Innovation in Near-Surface Geophysics

Innovation in Near-Surface Geophysics

Instrumentation, Application, and Data Processing Methods

Edited by

Raffaele Persico

Salvatore Piro

Neil Linford

ELSEVIER

Elsevier
Radarweg 29, PO Box 211, 1000 AE Amsterdam, Netherlands
The Boulevard, Langford Lane, Kidlington, Oxford OX5 1GB, United Kingdom
50 Hampshire Street, 5th Floor, Cambridge, MA 02139, United States

Library of Congress Cataloging-in-Publication Data
A catalog record for this book is available from the Library of Congress

British Library Cataloguing-in-Publication Data
A catalogue record for this book is available from the British Library

ISBN: 978-0-12-812429-1

For information on all Elsevier publications visit our website at https://www.elsevier.com/books-and-journals

Publisher: Candice Janco
Acquisition Editor: Marisa LaFleur
Editorial Project Manager: Carly Demetre
Production Project Manager: Omer Mukthar
Cover Designer: Christian J. Bilbow

Typeset by SPi Global, India

Contents

Contributors

Russell Abbey
Nuthampstead Airfield Museum, Bee Farm, Hertfordshire, United Kingdom

Paciello Antonella
ENEA, Rome, Italy

Adam D. Booth
School of Earth and Environment, University of Leeds, Leeds, United Kingdom

Ettore Cardarelli
Sapienza University of Rome—DICEA, Rome, Italy

Andrea Cataldo
University of Salento, Lecce, Italy

Ilaria Catapano
Institute for Electromagnetic Sensing of the Environment, National Research Council of Italy, Naples, Italy

Roger A. Clark
School of Earth and Environment, University of Leeds, Leeds, United Kingdom

Farrugia Daniela
Department of Geosciences, University of Malta, Malta

Zamuner Daniela
Institute for Technologies Applied to Cultural Heritage ITABC CNR, Roma, Italy

Egidio De Benedetto
University of Salento, Lecce, Italy

Giorgio De Donno
Sapienza University of Rome—DICEA, Rome, Italy

Lara De Giorgi
Institute for Archaeological and Monumental Heritage, National Research Council, Prov.le Lecce-Monteroni, Lecce, Italy

Philippe De Smedt
Department of Environment, Ghent University, Ghent, Belgium

Egil Eide
Department of Electronic Systems, Norwegian University of Science and Technology, Trondheim, Norway

Papale Enrico
Institute for Technologies Applied to Cultural Heritage ITABC CNR, Roma, Italy

Eileen G. Ernenwein
Department of Geosciences, East Tennessee State University, Johnson City, TN, United States

Simona Fontul
National Laboratory for Civil Engineering—LNEC, Lisboa, Portugal

Panzera Francesco
Department of Biological Geological and Environmental Sciences, University of Catania, Catania, Italy

Lombardo Giuseppe
Department of Biological Geological and Environmental Sciences, University of Catania, Catania, Italy

Chris M. Green
School of Earth and Environment, University of Leeds, Leeds, United Kingdom

Tomasz Herbich
Institute of Archaeology and Ethnology, Polish Academy of Sciences, Warsaw, Poland

Nathan Howland
Nuthampstead Airfield Museum, Bee Farm, Hertfordshire, United Kingdom

Kenneth L. Kvamme
Department of Anthropology, University of Arkansas, Fayetteville, AR, United States

Giovanni Leucci
Institute for Archaeological and Monumental Heritage, National Research Council, Prov.le Lecce-Monteroni, Lecce, Italy

Neil Linford
Geophysics Team, Historic England, Portsmouth, United Kingdom

Paul Linford
Geophysics Team, Historic England, Portsmouth, United Kingdom

Kucukdemirci Melda
Engineering Geophysical Department, University of Istanbul, Istanbul, Turkey

Jeremy G. Menzer
Environmental Dynamics Program, University of Arkansas, Fayetteville, AR, United States

Lara Pajewski
Department of Information Engineering, Electronics and Telecommunications, Sapienza University of Rome, Rome, Italy

Galea Pauline
Department of Geosciences, University of Malta, Malta

Andrew Payne
Geophysics Team, Historic England, Portsmouth, United Kingdom

Raffaele Persico
Institute for Archaeological and Monumental Heritage IBAM-CNR, Lecce; International Telematic University UNINETTUNO (UTIU), Rome, Italy

Dominic Pike
Nuthampstead Airfield Museum, Bee Farm, Hertfordshire, United Kingdom

Iannucci Roberto
Department of Earth Sciences, University of Rome "La Sapienza", Italy

Jacopo Sala
3D-Radar AS, Trondheim, Norway

Martino Salvatore
Department of Earth Sciences, University of Rome "La Sapienza", Italy

Piro Salvatore
Institute for Technologies Applied to Cultural Heritage ITABC CNR, Roma, Italy

D'Amico Sebastiano
Department of Geosciences, University of Malta, Malta

Francesco Soldovieri
Institute for Electromagnetic Sensing of the Environment, National Research Council of Italy, Naples, Italy

Mercedes Solla
Defense University Center, Spanish Naval Academy, Marín, Pontevedra, Spain

Veerle Vandeginste
School of Chemistry, University of Nottingham, University Park, Nottingham, United Kingdom

Lieven Verdonck
Department of Archaeology, Ghent University, Ghent, Belgium

Jeroen Verhegge
Department of Archaeology, Ghent University; Geosonda Environment NV, Ghent, Belgium

CHAPTER

1 Passive and active electric methods: New frontiers of application

Lara De Giorgi, Giovanni Leucci

Institute for Archaeological and Monumental Heritage, National Research Council, Prov.le Lecce-Monteroni, Lecce, Italy

CHAPTER OUTLINE

1 INTRODUCTION

In the field of applied geophysics, the geoelectric methods have been important for about a century, particularly for shallow and near-surface investigations. Geoelectric methods are used extensively to locate buried targets that are conductive and resistive in nature. The purpose of active geoelectric surveys is to determine the subsurface resistivity distribution by making measurements on the ground surface. The ground resistivity is related to various geologic parameters such as the mineral and fluid content, porosity, and degree of water saturation in the rock.

Geoelectric surveying has greatly improved and has become an important and useful tool in archaeological and monumental heritage conservation state studies and in hydrogeology, in environmental and engineering applications (e.g., Refs. [1–9]).

Schlumberger brothers introduce the resistivity method in the second decade of the twentieth century [10,11]. In the origin of this technique, the center point of the electrode array remains fixed, but the spacing between the electrodes is increased to

Innovation in Near-Surface Geophysics. https://doi.org/10.1016/B978-0-12-812429-1.00001-5

obtain more information about the deeper sections of the subsurface. The traditional horizontal layering techniques for interpreting geoelectric resistivity data are rapidly being replaced with two-dimensional (2D) and three-dimensional (3D) models of interpretations, especially in complex and heterogeneous subsurface media. Field techniques have advanced from the measurements made at separate and independent points to automated measuring systems with multielectrode array along the measurement profiles.

In the case of self-potential (SP) methods, where no subsurface energization using external artificial source is done, differences in natural ground potentials are measured between any two points on the ground surface. SP is the naturally occurring electric potential of the earth resulting from geologic, geochemical and hydrologic interactions that cause electric potentials to exist in the earth in the vicinity of the measurement point. Since 1830, the SP method has been employed in the search for minerals, but it is very rarely used in archaeological prospection because related phenomena are not very well known. The self-potentials are measured in millivolts (mv) relative to a "survey base," where the potential is arbitrarily assigned to be 0 V. Starting in the first years of the 1980s, the data acquisition was carried out manually, making labor-intensive and slow survey and often producing low-quality data. Successively, a range of automated multielectrode and multichannel data acquisition systems were developed. These methods allow fast and accurate data acquisition [12–18]. Since the electric techniques are regularly used by a vast number of geoscientists, authors use this chapter to expose the new research development reached in the field of instrumentation, data processing, and interpretation.

2 ABOUT OF SELF POTENTIAL METHOD

The self-potential (SP) measurements are performed using nonpolarizable electrodes. Several charge polarization mechanisms were proposed to explain SP anomalies [19]. They are associated with electrokinetic, electrochemical, thermoelectric, redox, and piezoelectric effects. The general equation for coupled flows can be written [20] as

$$J_i = \sum_j L_{ij} F_j \tag{1}$$

where the fluxes J_i (of charges, matter, heat, etc.) are related to the various forces F_j (gradients of electric potential, pressure, temperature, etc.) through the coupling coefficients L_{ij} ("phenomenological coefficients" or "conductivities") [21,22]. The L_{ij} can be considered as a symmetrical matrix. In this matrix, each flow is related to any combination of forces, and the diagonal elements are typical conductivities, such as those in Ohm's or Darcy's laws [19].

The total electric current density generated through coupling with one other phenomena in the Earth can be write in the simplest case as

$$J = -L_{11}\,\nabla F_1 - L_{1k}\,\nabla F_k = -\sigma\,\nabla\varphi - L_{1k}\,\nabla F_k \tag{2}$$

That represent a simplification of Eq. (1) [19].

The term $(-\sigma\nabla\varphi)$ is the conduction current density that flows throughout the Earth (σ represent the electric conductivity); the term $(-L_{1k}\nabla F_k)$ is a source current density related to the coupling process.

If the external sources of electric current are absent, the application of conservation equation [19] allows to affirm that the total current density is divergence-free (i.e., $\nabla \cdot J = 0$) and therefore

$$-\nabla \cdot \sigma \nabla \varphi = \nabla L \cdot \nabla F + L \nabla^2 F \quad (3)$$

The second term of Eq. (3) is related to the primary forcing process. It provides a source term for the self-potential signal [22,23] that is the main cause of the measured electric potential field in the ground surface. In order to resolve Eq. (3), typical boundary conditions require that $\phi \to 0$ as the distance from the source region becomes large, and the normal component of the electric current density is zero at the Earth's surface, that is, $n \cdot \sigma \nabla \phi = 0$ [24].

Sometimes, SP sources are localized within the subsoil resistivity structure [24a,24b]. In this case, SP data are related to the several physical gradients (thermal, chemical, and hydraulic) and to the resistivity and coupling coefficient structure of the medium (i.e., the electric potential gradients that are created by current flow through the subsoil generated by these sources).

2.1 SP FIELD DATA ACQUISITION

A simple equipment is require in order to acquired SP data. The nonpolarizing porous electrode are used (typically Pb-$PbCl_2$). They consist of a metal conductor in a saturated solution of its own salt that allows to slowly leak out the electrode and contact the ground (Fig. 1). The electrodes are connected by wire to a digital multimeter with a high input impedance capable of a reading accuracy of ± 0.0001 V. Schematically, the method of conventional SP data acquisition is showed in Fig. 2. One electrode represents the base station. The wire is unreeled to station 1 where the roving electrode is placed in a shallow hole ($\sim$10 cm deep) in the ground. The voltage between the base station electrode and the roving electrode is recorded with the defined sign convention. The base pot is always attached to the negative lead of the voltmeter. After making the voltage measurement, the roving electrode is picked up, and more wire is unreeled to station 2. The roving electrode is again placed in the earth, and another voltage measurement is made. This process is repeated until the survey is complete. The spacing between two successive measurement stations depends on the size of the anomalies we expect. The value of the electric resistance is also measured before each SP measurement to check the electric contact between the electrodes. A new reference station is established every time the end of the cable is reached or before (depending on the field conditions). The measurements are generally performed in a loop (closed profiles or profiles connected at both extremities to other profiles) in order to evaluate and correct the drift undergone during the acquisition of the data. This process requires two corrections of the raw SP data, the reference correction and closure correction (or loop correction). In a graph, SP data are represented as a function of the distance.

FIG. 1

Self-potential data acquisition electrode.

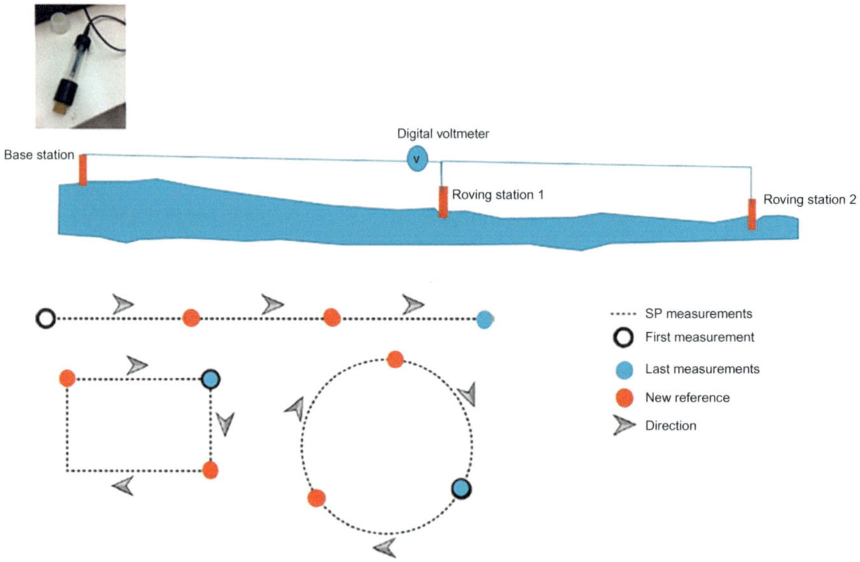

FIG. 2

Self-potential field data acquisition scheme.

However, there is also another method of measurements called dipole configuration (gradient configuration). It involves recording a measurement and then moving the electrodes along a survey line, maintaining a fixed spacing. The trailing electrode becomes the leading electrode as the electrodes are leapfrogged along the line. The positive voltmeter lead is always connected to the leading electrode to maintain proper polarity.

The current measurement method is slow, bulky, and requires a trained technician to gather reliable data. Therefore, an innovative system capable of performing multi-automatic SP measurements has been built. The syscal kid multichannel resistivity meter was modified in order to acquire SP in automatic mode.

In this case, the internal circuits were modified in order to obtain a reconfigurable instrumentation that allows the measurements of the electric resistivity, induced polarization, and self-potential.

First, the input impedance is increased in order to better assure a low contact resistance between soil and electrodes. Higher input impedances are desirable due also to the impedance reduction of air moisture.

In order to switch the instrumentation from resistivity to SP measurements circuit, a simple switch was added. The first idea for the electrode array capable of making SP measurements was to call for automating the rotation using solid-state switches. An electrode array design in the form of rectangular grid was considered. The connections between electrodes are controlled by PC via USB interface. In order to assure a good switch and the SP measurements an analog device cross-point switch was chosen because of its high density and simple programming interface. Being a cross-point switch, it can connect any of its 12 inputs to any of its 8 outputs, making an electrode array with 100 or more electrode feasible (Fig. 3). The device has a 300 MHz bandwidth and is capable of handling a 3 V input swing relative to analog ground.

2.2 SP DATA PROCESSING

The classical SP inverse problem consists of determining position and magnitude of SP sources. The first step in data processing is the reference correction.

This correction is made to join the different parts of a same SP profile, correcting the various changes of reference electrode.

The second step of SP data correction is the closure correction. In the case of a closed profile, the first point is identical to the last one, so the SP value measured should theoretically be the same. This would be true if no environmental perturbation occurred between the moments when the first and the last measurements were made. However, during the survey, the measurement conditions can change (e.g., soil moisture, soil temperature, and instrumental error), and a drift will be observed. The drift increases regularly from the first point to the last one, and it accumulates during the whole acquisition time. This drift is considered as parasitic, and it must be corrected.

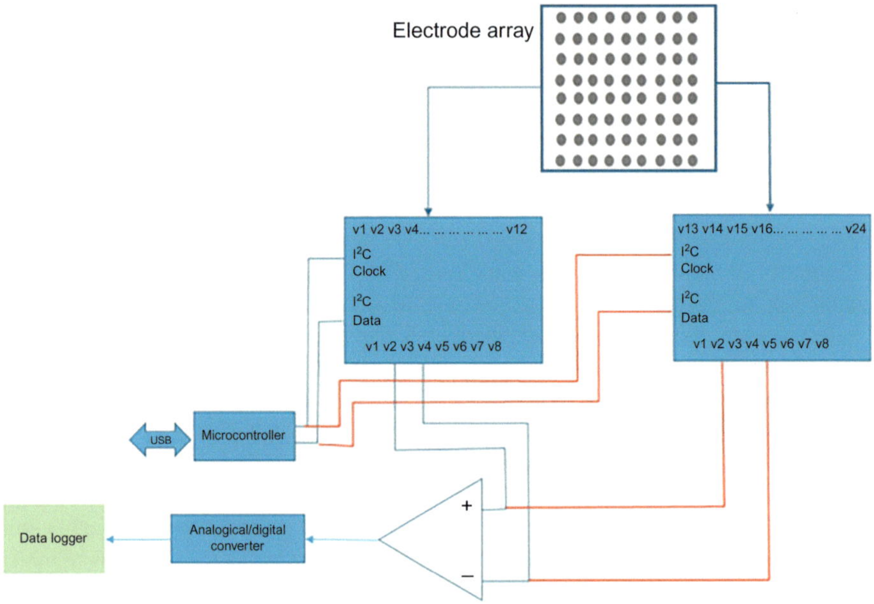

FIG. 3

The self-potential acquisition circuit scheme.

Successively, the estimation of the SP sources "**s**" could be done with the inverse problem. Obviously, it involves the calculation of **s** from a set of measured potentials. There are two principal difficulties that are addressed during the inversion. The first one involves the large variation in sensitivity of the data to different model parameters. The sensitivity is typically greatest near the measurement locations, though it is also a function of the resistivity structure. Sensitivity scaling is therefore applied by calculating the cumulative sensitivity of each model parameter to all the measured data. The second one is the nonunique solution, which is compounded by the limited set of available measurements. In order to address this nonuniqueness, a class of source solutions that have a desired spatial compactness could be selected. This choice amounts to the addition of prior knowledge about the properties of **s**, which may be dependent on the particular problem of interest. Starting from Eq. (4)

$$\nabla \cdot \sigma \nabla \varphi = s \tag{4}$$

and introducing the matrix $K = \nabla \cdot \sigma \nabla$.
it is possible to write

$$K\varphi = s \tag{5}$$

Therefore

$$K^{-1} s = \varphi \tag{6}$$

Eq. (6) states the basic inverse problem.

The sources **s** could be estimated "theoretically" if φ (the electric potential) is known. In practical, φ is sampled at a limited subset of locations that are often restricted to the surface of the earth. Therefore, it is possible to consider the N measured potential φ_d that is related to the sources through a subset of the K^{-1} matrix. This new operator O is a selector matrix $N \times M$ that consists of a single 1 on each row in the column that corresponds to the measurement location. Therefore, Eq. (6) can be written as

$$O K^{-1} s = \varphi_d \tag{7}$$

Each row of K^{-1} contains Green's function that multiplies the source vector and results in SP measurements. The solution that minimizes the squared difference between the predicted and observed SP is given in Eq. (8):

$$\left(K^{-1} O^T O K^{-1}\right) s = K^{-1} O^T \varphi_d \tag{8}$$

To estimate the depth of a buried structure from the SP data, it is possible to use a least-squares analysis method [25] based on normalizing the residual SP anomaly using three characteristic points and their corresponding distances on the anomaly profile; then, the depth for each horizontal position of the buried structure is determined using the least-squares method. The computed depths are plotted against the assumed horizontal positions on a graph. The solution for the depth and the horizontal position of the buried structure is read at the common intersection of the curves. Knowing the depth and the horizontal position and applying the least-squares method, the shape factor is determined using a simple linear equation. Procedures are also formulated to estimate the polarization angle and the electric dipole moment. The method is semiautomatic, and it can be applied to short or long residual SP anomaly profiles.

3 ABOUT ELECTRICAL RESISTIVITY METHOD

The physical parameter in geoelectric measurements is the specific electric resistivity ρ of the ground. The applications of 2D and 3D imaging are nowadays important for visualizing and interpreting complex archaeological structures at various depths [8]. The electric resistance, R, of a material is related to its physical dimension, cross-sectional area, S, and length, l, through the electric resistivity, ρ, by Reynolds [19]:

$$\rho = R \frac{S}{l} \tag{9}$$

Active resistivity method uses a dipole source of current injected into the subsurface through point electrodes, and the resulting potential difference is measured at other electrode positions in the neighborhood of the current flow (Fig. 4). Considering a completely homogeneous and isotropic medium for a semiinfinite conducting layer of uniform resistivity bounded by the ground surface, a current of strength I is injected at a point C_1 into the ground surface. The current I will flow away radially

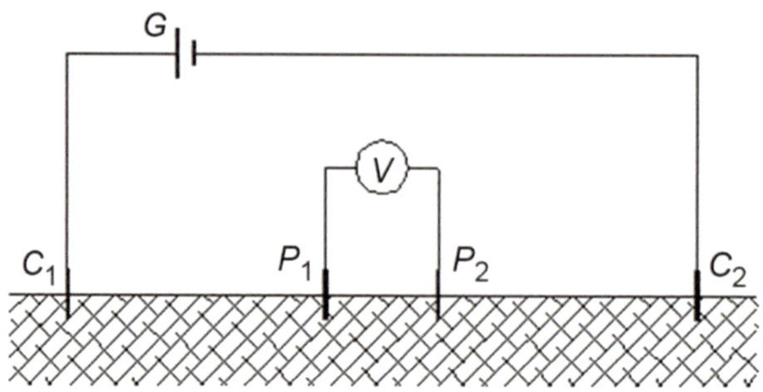

FIG. 4

Active resistivity method scheme.

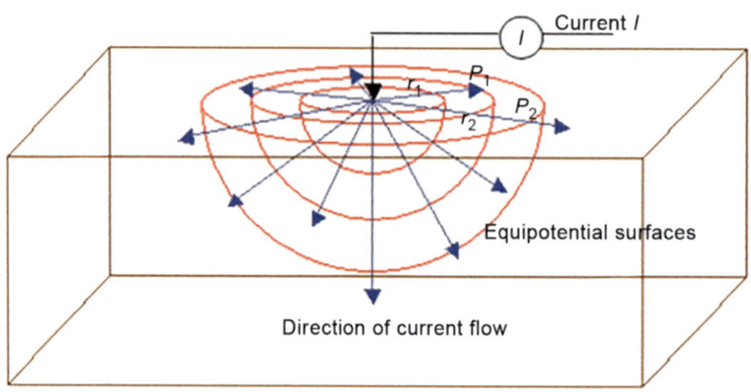

FIG. 5

Schematization of current flow from point source.

from the point of source, and its distribution will be uniform over a hemispheric shell of an underground region (Fig. 5) of resistivity ρ.

At a distance r of a point P in the medium from the point source, the surface area of the hemispheric shell is $2\pi r^2$ so that the potential for the homogeneous half-space is

$$V(r) = (\rho I / 2\pi r) + \cos t \tag{10}$$

The potential function, considering the two points P_1 and P_2 at distance r_1 and r_2 from the point source (Fig. 5), is obtained by applying Eq. (10):

$$\mathrm{V} = \frac{\rho I}{2\pi r}\left(\frac{1}{r_1} - \frac{1}{r_2}\right) \tag{11}$$

Thus, if the P_1 and P_2 are the potential electrodes and C_1 and C_2 are the current electrodes (Fig. 4), the potential difference between P_1 and P_2 can be written as

$$\Delta V = \frac{\rho I}{2\pi r}\left(\frac{1}{C_1P_1} - \frac{1}{C_1P_2} - \frac{1}{C_2P_1} + \frac{1}{C_2P_2}\right) \tag{12}$$

But subsurface isn't a homogeneous medium; it is typically heterogeneous. Therefore, the estimated electric resistivity must be considered as apparent electric resistivity ρ_a:

$$\rho_a = K\frac{\Delta V}{I} \tag{13}$$

Eq. (13) was obtained from Eq. (12) considering K as a geometric factor.

4 ELECTRICAL RESISTIVITY FIELD DATA ACQUISITION

The subsurface and the investigated medium are usually complex and multiscale, and spatial variation can change rapidly both laterally and vertically along survey profiles, since survey design must take into account the capability of the system for data acquisition; the target (shape and dimensions); and therefore the required resolution, the medium heterogeneity, the areal extension the cost, and the time required to complete the survey. Generally, as shown in Fig. 4, four electrodes are placed at arbitrary locations on the ground surface. Therefore, there are a number of electrode configurations that can be used in recording electric resistivity field data. Conventional arrays most commonly used include Wenner (alpha), Schlumberger, dipole-dipole, pole-pole, and pole-dipole arrays (Fig. 6), each one suitable for a particular geologic situation. The array choice depends on a number of factors related to the target to be delineated, heterogeneities of the medium, sensitivity of the resistivity meter system, the background noise level, electromagnetic coupling, sensitivity of the array to vertical and lateral resistivity variations, its depth of investigation, and the horizontal data coverage and signal strength of the array. The Wenner (alpha) and Schlumberger arrays are relatively sensitive to vertical variations in the resistivity below the center of the array but less sensitive to horizontal variations in the subsurface, and it has a moderate depth of investigation. A limitation of this array is the relatively poor horizontal coverage with increased electrode spacing [26].

Dipole-dipole is very sensitive to horizontal variations of resistivity and relatively insensitive to vertical variations of resistivity in the investigated medium. Thus, it is the most preferred array for mapping vertical structures like walls, dykes, tombs, and cavities. It is the most sensitive array to 3D structure [26]. The depth of investigation of the array depends on both the current electrode spacing a and the distance between the two dipoles n and is generally shallower than the Wenner array, but it has a better horizontal data coverage than Wenner array. It has a major disadvantage that consists in a decrease in signal strength with increasing distance between the dipole pair.

The pole-dipole array is an asymmetrical array; it has relatively good horizontal coverage and higher signal strength compared with dipole-dipole array. Repeating measurements with the electrodes arranged in the reverse order can eliminate the asymmetrical effect. The signal strength of the pole-dipole array is lower than that of Wenner and Schlumberger arrays and is very sensitive to vertical structures.

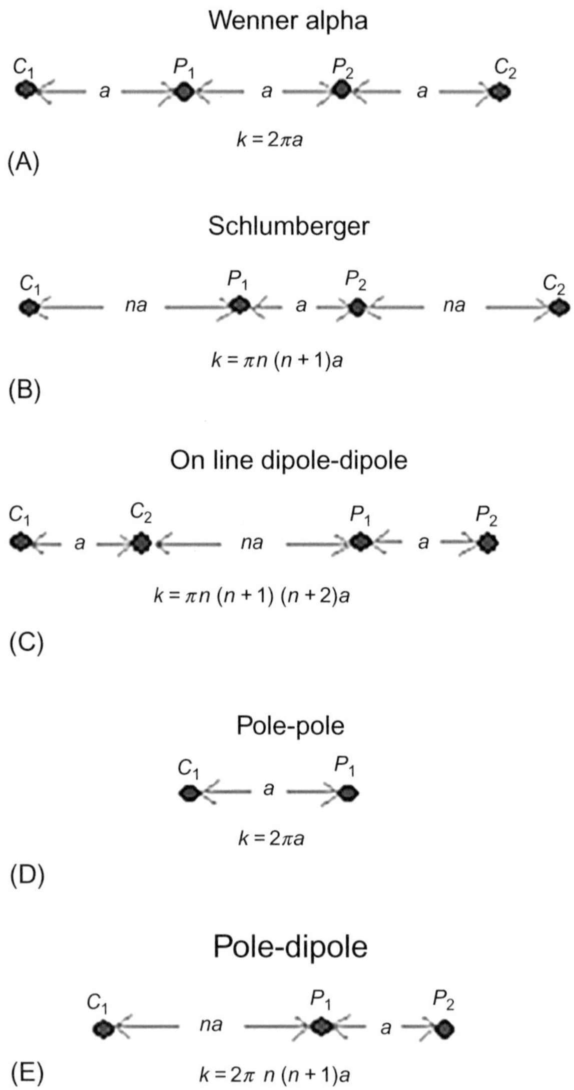

FIG. 6

The electrodes array commonly used in electric resistivity data acquisition. (A) Wenner alpha; (B) Schlumberger; (C) on line dipole-dipole; (D) pole-pole; (E) pole-dipole.

The pole-pole array consists of one current and one potential electrode with the second current and potential electrodes at infinite distances. Finding suitable locations for these electrodes so as to satisfy this theoretical requirement is often difficult. It has the widest horizontal coverage and the deepest depth of investigation but the poorest resolution. The resolution of the pole-pole array is very poor as subsurface structures tend to be smeared out in the inversion model [26].

The apparent-resistivity values observed by the different array types over the same structure can be very different.

Resistivity field data are acquired using several types of earth resistivity meter. The equipment is portable, light weight, and relatively cost-effective when compared with other geophysical data acquisition systems. A conventional setup of the earth resistivity meter (Fig. 7) basically consists of the following: a constant current source, commonly a battery pack connected to a unit control composed of (i) a commutated DC circuit to change polarity of the current source, (ii) an ammeter that measures the injecting current, and (iii) a very sensitive voltmeter that measures the response signal. Furthermore, a computer with an acquisition software allows to see in real time the resistivity data. Electrode is usually stainless steel. Two-dimensional (2D) resistivity imaging can be achieved by integrating the techniques of vertical electric sounding with that of electric profiling. They are usually carried out using large numbers of electrodes connected to multicore cables (Fig. 8). For a system with a limited number of electrodes, Dahlin and Bernstone [27] have developed a survey method in which the area covered by the survey can be extended along the survey line by moving the cables past one end of the line by several units of electrode spacing (roll-along method). The values of apparent resistivity observed are presented in pictorial form using pseudosection contouring that gives an approximate picture of the resistivity distribution in the measured medium. The shape of the contours depends on the type of array used in the investigation and the distribution of the true resistivity. Three-dimensional survey, in theory, gives the most accurate and reliable results especially in heterogeneous medium. Ideally, the

FIG. 7

The field electric resistivity data acquisition instrument.

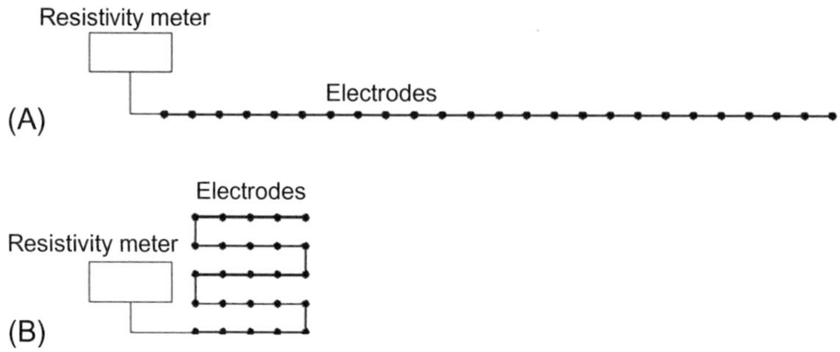

FIG. 8

The multielectrodes 2D (A) and 3D (B) acquisition scheme.

measurements of apparent-resistivity values that would constitute a complete 3D data set should be made in all possible directions.

In practice, instead of using the 3D measurement mode, 3D resistivity variations are recorded by recording a dense 2D measurement grid that is considered to be a more practical and economical approach for field data.

The 2D sets involve measuring parallel 2D lines with interline spacing equal to the interelectrode spacing and can take place along the *X* axis, *Y* axis, or along both axes. These 2D data are routinely interpreted with 2D algorithms, and the results are combined to generate pseudo 3D (x, y, and z) images. The obtained results are affected by artifacts due to the fact that 3D responses are related to 2D structures.

Electric resistivity inversion method is related to the finite element method (FEM) in order to solve the forward problem. The technique has been extensively described in many works, so only a brief description of the method is presented here.

The investigated medium is discretized into homogeneous triangular regions called elements. The potential within each element is approximated by a simple interpolation function, the basis function. To minimize the error between the approximated and real potential, the Galerkin minimization criterion [28] is applied. After applying the Galerkin minimization scheme to every element, the individual element equations can be assembled in to one global system that has the following form:

$$\mathbf{K}\mathbf{P} = \mathbf{S} \tag{14}$$

where **P** is the unknown transformed nodal potential vector, **S** is the vector describing the sources, and **K** is a matrix that is related to the nodal coordinates. After applying the homogeneous Dirichlet and Newman boundary conditions, the system of Eq. (14) is being solved, and the transformed nodal potential is obtained.

Eq. (14) is solved for several wave numbers, and applying the inverse Fourier transform the total potential is recovered. The knowledge of the nodal potential allows to obtain the potential differences and apparent resistivities point to point.

Now, the solution of the Poisson equation (the differential equation that governs the flow of the electric current in the medium) is sought by subdividing the area into hexahedral elements. The application of the Galerkin minimization scheme to every element allows to consider the global system that has the form of Eq. (14).

The final step is to solve the system of equations: for the 3D case that in general involves large systems of equations, an iterative technique is preferable. In this work, the conjugate gradient method for solving large sparse linear systems is used. For both the 2D and 3D case, the inversion core algorithm is the same. A nonlinear smoothness constrained inversion algorithm [29] is used.

Another way to acquire 3D data is related to the L geometry that consists of two perpendicular profiles. This process is discussed in detail by Tejero-Andrade et al. [30]. The array geometry allows various array types, including dipole-dipole (DD), Wenner (W), Schlumberger (S), Wenner-Schlumberger (WS), pole-pole (PP), and pole-dipole (PD). In general, any geometric configuration can be applied [30].

Another physical parameter that could be studied in electric tomography is the resistivity anomaly probability tomography (RAPT). The 3D RAPT imaging approach is very complicate, and therefore, a simple definition will be done at this point. It consists in a cross correlation procedure performed by a function "f" over the data called the tomospace, lying below the survey area S. In practice, an elementary cell is used with a positive resistivity anomaly of unitary strength to scan the whole tomospace and search where resistivity variations with respect to a reference function are placed in a probabilistic sense [31,32].

4.1 SOME FIELD APPLICATION: THE ARCHAEOLOGICAL SITE OF SAGALASSOS (TURKEY)

The archaeological site of Sagalassos is a very important settlement located in a magnificent mountain landscape, 7 km north from a little village named Ağlasun (province of Burdur, SW Turkey). Since 1990, the University of Leuven (Belgium) carries out an interdisciplinary archaeological research program that studies the uninterrupted occupation of more than 1000 years in all its aspects of daily life, from architecture to trade and its mechanisms and environmental conditions of the past. The ancient roman city is a site covered by layers of eroded soil and preserved many secrets that need to be revealed.

A geophysical campaign was planned along the south facing terraces of the mountain slopes, in order to highlight the structure of the city still covered. Site conditions (high slope, high grass, and several obstacles) and the need to investigate to depths in excess of 20 m led to the choice of geophysical methods. Therefore, passive and active electric resistivity tomographies were used. The results obtained in two different areas labeled, respectively, area 1 and area 2 were presented. Data were collected along nonconventional profiles using dipole-dipole array and variable electrode spacing. For area 1, two physical parameters were measured, the electric resistivity and self-potential in order to obtain their distribution in the subsurface in a three-dimensional volume. In the area 2, due to the high contact resistance between the electrodes and soil, a 3D distribution in the subsoil of resistivity tensor probability was calculated.

4.1.1 Area 1 Geophysical Results

In the area 1, ERT survey was performed in a 55 m by 95 m grid (Fig. 9). A combined full 3D and L geometry was used [33].

Electric resistivity tomography (ERT) data were then processed in 3D mode and presented as horizontal slices (parallel to the surface) through the ground (Fig. 10). Fig. 10 shows the electric resistivity model at 2 m depth. It is possible to note in the west side of the area a low-resistivity zone (100–150 Ωm) inside a dashed dark rectangles. The dimensions of this zone are about 2 m long × 1 m width, and the resistivity values could be related "probably" to voids filled with soil materials (tombs?). Some regular structures are clearly visible (*dashed dark line*). The resistivity values (ranging from 1000 to 1200 Ωm) suggest the likely presence of walls.

In this area, the analysis of SP data was performed using the algorithm proposed by Eppelbaum et al. [34,35]. It uses the similarities between magnetic and SP field. The results, developed in, show the horizontal depth slices (Fig. 11). At the depth of 0.5 and 1.0 m, probable traces of structure of archaeological interest are present in the form of positive values of SP (80–100 mV). At the depth of 2.0 m, it is possible to

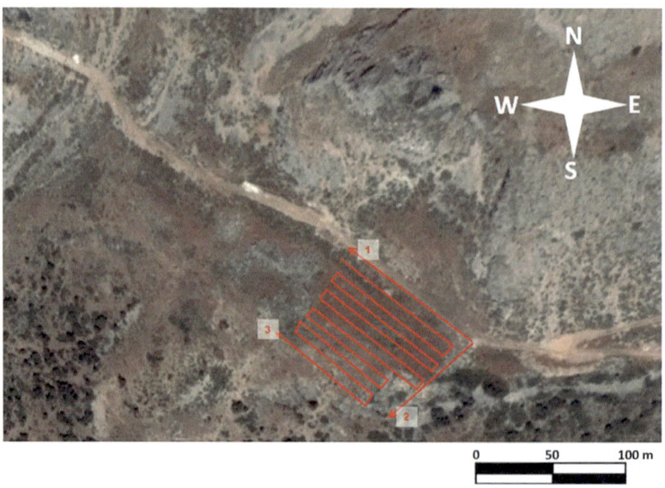

FIG. 9

The area 1 in Sagalassos: ERT data acquisition scheme: 1, 2, and 3 are the three ERT profiles.

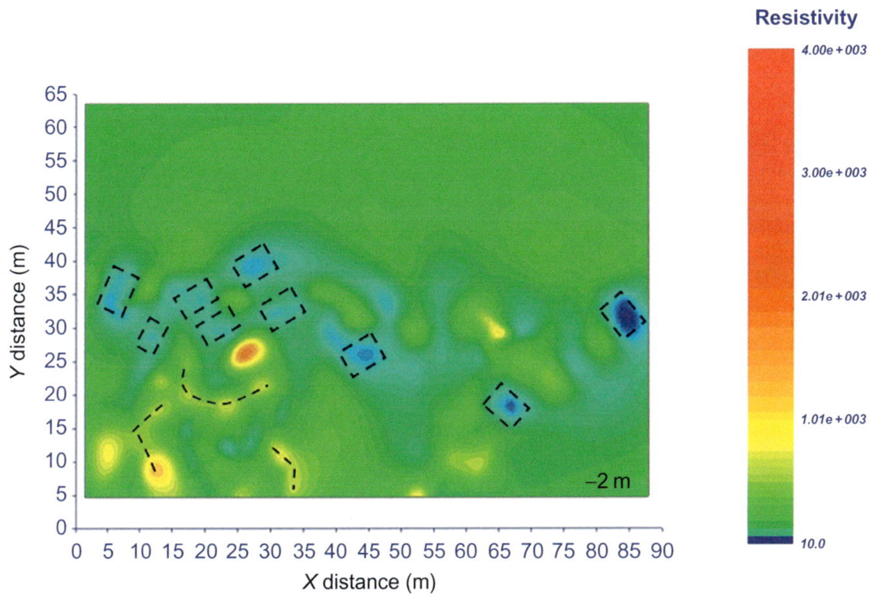

FIG. 10

Area 1: 3D electric resistivity distribution in the subsoil at a depth of 2 m.

compare with ERT depth slice at same depth (Fig. 11). Negative SP values could be related to probable traces of tombs. Probable traces of water flow from the positive SP values to negative SP values at about 9 m in depth (Fig. 11).

4.1.2 Area 2 Geophysical Results

In the area 2, ERT survey was performed in a 104 m by 128 m grid inside a depression area (Fig. 12). The measurements were collected along two lines [33]. The first line was an ERT profile that crosses the hill, and it is 115 m. The second line was acquired around the hill in roll-along mode in order to cover the whole area. It is 320 m long and consists of five lines, each 64 m long. For this line, the electrode spacing was variable from 2 to 3 m. This was due to the inaccessibility conditions of the site (bushes, stones, etc.). A total of 300,000 apparent-resistivity measurements were collected (from $n=1$ to $n=9$).

This area was chosen to test the applicability of the 3D resistivity anomaly probability tomography (RAPT) method in an actual archaeological context [31,32]. Apparent-resistivity tensor was estimated in 3D mode from the measured apparent-resistivity data through an algorithm written in Matlab. The results were presented as horizontal slices (parallel to the surface) through the ground (Fig. 13).

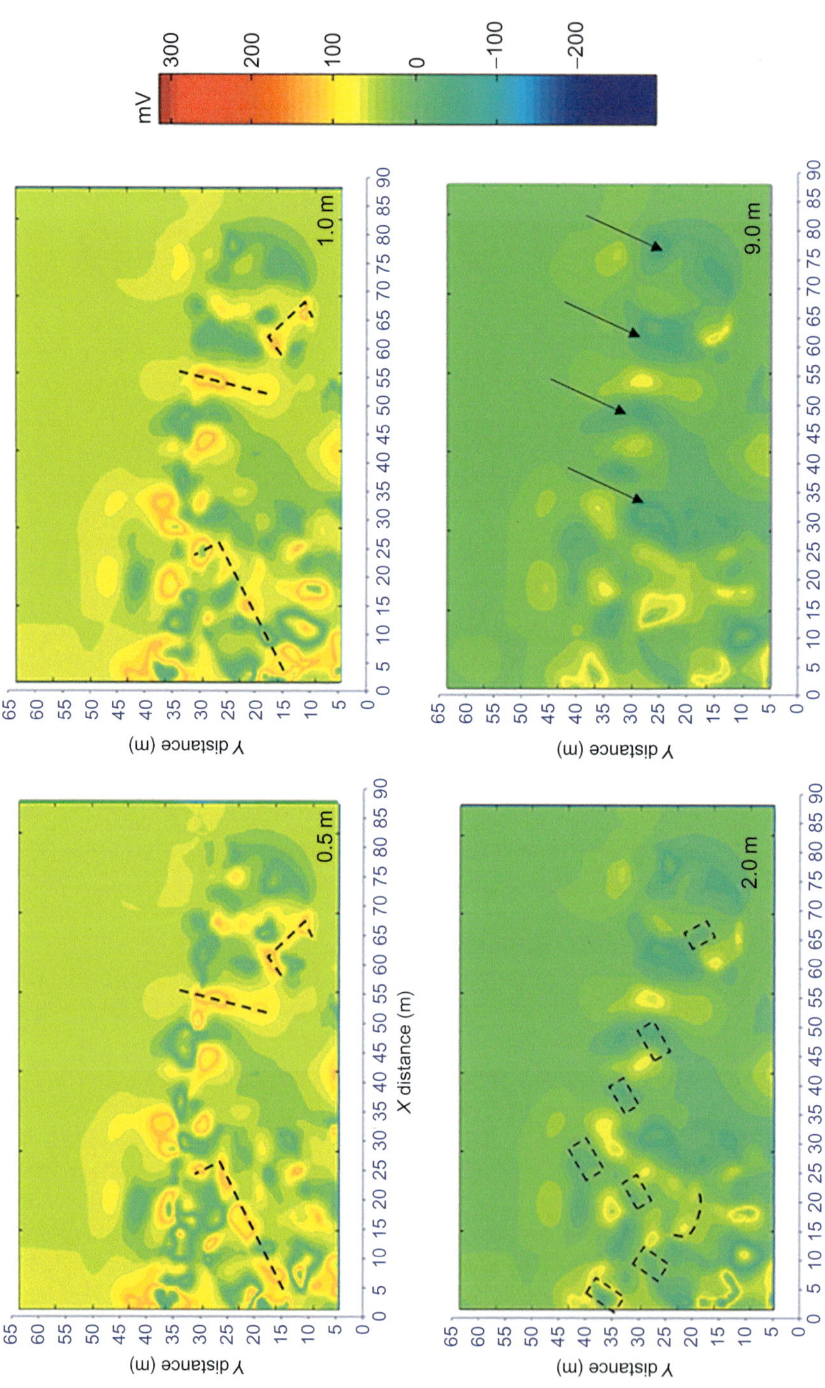

FIG. 11

Area 1: SP map at 1508 m asl.

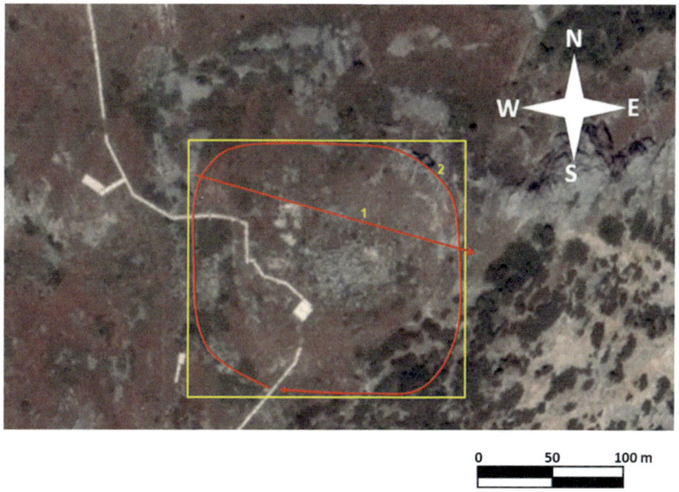

FIG. 12

Area 2: ERT profiles.

Fig. 13 shows the apparent-resistivity tensor model at four different depths. It is possible to note several anomalous zones (*dashed dark lines*) with a moderately high apparent-resistivity tensor (150–200 Ωm). The shape of these anomalies suggests a probable presence of structures of archaeological interest at a depth ranging between 0.5 and 2.0 m. At 4.0 m in depth, the apparent-resistivity tensor values increase until 250 Ωm. These values could be related to bedrock.

5 CONCLUSIONS

Some notions related to active and passive geoelectric methods were done in this chapter. A new instrumentation for self-potential (SP) data acquisition was presented. In this case, the SP data acquisition was automated and implemented in a multielectrode data acquisition system. This allows to perform 2D and 3D geoelectric active and passive data acquisition with a reduction in time and costs. Square, rectangular, and irregular grids of electrodes with unconstant electrode spacing beginning commonly used for 3D surveys yield large volume of independent data points. Nonconventional array survey method in which potential measurements are made at the electrodes along L geometry can reduce the number of measurements and the time required to carry out a 3D resistivity survey. Finally, the presented case studies show the good results that are possible to obtain.

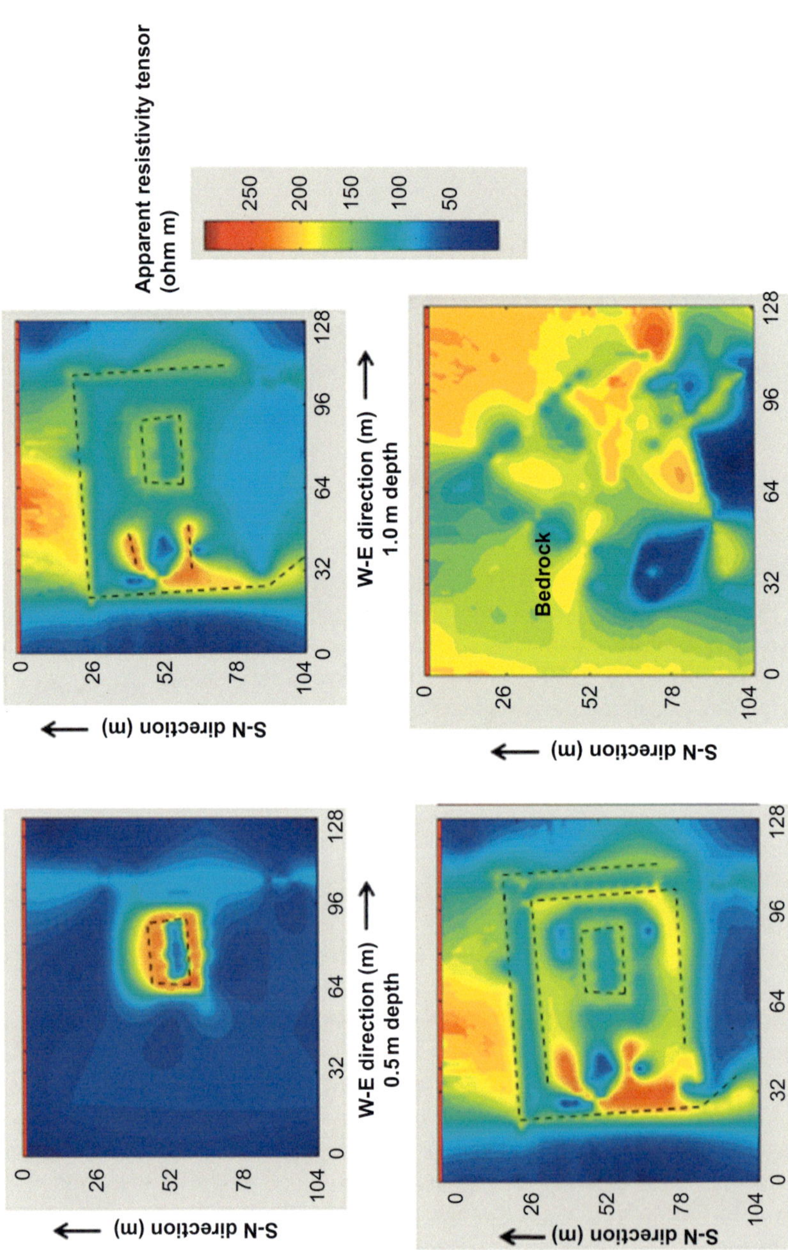

FIG. 13

Area 2: 3D apparent-resistivity tensor distribution in the subsoil.

REFERENCES

[1] S.A. Amidu, A.I. Olayinka, Environmental assessment of sewage disposal systems using 2D electrical resistivity imaging and geochemical analysis: a case study from Ibadan, Southwestern Nigeria, Environ. Eng. Geosci. 7 (3) (2006) 261–272.

[2] A.P. Aizebeokhai, A.I. Olayinka, V.S. Singh, Application of 2D and 3D geoelectrical resistivity imaging for engineering site investigation in a crystalline basement terrain, Southwestern Nigeria, Environ. Earth Sci. 61 (2010) 1481, https://doi.org/10.1007/s12665-010-0474-z.

[3] T. Dahlin, M.H. Loke, Quasi-3D resistivity imaging-mapping of three-dimensional structures using two-dimensional DC resistivity techniques, Proceedings of the 3rd Meeting of the Environ. Eng. Geophy. Soc., 1997, pp. 143–146.

[4] D.H. Griffiths, J. Turnbull, A.I. Olayinka, Two-dimensional resistivity mapping with a complex controlled array, First Break 8 (1990) 121–129.

[5] G. Leucci, Contribution of ground-penetrating radar and electrical resistivity tomography to identify the cavity and fractures under the main church in botrugno (Lecce, Italy). J. Archaeol. Sci. 33 (9) (2006) 1194–1204, https://doi.org/10.1016/j.jas.2005.12.009.

[6] G. Leucci, Geoscientific investigations for mapping the subsidence risk in an urban area, J. Geophys. Eng. 4 (2007) 317–331, https://doi.org/10.1088/1742-2132/4/3/S11.

[7] G. Leucci, L. De Giorgi, G. Scardozzi, Geophysical prospecting and remote sensing for the study of the San Rossore area in Pisa (Tuscany, Italy). J. Archaeol. Sci. 52 (2014) 256–276, https://doi.org/10.1016/j.jas.2014.08.028.

[8] S. Negri, G. Leucci, F. Mazzone, High resolution 3D ERT to help GPR data interpretation for researching archaeological items in a geologically complex subsurface, J. Appl. Geophys. 65, 3-4 (2008) 111–120, https://doi.org/10.1016/j.jappgeo.2008.06.004.

[9] G. Leucci, L. De Giorgi, F.T. Gizzi, R. Persico, Integrated geo-scientific surveys in the historical centre of Mesagne (Brindisi, Southern Italy), Nat. Hazards 86 (2017) 363–383, https://doi.org/10.1007/s11069-016-2645-x.

[10] M.H. Loke, J.E. Chambers, D.F. Rucker, O. Kuras, P.B. Wilkinson, Recent developments in the direct-current geoelectrical imaging method, J. Appl. Geophys. 95 (2013) 135–156, https://doi.org/10.1016/j.jappgeo.2013.02.017.

[11] V.S. Sarma, Electrical resistivity (ER), self potential (SP), induced polarisation (IP), spectral induced polarisation (SIP) and electrical resistivity tomography (ERT) prospection in NGRI for the past 50 years, J. Ind. Geophys. Union 18 (2) (2014) 245–272.

[12] E. Auken, L. Pellerin, N.B. Christensen, K. Sorensen, A survey of current trends in near-surface electrical and electromagnetic methods, Geophysics 71 (5) (2006) 249–260.

[13] R.D. Barker, The offset system of electrical resistivity sounding and its use with multicore cable, Geophys. Prospect. 29 (1981) 128–143.

[14] T. Dahlin, B. Zhou, Multiple-gradient array measurements for multichannel 2D resistivity imaging, Near Surv. Geophys. 4 (2) (2006) 113–123, https://doi.org/10.3997/1873-0604.2005037.

[15] M.H. Loke, P. Wilkinson, J. Chambers, Parallel computation of optimized arrays for 2-D electrical imaging, Geophys. J. Int. 183 (3) (2010) 1202–1315, https://doi.org/10.1111/j.1365-246X.2010.04796.x.

[16] R. Martorana, G. Fiandaca, A. Casas Ponsati, P.L. Cosentino, Comparative tests on different multi-electrode arrays using models in near-surface geophysics, J. Geophys. Eng. 6 (1) (2009) 1–20, https://doi.org/10.1088/1742-2132/6/1/001.

[17] R. Martorana, P. Capizzi, A. D'Alessandro, D. Luzio, Comparison of different sets of array configurations for multichannel 2D ERT acquisition, J. Appl. Geophys. 137 (2017) 34–48, https://doi.org/10.1016/j.jappgeo.2016.12.012.

[18] P. Stummer, H.R. Maurer, Real-time experimental design applied to high-resolution direct-current resistivity surveys, International Symposium on Optical Science and Technology, Expanded Abstracts, 2001, pp. 143–150.

[19] J.M. Reynolds, An Introduction to Applied and Environmental Geophysics, second ed., Wiley-Blackwell, Oxford, 2011, p. 696.

[20] J. Overbeek, Electrochemistry of the double layer, in: H.R. Kruyt (Ed.), Colloid Science: Irreversible Systems, vol. 1, Elsevier, New York, 1956, p. 115.

[21] S.R. de Groot, P. Mazur, Non-Linear Thermodynamics, Dover Publications, Amsterdam, 1983, p. 510.

[22] W.R. Sill, Self-potential modeling from primary flows, Geophysics 48 (1983) 76–86.

[23] A. Revil, The hydroelectric problem of porous rocks: thermodynamic approach and introduction of a percolation threshold, Geophys. J. Int. 151 (2002) 944–949.

[24] A. Dey, H.F. Morrison, Resistivity modeling for arbitrarily shaped three-dimensional structures, Geophysics 44 (4) (1979) 753–780.

[24a] J. Bigalke, E.W. Grabner, The geobattery model. A contribution to large scale electrochemistry, Electrochim. Acta 42 (1997) 3443–3452.

[24b] F. Timm, P. Moeller, The relation between electric and redox potential: evidence from laboratory and field measurements, J. Geochem. Explor. 72 (2) (2001) 115–128.

[25] M. Abdelazeem, M. Gobashy, Self-potential inversion using genetic algorithm, J. King Abdulaziz Univ. 17 (1) (2006) 83–101, https://doi.org/10.4197/Ear.17-1.5.

[26] M.H. Loke, Electrical Imaging Surveys for Environmental and Engineering Studies. A Practical Guide to 2-D and 3-D Surveys, 2001. on line. http://personales.upv.es/jpadin/lokenote.pdf.

[27] T. Dahlin, C. Bernstone, in: R.S. Bell (Ed.), A roll-along technique for 3D resistivity data acquisition with multi-electrode array, Proceedings of SAGEEP, EEGS, March, 1997, Reno, Nevada, vol. 2, 1997, pp. 927–935.

[28] I. Babuska, R. Tempone, G.E. Zouraris, Galerkin finite element approximations of stochastic elliptic partial differential equations, SIAM J. Numer. Anal. 42 (2) (2004) 800–825.

[29] Y. Sasaki, Resolution of resistivity tomography inferred from numerical simulation, Geophys. Prospect. 40 (1992) 453–464.

[30] A. Tejero-Andrade, G. Cifuentes, R.E. Chavez, A. Lopez Gonzalez, C. Delgado-Solorzano, "L" and "Corner" arrays for 3D electrical resistivity tomography: an alternative for urban zones, Near Surv. Geophys. 13 (2015) 1–13, https://doi.org/10.3997/1873-0604.2015015.

[31] B. Di Fiore, P. Mauriello, D. Monna, D. Patella, Examples of application of tensorial resistivity probability tomography to architectonic and archaeological targets, Ann. Geophys. 45 (2) (2002) 417–429.

[32] P. Mauriello, D. Patella, Introduction to tensorial resistivity probability tomography, (2005) 1–8, http://arXiv:physics/0512147v1.

[33] L. De Giorgi, G. Leucci, The archaeological site of Sagalassos (Turkey): exploring the mysteries of the invisible layers using geophysical methods, Explor. Geophys. (2017) (on line published). https://doi.org/10.1071/EG16154.

[34] L.V. Eppelbaum, B.E. Khesin, S.E. Itkis, Prompt magnetic investigations of archaeological remains in areas of infrastructure development: Israeli experience, Archaeol. Prospect. 8 (2001) 163–185, https://doi.org/10.1002/arp.167.

[35] L.V. Eppelbaum, B.E. Khesin, S.E. Itkis, Z. Ben-Avraham, Advanced analysis of self-potential data in ore deposits and archaeological sites, Proceedings of the 10th European Meeting of Environmental and Engineering Geophysics, Utrecht, The Netherlands, 2004, 4 pp.

FURTHER READING

[36] A.P. Aizebeokhai, 2D and 3D geoelectrical resistivity imaging: theory and field design, Sci. Res. Essays 5 (23) (2010) 3592–3605.

[37] L. Jouniaux, Electrokinetic techniques for the determination of hydraulic conductivity, in: L. Elango (Ed.), Hydraulic Conductivity-Issues, Determination and Applications, InTech Publisher, London, 2011, pp. 307–328.

[38] A. Revil, P.A. Pezard, P.W.J. Glover, Streaming potential in porous media. 1. Theory of the zeta potential, J. Geophys. Res. 104 (1999) 20021–20031.

CHAPTER

2 Advances in electric resistivity tomography: Theory and case studies

Ettore Cardarelli, Giorgio De Donno
Sapienza University of Rome—DICEA, Rome, Italy

CHAPTER OUTLINE

1 INTRODUCTION AND BACKGROUND

Electric direct current (DC) method is one of the oldest and most popular techniques among the noninvasive geophysical investigations, mainly because of its rapidity and cost-effectiveness. The first application of the electric method was made almost contemporary by Conrad Schlumberger (1912, France) and Frank Wenner (1915, United States). Since then, electric method has been extensively employed for hydrogeologic problems, mining studies, environmental and civil engineering, and archaeological investigations (see Refs. [1–5] for reviews of the application of electric method to near-surface targets).

The majority of electric surveys made until the 1980s were performed through the vertical electric sounding (VES) technique, under the assumption of one-dimensional soil layering [6]. Only few studies are considered 2D effects due to dipping discontinuities [7–9]. The VES technique has the major disadvantage to be a low resolution and relatively slow method. Therefore, the 1D approach was

Innovation in Near-Surface Geophysics. https://doi.org/10.1016/B978-0-12-812429-1.00002-7

progressively replaced by a multichannel and multielectrode acquisition system, thanks to the development of electronic multichannel devices (e.g., Refs. [10–13]) and algorithms for modeling and inversion (e.g., Refs. [14–16]). The two-dimensional systems can deal with automatic protocols for data acquisition, where current and potential electrodes are automatically switched along the investigated line in order to retrieve a tomographic reconstruction of the subsoil. This technique, called electric resistivity tomography (ERT), is analogous to the electric tomography employed for medical application on the human thorax. The ERT technique is currently the most popular in the geophysical community as demonstrated by the numerous applications during the last decades, involving both the geologic field, that is, aquifer characterization, cavities detection, landslide modeling, and soils layering (e.g., Refs. [17–22]), and the anthropogenic environment, that is, archaeological prospection, building characterization, landfills, and contaminated sites (e.g., Refs. [23–27]). Resolution of ERT investigation mainly depends on the electrode spacing, the array configuration (Wenner-Schlumberger, dipole-dipole, pole-dipole, gradient, etc.), the signal-to-noise ratio, and the type of algorithm used for inversion. However, 2D inverted models include spurious effects due to resistivity changes perpendicular to the investigated profile. Therefore, during the last 20 years, full 3D data acquisition systems have been increasing together with the corresponding algorithms for data processing and interpretation (e.g., Refs. [28–33]). This approach has been employed, whereas the reconstruction of three-dimensional targets is needed, particularly for small-scale surveys pertaining to dipping lateral contacts, cavities, buried structures, and safeguard of cultural heritage (e.g., Refs. [34–36]). However, there are still many limitations on the use of three-dimensional ERT, mainly due to the huge amount of measurements and to the increasing size of the investigated domains, compared with a 2D survey, that are straightforward indicators of the amount of memory required for data inversion. The ERT method is also used for monitoring, leading to the so-called time-lapse method, applicable for monitoring injections (resin, saline tracers, etc.), aquifers, and contaminated areas or for understating any changes due to the seasonality (e.g., Refs. [37–42]). Many researchers investigated also the induced polarization (IP) effect [43], linked to the capacitive response of the subsoil due to an external current injection. The capacitive response has been mostly analyzed in the time domain (e.g., Refs. [44,45]), even though frequency-domain investigations have been increased in the last decade (e.g., Refs. [46, 47]). Near-surface applications of the combined ERT/IP tomographic reconstruction often dealt with environmental engineering (e.g., Ref. [48]) and hydrogeophysics [3].

In this work, we aim to investigate the advances in ERT investigation made during the last years for near-surface applications, discussing the role played by the main parameters on data acquisition, forward modeling, and inversion and presenting strategies for removing the ambiguities arising in interpretation of geoelectric models. In Section 2.1, electric theory was briefly introduced considering electrodes both as point sources (commonly used approximation for ERT modeling) and as finite-length objects. The comparison between the two approaches will be discussed as a function of the ratio between electrode shape (length and size) and spacing.

Then, we study the role of the mesh coarseness for the accuracy of the forward solver. Data acquisition problems are discussed in Section 2.2, considering common measurement errors and exploring potential and limits of the 3D arrays. The role of inversion parameters within the ERT reconstruction and the strategies for removing ambiguities and for incorporating a priori and a posteriori information are presented in Section 2.3. Finally (Section 3), we illustrate some case studies concerning the application of ERT for archaeological purposes and soil characterization. Synthetic simulations and field data inversion presented hereafter are performed through the VEMI interface [49], developed by the authors for time-domain and frequency-domain data modeling and inversion of 2D and 3D ERT and IP data.

2 MODELING, DATA ACQUISITION, AND INVERSION

2.1 FORWARD MODELING

2.1.1 Theory

The resistive response of a subsoil has been described for years by Poisson's equation, under the hypothesis of an external point source [29]:

$$-\nabla\cdot[\sigma(\mathbf{r})\nabla\phi(\mathbf{r})]=I\delta(\mathbf{r}-\mathbf{r_s})\forall\mathbf{r}\in D \tag{1}$$

where σ is the electric conductivity, ϕ the electric potential, $\mathbf{r}\equiv\mathbf{r}(x,y,z)$ the position vector, I the injected current, δ Dirac's function, $\mathbf{r_s}\equiv\mathbf{r_s}(x_s,y_s,z_s)$ the position vector of the source, and D the domain.

The point-source hypothesis holds where the electrode spacing is large against electrode size and penetration depth that is often the case of classical ERT studies. However, during the last decade, the number of small-scale experiments grows up, particularly not only for near-surface archaeological and civil engineering applications, as mentioned before, but also for laboratory experiments (e.g., Refs. [50–53]). In these cases, we usually have limited electrode spacing compared with the electrode size. Therefore, the electrodes need to be modeled as finite-length objects. This condition may be assigned adding piecewise equations to the boundary conditions. As a consequence, we can employ Laplace's equation to define the electric response inside the domain, as follows:

$$\nabla\cdot[\sigma(\mathbf{r})\nabla\phi(\mathbf{r})]=0 \quad \forall\mathbf{r}\in D \tag{2}$$

The boundary conditions should include the finite-length behavior of the electrodes (contact impedances), the shunt effect, and the absence of current flow through the air-soil interface (Neumann condition):

$$\begin{cases} \phi+z_l\sigma\dfrac{\partial\phi}{\partial\nu}=U_l \quad \forall\mathbf{r}\in e_l;\ l=1,2,\ldots,L \\ \displaystyle\int_{e_l}\sigma\frac{\partial\phi}{\partial\nu}=I_l \quad \forall\mathbf{r}\in e_l;\ l=1,2,\ldots,L \\ \sigma\dfrac{\partial\phi}{\partial\nu}=0 \quad \forall\mathbf{r}\in D_s\backslash\left\{\bigcup_{l=1}^{L}e_l\right\} \end{cases} \tag{3a}$$

where z_l is the effective contact impedance between the l_{th} electrode and body, L the total number of electrodes, U_l the measured voltage on electrodes, e_l the subdomains representing electrodes, ν the outward normal, and D_S the air-soil interface.

Eqs. (2), (3a)–(3c) constitute the so-called complete electrode model (CEM), introduced by Cheng et al. [54] and widespread used for medical applications of the electric tomography (often called electric impedance tomography, see Ref. [55], for a review). However, for geophysical field investigations, we need to define also another boundary condition on the external boundaries D_∞, located "far enough" from the source, in terms of a Dirichlet boundary condition:

$$\phi = 0 \quad \forall \mathbf{r} \in D_\infty \tag{3b}$$

or, alternatively, through a Robin (mixed) condition:

$$\frac{\partial \phi}{\partial \nu} + \alpha \phi = 0 \quad \forall \mathbf{r} \in D_\infty \tag{3c}$$

where $\alpha = \boldsymbol{\nu} \cdot \mathbf{r}/|\mathbf{r}|^2$ [56].

When the point-source hypothesis is assumed, the boundary conditions consist only of the last of (3a) and of one between (3b) and (3c).

Many numerical approaches to the solution of Eqs. (1), (2), and (3a)–(3c) have been adopted during the last decades, both using the finite-element method (e.g., Ref. [57]) and the finite difference method (e.g., Ref. [29]). A widespread used technique is the Galërkin formulation of the finite-element method, applied by numerous authors in the last years (e.g., Refs. [58–60]).

At the end, for both cases, we need to solve the following linear system:

$$[\mathbf{Y}]\boldsymbol{\Phi} = \mathbf{I} \tag{4}$$

which requires $[\mathbf{Y}]$ to be invertible.

The latter system can be solved using the Cholesky or LU factorization [61]. Once the potential has been derived, we can compute the impedance for each quadrupole q, owing to the particular sequence of measurements [62]:

$$Z_q = \mathbf{M}_{p,c} \boldsymbol{\phi}_c \quad q = 1, 2, \ldots, N_Q \tag{5}$$

where $c = 1, 2, \ldots, C$ and $p = 1, 2, \ldots, P$ are the current and measurement patterns, q the corresponding quadrupole, $\mathbf{M}$ the measurement operator, and N_Q the number of quadrupoles. Finally, the predicted data (apparent resistivity) are given by

$$\rho_{a,q}^{pre} = K_q Z_q \tag{6}$$

being K the geometric factor.

2.1.2 Electrode effect

A proper choice of the electrode can play a nonnegligible role for near-surface ERT measurements. In fact, Rücker and Günther [60] demonstrated that significant errors in forward modeling (>2%) can arise where electrode spacing and electrode length are of the same order of magnitude. This error can be expressed in terms of electrode effect c and its deviation δ_C [60]:

$$\delta c = 100 \cdot (c - 1) \tag{7}$$

where $c = \frac{\phi_{CEM}}{\phi_{PEM}}$ and ϕ_{CEM} and ϕ_{PEM} are electric potential calculated using the CEM approach and the point-source approximation, respectively. The electrode effect can be positive or negative depending on the particular electrode type (rod, plate, etc.) and configuration (contact impedance, array, etc.).

An example of the electrode effect for a classical ERT investigation for geoexploration is depicted in Fig. 1A, where cylindrical rods having a diameter of 10 mm are grounded for a depth l and measurements were performed using a dipole-dipole array with an electrode spacing e and a dipole separation factor n. We set $l = (0.1, 0.2,$ and $0.4\,\text{m})$, $e = (0.2, 0.3, 0.5,$ and $1\,\text{m})$, and $n = (1, 2, 3,$ and $4)$, and each model is calculated using a contact impedance $z_l = 10^{-2}\,\Omega\,\text{m}^2$, a moderate value for the homogeneous conductivity equal to $10\,\Omega\,\text{m}$. Therefore, we can estimate for each model the electrode effect as a function of the $\frac{n \cdot e}{l}$ dimensionless ratio, which is a significant index of the mutual influence between a pair of electrodes. Fig. 1B shows a detail of the finite-element mesh, where the surface of the cylindrical electrode is highlighted.

Fig. 2 shows a good power-law correlation ($R^2 = 0.978$) between δc and the $\frac{n \cdot e}{l}$ ratio:

$$\delta c = -44.7 \left(\frac{n \cdot e}{l} \right)^{-1.2} \tag{8}$$

A huge effect ($\delta c > 10\%$) is seen for $\frac{n \cdot e}{l} \leq 5$, a case that often occurs in the current practice (e.g., electrodes grounded for 0.2 m, spaced <1 m using $n = 1$). On the contrary, negligible errors are made for large spacings and dipole separations, regardless of the penetration depth. These results, comparable with those after Rücker and

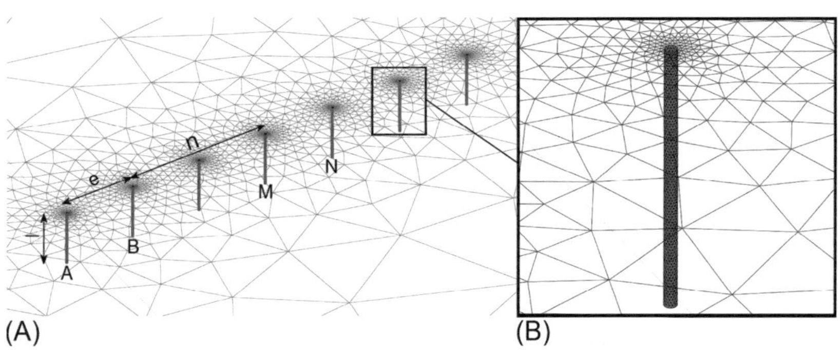

FIG. 1

(A) Finite-element discretization of a typical ERT investigation for geoexploration.
(B) Particular of the area within the *black rectangle* in (A).

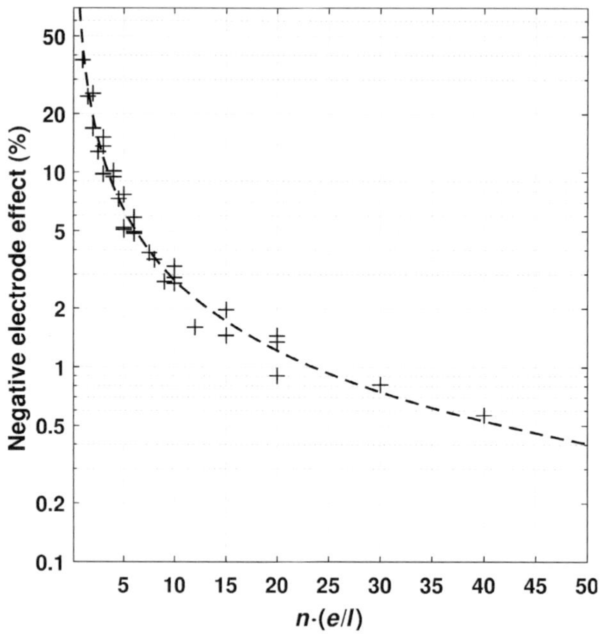

FIG. 2

Negative electrode effect as a function of the $n \cdot e/l$ ratio. The power correlation is marked with *dashed line.*

Günther [60], underline the importance of the CEM approach for small-scale surveys, where usually limited electrode spacing is employed.

However, in many archaeological or civil engineering applications of the ERT method, invasive electrodes (rods and nails) cannot be used, and surface noninvasive electrodes are often the only feasible option. In this respect, three examples of small-scale surveys for this kind of applications are presented in Fig. 3, performed using 10 mm square copper plates as electrodes.

In the case of surface electrodes, the influence of the electrode can be assessed through the ratio between the plate length d and the electrode spacing e. The respective finite mesh built using VEMI for surface electrodes is shown in Fig. 4A, where electric potential is computed for a homogenous conductive background (10 Ω m), with a contact impedance value $z_l = 10^{-2}\,\Omega\,\mathrm{m}^2$. Results are expressed in terms of electrode effect (positive) versus d/e.

Overall, the effect of the electrode is lower than for rods and negligible (<2%) for small d/e (<0.4). When the electrode spacing is small compared with the plate length (>0.5), the point-source approximation leads to significant errors in forward modeling (>3%), except for $n = 1$. The increase of the electrode effect at higher separations is due to the shorting effect of the unused electrodes between current carrying and

(A) (B) (C)

FIG. 3

Three examples of small-scale survey for civil engineering and archaeological prospection, using 100mm square plates as electrodes. (A) Monitoring an airport pavement. (B) Investigation of fractures on a damaged building. (C) Archaeological prospection for detecting foundation of a Roman "domus".

transmitter pairs, similarly to the results shown by Rücker and Günther [60]. In fact, since the electrodes are modeled as perfect conductors, they "short-circuit" the part of the boundary through which they are grounded and a proportion of the current applied to image the volume is, in this case, "shunted" through the electrodes [63].

2.1.3 Meshing

Generally, a finite-element mesh is built by making the mesh denser near the electrodes (where there are large potential gradient) and coarser elsewhere. The use of such criteria can help to avoid errors in the forward solution without drastically increase the allocated memory [64]. However, the accuracy of the forward solution can vary as a function of the number of elements and of the "quality" of the mesh. In the last decade, many researchers made use of adaptive mesh generators, for example, Netgen (https://ngsolve.org/), Distmesh [65], and Gmsh [66]. These generators are capable to build high-quality meshes—a mesh is rated good if the quality factor defined after Vauhkonen et al. [64] is higher than 0.7—progressively coarser toward the external boundary proportionally to a global grading factor *GF*. The global grading factor can be expressed as

$$GF = c_f \cdot e \tag{9}$$

where c_f is the coarseness factor and e the electrode spacing. Fig. 5 shows the influence of c_f for the accuracy of the forward solver both for 2D and for 3D modeling.

The finite-element meshes were built using Netgen within the VEMI interface, considering 48 electrodes spaced 5m apart with a dipole-dipole array (2D case) and a grid of 48 × 2 electrodes (spaced 5m apart in the x-direction and 20m in the y-direction) with a pole-dipole array (3D case). Electric potential was calculated

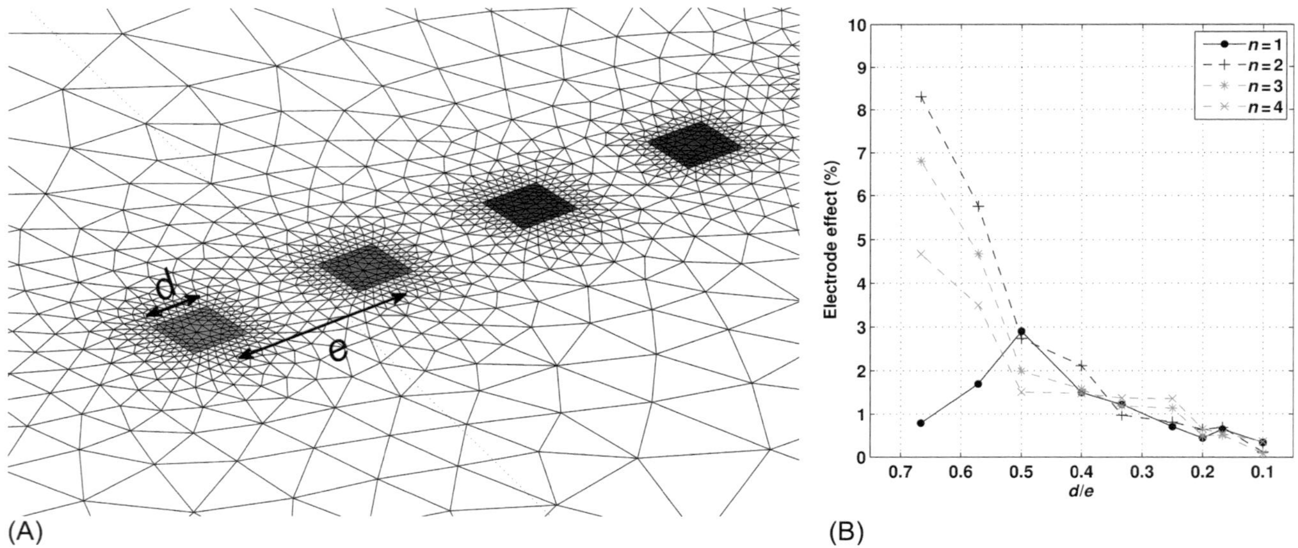

FIG. 4

(A) Finite-element discretization of surface electrodes. (B) Electrode effect as a function of the plate length versus electrode spacing ratio.

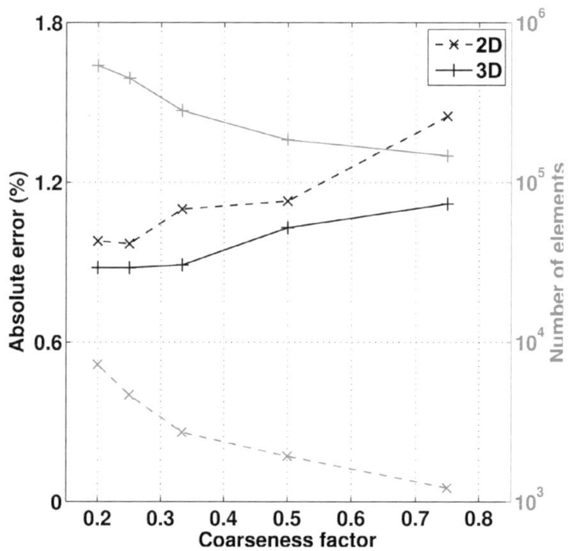

FIG. 5

Number of elements and absolute error between numerical and analytic solutions as a function of the coarseness factor for 2D and 3D geometries.

with the CEM approach for a homogenous model having a resistivity of $1\,\Omega$m, with electrodes modeled as cylinders having a diameter of 20mm and a length of 50mm with a contact impedance of $10^{-2}\,\Omega\text{m}^2$. Overall, the effect of the mesh coarseness on the accuracy is not significant (<1.5%) and noticeable only for $c_f > 0.5e$.

2.2 DATA ACQUISITION

In the time domain, we generally make use of a DC electric source, measuring the apparent resistivity ρ_a:

$$\rho_{a,q}{}^{obs} = K_q \frac{V_{p,q}}{I_q} \tag{10}$$

where V_p is the measured voltage during the application of the DC current I.

Recently, acquisition protocols have been evolved to avoid problems with electrode polarization: it is common to switch current electrodes frequently to dissipate charge buildup or use a plus-minus-plus injection cycle to minimize electrode polarization and cancel out spontaneous potential [67]. Another common practice is to perform reciprocal measurements by interchanging current and voltage electrodes to have a control on the noise level [68]. In addition to this, the influence of the electrode material has been analyzed for both AC [69] and DC [70] sources: errors due to a nonideal choice of the electrode material do not affect significantly the recorded dataset (<1%), differently from IP acquisitions.

However, one of the main drawbacks of ERT, particularly using noninvasive electrodes as plates, is surely the mitigation of the contact resistances, directly linked to the improvement of the signal-to-noise ratio. The contact resistance R_C can be written as the combination of a geometric (R_G) and interfacial (R_I) terms [71]:

$$R_C = R_G + R_I \tag{11}$$

where R_G thus describes the effect of the geometry of the electrode while R_I takes into account the interfacial resistance between electrode and soil due to the moisture conditions. Generally, contact resistance measured from surface electrodes is higher than those recorded from rods, steaks, and nails. This effect is mainly due to the interfacial component, since the contact area of surface and invasive electrodes is similar. By way of example, the square plate seen in Fig. 3 (100 × 100 mm) has a contact area comparable with that of a rod having a diameter of 20 mm inserted 150 mm into the ground, a common configuration for DC surveys. On the other hand, overall interfacial component of the contact resistance is higher for plates with respect to rods, because the surrounding medium (e.g., concrete, asphalt, and stones) has a lower moisture content due a lower porosity and/or indoor conditions [72]. Although a good signal-to-noise ratio can be ensured also for surface electrodes by improving the interfacial component, that is, using a conductive gel or water, in the presence of asphaltic bitumen-tar mixtures, the use of nondestructive electrodes is not feasible due to the high insulting nature of tar [72]. We present an example of ERT dataset acquired along a single profile on tarmac with square plates (100 × 100 mm) and steel rods (2 cm of diameter, grounded 15 cm below the tarmac). Data were acquired using a SyscalPro48 resistivity meter (IRIS Instruments) with 48 electrodes spaced 1 m apart and a dipole-dipole array using a constant maximum potential value at the transmission circuit ($V_{ab} = 800$ V).

We compare in Fig. 6 the two acquisitions in terms of signal-to-noise ratio (SNR), expressed as follows, for each quadrupole q:

$$SNR = \frac{\overline{R}_q}{s_q} \tag{12}$$

where $\overline{R}$ is the observed mean resistance value and s the corresponding standard deviation.

There are significant differences (about two decades) between the two datasets in terms of SNR. Since the noise level of the site can be assumed, the same for both acquisitions, the low signal-to-noise ratio is mainly due to the high contact resistances encountered using surface electrodes, as expected.

An additional aspect, which should be considered for planning an ERT survey, is surely the resolution needed for a specific array for retrieving a reliable reconstruction of the subsurface [73]. Nowadays, many archaeological and engineering applications require a high-resolution model of the subsurface, where the expected targets need to be reconstructed with a high degree of accuracy. To this end, three-dimensional ERT is often the only feasible option to investigate the hidden subsoil in order to reconstruct complex targets (e.g., foundation and cavity net). The

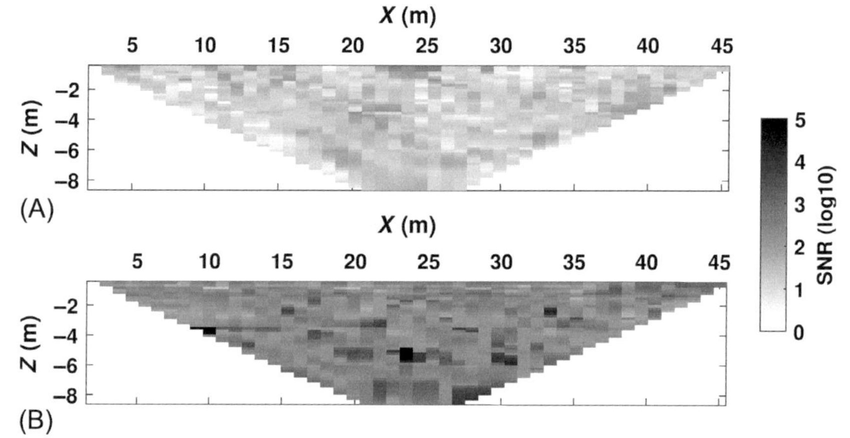

FIG. 6

Comparison between plates and rods for ERT acquisition on tarmac given in terms of signal-to-noise ratio for plates (A) and rods (B).

three-dimensional acquisition is relatively time consuming and often requires customized site-specific arrays [74]. The role of the three-dimensional array for the reliability of an ERT reconstruction can be assessed in terms of cumulative sensitivity [75]. In fact, the cumulative sensitivity distribution shows how "sensitive" are the model cells to any changes of resistivity occurring in the subsoil and consequently the zones where the reconstruction is reliable. It can be simply derived as follows [25]:

$$S_k = \frac{\sum_{q=1}^{N_Q} |J_{q,k}|}{N_Q} \quad k = 1,2,\ldots,N_E \tag{13}$$

where $\mathbf{J}$ is the Jacobian matrix related to the inversion process described in Section 3 and N_E is the number of finite elements. The absolute value of $\mathbf{J}$ is given to avoid situations, where a negative sensitivity value for one measurement cancels out the positive value for another measurement.

An example of cumulative sensitivity of two different 3D arrays is given in Fig. 7.

The aim of the study was the detection of the foundation of a Roman building, named "Casa di Diana" and located at the Ostia Antica archaeological site (Rome, Italy). Two different arrays were designed to achieve this goal, both using 48 electrodes spaced 3 m apart: a perimeter array (Fig. 7A) and a combination of U- and M-shaped arrays, where electrodes were positioned also inside the house (Fig. 7B). The corresponding cumulative sensitivity maps, shown in Fig. 7C and D, display substantial differences in terms of resolution achieved in the three main directions, stressing once more the importance to correctly plan an ERT survey as a function of the expected target.

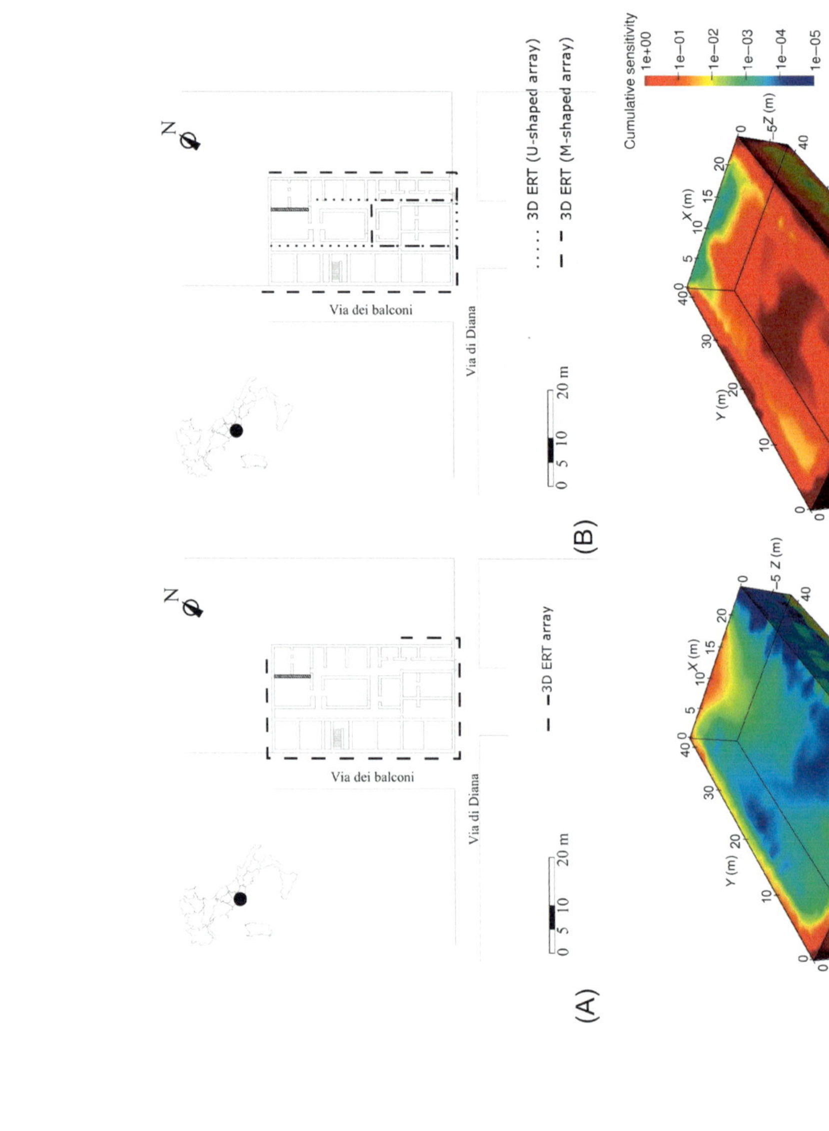

FIG. 7

Three-dimensional array for detecting foundation of the "Casa di Diana." (A) Perimeter array. (B) Combination of U- and M-shaped arrays. (C) Cumulative sensitivity for the perimeter array. (D) Cumulative sensitivity for the U- and M-shaped arrays.

2.3 INVERSION

During the last decades, inversion of electric resistivity tomography data has been performed almost only using local minimization search algorithms via a Tikhonov regularization [76–78]. In fact, the high computational effort, required for the solution of the nonlinear forward problem, prevents the application of global inversion methods, where the minimum number of random steps for achieving convergence is often higher than 10^4. However, the rapid improvement in the performance of processors could allow the application of global inversion for the majority of end users in the next decade.

During the last years, many research groups have developed software for geophysical data inversion. Commonly known closed-source codes developed for ERT data inversion are RES2DINV/RES3DINV (Geotomo Software, [15]), AarhusInv [79], ERTLab (Geostudi Astier and Multiphase Technologies), BERT [80], and COMSOL Multiphysics (Comsol Inc.). Although these are all reliable package for ERT data inversion, many researchers have been focused on developing open-source codes, with the primary aim to share data and promote collaboration between different groups and to have a more flexible tool compared with the available commercial software. A Matlab tool for 3D ERT data inversion is RESINVM3D [81], where inversion is implemented through an inexact Gauss-Newton solver, with the model update being calculated using a preconditioned conjugate gradient algorithm. The EIDORS package [55] is able to perform 3D tomographic inversion of resistivity data with a Gauss-Newton inversion. It was firstly developed for medical application on human thorax (cylindrical domains) and progressively adapted for geophysical purposes. It can be also used in combination with the VEMI interface [49] in order to deal with the whole range of geophysical data (time and frequency domain). Karaoulis et al. [82] recently developed a software in Matlab (IP4DI) to perform a tomographic time-lapse inversion of ERT data with an active time-constrained approach. Actually, inversion of ERT data is now a well-established and tested technique. Therefore, using different codes from the abovementioned list often leads to slight differences in the final inverted resistivity model. In the following section, we analyze the choice of the inversion parameters for the Gauss-Newton inversion scheme, widely employed in many case studies.

2.3.1 Inversion parameters

Generally, inversion of resistivity data is accomplished by minimizing an augmented least square (l_2-norm) functional, yielding the so-called Tikhonov regularization:

$$\operatorname{argmin}\left\{ \|\Phi_d(\mathbf{m})\|_2^2 + \alpha \|\Phi_m(\mathbf{m})\|_2^2 \right\} \tag{14a}$$

where $\Phi_d(\mathbf{m})$ and $\Phi_m(\mathbf{m})$ are data misfit and model roughness and $\alpha > 0$ is the regularization parameter. Eq. (14a) can be expressed as follows:

$$\operatorname{argmin}\left\{ \|\mathbf{W}(\mathbf{d} - g(\mathbf{m}))\|_2^2 + \alpha \|\mathbf{L}(\mathbf{m} - \mathbf{m}_0)\|_2^2 \right\} \tag{14b}$$

where **d** is the observed data vector ($d_q = \ln\left(\rho_{a\,q}{}^{obs}\right)$), $g(\mathbf{m})$ the predicted data vector ($g_q(\mathbf{m}) = \ln\left(\rho_{a\,q}{}^{pre}\right)$), **L** is an appropriate regularization matrix (e.g., Refs. [80, 81, 83]), **m** the model vector ($m_k = \ln(\rho_k)$), and $\mathbf{m}_0$ the initial guess for **m**.

The optimization problem in Eq. (14b) is often solved by means of an iterative Gauss-Newton scheme, yielding

$$\delta\mathbf{m}_h = \left[\mathbf{J}(\mathbf{m}_h)^{\mathrm{T}}\mathbf{W}^{\mathrm{T}}\mathbf{W}\,\mathbf{J}(\mathbf{m}_h) + \alpha_h\mathbf{L}^{\mathrm{T}}\mathbf{L}\right]^{-1}\cdot\left\{\mathbf{J}(\mathbf{m}_h)^{\mathrm{T}}\mathbf{W}^{\mathrm{T}}\mathbf{W}[\mathbf{d} - g(\mathbf{m}_h)] - \alpha_h\mathbf{L}^{\mathrm{T}}\mathbf{L}(\mathbf{m}_h - \mathbf{m}_0)\right\} \tag{15}$$

where $\delta\mathbf{m}$ is the model update vector, **J** the Jacobian matrix ($J_{q,k} = \frac{\partial d_q}{\partial m_k}$), **W** the data weight matrix ($W_q = \mathrm{diag}\left(\frac{1}{s_q/100}\right)$), s the observed standard deviation expressed in %, h the outer iteration, k the finite element, and T the transpose matrix.

However, since the least-squares method uses the square of the data misfit, it results particularly sensitive to bad data points [84]. In this case, the minimization of the l_1-norm (the so-called robust inversion) is more appropriate, yielding

$$\begin{aligned}\delta\mathbf{m}_h = &\left[\mathbf{J}(\mathbf{m}_h)^{\mathrm{T}}\mathbf{W}\mathbf{R}_{\mathbf{d}}{}^{\mathrm{T}}\mathbf{W}\,\mathbf{J}(\mathbf{m}_h) + \alpha_h\mathbf{L}^{\mathrm{T}}\mathbf{R}_{\mathbf{m}}\mathbf{L}\right]^{-1}\\ &\cdot\left\{\mathbf{J}(\mathbf{m}_h)^{\mathrm{T}}\mathbf{W}^{\mathrm{T}}\mathbf{R}_{\mathbf{d}}\mathbf{W}[\mathbf{d} - g(\mathbf{m}_h)] - \alpha_h\mathbf{L}^{\mathrm{T}}\mathbf{R}_{\mathbf{m}}\mathbf{L}(\mathbf{m}_h - \mathbf{m}_0)\right\}\end{aligned} \tag{16}$$

where $\mathbf{R}_{\mathbf{d}}$ and $\mathbf{R}_{\mathbf{m}}$ are weighting matrices that give weights in the inversion process to data misfit and model roughness vectors, respectively [85].

The iterative procedure is stopped when the root mean square (RMSE) or the absolute error (AE) is below a certain threshold (e.g., error difference between two consecutive iterations under 1%), for the standard and robust procedure, respectively:

$$RMSE = \sqrt{\frac{\sum_{q=1}^{N_Q}\left(\frac{\rho_{a\,q}^{obs} - \rho_{a\,q}^{pre}}{\frac{s_q}{100}\rho_{a\,q}^{obs}}\right)^2}{N_Q}};\quad AE = \frac{\sum_{q=1}^{N_Q}\left(\frac{\left|\rho_{a\,q}^{obs} - \rho_{a\,q}^{pre}\right|}{\frac{s_q}{100}\left|\rho_{a\,q}^{obs}\right|}\right)}{N_Q} \tag{17}$$

We compare in Fig. 8 resistivity models resulting from standard least squares (Figs. 8A and C) and robust (Fig. 8B and D) inversion procedures using a fixed damping value (0.1) or an optimized damping, chosen at each iteration as the value that minimizes the RMSE [50].

The ERT survey, described in Cardarelli and Fischanger [86], aimed to detect a buried tomb in the necropolis of Montelibretti (Rome, Italy): it consists of a single ERT line with 28 electrodes spaced 0.5 m apart with a Wenner array. The true position of the tomb is marked with a black rectangle in Fig. 8. The high-resistivity target related to the tomb is always detected, even though with higher amplitude response and lower RMSE for optimized damping than for a fixed one. In this sense, minor changes can be noticed also between robust and standard inversion.

However, only achieving a reasonably low error comparable with the recorded noise level is not sufficient to rate "good" a geoelectric reconstruction. In fact, the inverted resistivity model could be of nonunivocal interpretation, both referring

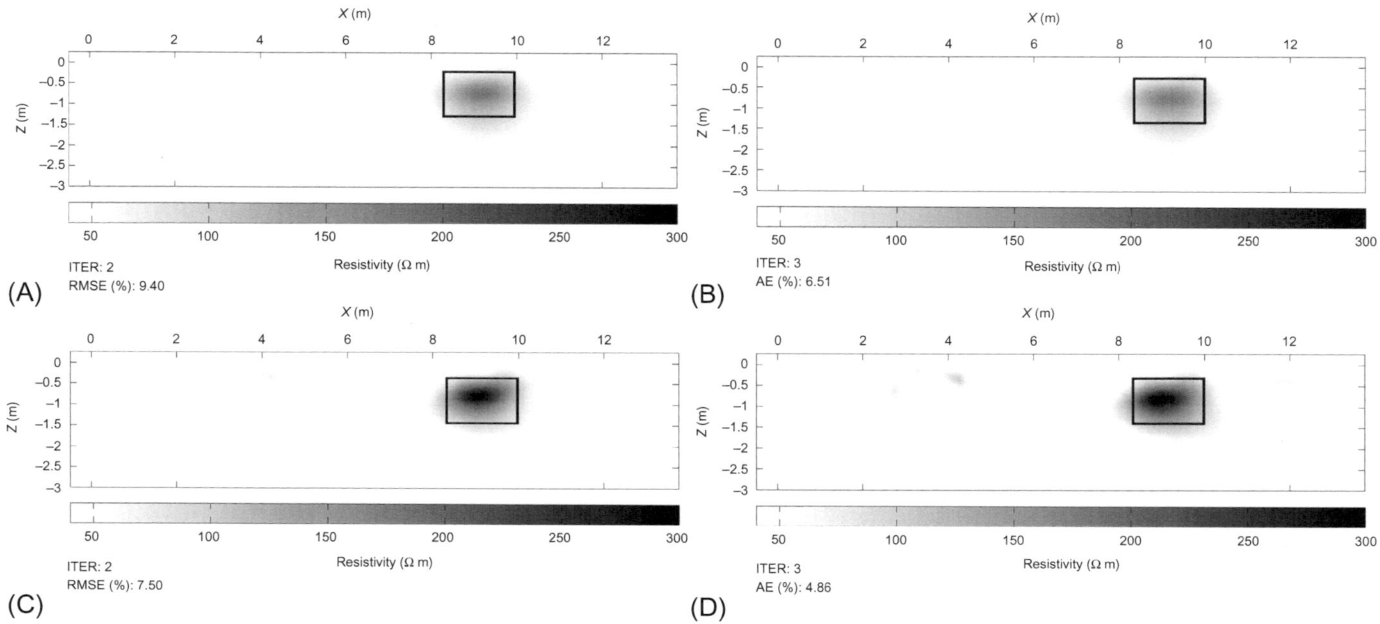

FIG. 8

Inversion of ERT data acquired for detecting an Etruscan tomb [86]. (A) Standard inversion with fixed damping=0.1. (B) Robust inversion with fixed damping=0.1. (C) Standard inversion with optimized damping. (D) Robust inversion with optimized damping. The true position of the tomb is marked with a *black rectangle*.

to the effective size and to the depth of the detected targets. Firstly, since the sensitivity (that is the Jacobian matrix **J**) rapidly decreases in depth, any sharp variation occurring at the deeper zones could not be followed by a rapid change in resistivity. Therefore, the speculative interpretation of the inverted model may lead to important errors both for the geometry of the anomaly and for the resistivity value. Secondly, in case of numerous resistive-to-conductive transitions, the effective thickness of the detected layers can be biased. In light of this, the minimization of these ambiguities represents a key factor for many small-scale applications, where the geometric and physical properties of a geophysical target may be required with a high accuracy.

2.3.2 Model appraisal

We see in Section 2.2 that a popular method to assess the reliability of an ERT reconstruction is the calculation of the cumulative sensitivity matrix that shows how "sensitive" are the model cells to any changes of resistivity occurring in the subsoil and consequently the maximum depth where the reconstruction is reliable. However, there are other methods employed during the last decades for this purpose, starting from early works addressing model calibration and uncertainty prediction [87]. A rapid method is the calculation of the model resolution matrix [88], defined as the ratio between the model parameter vector obtained by the inversion and the vector of true (unknown) parameters [3]. Since it often requires a high computational effort, other authors proposed different indexes for the evaluation of the reliability of the inversion. Oldenburg and Li [89] introduced the depth of investigation (DOI) index, as a function of two independent inversions $m_{i,j}^{(1)}$ and $m_{i,j}^{(2)}$ performed on different reference models $\rho_b^{(1)}$ and $\rho_b^{(2)}$:

$$DOI_{i,j} = \frac{m_{i,j}^{(1)} - m_{i,j}^{(2)}}{\rho_b^{(1)} - \rho_b^{(2)}} \tag{18}$$

where i and j are the indexes of the elements in the x- and z-directions, respectively, (for a 2D case) and all variables are expressed as log-transformed quantities. The two reference models are usually taken as 0.1 and 10 times the mean apparent resistivity of the field dataset. If the DOI index is close to zero, the reconstruction is reliable, as the inversion does not depend on the background value. On the contrary, low credibility would be given to the regions where DOI approaches the unity. Marescot et al. [90] considered a normalized DOI index for its maximum value, while Deceuster et al. [91] used the median DOI value of the deepest row of the model cells for the normalization.

Two examples of model appraisal are given in Fig. 9 with reference on different applications: deep bedrock detection at the Rieti Plain in Central Italy (Fig. 9A, C, and E) and detection of buried voids and cavities in the urban area of Rome, Italy (Fig. 9B, D, and F). The survey geometry consists of a pole-dipole array applied on a single ERT line with 48 electrodes spaced 5 and 4 m apart.

The bedrock detected by ERT at a depth of 45 m (resistivity $>100\ \Omega$m in Fig. 9A) cannot be inferred with a sufficient degree of confidence, given the low cumulative

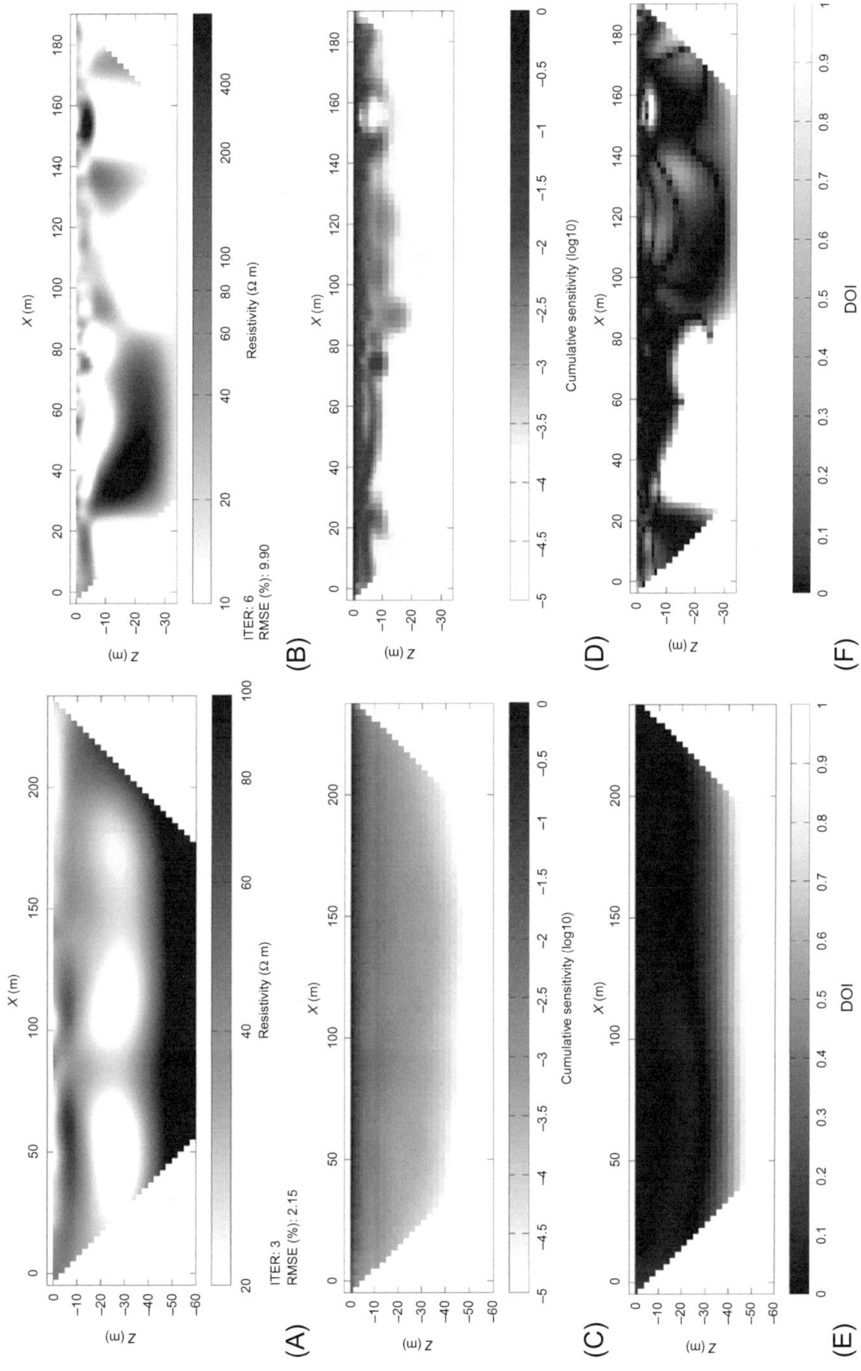

FIG. 9 Inversion of ERT data acquired for bedrock (A) and cavity (B) detection. Cumulative sensitivity (C and D) and DOI index (E and F) calculated for both examples.

sensitivity ($<10^{-5}$ in Fig. 9C) and the high DOI value (>1 in Fig. 9E) related to the bedrock. The same principle holds for the detected cavity (resistivity higher than 500 Ωm in Fig. 9B). We have low sensitivities ($<10^{-5}$ in Fig. 9C) and high DOI values (>1 in Fig. 9E) in correspondence of the detected anomaly. Therefore, for both cases, we cannot associate the detected anomalies to the corresponding geologic features (bedrock and cavity) with a degree of uncertainty, compatible with the engineering applications (drilling a water well within the bedrock and drilling a borehole for safeguarding of cavities).

In this sense, the use of a more quantitative procedure can help for reducing the degree of uncertainty. Here, we present a new iterative procedure, called model validation, early proposed by Narayan et al. [92] and fully described by De Donno and Cardarelli [25], developed with the aim to give a comparison between the field data inversion and piecewise synthetic models (Fig. 10).

Given the model $\mathbf{m}_F^{inv}$, resulting from inversion of field data $\mathbf{d}_F$, the validation proceeds as follows:

1. Create a piecewise synthetic model at the first iteration $\mathbf{m}_S(h)$, on the basis of the field inverted model (a posteriori information).
2. Generate the corresponding synthetic dataset $\mathbf{d}_S(h)$, adding a zero-mean Gaussian noise having a standard deviation comparable with the mean standard deviation value of the recorded field dataset.
3. Invert the synthetic dataset to achieve an inverted model $\mathbf{m}_S^{inv}(h)$.
4. Compare $\mathbf{m}_F^{inv}$ and $\mathbf{m}_S^{inv}(h)$ in terms of RMSE or AE of the model cells (finite elements).
5. Assess if the RMSE or the AE is lower than a given tolerance (e.g., 10%). If yes, $\mathbf{m}_S(h)$ is the final resistivity model of the investigated site; otherwise, we need to create an updated synthetic model and to iterate the above procedure.

At this stage, the model validation algorithm uses a trial and error procedure for the choice of the updated synthetic model. In this sense, the role of the operator is still

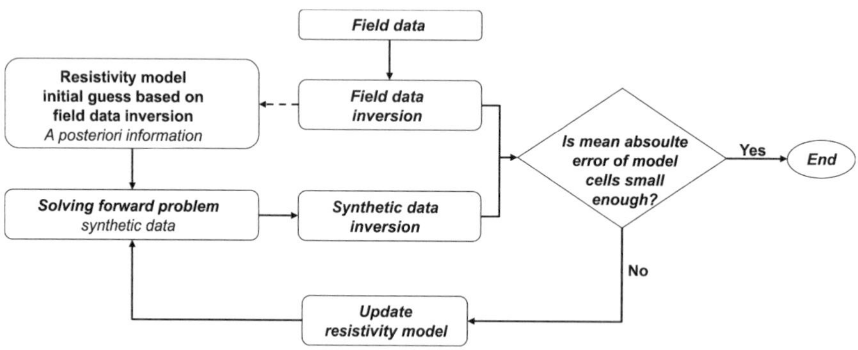

FIG. 10

Flow chart of the model validation procedure.

crucial in order to steer the search toward the "best" model and consequently to reduce the number of trials. Nevertheless, an easily implementable improvement of this algorithm could be the insertion of an inner loop that automatically minimizes the misfit between $\mathbf{m}_F^{inv}$ and $\mathbf{m}_S^{inv}$, using the well-known optimization procedures.

An example of application of the model validation procedure is given in Fig. 11. It represents an old tuff quarry used for years as illegal waste disposal and now covered. Electric resistivity tomography investigation was executed on a 2D line with 48 electrodes spaced 3 m apart with a pole-dipole array, in order to confirm the presence of the waste disposal and to detect the size of the waste mass.

Results of field data inversion are shown in Fig. 11A. Using the model validation procedure, we attain a final resistivity model (Fig. 11C), in which we can evaluate the maximum depth and the particular shape of the waste disposal. In fact, the corresponding inverted model (Fig. 11B) related to synthetic data generated using the model in Fig. 11C displays a low mean absolute error (1.96%), evaluated in terms of resistivity of the model cells with respect to the field model (Fig. 11A).

2.3.3 Using a priori information

An important role for ERT data inversion is played by the use of a priori information. In fact, information about geology, shape, or type of expected target may be available at the time when a geoelectric survey has to be planned. Therefore, it could be pivotal to use appropriately this information, possibly embedding it within the inversion process. A way to consider this additional information is to insert it as inequality constraints on the model parameters [92a]:

$$\mathbf{m}' = \ln\left(\frac{\mathbf{m}-\mathbf{a}}{\mathbf{b}-\mathbf{m}}\right) \tag{19}$$

where $\mathbf{m}'$ is the new model vector and $\mathbf{a}$ and $\mathbf{b}$ are vectors containing minimum and maximum acceptable values for the study parameter. This procedure can be applied only within a particular subdomain or in the whole domain. The benefit of this procedure should be evaluated in terms of error and of speed of convergence. An example of application of inequality constraints is given in Fig. 12 for the simulation of an ERT investigation (48 electrodes spaced 3 m apart with a dipole-dipole array) of a cutoff wall (resistivity = 5000 Ωm) extended up to a low-permeability conductive layer.

Data inversion without adding inequality constraints is shown in Fig. 12B, where RMSE is equal to 9.81% at the fourteenth iteration. If a priori information is considered (resistivity can vary only between 4000 and 6000 Ωm within the wall), the speed of convergence grows up substantially, and the RMSE is lower (5.68% at the fourth iteration), close to the noise level added to the simulation (zero-mean Gaussian noise with 5% of standard deviation). When the RMSE at each iteration is plotted against the number of iteration, we can have a confirmation of the step forward made using a priori information in terms of convergence and error level.

Therefore, for such case, we can execute a preliminary inversion without any constraints to assess the geologic setting (position of low-permeability layer) and the

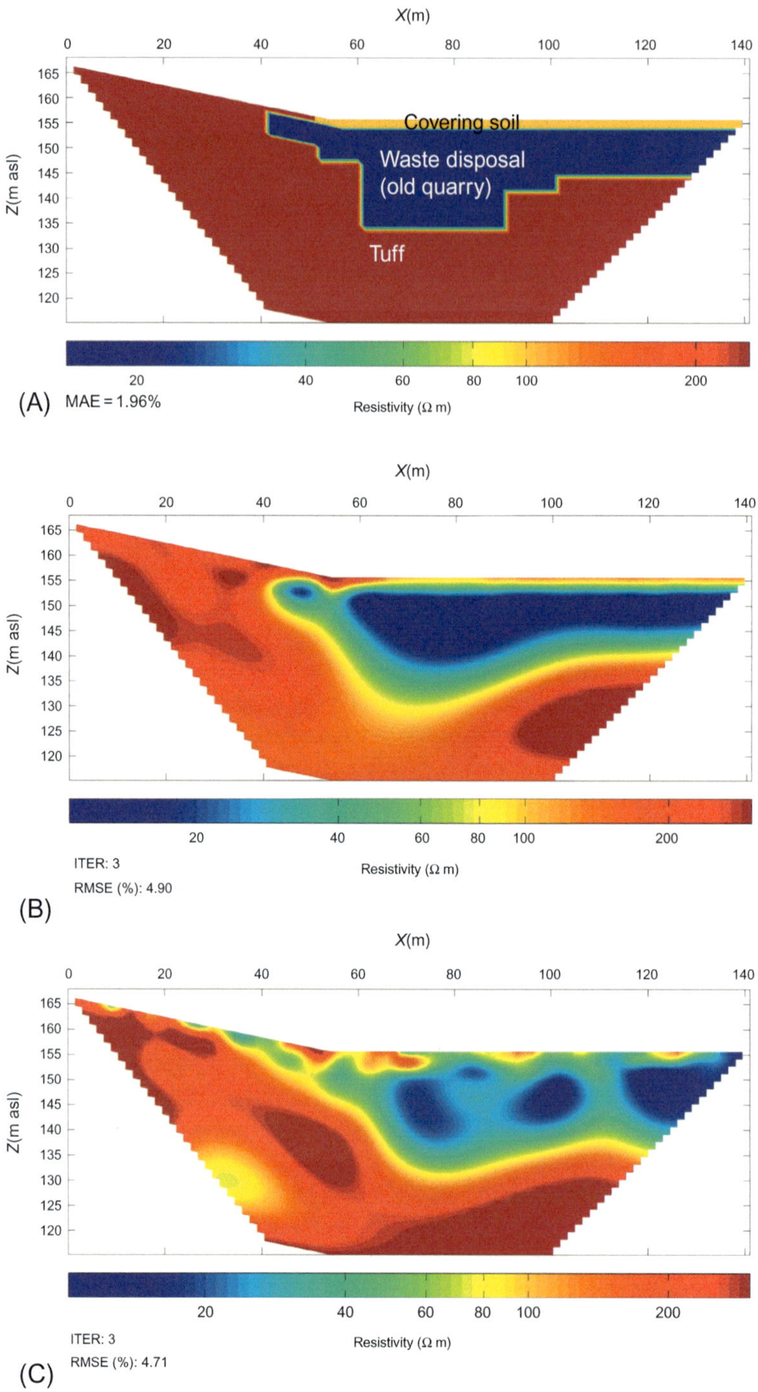

FIG. 11

Model validation procedure applied for the detection of an unauthorized waste disposal. (A) Inversion of field data, (B) inversion of synthetic data at last iteration, and (C) final resistivity model. The mean absolute error of model cells is 1.96%.

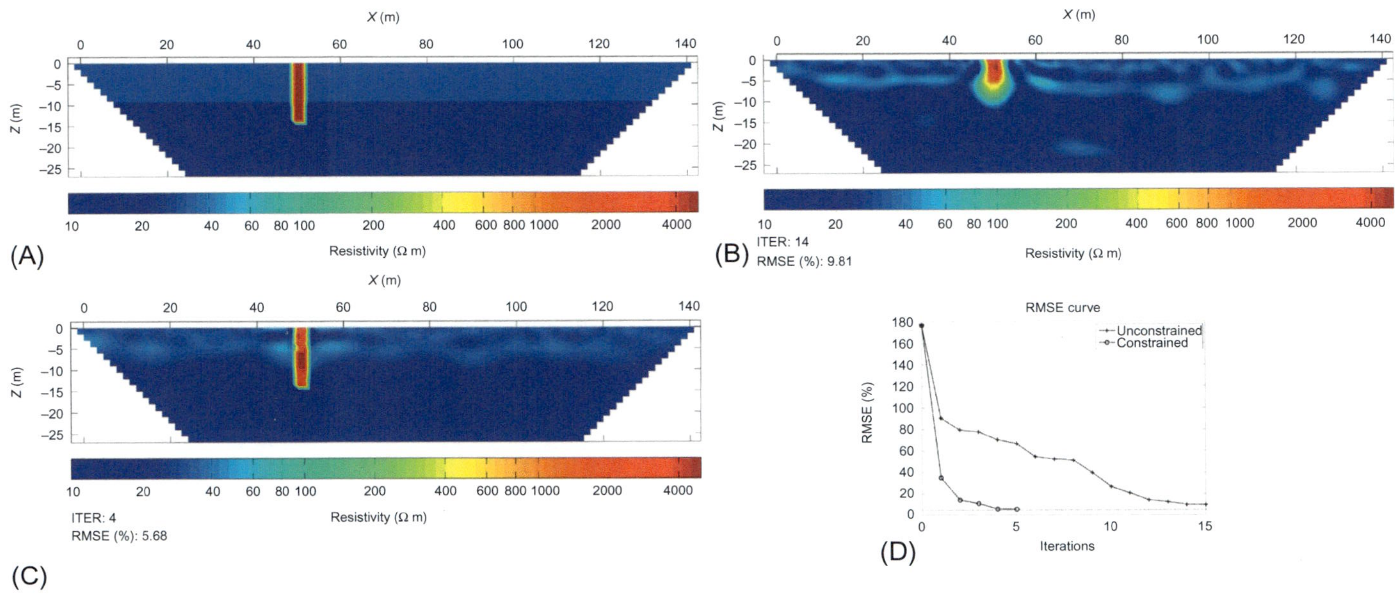

FIG. 12

(A) Synthetic model of a cutoff wall extended up to a low-permeability conductive layer. (B) Inverted model without using a priori information. (C) Inverted mode using a priori information. (D) Comparison of RMSE curves for the two methods.

approximate position of the wall. Then, knowing the usual dimension and shape of the cutoff walls, we can perform a further inversion using inequality constraints on the resistivity values of the wall in order to estimate the improvement on final error and speed of convergence. The final model can be also used to evaluate inhomogeneities of the wall seen as resistivity changes within the area where inequality constraints are applied.

In the following section, we present two examples of application of the above-mentioned procedures (model appraisal and validation) related to archaeological prospection and engineering geology, where the whole geoelectric processing (modeling, inversion, and appraisal) is managed through the VEMI interface.

3 CASE STUDIES

3.1 ARCHAEOLOGICAL PROSPECTION

3.1.1 Site description and acquisition parameters

Before the Great Jubilee of the Roman Catholic Church (2000), the Municipality of Rome planned a restoration of the Basilica of Maxentius and Constantine, one of the most important monuments of the Roman age, located close to the Colosseum and the Roman Forum. To this end, geoelectric and seismic investigations were executed both inside and outside the monument for the characterization of the geologic background and for detection of foundation. The first phase of the survey encompassed the use of VES, 2D ERT, and seismic refraction, while during the second phase, four 2D ERT parallel lines were acquired inside the Basilica. Results of the first phase were fully described in Cardarelli and Di Filippo [93], where foundation walls (strip footings) were identified from a depth of few centimeters from the pavement. In the same paper, the authors present some preliminary results of the second phase that highlight slightly difference of foundation depths. In fact, foundation of the northern aisle of the Basilica seems to be deeper with respect to the southern aisle, due to an increase of the bedrock depth in that zone.

In the following section, we perform a pseudo-3D inversion of the four ERT profiles acquired in the second phase (L1, L2, L3, and L4 in Fig. 13) in order to confirm the previous evidences by a volumetric reconstruction of the foundation wall.

ERT lines were acquired through the resistivity meter SyscalPro48 (IRIS Instruments) with 48 electrodes spaced 1.5 m apart using a Wenner array ($a_{max} = 15$) to ensure a good signal strength. The electrodes are small nails (10 cm long with a diameter of 5 mm), grounded in the interspace between adjacent paving slabs. The contact resistance values are always between 5 and 15 kΩ. Plate electrodes are not applicable here due to very high contact resistance values (>100 kΩ).

Forward computation is based on the CEM assumption using a dual reconstruction with $c_f = 0.3$. Inversion is achieved through the minimization of the L1-norm using the optimized damping procure (initial damping value = 0.1), without adding any a priori information on the resistivity model.

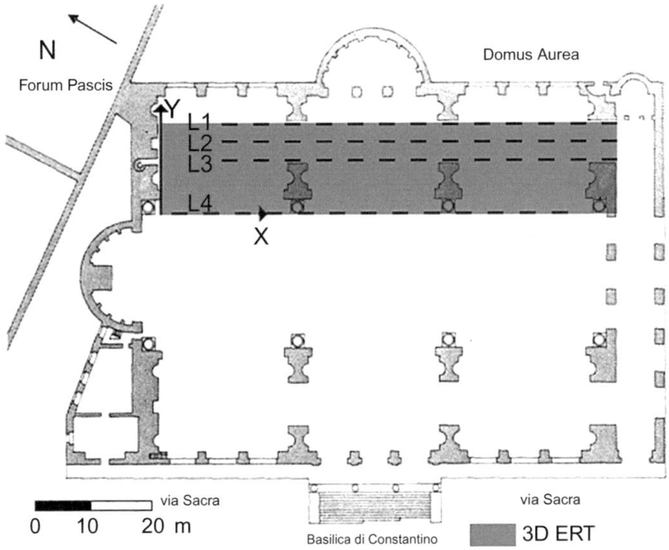

FIG. 13

Basilica of Maxentius and Constantine (Rome). ERT investigation is marked with a *gray area*.

3.1.2 Results

Results of pseudo-3D inversion achieved by the VEMI algorithm are shown in Fig. 14 in terms of horizontal slices at different depths (1.5, 3, 5, and 7 m from the pavement).

The high-resistivity zones are located in correspondence to the pillars of the north-eastern aisle (K-K′, L-L′, and M-M′ in Fig. 14), while elsewhere, the resistivity is lower due to the presence of the clayey soil. The shape of the anomalies changes as a function of depth: In the northern part of the model (pillars K′, L′, and M′), high resistivity is seen up to the deepest zones (depth = 7 m, Fig. 14D), while at the same depth, the effect of the foundation of the southern pillars (K, L, and M) vanishes. These anomalies, highlighted also in the first phase of the survey, were further confirmed by archaeological excavation to be the foundation of the Basilica (strip footings). The cumulative sensitivity of the whole survey (Fig. 15) is quite homogenous, except at lateral boundaries and at the zone where pillars are present ($y = 3$–6 m). Therefore, we can interpret the anomalies to reconstruct a quantitative model for the foundations.

The early hypothesis for which foundations have different thicknesses depending on the position of pillars is confirmed by the volumetric reconstruction of resistivity values higher than 50 Ω m given in Fig. 16A. The investigated volume is superimposed on the archaeological model made in the first phase of survey (Fig. 16B).

It is clearly demonstrated that resistivity anomalies are deeper for northern pillars (K′, L′, and M′) with respect to the southern ones (K, L, M). In the first phase, this

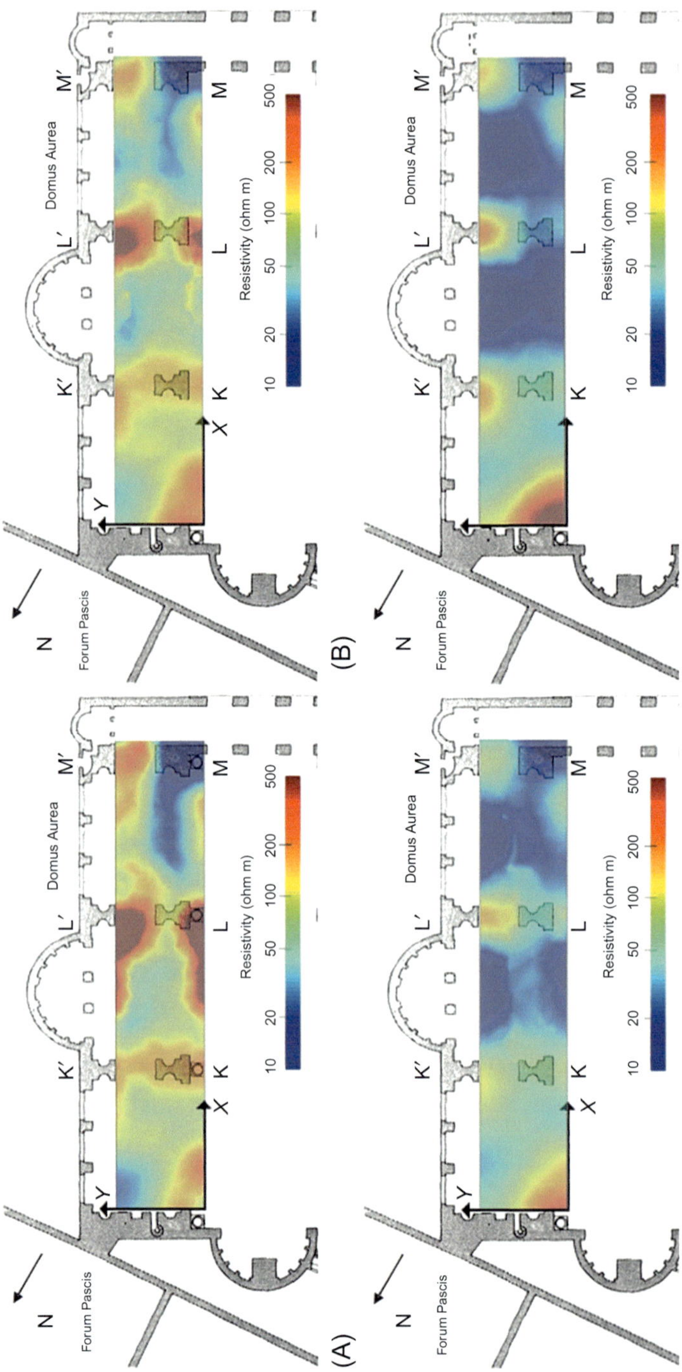

FIG. 14

Basilica of Maxentius and Constantine (Rome). Inversion of 3D ERT data. Horizontal slices are drawn at $z=1.5$ (A), 3 (B), 5 (C), and 7 m (D) from surface. RMSE = 12.3%.

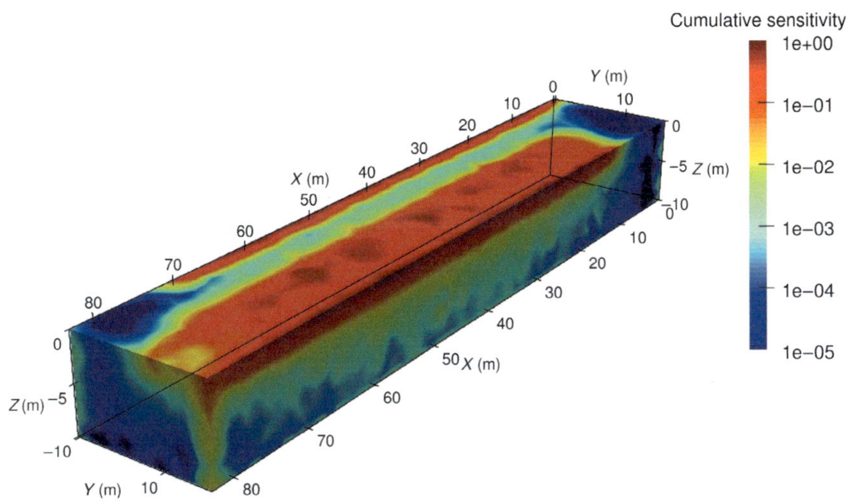

FIG. 15

Basilica of Maxentius and Constantine (Rome). Cumulative sensitivity of pseudo-3D ERT.

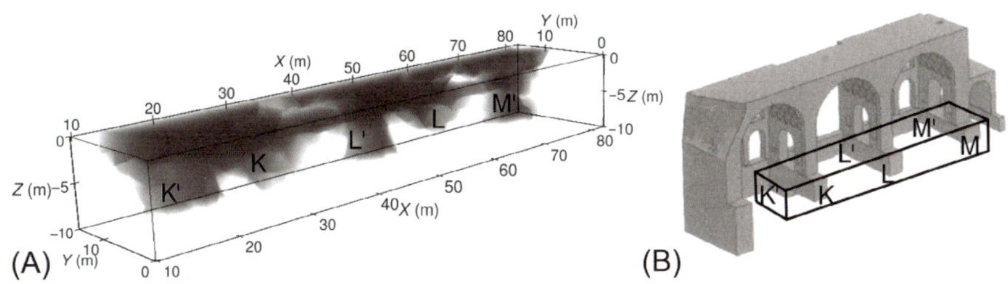

FIG. 16

Basilica of Maxentius and Constantine (Rome). Inversion of 3D ERT data. Volumetric representation of resistivity values higher than 50 Ω m. View from the western (A) corners and sketch of the building structures and buried foundations (B).

effect seemed only visible on K-K′ pillars (Fig. 16B), while as a matter of fact, the three foundations have similar shapes (deeper in the north-eastern direction and shallower in the south-western one), as can be expected from the archaeological and geologic information.

3.2 ENGINEERING GEOLOGY

3.2.1 Site description and acquisition parameters

This case study is an urban park (Villa de Sanctis) located in the fifth submunicipality of Rome. Many evidences of the presence of an anthropogenic dense cavity net have been found for years by geologic and speleological investigations, even though an extensive mapping of the net throughout the park is still underway. The cavity net is located within the deep volcanic deposits (pozzolane nere), whose roof was estimated at a depth of 10m, by geologic information [94]. The overlying volcanic deposits are the pozzolanelle (0–3m), a thin layer of tuff (Tufo Lionato, 3–5m), a paleosol stratum (5–9m), and the pozzolane nere (from 9m). Two-dimensional ERT profiles were acquired on a selected area (around 100 × 100m), using the SyscalPro48 resistivity meter with 48 electrodes spaced 2m apart and a dipole-dipole array with $a_{max} = 5$ and $n_{max} = 6$ (Fig. 17).

The electrodes (40cm-long stainless steel rods having a diameter of 1.5cm) were grounded for 20cm in the subsoil, yielding very low contact resistance values (<3kΩ).

The point-source hypothesis was used for forward modeling together with a dual reconstruction process with $c_f = 0.25$.

Results of ERT inversion, fully described by Bianchi Fasani et al. [94], highlighted that the cavity roof seems to be deeper than expected from the geologic model (15m instead of 10m). Below, we report the results of a 2D ERT inversion for the L15 ERT line selected because it crossed the cavity net in three well-defined zones (C1, C2, and C3 in Fig. 17). In fact, theoretically, a 2D inversion can generate artifacts or lead to a biased position for voids, due to the three-dimension shape of the cavity net. However, the particular position of the L15 line with respect to the cavity net (Fig. 17) helped us to minimize the spurious effect of the adjacent cavities, running parallel to the ERT line.

Standard inversion (minimization of L2-norm) of ERT data employed the optimized damping procurement (initial damping value = 0.1), without adding any a

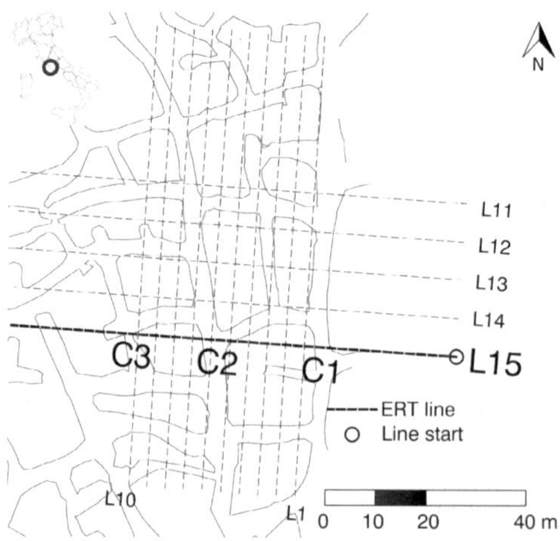

FIG. 17

Villa De Sanctis urban park (Rome), where ERT lines *(dotted lines)* are superimposed to the cavity net *(black lines)*. The studied alignment (L15) is marked with dashed line. C1, C2, and C3 indicate zones where the presence of cavities is likely.

priori information on the resistivity model. Then, starting from the inversion results, we try to reduce the ambiguity in the interpretation of ERT model using the model validation procedure.

3.2.2 Results

The inverted resistivity model of the L15 line is represented in Fig. 18A. The inverted model displays a three-layer configuration (conductive-resistive-conductive) as expected from the geologic information. The cavity net can be attributed to the high-resistivity anomalous zone detected from a depth of 15 m, even though the geometric layout of the net is unclear from the speculative interpretation of the model. We are not able to detect the three cavities (C1, C2, and C3) with a sufficient degree of confidence. In this sense, the cumulative sensitivity of the deepest part of the

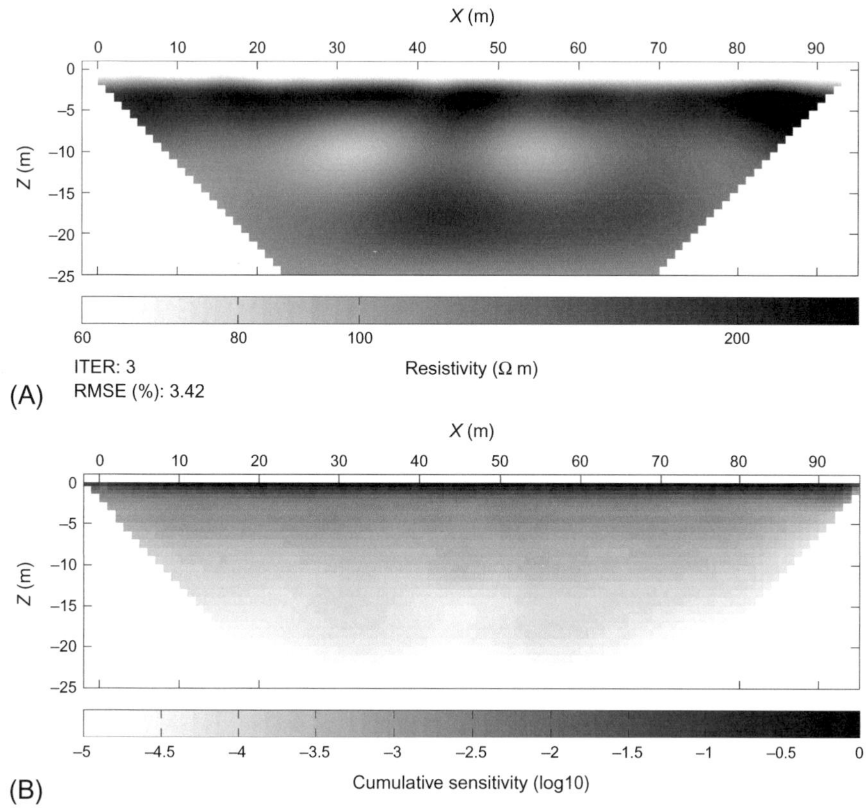

FIG. 18

(A) Inverted resistivity model of the L15 line. (B) Cumulative sensitivity at last iteration.

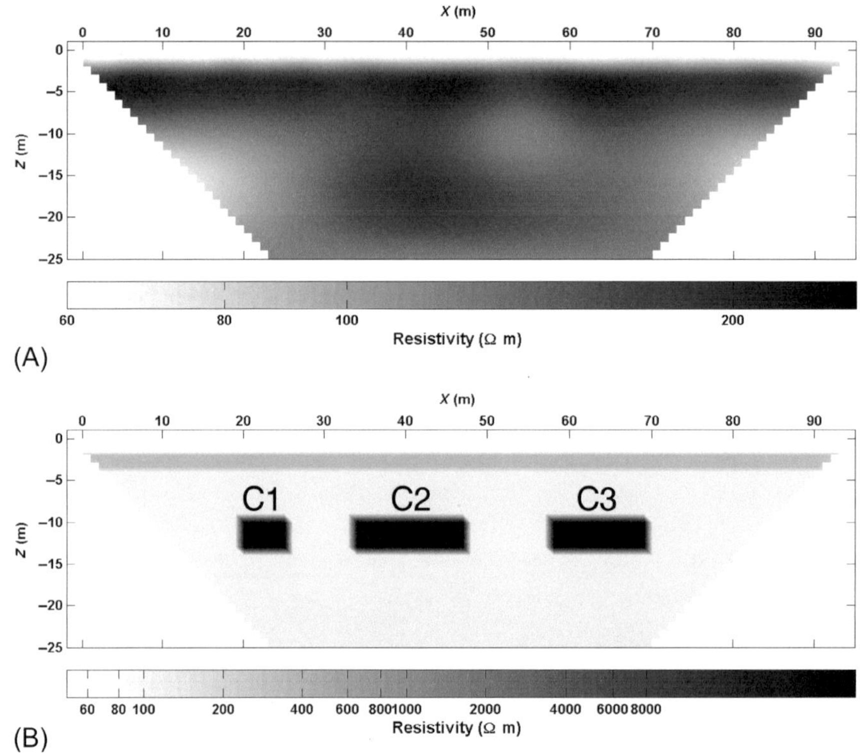

FIG. 19

(A) Inverted resistivity model at last iteration of the model validation procedure. (B) Final resistivity model of the cavity net.

model is very low ($<10^{-4}$ in Fig. 18B), and therefore, low credibility should be given to these zones.

In order to reduce this ambiguity, we apply the model validation procedure to the L15 line. Results are shown in Fig. 19 in terms of synthetic inverted model at the last iteration (Fig. 19A) and final resistivity model of the subsoil (Fig. 19B).

The mean absolute error between synthetic (Fig. 19A) and field (Fig. 18A) inverted model is 7.6%. The final results clearly demonstrate that (i) the effective depth of the roof is around 10m according to the geologic information, (ii) the three cavities can be distinguished, and (iii) the position of the three cavities is the same of that inferred from the topographic plan. Therefore, the cavities appear distorted in the field data inversion as an effect of the low sensitivity of the deepest zones, and the location of the cavity net carried out only with a speculative interpretation of the ERT inverted model can be misleading.

4 CONCLUSIONS

Nowadays, the reliability of an electric resistivity tomography investigation needs to be addressed quantitatively, particularly for near-surface applications. Therefore, in this work, we give a special focus on the role played by forward and inversion parameters within the data processing, on the reliability of collected dataset, and on piecewise strategies that should be implemented for removing ambiguities often arising in interpretation of ERT inverted models.

To these aims, we firstly analyzed the effect of the finite length of the electrodes for a correct forward reconstruction for both invasive (rods) and noninvasive (plates) electrodes. The influence of the electrode size was evaluated for rods by relating the electrode effect (ratio between the forward solution achieved using the point electrode model and the complete electrode model) to the $\frac{n \cdot e}{l}$ dimensionless ratio that is a significant index of the mutual influence between a pair of electrodes. Results show a good power-law correlation with a strong effect of the electrode size visible for $\frac{n \cdot e}{l} < 5$, a frequent case often occurring in the current practice. For surface electrodes, the electrode effect was plotted as a function of the ratio between the electrodes size and their separation $\left(\frac{d}{e}\right)$. Results demonstrate that the electrode effect is lower than for rods and negligible (<2%) for small-size length separation ratio (<0.4). However, when the electrode spacing is small compared with the plate length (>0.5), the point source approximation leads to a significant error in forward modeling (>3%).

Compared with the electrode effect, the mesh coarseness plays a minor role for the accuracy of an ERT forward solver, mainly because of the high optimization reached by mesh generators. The maximum error due to the mesh choice, evaluated as a function of a coarseness factor c_f for both 2D and 3D domains, is relatively small and always lower than 1%. The choice between invasive and surface electrodes should be constrained not only by the need to preserve places of architectural or heritage value but also by the type of the investigated surface: we demonstrated that acquisition on tarmac is unfeasible using surface electrodes because of the high contact resistances and consequently of the low signal-to-noise ratio.

For engineering and archaeological applications, 3D ERT represents the only practical tool able to detect foundation or buried structures with a sufficient degree of accuracy. For such cases, it is recommended to design site-specific arrays and, where feasible, to mix together more than a 3D array, using cumulative sensitivity as the parameter for assessing the spatial coverage of the electrode configuration. If cumulative sensitivity is very low ($<10^{-5}$), the 3D array should be redesigned, or if not feasible, inverted models should be analyzed in any event with this in mind.

For inversion, we investigated the effect of the damping choice (fixed vs optimized) and of the type of inversion (least squares vs robust), among the numerous inversion parameters involved within an iterative Gauss-Newton procedure, almost only used during the last years for ERT inversion. The comparison is performed on a field example for the detection of a buried tomb, whose true size and position are known. The resistivity models resulting from a robust inversion and an optimized damping procedure show a better reconstruction of the tomb and lower RMSE values compared with standard procedures. However, achieving a low RMSE value, it is not sufficient to rate "good" an ERT reconstruction, both due to the decrease of resolution in depth and to the possible bias caused by multiple succession of conductive and resistive layers. Therefore, each inverted model should always be analyzed in terms of model resolution matrix, cumulative sensitivity, or DOI index, which are diagnostic parameters to evaluate quantitatively the model uncertainty. However, only quantifying the uncertainty does not allow to recognize which is the actual model of the subsoil. In this sense, we present a new procedure, called model validation and based on the minimization of the error between the field data inversion and piecewise synthetic models, generated starting from an initial guess based on field data inversion (a posteriori information). Finally, we discuss the importance of inserting a priori information, where available, during the inversion process using a synthetic example, in which the aid of a priori information results in an increase of the speed of convergence together with a lowering of the RMSE. The whole features have been implemented during the last years by the authors within a flexible numerical tool (VEMI), able to perform ERT modeling and inversion for 2D and 3D cylindrical and prismatic geometries both for time- and for frequency-domain data.

At the end, we employed the VEMI interface for ERT reconstruction on two case studies related to archaeological prospecting and engineering geology. The first example is the study of the foundation of the Basilica of Maxentius and Constantine, achieved by a pseudo-3D inversion where buried structures were detected with high accuracy. We found that foundations have different depths depending on the peculiarity of the foundation soil, deeper in the northern part of the Basilica. Cumulative sensitivity calculated for the ERT array shows quite homogeneous high values confirming the robustness of these results, successively validated by archaeological excavations.

The second case study aimed to detect a cavity net located within an urban park in Rome using 2D ERT. However, we demonstrated that the speculative interpretation of inverted models is not adequate here (depth of the cavity estimated at 15 m instead of 10 m that is the true depth), because of low values of sensitivity at growing depths. Therefore, we made use of the model validation procedure to retrieve a more realistic resistivity model. Results clearly confirm that the depth of the cavity roof is around 10 m, in agreement with the geologic information. In light of this, the model validation can be an effective tool for a more accurate ERT reconstruction in cases of deep targets and low sensitivities. However, this tool is still very time consuming and needs to be optimized. In fact, the optimum model is currently found by trial and error, and hopefully, it should be selected through the automatic minimization of the absolute difference between the inverted models of the synthetic and field datasets.

Owing to the advances made on forward modeling and the rapid increase in the development of processors and memories, we think that future works would involve the development of global inversion approaches for ERT reconstruction of near-surface targets.

REFERENCES

[1] A.P. Aizebeokhai, 2D and 3D geoelectrical resistivity imaging: theory and field design, Sci. Res. Essays 5 (23) (2010) 3592–3605.

[2] E. Auken, L. Pellerin, N.B. Christensen, K. Sørensen, A survey of current trends in near-surface electrical and electromagnetic methods, Geophysics 71 (5) (2006) 249–260.

[3] A. Binley, A. Kemna, DC resistivity and induced polarization methods, Hydrogeophysics, Springer, Netherlands, 2005, pp. 129–156.

[4] M.H. Loke, J.E. Chambers, D.F. Rucker, O. Kuras, P.B. Wilkinson, Recent developments in the direct-current geoelectrical imaging method, J. Appl. Geophys. 95 (2013) 135–156.

[5] A. Samouëlian, I. Cousin, A. Tabbagh, A. Bruand, G. Richard, Electrical resistivity survey in soil science: a review, Soil Tillage Res. 83 (2) (2005) 173–193.

[6] R.A. Van Overmeeren, I.L. Ritsema, Continuous vertical electrical sounding, First Break 6 (10) (1988) 313–324.

[7] M. Bernabini, E. Cardarelli, Geoelectrical surveys of dipping structures, Geophys. Prospect. 39 (1991) 953–966.

[8] M. Bernabini, E. Cardarelli, Geoelectrical surveys of structures closed in depth by two dipping discontinuities, J. Appl. Geophys. 32 (1994) 537–548.

[9] J. Chastenet De Gery, G. Kunetz, Potential and apparent resistivity over dipping beds, Geophysics 21 (3) (1956) 780–793.

[10] R.D. Barker, The offset system of electrical resistivity sounding and its use with a multicore cable, Geophys. Prospect. 29 (1) (1981) 128–143.

[11] D.H. Griffiths, J. Turnbull, A multi-electrode array for resistivity surveying, First Break 3 (1985) 16–20.

[12] D.H. Griffiths, J. Turnbull, A.I. Olayinka, Two-dimensional resistivity mapping with a complex controlled array, First Break 8 (1990) 121–129.

[13] K.I. Sørensen, Pulled array continuous electrical profiling, First Break 14 (1996) 85–90.

[14] A. Dey, H.F. Morrison, Resistivity modelling for arbitrary shaped two-dimensional structures, Geophys. Prospect. 27 (1979) 1020–1036.

[15] M.H. Loke, R.D. Barker, Rapid least-squares inversion of apparent resistivity pseudosections by a quasi-Newton method, Geophys. Prospect. 44 (1) (1996) 131–152.

[16] P. Stummer, H. Maurer, A. Green, Experimental design: electrical resistivity data sets that provides optimum subsurface information, Geophysics 69 (2004) 120–139.

[17] E. Cardarelli, M. Cercato, G. De Donno, G. Di Filippo, Detection and imaging of piping sinkholes by integrated geophysical methods, Near Surv. Geophys. 12 (2014) 439–450.

[18] F.D. Day-Lewis, E.A. White, C.D. Johnson, J.W. Lane, Continuous resistivity profiling to delineate submarine groundwater discharge—examples and limitations, Lead. Edge 25 (2006) 724–728.

[19] J. Dunbar, P. Allen, J. White, R. Neupane, T. Xu, J. Wolfe, J. Arnold, Characterizing a shallow groundwater system beneath irrigated sugarcane with electrical resistivity and radon (222Rn), Puunene, Hawaii, J. Environ. Eng. Geophys. 20 (2) (2015) 165–181.

[20] H.L. Hsu, B.J. Yanites, C.C. Chen, Y.G. Chen, Bedrock detection using 2D electrical resistivity imaging along the Peikang River, Central Taiwan, Geomorphology 114 (3) (2010) 406–414.

[21] Y. Ratnakumari, S.N. Rai, S. Thiagarajan, K. Dewashish, 2D electrical resistivity imaging for delineation of deeper aquifers in a part of the Chandrabhaga river basin, Nagpur District, Maharashtra, Curr. Sci. 102 (1) (2012) 61–69.

[22] M. van Schoor, Detection of collapse sinkholes using 2D resistivity imaging, J. Appl. Geophys. 50 (2002) 393–399.
[23] E. Cardarelli, F. Fischanger, S. Piro, Integrated geophysical survey to detect buried structures for archeological prospecting. A case-history at Sabine Necropolis (Rome, Italy), Near Surv. Geophys. 1 (2008) 15–20.
[24] J.E. Chambers, O. Kuras, P.I. Meldrum, R.D. Ogilvy, J. Hollands, Electrical resistivity tomography applied to geologic, hydrogeologic, and engineering investigations at a former waste-disposal site, Geophysics 71 (6) (2006) B231–B239.
[25] G. De Donno, E. Cardarelli, Tomographic inversion of time-domain resistivity and chargeability data for the investigation of landfills a priori information, Waste Manage. 59 (2017) 302–315.
[26] I.F. Louis, V.K. Karastatis, A.P. Vafidis, F.I. Louis, Resistivity modelling and imaging methods for mapping near surface features: application to a site characterization at the ancient Temple Olympian Zeus in Athens, J. Geol. Soc. 5 (2002) 135–144.
[27] G.N. Tsokas, P.I. Tsourlos, G. Vargemezis, M. Novack, Non-destructive electrical resistivity tomography for indoor investigation: the Case of Kapnikarea Church in Athens, Archaeol. Prospect. 15 (2008) 47–61.
[28] T. Dahlin, C. Bernstone, A roll-along technique for 3D resistivity data acquisition with multi-electrode arrays, Symposium on the Application of Geophysics to Engineering and Environmental Problems, Society of Exploration Geophysicists, 1997, pp. 927–935.
[29] A. Dey, H.F. Morrison, Resistivity modeling for arbitrarily shaped three-dimensional shaped structures, Geophysics 44 (1979) 753–780.
[30] R.G. Ellis, D.W. Oldenburg, The pole-pole 3-D DC-resistivity inverse problem a conjugate gradient approach, Geophys. J. Int. 119 (1994) 187–194.
[31] Y. Li, D.W. Oldenburg, Inversion of 3D DC resistivity data using an approximate inverse mapping, Geophys. J. Int. 116 (1994) 527–537.
[32] M.H. Loke, R.D. Barker, Practical techniques for 3D resistivity surveys and data inversion, Geophys. Prospect. 44 (1996) 499–524.
[33] P.I. Tsourlos, R. Ogilvy, An algorithm for the 3-D inversion of tomographic resistivity and induced polarisation data: preliminary results, J. Geol. Soc. 2 (2) (1999) 30–45.
[34] E. Cardarelli, G. De Donno, I. Oliveti, C. Scatigno, Three-dimensional reconstruction of a masonry building through electrical tomography integrated by chemical analysis. Near Surv. Geophys. 16 (1) (2018), https://doi.org/10.3997/1873-0604.2017040.
[35] R. Ogilvy, P. Meldrum, J. Chambers, Imaging of industrial waste deposits and buried quarry geometry by 3D tomography, Eur. J. Environ. Eng. Geophys. 3 (1999) 103–113.
[36] N.G. Papadopoulos, M. Jong Yi, J. Ho Kim, G.N. Tsokas, P.I. Tsourlos, A. Sarris, Tumuli exploration using surface 3D electrical resistivity tomography, Arch. Sci. 33 (2009) 129–132.
[37] J.E. Chambers, D.A. Gunn, P.B. Wilkinson, P.I. Meldrum, E. Haslam, S. Holyoake, M. Kirkham, O. Kuras, A. Merritt, J. Wragg, 4D electrical resistivity tomography monitoring of soil dynamics in an operational railway embankment, Near Surv. Geophys. 12 (2014) 61–72.
[38] F. Fischanger, G. Morelli, G. Ranieri, G. Santarato, M. Occhi, 4D cross-borehole electrical resisitivity tomography to control resin injection for ground stabilization: a case history in Venice (Italy), Near Surv. Geophys. 11 (2013) 41–50.
[39] V. Leroux, T. Dahlin, Time-lapse resistivity investigations for imaging saltwater transport in glaciofluvial deposits, Environ. Geol. 49 (2006) 347–358.
[40] S. Park, Fluid migration in the vadose zone from 3D inversion of resistivity monitoring data, Geophysics 63 (1998) 41–51.

[41] C. Power, J.I. Gerhard, M. Karaoulis, P. Tsourlos, A. Giannopoulos, Evaluating four-dimensional time-lapse electrical resistivity tomography for monitoring DNAPL source zone remediation, J. Contam. Hydrol. 162–163 (2014) 27–46.

[42] K. Singha, F.D. Day-Lewis, T. Johnson, L.D. Slater, Advances in interpretation of subsurface processes with time-lapse electrical imaging, Hydrol. Process. 29 (6) (2015) 1549–1576.

[43] H.O. Seigel, Mathematical formulation and type curves for induced polarization, Geophysics 24 (3) (1959) 547–565.

[44] G. Fiandaca, E. Auken, A.V. Christiansen, A. Gazoty, Time-domain-induced polarization: full-decay forward modeling and 1D laterally constrained inversion of Cole-Cole parameters, Geophysics 77 (3) (2012) E213–E225.

[45] G. Fiandaca, J. Ramm, A. Binley, A. Gazoty, A.V. Christiansen, E. Auken, Resolving spectral information from time domain induced polarization data through 2-D inversion, Geophys. J. Int. 192 (2) (2013) 631–646.

[46] G. Cassiani, A. Kemna, A. Villa, E. Zimmermann, Spectral induced polarization for the characterization of free-phase hydrocarbon contamination of sediments with low clay content, Near Surv. Geophys. 7 (5–6) (2009) 547–562.

[47] A. Kemna, A. Binley, G. Cassiani, E. Niederleithinger, A. Revil, L. Slater, K. H. Williams, A.F. Orozco, F.-H. Haegel, A. Hördt, S. Kruschwitz, V. Leroux, K. Titov, E. Zimmermann, An overview of the spectral induced polarization method for near-surface applications, Near Surv. Geophys. 10 (6) (2012) 453–468.

[48] T. Dahlin, H. Rosqvist, V. Leroux, Resistivity-IP mapping for landfill applications, First Break 28 (8) (2010) 101–105.

[49] G. De Donno, E. Cardarelli, VEMI: a flexible interface for 3D tomographic inversion of time- and frequency-domain electrical data in EIDORS, Near Surv. Geophys. 15 (1) (2017) 43–58.

[50] G. De Donno, 2D tomographic inversion of complex resistivity data on cylindrical models, Geophys. Prospect. 61 (2013) 586–601.

[51] D.J. LaBrecque, A.L. Ramirez, W.D. Daily, A.M. Binley, S.A. Schima, ERT monitoring of environmental remediation processes, Meas. Sci. Technol. 7 (3) (1996) 375.

[52] L. Slater, A.M. Binley, W. Daily, R. Johnson, Cross-hole electrical imaging of a controlled saline tracer injection, J. Appl. Geophys. 44 (2000) 85–102.

[53] L. Slater, A.M. Binley, R. Versterg, G. Cassiani, R. Birken, S. Sandberg, A 3D ERT study of solute transport in a large experimental tank, J. Appl. Geophys. 49 (2002) 85–102.

[54] K.S. Cheng, D. Isaacson, J.N.D. Gisser, Electrode model for electric current computed tomography, IEEE Trans. Biomed. Eng. 36 (1989) 918–924.

[55] A. Adler, W.R.B. Lionheart, Uses and abuses of EIDORS: an extensible software base for EIT, Physiol. Meas. 27 (2006) S25–S42.

[56] C. Rücker, T. Günther, K. Spitzer, Three-dimensional modelling and inversion of DC resistivity data incorporating topography—I. Modelling, Geophys. J. Int. 166 (2) (2006) 495–505.

[57] D. Pridmore, G.W. Hohmann, S.H. Ward, W. Sill, An investigation of finite element modeling for electrical and electromagnetic data in three dimensions, Geophysics 46 (1981) 1009–1024.

[58] J.P.A. Bastos, N. Sadowski, Electromagnetic modeling by finite element method, Electrical Engineering and Electronic Series, vol. 117, Marcel Dekker, New York, 2003. ISBN 0-8247-4269-9.

[59] M. Vauhkonen, Electrical Impedance Tomography and Prior Information (Ph.D. Thesis). Department of Applied Physics, Kuopio University, Finland, 1997.

[60] C. Rücker, T. Günther, The simulation of finite ERT electrodes using the complete electrode model, Geophysics 76 (2011) F227–F238.
[61] N. Polydorides, W. Lionheart, A Matlab toolkit for three-dimensional electrical impedance tomography: a contribution to the Electrical Impedance and Diffuse Optical Reconstruction Software project, Physiol. Meas. 13 (2002) 1871–1883.
[62] G. De Donno, E. Cardarelli, 3D complex resistivity tomography on cylindrical models using EIDORS, Near Surv. Geophys. 12 (5) (2014) 587–598.
[63] K. Paulson, W. Breckon, M. Pidcock, Electrode modelling in electrical impedance tomography, SIAM J. Appl. Math. 52 (4) (1992) 1012–1022.
[64] P.J. Vauhkonen, M. Vauhkonen, T. Savolainen, J.P. Kaipio, Three-dimensional electrical impedance tomography based on the complete electrode model, IEEE Trans. Biomed. Eng. 46 (9) (1999) 1150–1160.
[65] P.-O. Persson, G. Strang, A simple mesh generator in MATLAB, SIAM Rev. 46 (2) (2004) 329–345.
[66] C. Geuzaine, J.F. Remacle, Gmsh: a 3-D finite element mesh generator with built-in pre- and post-processing facilities, Int. J. Numer. Methods Eng. 79 (11) (2009) 1309–1331.
[67] T. Dahlin, Short note on electrode charge-up effects in DC resistivity data acquisition using multi-electrode arrays, Geophys. Prospect. 48 (1) (2000) 181–187.
[68] D.J. LaBrecque, M. Miletto, W. Daily, A. Ramirez, E. Owen, The effect of noise on Occam inversion of resistivity tomography data, Geophysics 6 (1996) 538–548.
[69] D. LaBrecque, W. Daily, Assessment of measurement errors for galvanic-resistivity electrodes of different composition, Geophysics 73 (2) (2008) F55–F64.
[70] G. De Donno, E. Cardarelli, Assessment of errors from different electrode materials and configurations for electrical resistivity and time-domain IP data on laboratory models, Boll. Geofis. Teor. Appl. 52 (2) (2011) 211–223.
[71] T. Ingeman-Nielsen, S. Tomaškovičová, T. Dahlin, Effect of electrode shape on grounding resistances—part 1: the focus-one protocol, Geophysics 81 (1) (2016) WA159–WA167.
[72] E.N. Athanasiou, P.I. Tsourlos, G.N. Vargemezis, C.B. Papazachos, G.N. Tsokas, Nondestructive DC resistivity surveying using flat-base electrodes, Near Surv. Geophys. 5 (2007) 263–271.
[73] A. Furman, T.P.A. Ferre, G.L. Heath, Spatial focusing of electrical resistivity surveys considering geologic and hydrologic layering, Geophysics 72 (2007) 65–73.
[74] P.B. Wilkinson, P.I. Meldrum, J.E. Chambers, O. Kuras, R.D. Ogilvy, Improved strategies for the automatic selection of optimized sets of electrical resistivity tomography measurement configurations, Geophys. J. Int. 169 (2006) 1119–1126.
[75] S.K. Park, G.P. Van, Inversion of pole-pole data for 3D resistivity structure beneath arrays of electrodes, Geophysics 56 (1991) 951–960.
[76] S.C. Constable, R.L. Parker, C.G. Constable, Occam's inversion: a practical algorithm for generating smooth models from electromagnetic sounding data, Geophysics 52 (3) (1987) 289–300.
[77] C. de Groot-Hedlin, S.C. Constable, Occam's inversion to generate smooth two- dimensional models from magnetotelluric data, Geophyiscs 55 (1990) 1613–1624.
[78] S.A. Greenhalgh, Z. Bing, A. Green, Solutions, algorithms and inter-relations for local minimization search geophysical inversion, J. Geophys. Eng. 3 (2) (2006) 101.
[79] E. Auken, A.V. Christiansen, C. Kirkegaard, G. Fiandaca, C. Schamper, A. A. Behroozmand, A. Binley, E. Nielsen, F. Effersø, N.B. Christensen, K. Sørensen, N. Foged, G. Vignoli, An overview of a highly versatile forward and stable inverse

algorithm for airborne, ground-based and borehole electromagnetic and electric data, Explor. Geophys. 46 (3) (2014) 223–235.
[80] T. Günther, C. Rücker, K. Spitzer, Three-dimensional modelling and inversion of DC resistivity data incorporating topography—II. Inversion, Geophys. J. Int. 166 (2) (2006) 506–517.
[81] A. Pidlisecky, E. Haber, R. Knight, RESINVM3D: a 3D resistivity inversion package, Geophysics 72 (2) (2007) H1–H10.
[82] M. Karaoulis, A. Revil, P. Tsourlos, D.D. Werkema, B.J. Minsley, IP4DI: a software for time-lapse 2D/3D DC-resistivity and induced polarization tomography, Comput. Geosci. 54 (2013) 164–170.
[83] Y. Sasaki, Resolution of resistivity tomography inferred from numerical simulation, Geophys. Prospect. 40 (1992) 453–464.
[84] C.G. Farquharson, D.W. Oldenburg, Non-linear inversion using general measures of data misfit and model structure, Geophys. J. Int. 134 (1) (1998) 213–227.
[85] M.H. Loke, I. Acworth, T. Dahlin, A comparison of smooth and blocky inversion methods in 2D electrical imaging surveys, Explor. Geophys. 34 (3) (2003) 182–187.
[86] E. Cardarelli, F. Fischanger, 2D data modelling by electrical resistivity tomography for complex subsurface geology, Geophys. Prospect. 54 (2006) 121–133.
[87] K. Beven, A. Binley, The future of distributed models: model calibration and uncertainty prediction, Hydrol. Process. 6 (3) (1992) 279–298.
[88] W. Menke, Discrete Inverse Theory, Academic Press, San Diego, 1989, p. 330.
[89] D.W. Oldenburg, Y. Li, Estimating depth of investigation in dc resistivity and IP surveys, Geophysics 64 (1999) 403–416.
[90] L. Marescot, M.H. Loke, D. Chapellier, R. Delaloye, C. Lambiel, E. Reynard, Assessing reliability of 2D resistivity imaging in mountain permafrost studies using the depth of investigation index method, Near Surv. Geophys. 1 (2) (2003) 57–67.
[91] J. Deceuster, A. Etienne, T. Robert, F. Nguyen, O. Kaufmann, A modified DOI-based method to statistically estimate the depth of investigation of dc resistivity surveys, J. Appl. Geophys. 103 (2014) 172–185.
[92] S. Narayan, M.B. Dusseault, D.C. Nobes, Inversion techniques applied to resistivity inverse problems, Inverse Prob. 10 (3) (1994) 669.
[92a] H.J. Kim, Y. Song, K.H. Lee, Inequality constraint in least-squares inversion of geophysical data, Earth Planets Space 51 (1999) 255–259.
[93] E. Cardarelli, G. Di Filippo, Integrated geophysical methods for the characterisation of an archaeological site (Massenzio Basilica—Roman forum, Rome, Italy), J. Appl. Geophys. 68 (4) (2009) 508–521.
[94] G. Bianchi Fasani, F. Bozzano, E. Cardarelli, M. Cercato, Underground cavity investigation within the city of Rome (Italy): a multi-disciplinary approach combining geological and geophysical data, Eng. Geol. 152 (1) (2013) 109–121.

FURTHER READING

[95] W. Daily, E. Owen, Cross-borehole resistivity tomography, Geophysics 56 (1991) 1228–1235.

CHAPTER

3 Time-domain reflectometry: Current uses and new possibilities

Raffaele Persico*,†, Andrea Cataldo‡, Egidio De Benedetto‡

Institute for Archaeological and Monumental Heritage IBAM-CNR, Lecce, Italy International Telematic University UNINETTUNO (UTIU), Rome, Italy† University of Salento, Lecce, Italy‡*

CHAPTER OUTLINE

1 OVERVIEW OF TDR APPLICATIONS

TDR is largely used for diagnostic and monitoring purposes in different fields, and this is due to its high versatility, accuracy, and relatively low implementation costs, and also, thanks to the possibility of carrying out continuous real-time measurements. An additional advantage is that TDR can be automated (so as to be remotely controlled) and through multiplexers it is possible to employ simultaneously several probes with a single measurement instrument. For example, the TDR200 reflectometer can control up to 512 probes through multiplexers. One of the first TDR applications relates to fault diagnosis in wires. In particular, TDR has been used for fault detection in electric wires in buildings, aircraft, and transportation systems; additionally, TDR has been used to identify bad splices in telephone cables and to characterize electronic devices/components and antennae.

Innovation in Near-Surface Geophysics. https://doi.org/10.1016/B978-0-12-812429-1.00003-9

Dielectric characterization of materials is another major application field of TDR. Furthermore, impedance measurements; failure in interconnection mechanisms; and qualitative, quantitative, and/or structural controls on materials are only a few of the domains in which TDR has gained particular interest. TDR has also been largely employed in moisture content measurements. In particular, TDR is widely used for routine field monitoring of water content in soil. In fact, over the years, several theoretical (e.g., Maxwell-De Lor) and empirical models (e.g., fitting exponent mixing model) have been developed to infer soil moisture content from TDR measurements.

Moreover, TDR has been adopted for measuring water content of agrofood materials (e.g., soybean, coffee, bran, and corn), thus inferring their qualitative status, and also for studying the microwave dielectric properties of solid and liquid foods. TDR is also used in geotechnical engineering for assessing distributed pressure profiles (useful for evaluating the strength of the ground for constructions), for monitoring liquefaction of soils, for the detection of organic pollutants in sandy soils, etc. Additionally, TDR has also been used for the investigation of contaminated land and in particular of leachate-contaminated soil. Another well-established application of TDR relates to dielectric characterization of soil.

TDR has also been used for landslide monitoring, for example, by using coaxial cables grouted in rock or soil mass and watching for reflections due to cable deformity induced by the ground deformation. Recently, TDR technology has been implemented for measurement of slope underground movements, with the intent of providing a geotechnical monitoring of road embankment in landslide areas. TDR has become a well-established technique also for monitoring the static electric conductivity of materials. In this regard, TDR becomes particularly useful when dealing with porous media (e.g., soil). In this case, traditional methods such as conductive meters (in which an alternating current is applied to two electrodes and the voltage is measured to retrieve the conductance) pose problems with regard to electrode polarization; on the other hand, the principle behind TDR makes this technique less sensitive to this phenomenon.

Finally, TDR-based techniques are also providing promising results for monitoring applications in civil engineering for dynamic structure analysis.

In the following section, we provide the basic principles of time-domain reflectometry; then, the common use of TDR in the time domain (Section 3), in a combination of time and frequency domain (Section 4), and in the "pure" frequency domain (Section 5) is introduced. In Section 6, we introduce the transmission line model, starting from an infinite line, in order to insert TDR probes in a wider electromagnetic theory allowing to look for both dielectric and magnetic properties of the propagation media. In Section 7, finite (i.e., real) transmission lines are introduced, and in Section 8, we propose a prompt on the problem of the higher-order modes. In Section 9, an investigation of lossless and dispersionless materials is proposed, whereas in Section 10, the investigation of lossy and/or dispersive materials is presented. These two sections show also numerical examples. Conclusions follow.

2 BASIC PRINCIPLES

Generally, in microwave reflectometry, a low-power electromagnetic signal is propagated into the system under test. Through the analysis of the reflected signal together with specific data processing, the desired information on the system under test can be retrieved.

Two main elements are involved in microwave reflectometry measurements:

1. The instrument for generating/receiving the electromagnetic (EM) signal.
2. The measurement cell, which includes the sensing element (or probe) and the material (more in general the object) under test.

Microwave-reflectometry-based measurements can be performed either in time domain (time-domain reflectometry (TDR)) or in frequency domain (frequency-domain reflectometry (FDR)). Depending on the specific application, one of the two approaches may be more suitable than the other. TDR instrumentation is usually less expensive than instruments operating in frequency domain. Nevertheless, instruments operating directly in frequency domain, such as vector network analyzers (VNAs), usually provide higher measurement accuracy. A good trade-off (in terms of low cost and measurement accuracy) can be achieved through a time-domain/frequency-domain (TD/FD) combined approach. In the latter case, measurements are carried out in TD, and the corresponding FD information is retrieved through appropriate processing of the acquired data.

One of the greatest advantages of operating in FD relates to the possibility of performing calibration procedures, thus enhancing the final measurement accuracy. This procedure can also be implemented in the TD/FD approach. Consequently, the combination of the two approaches is considered as a powerful tool for guaranteeing relatively low cost of the system and adequate measurement accuracy.

In TDR measurements, the EM stimulus is usually a steplike signal that propagates along the probe, through the system under test. It consists of a step pulse generator, a sampling scope, and a signal analyzer. The step generator produces a positive-going incident wave that is applied to the system under test. The steplike pulse travels down the transmission line at the velocity of propagation of the line. Any impedance variation causes the partial reflection of the propagating signal. If the load impedance is equal to the characteristic impedance of the line, no wave is reflected, and the oscilloscope displays only the incident voltage step. If a mismatch exists at the load, part of the incident wave is reflected [1]:

$$\rho(t) = \frac{v_{refl}(t)}{v_{inc}(t)} \tag{1}$$

The reflected signal is acquired by the oscilloscope, and its voltage amplitude is displayed as a function of time (or as a function of the traveled electric distance). The ratio between the amplitude of the reflected signal, $v_{refl}(t)$, and the amplitude of the generated signal, $v_{inc}(t)$, gives the value of the reflection coefficient in time domain, $\rho(t)$, where $-1 \leq \rho(t) \leq +1$ [2].

Generally, the TDR signal generator and the oscilloscope are integrated into a single measurement instrument, which can be either portable or benchtop. Clearly, the behavior of $\rho(t)$ is strictly associated with the impedance variations along the electric path traveled by the EM signal.

3 TYPICAL TDR MEASUREMENTS

When the TDR signal propagates along a transmission line, impedance mismatches cause the reflection of a portion of the signal. The TDR signal's propagation velocity c is related to the relative dielectric permittivity ε_r of the medium, which in the first moment is assumed to be lossless or at least with negligible conductivity, and to the relative magnetic permeability μ_r, which is equal to 1 for many materials but not for all of them:

$$c = \frac{c_o}{\sqrt{\varepsilon_r \mu_r}} \tag{2}$$

where c_o is the speed of light in free space ($c_o \approx 3 \times 10^8$ m/s). To give an example, let us consider a three-rod probe (see Fig. 1), whose electrodes have length L. A typical TDR waveform for such a probe, immersed in a generic homogeneous dielectric, is represented in Fig. 2.

The analysis of TDR waveform directly leads to the evaluation of the apparent length L_{app} (also called electric distance), which can be directly associated with the dielectric characteristic of the medium. In fact, L_{app} can be considered as the distance that would be traveled by the EM signal, in the same interval of time, if the signal was propagating at a speed c_o. Typically, in TDR analysis, the apparent distance is evaluated through the identification of the beginning and end of the probe;

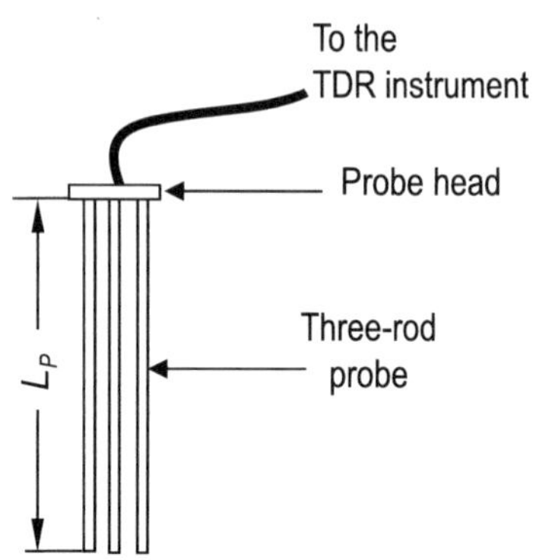

FIG. 1

Schematic of a three-rod TDR probe: the electromagnetic waves propagate guided along the three rods and are reflected at the end.

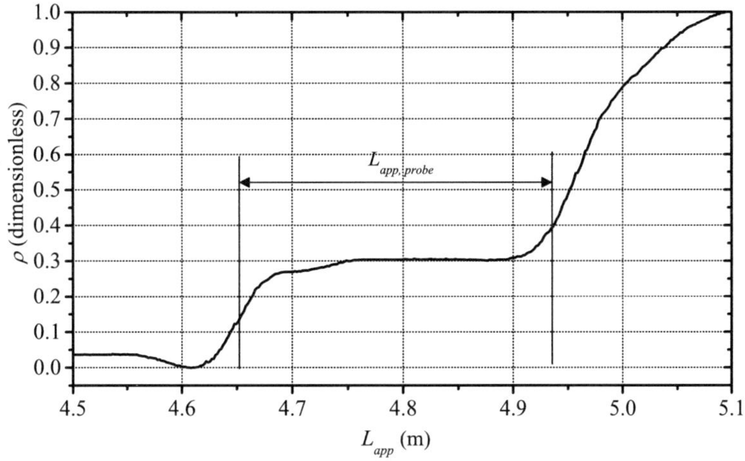

FIG. 2

Example of a TDR waveform acquired with a three-rod probe inserted into a homogenous dielectric material. The portion of the TDR waveform corresponding to the probe is also indicated.

thus, it can be associated with the physical length, L, through the following equation (meant for nonmagnetic materials):

$$L_{app} = L\sqrt{\varepsilon_{app}} = \frac{c_o \times t_t}{2} \tag{3}$$

where t_t is the travel time (i.e., the round-trip time taken by the signal to travel between the beginning and end points) and ε_{app} is referred to as apparent dielectric permittivity of the material in which the probe is inserted [3]. Indeed, the apparent permittivity depends on the frequency, and it can be shown that a more rigorous expression for it is given by [4]

$$\varepsilon_{app}(f) = \frac{\varepsilon_r'(f)}{2}\left(1 + \sqrt{1 + \left\{\frac{\varepsilon_r''(f) + \dfrac{\sigma}{2\pi f \varepsilon_o}}{\varepsilon_r'(f)}\right\}^2}\right) \tag{4}$$

where $\varepsilon_o = 8.854 \times 10^{-12}$ F/m is the dielectric permittivity of free space, f is the frequency, $\varepsilon_r'(f)$ describes energy storage, $\varepsilon_r''(f)$ accounts for the dielectric losses, and σ_0 is the static electric conductivity.

However, when dealing with low-loss materials, the imaginary part of the complex permittivity can be neglected so that the apparent relative permittivity reduces to $\varepsilon_r'(f)$. Additionally, considering low dispersive materials, the dependence of $\varepsilon_r'(f)$ on frequency can be considered negligible. Under these circumstances, $\varepsilon_{app}(f)$ can be considered to be approximately constant.

On the basis of these theoretical considerations, TDR-based dielectric measurements simply rely on the identification (in the TDR waveform) of the points corresponding to the beginning and to the end of the probe. To identify these two points, the so-called tangent method can be used. However, L_{app} can be easily evaluated through the derivative of the TDR waveform, which typically exhibits prominent peaks in correspondence of the probe-beginning and probe-end sections. Indeed, not only this "derivative method" is simple and quick, but also it proves to be particularly advantageous when significant impedance changes are masked by other impedance variations (e.g., by the transition coaxial cables/rods).

4 TD/FD COMBINED APPROACH

The adoption of the TD/FD combined approach leads to several advantages. In particular, estimating the frequency-dependent scattering parameter starting from TDR measurements can help retrieving some useful information that is masked in time domain (e.g., single or multiple dielectric relaxation).

This strategy is regarded as a powerful tool for guaranteeing simultaneously low cost and portability of the instruments and measurement accuracy; in fact, low-cost portable TDR instruments are readily available on the market. The $S_{11}(f)$ of the monitored system can be calculated as the ratio between the discrete Fourier transform (DFT) of the reflected signal and that of the input signal [5]:

$$S_{11}(f) = \frac{\mathrm{DFT}(r(t))}{\mathrm{DFT}(v_o(t))} \tag{5}$$

where $v_o(t)$ is the input function or incident signal, which should be the unaltered signal generated by the TDR instrument and propagating (along the cable) up to the section where the $S_{11}(f)$ is being measured. This approach provides reliability and adequate measurement accuracy, although relying on a relatively inexpensive TDR instrumentation.

Customarily, when using a TDR probe in frequency domain, some dispersion law should be known a priori, so that the data provide the information needed to estimate some parameter of this dispersion law (usually of Debye kind [6–8]). However, in order to propose some possible methodological innovation, here, we propose an approach where the dispersion law is not necessarily known a priori, and also, possible magnetic properties of the investigated material will be admitted and looked for. Conceptually, this inserts the proposed dealing within the Nicolson-Ross-Weir method for the measurement of both dielectric and magnetic properties of the media [9–11]. In particular, the Nicolson-Ross-Weir theory enables the indirect measurement of both dielectric permittivity and magnetic permittivity of a probed material from the S parameters linked to the two-port microwave vector network generated by a parallelepiped of this material inserted in a rectangular waveguide [12]. The approach proposed here shares with the Nicolson-Ross-Weir approach the fact that the measurements are performed in frequency domain and that they are based on the guided propagation with only the fundamental mode propagating. Moreover, the proposed approach is based on the measurement of the reflection coefficient at the air-soil interface, which means the S_{11} parameter. Clearly, we cannot use the entire S matrix because TDR relies on reflection measurements (and not transmission measurements). On the other hand, a TDR probe is easily brought in the field and allows to gather data without perturbing the compactness, the moisture level, or the temperature of the investigated material. Moreover, it does not require particular attention about the fact that the section of the waveguide is properly filled up with the material under test (void parts would generate a meaningful contribution from higher-order modes, dealt with in the following). However, the insertion of a TDR probe in a hard material, for example, a rock, might be a hard task, and in such cases, it could be more practical to bring a piece of material in laboratory. In particular, unlike a sample of loose soil, a piece of rock can be immediately and suitably preserved and does not lose immediately its moisture content. It can be easily transported for measurements in waveguide (in some cases, its characteristics can also be retrieved from capacity measurements at radio frequency [13]).

5 TDR IN FREQUENCY DOMAIN

In the following, we will focus on TDR probes in frequency domain. Rather than transforming time-domain data, frequency-domain data might be obtained also directly with a stepped sinusoidal generator constituted, for example, by a network analyzer. The comprehension of the physics underlying the use of a TDR probe in frequency domain requires a specific dealing of the TDR probe in frequency domain in the framework of the transmission line theory.

In particular, from a physical-mathematical point of view, there is some extra information that measurements in frequency domain allow to retrieve with respect to measurements gathered in time domain, namely, the information relative to the intrinsic impedance of the equivalent transmission line, as described in more detail in the following sections. Indeed, the measurement of the return time of a pulse along the rods of the probe allows to retrieve the propagation velocity of the electromagnetic waves (somehow averaged in the frequency band of the propagating pulse) in the probed medium. As can be easily worked out in Eq. (2), the return time is linked to the product between the relative dielectric permittivity and the relative magnetic permeability, so it cannot allow to discriminate the two quantities from each other.

Indeed, the intrinsic impedance of the transmission line where the TEM fundamental mode propagates is linked to the ratio between μ_r and ε_r, and this “does the trick” as it will be shown.

In lossy cases, the relative dielectric permittivity and the relative magnetic permeability are complex quantities, and in general, the involved unknowns are functions of the frequency. So, there is a theoretical impossibility in frequency domain for the solution. In fact, adding information by increasing the number of exploited frequencies also makes larger the number of unknowns looked for (that are theoretically

4 for each frequency). In order to make the problem resolvable, a constitutive relationship between the parameters of the medium at different frequencies is needed. Such a relationship is provided by a dispersion law that is imposed on the data thanks to some a priori information. Possibly, the medium is a priori assumed to be nondispersive, which is equivalent to impose a nondispersive law. In any case, a general question about the reliability of the available or adopted dispersion law arises [14].

In general, the actual dispersion law depends on the chemical composition and on the water content of the material, and also other parameters as the mixture of different mineral components, the porosity, and the temperature have or at least can have a meaningful influence. As a consequence, if the probed material is homogeneous (a liquid or a stock of soybeans), then it is reasonable to suppose that one might have reliable a priori information about the dispersion law. However, in the case of a natural soil, it is often not easy to have the possibility to measure all the listed quantities reliably in the field, so it can be of interest also to have at disposal an approach where the dispersion law has not to be assumed a priori. This can be provided by a TDR probe with extensible rods, as it will be detailed in the next sections.

6 INFINITE TRANSMISSION LINES

As discussed in the previous section, some limits of the current TDR methods can be overcome by using a transmission line model in frequency domain. Therefore, we now recall some details about the transmission line theory. For an in-depth discussion on this topic, the reader is referred to the general literature on electromagnetism [12,15].

In principle, a transmission line is constituted by two or more long perfect electric conductors,[1] parallel to each other. First, let us consider an infinitely long transmission line in free space (or in air, which can be assimilated to the free space), and let us apply a voltage at fixed frequency at the beginning of the line, as shown in Fig. 3. Moreover, let us suppose that the transverse cross section of the line is small with

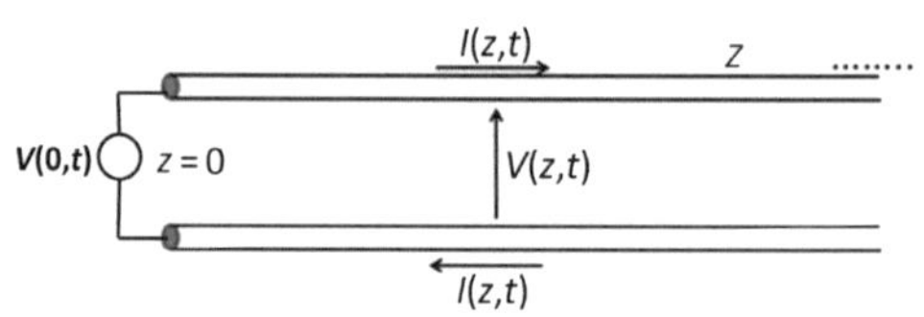

FIG. 3

Scheme for an infinite transmission line.

[1] In the theory of the transmission lines, also the case of lossy conductors is considered. Here, we will consider only the case of possibly lossy propagation media considering the conductors in any case perfect. In other words, we will assume that the losses in the conductors are negligible with respect to those in the medium.

respect to the wavelength. In these conditions, it can be shown that the voltage at any abscissa z and any time instant t along the line can be expressed as

$$V(z,t) = V_o \cos(2\pi f t - k_o z + \phi) \tag{6}$$

where f is the frequency, k_o is a constant called wave number, V_o is the maximum amplitude of the propagating wave, and ϕ is the initial phase, linked to the value of the voltage at the zero abscissa at the initial instant ($V_{in} = V(0,0) = V_o \cos(\phi)$).

Since the line is in air, the wave number and the frequency are linked to each other by the following equation:

$$\frac{2\pi}{k_o} = \frac{c_o}{f} \tag{7}$$

In turn, the quantity $\frac{2\pi}{k_o} = \lambda_o$ is the wavelength of the signal propagating along the line in air. Eq. (6) represents a double sinusoidal signal, in the sense that an observer placed at any fixed abscissa will record a sinusoidal function of the time (whose initial phase $-k_o z + \phi$ depends on the position) and an observer that gathers all the voltage signals along the line at any fixed instant time (by means of an ideal array of voltmeters) will record a sinusoidal function of the abscissa (whose initial phase $-2\pi f t - \phi$ depends on the time). The temporal period of the sinusoid versus the time in any spatial position is the same and is given by the inverse of the frequency $T = \frac{1}{f}$. The spatial period of the sinusoid versus the abscissa is the same for any time instant and is given by the wavelength λ_o. The voltage along the line is therefore a sinusoidal wave, which means that it is provided by a sinusoid "rigidly translating" along the line. In order to deliver correctly the sense of Eq. (6), it is to be specified that Eq. (6) represents a steady regime solution. In other words, if we switch on the voltage at the zero time, what really happens at the abscissa $z = 0$ is that the signal follows a sinusoidal law versus time after a brief transient time. Let us call it tr. At any abscissa $z > 0$, the signal is identically equal to zero until the wave front of the signal propagates up to that point, and then, it becomes sinusoidal after a further time interval tr. So, whatever fixed abscissa we consider after a transitory time (different from point to point), the voltage is well described by Eq. (6). Let us note that, consequently, the zero time in Eq. (6) is an equivalent zero time achieved back prolonging the steady regime solution and should be not meant as the time instant at which the voltage generator was switched on.

That said, accounting for the verses expressed in Fig. 3, along the rods of the transmission line, a current along the line flows (equal and opposite along the two parallel rods) whose expression is given by

$$I(z,t) = I_o \cos(2\pi f t - k_o z + \phi) \tag{8}$$

The ratio between the voltage and the current at any fixed abscissa is a constant given by

$$Z_{lo} = \frac{V_o}{I_o} \tag{9}$$

Z_{lo} has the physical dimensions of a resistance and is called the intrinsic impedance of the line in air. If we mask the line with a screen along the cross section at the zero abscissa, indeed, an observer able to measure both voltage and current at the zero abscissa will be not able to distinguish whether beyond the screen there is a resistance Z_{lo} or an infinite line with intrinsic impedance Z_{lo}. As a consequence, the average power absorbed by an infinite line with intrinsic impedance Z_{lo} is the same as that absorbed by a resistance Z_{lo} connected to an ideal sinusoidal voltage generator with maximum amplitude V_o that is given (as well known) by

$$P = \frac{1}{2}\frac{V_o^2}{Z_{lo}} = \frac{1}{2}V_o I_o = \frac{1}{2}Z_{lo}I_o^2 \tag{10}$$

The physical difference between a lumped resistance Z_{lo} and an infinite line with intrinsic impedance Z_{lo} is that the first one is a device that transforms the absorbed energy into heat, whereas the second one stores the absorbed energy indefinitely (the line represents an infinite tank for the energy because of its infinite length, and the absorbed energy propagates indefinitely along the line). It is worth noting that the line is a passive device, and consequently, the average power that it absorbs from the generator is necessarily positive; otherwise, it would transfer energy to the generator. As a result, the intrinsic impedance of a transmission line in air is necessarily positive. Before closing this section, it is worth noting the symbolic formalism of the phasors that will be particularly useful in the next sections. In particular, as well known, we can issue a biunivocal relationship between the sinusoidal function $\cos(2\pi ft - k_o z + \phi)$ and the complex function $\exp[j(-k_o z + \phi)] = \exp(-jk_o z)\exp(j\phi)$. In fact, starting from the cosine, the phasor $\exp(-jk_o z)\exp(j\phi)$ is achieved by dropping the time factor $2\pi ft$ (the frequency is fixed) and replacing the cosine with the imaginary exponential (or, which is the same, adding the quantity $j\sin(-k_o z + \phi)$ to $\cos(-k_o z + \phi)$). Reversely, starting from the phasor $\exp[j(-k_o z + \phi)]$, we can multiply it by the factor $\exp(j2\pi ft)$ and can take the real part, thus obtaining $\cos(2\pi ft - k_o z + \phi)$. Therefore, the steady regime voltage can be expressed in phasor terms as

$$\hat{V}(z) = V_o e^{j\phi} e^{-jkz} \tag{11}$$

where the circumflex accent is used in order to distinguish the phasor from its corresponding quantity in time domain. Analogously, the phasor of the current is given by

$$\hat{I}(z) = I_o e^{j\phi} e^{-jkz} \tag{12}$$

The ratio between the phasor of the voltage and the phasor of the current is the impedance along the line seen at the considered abscissa looking toward the direction far from the source. It is given by

$$Z(z) = \frac{\hat{V}(z)}{\hat{I}(z)} \tag{13}$$

In the case of an infinite line in air, this impedance is independent of the abscissa and coincides with the intrinsic impedance of the line, but in general, these two quantities are not the same, and the impedance at a given abscissa is in general not constant if the line is not infinitely long. This point will be more deeply discussed in the next section.

In terms of phasors, it can be shown that the average power absorbed by the line at a given abscissa can be expressed as

$$P = \mathrm{Real}\left\{\frac{1}{2}\hat{V}(z)\hat{I}^*(z)\right\} = \frac{1}{2}\mathrm{Real}\{Z(z)\}|I(z)|^2 \tag{14}$$

where the symbol "*" indicated the conjugate. It can be easily shown that Eq. (14) reduces to Eq. (10) in the case of a lossless line. Eq. (14) represents a more general form of Eq. (10) and holds also when the line is embedded in a lossy medium. In the latter case, expression (14) also indicates that the power absorbed by the line is not the same at any abscissa in general, just because the losses make the electromagnetic energy propagating along the line progressively transformed into heat. In particular, Eq. (14) implicates that the real part of the impedance along the line looking far from the source is necessarily a nonnegative quantity; otherwise, the line would supply energy to the generator.

At this point, let us consider a line embedded in a propagation medium that shows a relative dielectric permittivity ε_r and a relative magnetic permeability μ_r. Both ε_r and μ_r are in general complex quantities in lossy cases, whereas they are real in lossless cases. In particular, in lossy cases, the imaginary part of the relative dielectric permittivity ε_r also accounts for a possible electric conductivity. In lossy cases at fixed frequency, the general expression of the steady regime voltage is given by

$$V(z,t) = V_o \exp(-\alpha z)\cos(2\pi f t - \beta z + \phi_V) \tag{15}$$

whereas the expression of the current is given by

$$I(z,t) = I_o \exp(-\alpha z)\cos(2\pi f t - \beta z + \phi_I) \tag{16}$$

As can be seen, the effect of the losses is that the steady regime solution is a sinusoidal wave that attenuates exponentially while propagating through the line (α is a nonnegative number, as well as β). The wave is a sinusoidal function of the time at any abscissa, and the amplitude and phase of this sinusoid depend on z (in the lossless case, the phase depended on z, but the amplitude was the same at any abscissa along the line). For any fixed time instant, the voltage (or the current) along the line is a sinusoidal function multiplied by an exponential. Hence, the function is no longer periodic along z for any fixed instant. Notwithstanding, the spatial distance between the zeros for the function is still constant, and this enables the possibility of identifying the wavelength. In particular, the wavelength is defined as

$$\lambda = \frac{2\pi}{\beta} \tag{17}$$

and corresponds to twice the spatial distance between two consecutive zeros of the voltage along the transmission line at any fixed time instant. It is particularly worth noting that, in lossy cases, the phase term ϕ_V in Eq. (15) and the phase term ϕ_I in Eq. (16) are in general different from each other, which means that the voltage and the current at any fixed abscissa are in general not in phase with each other.

In phasor terms, for an infinitely long transmission line embedded in a lossy medium, we have that

$$\hat{V}(z) = V_o \exp(-\alpha z) \exp[j(-\beta z + \phi_V)] = V_o \exp\{-j[(\beta - j\alpha)z - \phi_V]\} \tag{18}$$

$$\hat{I}(z) = I_o \exp(-\alpha z) \exp[j(-\beta z + \phi_I)] = I_o \exp\{-j[(\beta - j\alpha)z - \phi_I]\} \tag{19}$$

and the impedance along the line is given by

$$Z(z) = \frac{\hat{V}(z)}{\hat{I}(z)} = \frac{V_o}{I_o} \exp[j(\phi_V - \phi_I)] \tag{20}$$

Since we are considering an infinite line, the impedance along the line is the same at any abscissa, and it corresponds to the intrinsic impedance of the line, which is complex in lossy cases. In particular, the power absorbed along the line is given again by Eq. (14), which can be rewritten as

$$P = \text{Real}\left\{\frac{1}{2}\hat{V}(z)\hat{I}^*(z)\right\} = \frac{1}{2}\text{Real}\{Z(z)\}I_o^2 \exp(-2\alpha z) \tag{21}$$

which indicates that the real part of the intrinsic impedance of the line $\text{Real}\{Z(z)\} = \frac{V_o}{I_o}\cos(\phi_V - \phi_I)$ has to be nonnegative.

The intrinsic impedance of the line depends on the geometry of the line, and its calculation requires a deeper dealing based on Maxwell's equation, which is beyond the purposes of this chapter (the interested reader can consult [12,15], and references therein). So, we will just enounce some results without demonstrating them. In particular, the transmission line is characterized by the fact that there are two or more conductors (a single conductor does not constitute a transmission line, even if it can constitute a guiding structure) detached from each other, developing along the same direction, and placed at a small reciprocal distance in terms of wavelength. If the line is constituted by two parallel cylindrical rods with the same radius, its intrinsic impedance is given by

$$Z_l = \frac{1}{\pi}\sqrt{\frac{\mu_r}{\varepsilon_r}}\sqrt{\frac{\mu_o}{\varepsilon_o}}\ \ln\left(\frac{d}{r}\right) \tag{22}$$

where d is the distance between the two conductors and r is their radius. Moreover, μ_o is the magnetic permeability of the free space, equal to about 12.56×10^{-7} H/m (as already said, $\varepsilon_o = 8.854 \times 10^{-12}$ F/m is the absolute permittivity of the free space).

The ratio $\sqrt{\frac{\mu_o}{\varepsilon_o}}$ has the physical dimensions of a resistance, and its numerical value is about 377 Ω. It is also called the wave impedance of the free space. The constant $Z = \frac{1}{\pi}\sqrt{\frac{\mu_o}{\varepsilon_o}}$ is about equal to 120 Ω, so it is also reported that the wave impedance (or intrinsic impedance) of a bifilar line is equal to $120\sqrt{\frac{\mu_r}{\varepsilon_r}}\ \ln\left(\frac{d}{r}\right)$ (Ω) ($\frac{120}{\sqrt{\varepsilon_r}}\ \ln\left(\frac{d}{r}\right)$ (Ω) for nonmagnetic materials).

If the probe is constituted by a coaxial cable (often used, e.g., for monitoring liquid or possibly powdered substances), it can be shown that the intrinsic impedance is given by

$$Z_l = \frac{1}{2\pi}\sqrt{\frac{\mu_r}{\varepsilon_r}}\sqrt{\frac{\mu_o}{\varepsilon_o}}\ \ln\left(\frac{b}{a}\right) \tag{23}$$

where b is the inner diameter of the outer conductor and a is the outer diameter of the inner conductor.

Analogously to the intrinsic impedance of a bifilar line, it is often reported that the wave impedance of a coaxial cable is equal to $60\sqrt{\frac{\mu_r}{\varepsilon_r}}\ \ln\left(\frac{b}{a}\right)$ (Ω) ($\frac{60}{\sqrt{\varepsilon_r}}\ \ln\left(\frac{b}{a}\right)$ (Ω) for nonmagnetic materials [16,17]).

For soil monitoring and for granular materials monitoring in general, the three-rod configuration has become a widespread solution since its electric behavior resembles coaxial cells. However, unfortunately, for configurations different from a coaxial line, models are not always available, and the evaluation of the probe impedance is not straightforward, especially for unbalanced probes. However, for multirod rotationally symmetrical probes, an analytic formulation relating the impedance of the probe in air, Z_p, to the geometric characteristics of the line has been found as follows [18]:

$$Z_l = \frac{1}{2\pi(n-1)}\sqrt{\frac{\mu_r}{\varepsilon_r}}\sqrt{\frac{\mu_o}{\varepsilon_o}}\ \ln\left(H+\sqrt{H^2+2}\right) \tag{24}$$

where H is given by

$$H = \frac{\left(s^2-a_o^2\right)^{n-1} - a_i^{2(n-1)}}{a_i^{n-1}\left[\left(s+a_o\right)^{n-1} - \left(s-a_o\right)^{n-1}\right]} \tag{25}$$

In the previous equations, n is the number of rods of the probe, a_i is the radius of the central conductor, a_0 is the radius of the outer conductors, and s is the distance between the middle of the center conductor and the middle of the outer conductor. Fig. 4 shows a sketch of the cross section of a four-conductor probe.

In Eq. (24), the impedance is to be meant as the ratio between the voltage phasor and a current phasor defined as follows: The voltage phasor is relative to the difference of potential between the central rod and any of the peripheral ones (due to the geometric symmetry on the line and to the symmetry of the alimentation that we implicitly assume, this quantity does not depend on which peripheral rod is specifically chosen). The current is that flowing along the central rod. In particular, for symmetry, the currents flowing along the peripheral rods are equal to each other and have the same verse. Moreover, they are opposite to that flowing along the central rod, and the current comprehensively flowing at any abscissa is equal to zero.

A simplification of the previous equations for three-rod probes is the following [19]:

$$Z_l = \frac{1}{4\pi}\sqrt{\frac{\mu_r}{\varepsilon_r}}\sqrt{\frac{\mu_o}{\varepsilon_o}}\ \ln\left(\frac{1-\left(\frac{g}{s}\right)^4}{2\left(\frac{g}{s}\right)^3}\right) \tag{26}$$

where s is the center-to-center rod spacing and g is the rod radius (see Fig. 5).

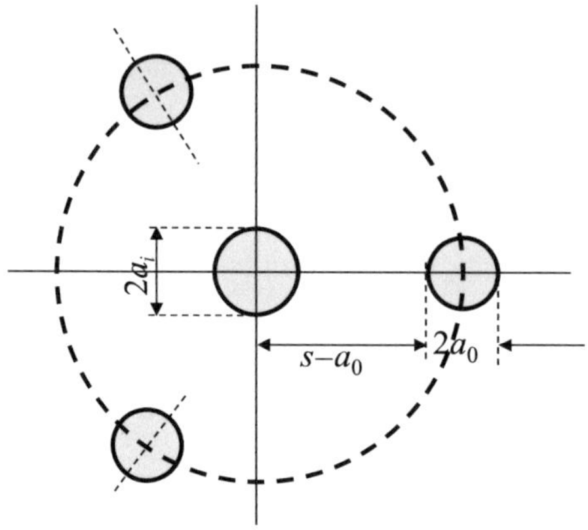

FIG. 4

Schematic of a multirod transmission line.

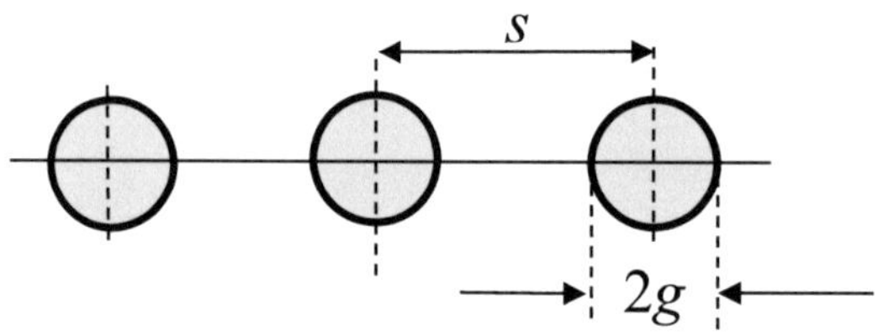

FIG. 5

Schematic of a three-rod transmission line.

As shown earlier in this chapter, in any case, the intrinsic impedance is proportional to the ratio $\sqrt{\frac{\mu_r}{\varepsilon_r}}$ and is equal in particular to the product between the intrinsic impedance that the same line would show in air times the ratio $\sqrt{\frac{\mu_r}{\varepsilon_r}}$. This dependence will be exploited in the next sections.

7 FINITE TRANSMISSION LINES

Now, we remove the hypothesis of infinite line. This will allow to deal with a model that can describe a TDR probe. In order to do this, we have also to take into account the fact that the voltage at the beginning of the line cannot be applied by means of an ideal generator, because ideal generators do not exist. A widely accepted model for the generator is that of an ideal generator with its internal impedance Z_g in series. We will add a piece of line with intrinsic impedance Z_g in air, that is, a piece of line matched to the generator. This part of the probe remains in air, but it is connected with the rods that are knocked in the material to be investigated. The knocked rods

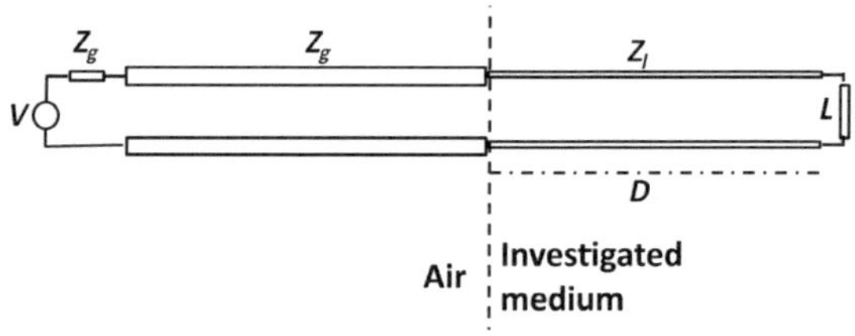

FIG. 6

Scheme for a finite transmission line applicable to a TDR probe knocked in the medium to be investigated.

constitute a transmission line with intrinsic impedance Z_l closed on a final load impedance L placed at distance D from the air-soil interface, according to the scheme depicted in Fig. 6.

In Fig. 6, the length of the line inserted in the soil is D; the length of the part in air is irrelevant. The effect of the load L and more in general the effect of any discontinuity along the line are the generation of a reflected voltage wave and of a reflected current wave that backpropagates from the load to the generator and that is added to the incident waves as described in the previous section. In the presence of a lossy material, the incident wave attenuates progressively toward the load, whereas the reflected wave attenuates progressively toward the generator, because each wave attenuates along its own direction of propagation. In order to describe the voltage and current waves, it is convenient now to choose a reference system with the zero of the z-axis on the load instead of on the generator as done before. In this way, the air-soil interface is placed at the negative abscissa $-D$. The voltage along the piece of line embedded in the material to be investigated can be expressed in the domain of the phasors, as

$$\hat{V}(z) = V^+ \exp\left(-jk_o\sqrt{\varepsilon_r\mu_r}z\right) + V^- \exp\left(jk_o\sqrt{\varepsilon_r\mu_r}z\right) \tag{27}$$

V^+ and V^- are complex quantities, as they include the amplitude and the initial phase of the incident and reflected voltage, respectively. The signs of the exponentials reflect the verses of the propagation. The possibility of losses is included in the possibly complex relative dielectric permittivity and relative magnetic permeability in Eq. (27). The current along the line is expressed by the sum of an incident and a reflected wave according to the following expression:

$$\hat{I}(z) = I^+ \exp\left(-jk_o\sqrt{\varepsilon_r\mu_r}z\right) - I^- \exp\left(jk_o\sqrt{\varepsilon_r\mu_r}z\right) \tag{28}$$

The minus sign before the reflected term in Eq. (28) is due to the fact that the voltage wave brings a current conventionally oriented toward the propagation direction of the voltage wave on the conductor at conventionally higher potential level. Consequently, the minus sign in Eq. (28) is due to the fact that the reflected voltage propagates toward the negative verse of the z-axis. From Eqs. (27), (28), it is immediate to see that, on the load, we have

$$\begin{cases} \hat{V}(0) = V^+ + V^- \\ \hat{I}(0) = I^+ - I^- \end{cases} \tag{29}$$

Moreover, on the load we have

$$Z(0) = \frac{\hat{V}(0)}{\hat{I}(0)} = L \tag{30}$$

Moreover, it is worth noting that it results

$$Z_l = \frac{V^+}{I^+} = \frac{V^-}{I^-} \Rightarrow I^+ - I^- = \hat{I}(0) = \frac{V^+ - V^-}{Z_l} \Rightarrow V^+ - V^- = Z_l \hat{I}(0) \tag{31}$$

Eq. (31) is derived from the fact that the incident and the reflected waves propagate along the same line that has a symmetrical structure either from the left-hand side to the right-hand side or vice versa and therefore shows the same intrinsic impedance whatever is the verse of propagation of a guide wave along it.

At this point, with some manipulation on Eqs. (27), (28) and accounting for Eqs. (29), (30), (31), we achieve

$$\begin{aligned} \hat{V}(z) &= V^+ \cos\left(k_o\sqrt{\varepsilon_r\mu_r z}\right) - jV^+ \sin\left(k_o\sqrt{\varepsilon_r\mu_r z}\right) + V^- \cos\left(k_o\sqrt{\varepsilon_r\mu_r z}\right) + jV^- \sin\left(k_o\sqrt{\varepsilon_r\mu_r z}\right) \\ &= (V^+ + V^-)\cos\left(k_o\sqrt{\varepsilon_r\mu_r z}\right) - j(V^+ - V^-)\sin\left(k_o\sqrt{\varepsilon_r\mu_r z}\right) \\ &= \hat{V}(0)\cos\left(k_o\sqrt{\varepsilon_r\mu_r z}\right) - j\hat{I}(0)Z_l \sin\left(k_o\sqrt{\varepsilon_r\mu_r z}\right) \end{aligned} \tag{32}$$

and analogously

$$\begin{aligned} \hat{I}(z) &= I^+ \cos\left(k_o\sqrt{\varepsilon_r\mu_r z}\right) - jI^+ \sin\left(k_o\sqrt{\varepsilon_r\mu_r z}\right) - I^- \cos\left(k_o\sqrt{\varepsilon_r\mu_r z}\right) - jI^- \sin\left(k_o\sqrt{\varepsilon_r\mu_r z}\right) \\ &= (I^+ - I^-)\cos\left(k_o\sqrt{\varepsilon_r\mu_r z}\right) - j(I^+ + I^-)\sin\left(k_o\sqrt{\varepsilon_r\mu_r z}\right) \\ &= \hat{I}(0)\cos\left(k_o\sqrt{\varepsilon_r\mu_r z}\right) - j\frac{\hat{V}(0)}{Z_l}\sin\left(k_o\sqrt{\varepsilon_r\mu_r z}\right) \end{aligned} \tag{33}$$

In Eqs. (32), (33), the exponentials have been replaced with sine and cosine functions. However, the arguments of the trigonometric functions are meant as complex in lossy cases. The ratio between Eq. (32), (33) provides the impedance seen at the generic abscissa along the line, looking toward the final load.

This is given by

$$\begin{aligned} Z(z) &= \frac{\hat{V}(z)}{\hat{I}(z)} = \frac{\hat{V}(0)\cos\left(k_o\sqrt{\varepsilon_r\mu_r z}\right) - j\hat{I}(0)Z_l \sin\left(k_o\sqrt{\varepsilon_r\mu_r z}\right)}{\hat{I}(0)\cos\left(k_o\sqrt{\varepsilon_r\mu_r z}\right) - j\frac{\hat{V}(0)}{Z_l}\sin\left(k_o\sqrt{\varepsilon_r\mu_r z}\right)} \\ &= \frac{\hat{V}(0) - j\hat{I}(0)Z_l tg\left(k_o\sqrt{\varepsilon_r\mu_r z}\right)}{\hat{I}(0) - j\frac{\hat{V}(0)}{Z_l} tg\left(k_o\sqrt{\varepsilon_r\mu_r z}\right)} = \frac{L - jZ_l tg\left(k_o\sqrt{\varepsilon_r\mu_r z}\right)}{1 - j\frac{L}{Z_l} tg\left(k_o\sqrt{\varepsilon_r\mu_r z}\right)} \\ &= Z_l \frac{L - jZ_l tg\left(k_o\sqrt{\varepsilon_r\mu_r z}\right)}{Z_l - jL tg\left(k_o\sqrt{\varepsilon_r\mu_r z}\right)} \end{aligned} \tag{34}$$

As can be seen, the impedance seen looking toward the load (the so-called transported impedance) is a function of position. For $z = 0$, Eq. (34) gives back the impedance of the load. It is also immediate to see that if the load impedance is equal to the intrinsic impedance of the line, then in any position the impedance seen looking toward the load is equal to the intrinsic impedance of the line. At the abscissa of the air-soil interface $z=-D$, the impedance seen looking toward the load is given by

$$Z(-D)=Z_l\frac{L+jZ_l tg\left(k_o\sqrt{\varepsilon_r\mu_r}D\right)}{Z_l+jLtg\left(k_o\sqrt{\varepsilon_r\mu_r}D\right)} \tag{35}$$

If the load is an open circuit, its impedance is infinite, because it prevents the passage of current whatever is the applied voltage. In this case, the open-circuit impedance Z_{oc} seen from the beginning of the line can be calculated as the limit for $L\rightarrow\infty$ of Eq. (35), and it is equal to

$$Z_{oc}(-D)=Z_l\frac{1}{jtg\left(k_o\sqrt{\varepsilon_r\mu_r}D\right)}=-jZ_l\cot\left(k_o\sqrt{\varepsilon_r\mu_r}D\right) \tag{36}$$

This impedance is purely imaginary if the line is lossless; otherwise, it will have a real and an imaginary part. This result is of specific interest for the problem at hand, because a TDR probe is an open-ended transmission line, and Eq. (37) shows the impedance seen at the air-soil interface looking underground (more generally at the interface between the air and the investigated medium looking into the medium). This impedance contains information on the product $\sqrt{\varepsilon_r\mu_r}$ through the wave number and contains information on the ratio $\sqrt{\frac{\mu_r}{\varepsilon_r}}$ through the intrinsic impedance of the line, which allows in principle to discriminate the magnetic from the dielectric characteristics of the medium under investigation. However, the impedance along the line is a quantity not easy to measure directly. So, we will formulate the problem on the basis of reflection coefficient data, which is a quantity more easily measurable. The reflection coefficient is evaluated at the air-soil interface, on the air side. The reflection coefficient Γ at the abscissa z is defined as the ratio between the reflected voltage wave and the incident voltage wave at that abscissa. Moving from the load to the generator, until no discontinuity is met along the line, this quantity is therefore given by

$$\Gamma(z)=\frac{V^-\exp(jkz)}{V^+(\exp-jkz)}=\frac{V^-}{V^+}\exp(2jkz)=\Gamma(0)\exp(2jkz) \tag{37}$$

The ratio $\frac{V^-}{V^+}$ is the reflection coefficient on the load. On the basis of Eqs. (29), (30), (31), we can express this ratio versus the impedance of the load and the intrinsic impedance of the line. In fact,

$$\begin{cases}\hat{V}(0)=V^++V^-=V^+(1+\Gamma(0))\\ \hat{I}(0)=\dfrac{V^+-V^-}{Z_l}=\dfrac{V^+}{Z_l}(1-\Gamma(0))\end{cases}\Rightarrow$$

$$\frac{\hat{V}(0)}{\hat{I}(0)}=L=Z_l\frac{1+\Gamma(0)}{1-\Gamma(0)}\Rightarrow(1-\Gamma(0))\frac{L}{Z_l}=1+\Gamma(0)\Rightarrow \tag{38}$$

$$\frac{L}{Z_l}-1=\Gamma(0)\left(1+\frac{L}{Z_l}\right)\Rightarrow\Gamma(0)=\frac{L-Z_l}{L+Z_l}$$

Replacing Eq. (38) into Eq. (37), we can easily calculate the reflection coefficient up to air-soil (air-material) interface on the soil side. However, passing from the soil to the air, there is a discontinuity along the line, provided by the different intrinsic impedances of the two connected pieces of lines. In particular, along the piece of line in air, the voltage is again composed of an incident plus reflected sinusoidal waves. However, the amplitude and phases of the two waves are different from those relative to the voltage waves propagating along the piece of line in the material to be investigated.

Consequently, the reflection coefficient shows a first kind of discontinuity, and its value after and before the abscissa $z=-D$ is not the same. In particular, the quantity of our interest is the reflection coefficient at the air-soil interface at the air side ($z=-D^-$), where the measurement can be more easily performed. In order to calculate it, we can exploit the formula of the transport of the impedance for an open-ended line (Eq. 36). In particular, the piece of line from the abscissa $z=-D$ up to the open end (i.e., the infinite load) can be viewed as a load given by the right-hand member of Eq. (36) placed at the abscissa $z=-D^-$. Consequently, rearranging the reference system and making use of Eq. (38) for the calculation of the new initial reflection coefficient, we have that, at $z=-D^-$, the reflection coefficient is given by

$$\begin{aligned}
\Gamma(-D^-) &= \frac{Z_{oc}(-D)-Z_g}{Z_g+Z_{oc}(-D)} = \frac{-jZ_l\cot\left(k_o\sqrt{\varepsilon_r\mu_r}D\right)-Z_g}{Z_g-jZ_l\cot\left(k_o\sqrt{\varepsilon_r\mu_r}D\right)} \\
&= \frac{-Z_g\sin\left(k_o\sqrt{\varepsilon_r\mu_r}D\right)-jZ_l\cos\left(k_o\sqrt{\varepsilon_r\mu_r}D\right)}{Z_g\sin\left(k_o\sqrt{\varepsilon_r\mu_r}D\right)-jZ_l\cos\left(k_o\sqrt{\varepsilon_r\mu_r}D\right)} \\
&= \frac{-Z_g\dfrac{\exp\left(jk_o\sqrt{\varepsilon_r\mu_r}D\right)-\exp\left(-jk_o\sqrt{\varepsilon_r\mu_r}D\right)}{2j}-jZ_l\dfrac{\exp\left(jk_o\sqrt{\varepsilon_r\mu_r}D\right)+\exp\left(-jk_o\sqrt{\varepsilon_r\mu_r}D\right)}{2}}{Z_g\dfrac{\exp\left(jk_o\sqrt{\varepsilon_r\mu_r}D\right)-\exp\left(-jk_o\sqrt{\varepsilon_r\mu_r}D\right)}{2j}-jZ_l\dfrac{\exp\left(jk_o\sqrt{\varepsilon_r\mu_r}D\right)+\exp\left(-jk_o\sqrt{\varepsilon_r\mu_r}D\right)}{2}} \\
&= \frac{(Z_g-Z_l)\exp\left(jk_o\sqrt{\varepsilon_r\mu_r}D\right)-(Z_l+Z_g)\exp\left(-jk_o\sqrt{\varepsilon_r\mu_r}D\right)}{-(Z_l+Z_g)\exp\left(jk_o\sqrt{\varepsilon_r\mu_r}D\right)-(Z_l-Z_g)\exp\left(-jk_o\sqrt{\varepsilon_r\mu_r}D\right)} \\
&= \frac{(Z_l-Z_g)+(Z_l+Z_g)\exp\left(-j2k_o\sqrt{\varepsilon_r\mu_r}D\right)}{(Z_l+Z_g)+(Z_l-Z_g)\exp\left(-j2k_o\sqrt{\varepsilon_r\mu_r}D\right)} \\
&= \frac{\left(Z_{lo}\sqrt{\dfrac{\mu_r}{\varepsilon_r}}-Z_g\right)+\left(Z_{lo}\sqrt{\dfrac{\mu_r}{\varepsilon_r}}+Z_g\right)\exp\left(-j\dfrac{4\pi f}{c_o}\sqrt{\varepsilon_r\mu_r}D\right)}{\left(Z_{lo}\sqrt{\dfrac{\mu_r}{\varepsilon_r}}+Z_g\right)+\left(Z_{lo}\sqrt{\dfrac{\mu_r}{\varepsilon_r}}-Z_g\right)\exp\left(-j\dfrac{4\pi f}{c_o}\sqrt{\varepsilon_r\mu_r}D\right)}
\end{aligned} \tag{39}$$

From a theoretical point of view, $\Gamma(f)$ as expressed in Eq. (39) and $S_{11}(f)$ as expressed in Eq. (5) are the same. Experimentally, working at fixed frequency the measurement of the reflection coefficient can be more reliable because it does not require the a priori knowledge of the incident wave but can be based on the wave standing ratio (WSR) and on the position of the minima of the comprehensive (incident plus reflected) voltage [15].

Eq. (39) depends on both the product $\sqrt{\varepsilon_r \mu_r}$ and the ratio $\sqrt{\frac{\mu_r}{\varepsilon_r}}$, and this allows in principle to retrieve both the relative dielectric permittivity and the relative magnetic permeability of the material under test by a suitable least-squares minimization. Measurements based only on time-domain data customarily do not exploit the intrinsic impedance of the line but only the return time, and so, they cannot allow to discriminate between ε_r and μ_r, because they depend only on the product $\sqrt{\varepsilon_r \mu_r}$.

In lossy cases, two complex square roots appear in Eq. (38), but the ambiguity about the determination to choose is easily solved on the basis of the previous dealing. In fact, the intrinsic impedance of the line "knocked" in the soil is proportional to $\sqrt{\frac{\mu_r}{\varepsilon_r}}$, but (see Eq. (21)) the real part of the intrinsic impedance of a lossy line has to be nonnegative. Consequently, the determination of $\sqrt{\frac{\mu_r}{\varepsilon_r}}$ with nonnegative real part has to be retained. Moreover, the wave number is proportional to $\sqrt{\varepsilon_r \mu_r}$, but the condition that the fields (and consequently the power) propagating along the line decrease and eventually vanish far from the sources (see again Eq. 21) drives to the condition that the imaginary part of the wave number (and consequently the imaginary part of $\sqrt{\varepsilon_r \mu_r}$) has to be nonpositive. In particular, the complex wave number (see Eqs. 18, 19) in lossy cases has been put in the form $k = \beta - j\alpha$ with both α and β nonnegative and with $\beta - j\alpha = k_o \sqrt{\varepsilon_r \mu_r} = \frac{2\pi f}{c_o} \sqrt{\varepsilon_r \mu_r}$.

8 HIGHER ORDER MODES

It is important to mention the issue of higher-order modes so as to make clearer the subsequent discussion. For an adequate dealing of the higher-order modes, again, the interested reader is referred to books of electromagnetism, for example [12,15], and references therein.

For our purposes, the problem can be exposed as follows: in the proposed dealing, we have implicitly assumed that the electromagnetic field and consequently the electromagnetic energy propagates just along the path of the transmission line. This is intuitive, but we do not have any physical assurance that this has to be the path of the energy. Indeed, things work in this way as far as the transmission line is constituted by at least two or more conductors, and its cross section is small in terms of wavelength. If we consider a progressively increasing frequency, at a certain point part of the energy still follows precisely the path of the line, but another part propagates, let say, wrapping around the line. In other words, part of the energy, even if it still follows the line, does not follow exactly its path. This alters the propagation time of part of the energy and consequently the phase of the retrieved signal in frequency domain. In particular, it is as if there were two incident waves and two reflected waves at the same time, with different amplitudes and phases and also different wave numbers and consequently also with different intrinsic impedances and reflection coefficients at any abscissa.

In particular, the phasor of the voltage[2] signal along the line can be expressed as

$$\hat{V}(z) = V_1^+ \exp(-jk_1 z) + V_1^- \exp(jk_1 z) + V_2^+ \exp(-jk_2 z) + V_2^- \exp(jk_2 z) \tag{40}$$

When this happens, it is said that the first higher-order mode propagates. When the frequency still increases, at a certain point, part of the energy will follow a third path along the line. An additional term is to be included in Eq. (40) in this case, representing the second higher-order mode. The mechanism goes on the same for progressively higher frequencies. The higher-order modes are infinite, and increasing the frequency, we progressively "excite" them sequentially. From the applicative point of view, the multimodal propagation is undesired because it generates ambiguous echoes in the retrieved signal and makes the interpretation of the returned signal trickier. This means that the cross section of the transmission line should be narrow enough with respect to the wavelength, let say of the order of $\frac{\lambda}{2}$ at most. In particular, for a coaxial cable, the first higher-order propagation mode is excited at the frequency $f = \frac{2c_o}{(a+b)\pi\sqrt{\varepsilon_r \mu_r}}$. It is important to outline also that, when a multiwire transmission line is exploited, the fundamental mode propagates at any frequency, even if very low. The fundamental mode is also called the transverse electromagnetic (TEM) mode because the electric and magnetic fields are both necessarily orthogonal to the direction along which the transmission line develops. Another feature worth outlining is that the intrinsic impedance of the line associated with the TEM mode does not depend on the frequency if the embedding material is not dispersive. Instead, with higher-order modes, the intrinsic impedance depends on the frequency even if the material is not dispersive.

In some cases, it is well advised to exploit a structure with only one conductor. This is as in the case of measurements in rectangular waveguide, typical for the Nicholson-Ross-Weir method (for which the fundamental mode is transverse electric (TE) but not transverse magnetic), or is also the case of some single-rod TDR probes recently exploited for monitoring soil moisture, for which the fundamental mode is transverse magnetic (TM) but not transverse electric. This kind of probe is simple to realize and easy to insert; nevertheless, the analysis of the TDR waveform is somewhat more complicated, and the region where the fields are meaningful is relatively large.

Before closing this section, it is worth mentioning one last important aspect, namely, the fact that the higher-order modes can have an effect even when they are in cutoff (i.e., the frequency is too low to allow their propagation). In fact, the cutoff does not imply the absence of higher-order modes but more precisely the fact that they attenuate exponentially (and quite rapidly if the work frequency is not very close to the cutoff frequency) along the line instead of propagating. This means that, especially close to discontinuities of any kind, there is in general some contribution to the field due to higher-order modes (this is theoretically needed in order to satisfy

[2]Indeed, it should be not called any longer "voltage," but we skip over about this further aspect, and for simplicity, we will continue to use this now not any longer rigorous term.

the boundary conditions), but this contribution decreases versus z along both the positive and negative verses. A detailed discussion on this matter goes beyond the scope of this chapter, but the reader should be aware that a solution that neglects (as often happens and as we also will do this way) the higher-order modes is not exact in any case, because it is not based on a complete (full-wave) model of the electromagnetic propagation.

9 INVESTIGATION OF LOSSLESS AND DISPERSIONLESS MATERIALS

Let us exploit Eq. (39) for the investigation of the dielectric permittivity and the magnetic permeability of a lossless and dispersionless material. This simplified model is not realistic in many cases, but allows us to introduce progressively the difficulties and allows to present some important physical features of the phenomenon in an easy way. In the lossless case, both ε_r and μ_r are real quantities. We will also assume (as usually is) that Z_g is real (i.e., we suppose that the generator does not show meaningful internal inductances and or capacitances); then, $\Gamma(-D^-)$ is a complex quantity with unitary modulus, that is, it is a pure phase term. This is due to the fact that, neglecting any loss, all the incident power is reflected back toward the generator and dissipated in its internal resistance Z_g. The problem at single frequency can be recast as the least-squares minimization of a cost function in two variables defined as follows:

$$F(x,y) = \left| \Gamma_m - \frac{(Z_{lo}x - Z_g) + (Z_{lo}x + Z_g)\exp\left(-j\frac{4\pi fD}{c_o}y\right)}{(Z_{lo}x + Z_g) + (Z_{lo}x - Z_g)\exp\left(-j\frac{4\pi fD}{c_o}y\right)} \right|^2 \tag{41}$$

where Γ_m is the measured value of $\Gamma(-D^-)$, $x = \sqrt{\frac{\mu_r}{\varepsilon_r}}$, and $y = \sqrt{\varepsilon_r \mu_r}$. Once one has retrieved the auxiliary unknown x and y, the relative dielectric permittivity and the relative magnetic permeability are given by

$$\begin{cases} \varepsilon_r = \dfrac{y}{x} \\ \mu_r = xy \end{cases} \tag{42}$$

Considering Eq. (41), however, it can be noted that the cost function is periodic along y, because the imaginary exponential is a periodic function. It is easily recognized that the period of the cost function is given by $P = \frac{c_o}{2fD}$. Hence, a single-frequency measurement provides in this case an ambiguous result, because there is a sequence of global minima, mathematically equivalent to one another, distant P from one another along the y-axis, and there is no means to discriminate which one among them provides the actual couple (x, y). In order to introduce some diversity and enhance the available information, multifrequency data can be exploited, starting from a minimum frequency f_o and making use of M frequencies uniformly spaced with a frequency step Δf. In this way, the problem can be recast as the minimization of a multifrequency (MF) cost function given by

$$F_{MF}(x,y)=\sum_{k=1}^{M}\left|\Gamma_{mk}-\frac{(Z_{lo}x-Z_g)+(Z_{lo}x+Z_g)\exp\left\{-j\frac{4\pi[f_o+(k-1)\Delta f]Dy}{c_o}\right\}}{(Z_{lo}x+Z_g)+(Z_{lo}x-Z_g)\exp\left\{-j\frac{4\pi[f_o+(k-1)\Delta f]Dy}{c_o}\right\}}\right|^2 \tag{43}$$

where Γ_{mk} is the reflection coefficient measured at the kth exploited frequency. From Eq. (43), we appreciate immediately that the cost function is still periodic along the y-axis if the minimum exploited frequency is a multiple integer of the frequency step. In fact, in this case, we have that an integer number N exists such as $f_o=N\Delta f$, which when replaced into Eq. (43) provides

$$F_{MF}(x,y)=\sum_{k=1}^{M}\left|\Gamma_{mk}-\frac{(Z_{lo}x-Z_g)+(Z_{lo}x+Z_g)\exp\left\{-j\frac{4\pi[N+(k-1)]\Delta fDy}{c_o}\right\}}{(Z_{lo}x+Z_g)+(Z_{lo}x-Z_g)\exp\left\{-j\frac{4\pi[N+(k-1)]\Delta fDy}{c_o}\right\}}\right|^2 \tag{44}$$

In Eq. (44), the generic kth addend is a periodic function versus y, and the period is

$$P_k=\frac{c_o}{2[N+(k-1)]D\Delta f} \tag{45}$$

Consequently, the quantity

$$P_{MF}=\frac{c_o}{2D\Delta f} \tag{46}$$

is an integer multiple of P_k for any value of k, that is, all the addends can be also seen as periodic quantities with a common period P_{MF}, which is therefore also the period of the entire multifrequency cost function in Eq. (44). Of course, the multifrequency period is N times larger than the period achieved when making use of the only minimal frequency f_o, and this reduces the ambiguity significantly. Even if the periodicity is not eliminated, a priori information can help in identifying the right global minimum. In particular, a range R in y can be chosen a priori, and in order to guarantee that only one global minimum is present in this range, the frequency step should be decreased enough such that $P_{MF}>R$, which means

$$\Delta f<\frac{c_o}{2DR} \tag{47}$$

An estimation of where this range could be advantageously centered can be retrieved from a common TDR measurement in time domain, thanks to which a first estimation of the $y=\sqrt{\varepsilon_r\mu_r}$ can be retrieved from the return time of a pulse radiated along the rods. That said, a range has to be chosen also with regard to the quantity $x=\sqrt{\frac{\mu_r}{\varepsilon_r}}$, and we will make use of the range 0.1–4, extremes included. For natural materials in the radio frequency or microwave range, it is definitely rare (even if not physically impossible) to find materials for which $\sqrt{\frac{\mu_r}{\varepsilon_r}}$ lies beyond this range. The minimization that we will apply here is exhaustive, because the unknowns looked for are only two, and therefore, an exhaustive minimization does not present particular computational problems. On the other hand, as discussed later on in this chapter, the cost function possibly exhibits many local minima [20], also beyond the problem of the periodicity, and so, an exhaustive minimization mitigates the risk of being trapped into a false solution.

Let us now remove the hypothesis that the minimum frequency is an integer multiple of the frequency step. It is easily recognized that the cost function is still periodic if the ratio between f_o and Δf is a rational number. However, this feature is not much of interest for us. It is more relevant to observe that, if (as reasonable) we choose $\Delta f \ll f_o$, then we can approximate the ratio $\frac{f_o}{\Delta f} \gg 1$ with its integer part, in the sense that the cost function achieved with the real value of Δf and that achieved from its approximation as an integer subpart of f_o becomes necessarily progressively similar to each other. This means indeed that in this limit, the cost function tends to show either a periodicity or a pseudoperiodicity along the y-axis, and P_{MF} provided in Eq. (46) is either the period or the pseudoperiod. Due to the fact that in real cases, the data are necessarily noisy, some model error is present for the neglecting of the higher-order modes, and possibly, some losses not accounted for in this preliminary model are present; the pseudoperiodicity is substantially equivalent to the periodicity.

Before showing some examples, let us also note that in Eq. (43), the frequency appears explicitly only multiplied by D, that is, times the length of the rods of the probe. Consequently, the reasoning proposed with regard to the multifrequency could be repeated with regard to an extensible probe where the rods can be somehow prolonged (at fixed frequency f_o) starting from a minimal length D_o with uniform length step ΔD. We can so define a multilength (ML) cost function given by

$$F_{ML}(x,y) = \sum_{k=1}^{M} \left| \Gamma_{mk} - \frac{(Z_{lo}x - Z_g) + (Z_{lo}x + Z_g)\exp\left\{-j\dfrac{4\pi f_o[D_o + (k-1)\Delta D]y}{c_o}\right\}}{(Z_{lo}x + Z_g) + (Z_{lo}x - Z_g)\exp\left\{-j\dfrac{4\pi f_o[D_o + (k-1)\Delta D]y}{c_o}\right\}} \right|^2 \tag{48}$$

where now Γ_{mk} is the reflection coefficient measured in correspondence of the kth exploited length of the probe, namely, $D_k = D_o + (k-1)\Delta D$. It is immediately verified that the multifrequency cost function with M frequencies at fixed length D_o is mathematically equivalent to the multilength functional at fixed frequency f_o if $\frac{\Delta D}{D_o} = \frac{\Delta f}{f_o}$, that is, if the length increment exploited in the multilength case is the same in percentage as the frequency increment in the multifrequency case.

The multilength solution is not exploited in practice to the best of our knowledge, because it is less practical than a multifrequency solution. However, we deem it worth considering because it allows to achieve a diversity fully equivalent to that achievable from multifrequency data without increasing the cross section of the line in terms of wavelength. This prevents the excitation of higher-order modes that instead might occur within a multifrequency approach. Moreover, when some dispersion is present (even if we are not considering it here), the multilength approach allows to achieve information at fixed frequency, so allowing to measure the dispersion law or allowing to test a supposed or suspected dispersion law.

At this point, let us show a first example. Let us suppose to have a bifilar TDR probe. The length of the rods is 25 cm, the radius of the conductors is 2 mm, and their reciprocal distance is 2.5 cm. We exploit initially a multifrequency configuration, and we propose the results achieved at a single frequency equal to 200 MHz and those achieved adding progressively up to 18 frequencies, with a uniform frequency step equal to 19 MHz. The actual value of the relative dielectric permittivity is $\varepsilon_r = 5$, and the actual value of the relative magnetic permeability is $\mu_r = 2$. These values drive to an exact value of the quantity $y = \sqrt{\varepsilon_r \mu_r} = 3.74$. Let us suppose to have a rough estimation, derived, for example, from time-domain measurements. In particular, we assume as first estimation the quantity $y_o = 5$, with an error of 33% with respect to the actual value, and so, we choose as range of variability for x the interval [1,10], that is, an interval of amplitude about twice larger than 5, about centered on 5, excluding the values $y < 1$ that are physically impossible. For x, instead, the range is [0.1,4].

The minimization is achieved exhaustively on 801×801 pixels. The cost function in all the considered cases is represented in Fig. 7.

As can be seen, the cost function becomes progressively cleaner when increasing the number of frequencies, but local minima do not disappear completely, even with 18 frequencies. On the other hand, the local minima become eventually quite shallow compared with the global minimum and do not constitute a problem in the

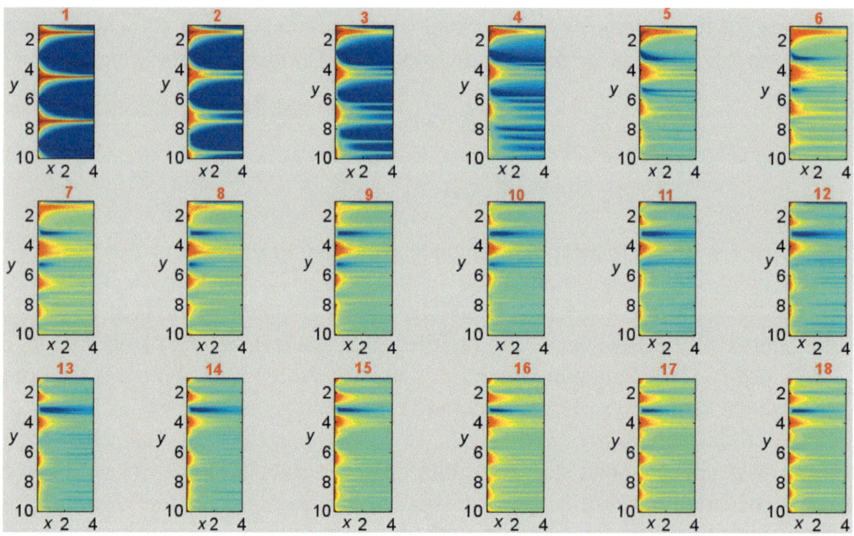

FIG. 7

Cost function with $\varepsilon_r = 5$, $\mu_r = 2$, $f_o = 200$ MHz, $\Delta f = 19$ MHz. The number of frequencies is shown in red on the top of each figure. The data are noiseless.

framework of an exhaustive minimization. The retrieved value for the relative dielectric permittivity with 18 frequencies is 5.0049, and the retrieved value for the relative magnetic permeability with 18 frequencies is 1.9951. The percentage error for the relative dielectric permittivity and the relative magnetic permeability is shown in Figs. 8 and 9, respectively. The error on both the permittivity and the permeability

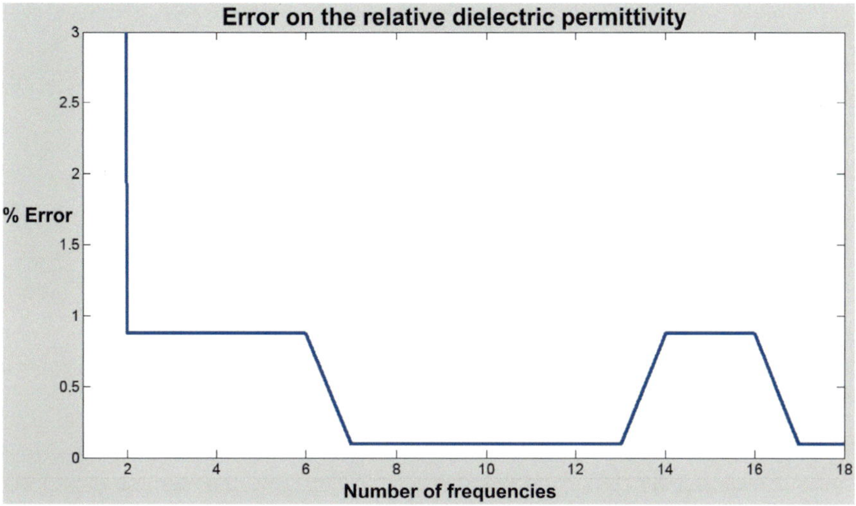

FIG. 8

Percentage error in the evaluation of the dielectric permittivity, $\Delta f = 19$ MHz. Noiseless data.

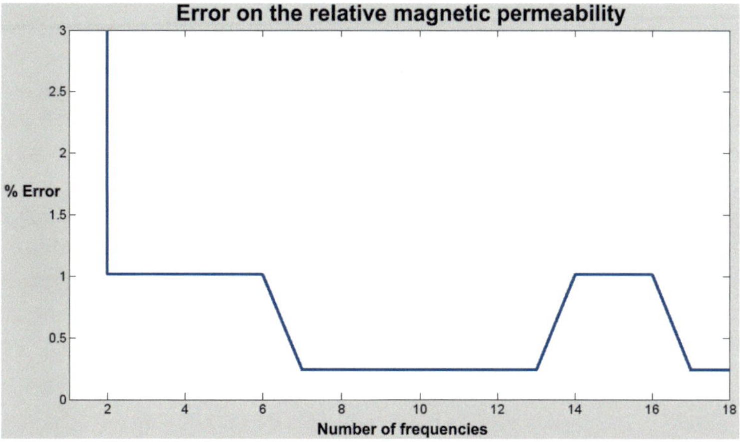

FIG. 9

Percentage error in the evaluation of the magnetic permeability, $\Delta f = 19$ MHz. Noiseless data.

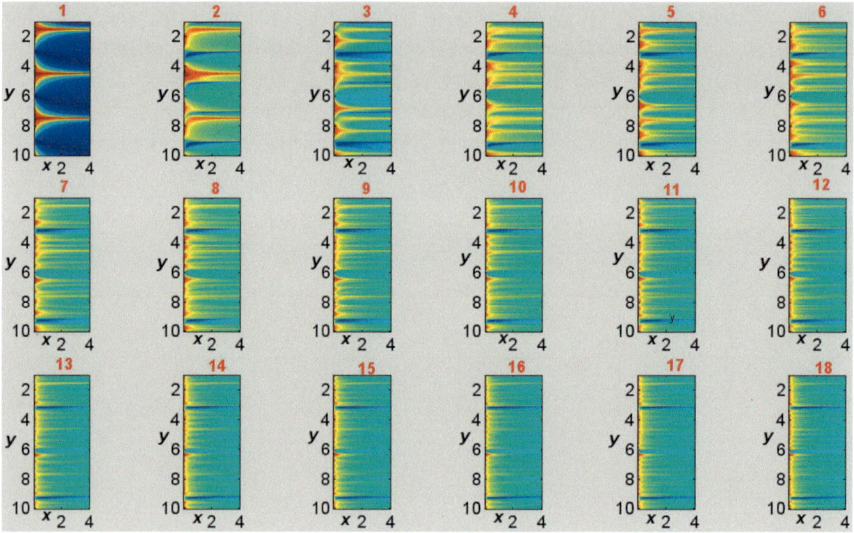

FIG. 10

Cost function with $\varepsilon_r=5$, $\mu_r=2$, $f_o=200\,\text{MHz}$, $\Delta f=98\,\text{MHz}$. The number of frequencies is shown in *red* on the top of each figure. The data are noiseless.

with only one frequency is quite high because in that case, the function is periodic along the y-axis, and as can be seen from the first image in Fig. 7, three global minima of the cost function are included in the considered range along y. For minimal numerical variations, the algorithm chooses substantially in a casual way on them. The behavior of the error is not monotonically decreasing versus the number of frequencies, but even with only two frequencies, the errors reach values of the order of 1% and keep below this threshold. In order to show the effects of a large frequency step, in Fig. 10, we show the cost function achieved with a frequency step of 98 MHz. This frequency step (as well as the previous step of 19 MHz) was chosen in order not to have an integer submultiple of the minimum frequency 200 MHz. All the other parameters are the same with respect to the graphs presented in Fig. 7. As can be seen, now, the cost function tends to show an evident pseudoperiodicity, and beyond the right global minimum occurring at $x=0.646$, $y=3.16$, there is another deep minimum at about $x=0.6$, $y=9.25$.

The vertical distance between the y-coordinates of these two minima is given by $\Delta y=9.25-3.16=6.09$ and corresponds with a good approximation to the value of the pseudoperiodicity given in Eq. (46); in fact, in the case at hand, we have $P_{MF}=\frac{c_o}{2D\Delta f}=\frac{3\times10^8\,\text{m/s}}{2\times0.25\,\text{m}\times98\times10^6\,\text{s}^{-1}}=6.12$. In particular, the pseudoperiod is smaller than the amplitude of the chosen range along the y-axis that is equal to 9 (the range along y extends in fact from 1 to 10). Incidentally, the pseudoperiod with $\Delta f=19\,\text{MHz}$ was equal to $P_{MF}=\frac{c_o}{2D\Delta f}=\frac{3\times10^8\,\text{m/s}}{2\times0.25\,\text{m}\times19\times10^6\,\text{s}^{-1}}=31.57$, so it was larger than 9,

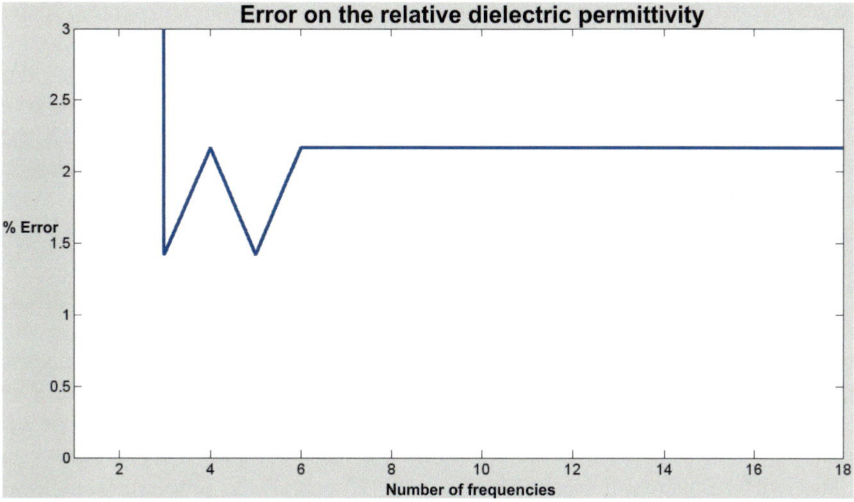

FIG. 11

Percentage error in the evaluation of the dielectric permittivity, $\Delta f = 98\,\text{MHz}$. Noiseless data.

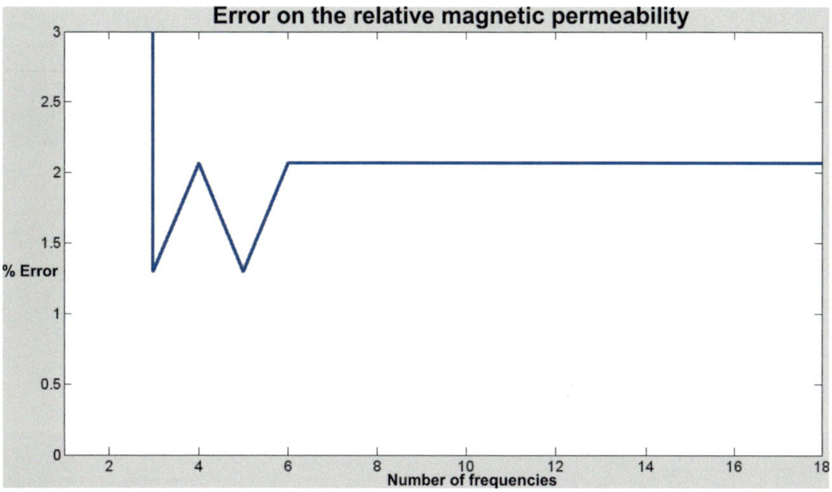

FIG. 12

Percentage error in the evaluation of the magnetic permeability, $\Delta f = 98\,\text{MHz}$. Noiseless data.

and the first minimum relative to the pseudoperiods occurred beyond the chosen range. In Figs. 11 and 12, the graphs of the percentage errors for the relative dielectric permittivity and the relative magnetic permeability are provided, respectively.

As can be seen, the errors are on a value slightly higher than 2% both with regard to the dielectric permittivity and the magnetic permeability. The "definitive" error is slightly higher than it was with $\Delta f = 19\,\text{MHz}$, probably for numerical approximation errors. However, the data were noiseless (and in particular, they were not full-wave data but data based on the propagation of the only fundamental propagation mode). Let us now add some noise to the data. In particular, white Gaussian noise has been added to each reflection coefficient (for 1, 2 ... up to 18 frequencies), so that SNR = 20 dB (10 in natural scale) is guaranteed. In particular, at each frequency, one more noisy datum is added to those already considered up to that point. Since the reflection coefficient is a complex quantity, noise has been separately added both to the real part and on the imaginary part of the reflection coefficient.

The percentage error on the relative dielectric permittivity with a frequency step $\Delta f = 19\,\text{MHz}$ is shown in Fig. 13, whereas the error on the relative magnetic permeability with $\Delta f = 19\,\text{MHz}$ is shown in Fig. 14. Moreover, the homologous quantities with $\Delta f = 98\,\text{MHz}$ and with the same kind of noise and the same SNR are shown in Figs. 15 and 16, respectively. As can be seen, the algorithm proves to be quite robust against the noise, and in the cases at hand, it seems also quite robust with respect to the pseudoperiodicity. However, the problem is enclosed in a probabilistic framework, and therefore, it is in any case well advised to choose a frequency step according to Eq. (47); otherwise, the pseudoperiodicity (or even more the periodicity in the case that the minimum frequency is an integer multiple of the frequency step) can represent a risk even for an exhaustive minimization procedure.

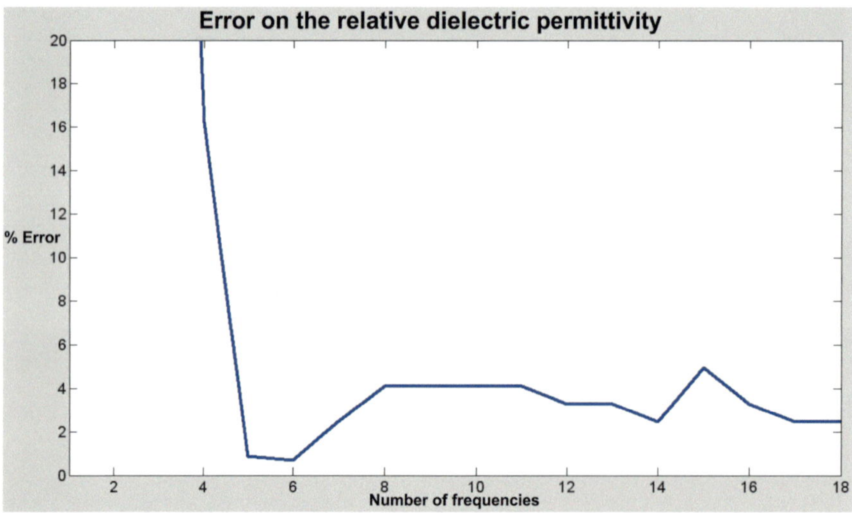

FIG. 13

Percentage error in the evaluation of the dielectric permittivity, $\Delta f = 19\,\text{MHz}$. Gaussian noise on the data, SNR = 20 dB.

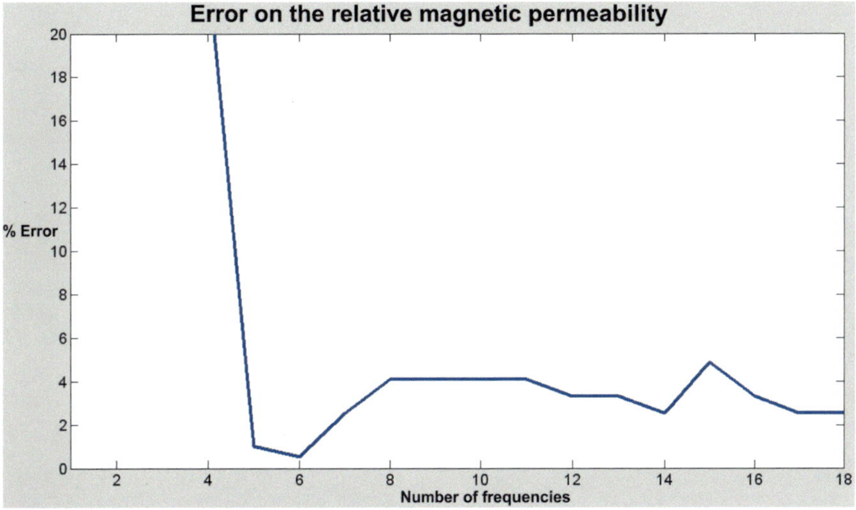

FIG. 14

Percentage error in the evaluation of the magnetic permeability, $\Delta f = 19\,\text{MHz}$. Gaussian noise on the data, $SNR = 20\,\text{dB}$.

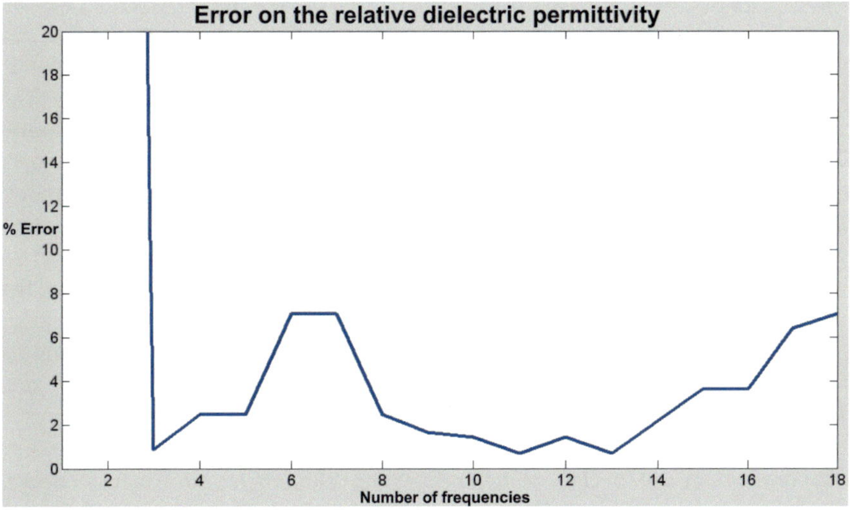

FIG. 15

Percentage error in the evaluation of the dielectric permittivity, $\Delta f = 98\,\text{MHz}$. Gaussian noise on the data, $SNR = 20\,\text{dB}$.

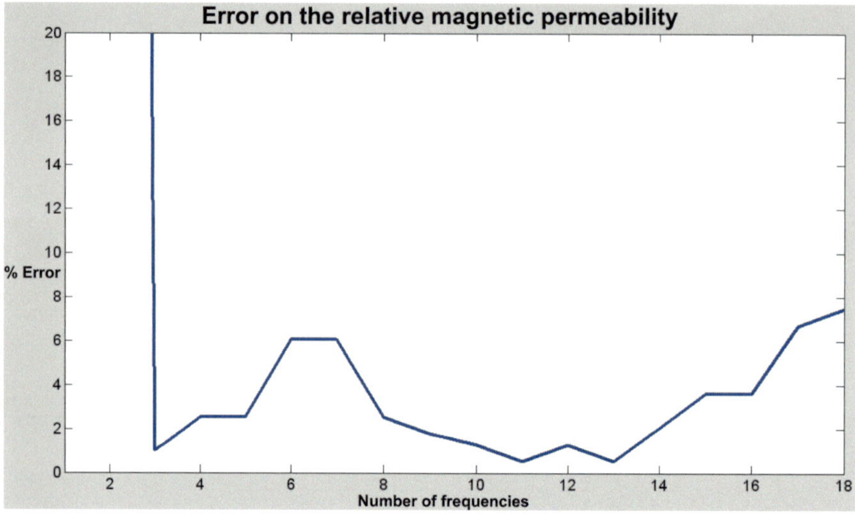

FIG. 16

Percentage error in the evaluation of the magnetic permeability, $\Delta f = 98$ MHz. Gaussian noise on the data, SNR = 20 dB.

Figs. 17–20 are homologous to Figs. 13–16 but with a uniformly distributed noise instead of a Gaussian noise. As can be seen, the kind of noise has some effect too, and in particular, the error tends to be slightly larger with a uniform noise on the data than with a Gaussian noise. From Figs. 8–9 and 11–20, we appreciate also the fact that the information added by the frequencies progressively added does not necessarily drive toward an improved solution, because the graphs of the errors are not uniformly decreasing with the number of frequencies. Indeed, the problem should be coped within a probabilistic framework, because the noise is a random quantity. In particular, the positive influence of the addition of more frequencies is expected only after averaging the results of repeated measurements, but such an analysis is beyond our purposes here. However, even within the limits of the presented analysis, we see that the addition of more frequencies tends to make and keep low the error. With a few frequencies, in particular, we see that the error can be very high because even the exhaustive minimization algorithm is "deceived" by the presence of "deep" local minima. When the discretization available information is too poor, even the small error due to the discretization can drive the algorithm toward a wrong solution. This is the reason why the result is completely wrong with one or two frequencies even with noiseless data, as can be seen from Figs. 11 and 12. We also see that the narrower frequency step provides the better results. Indeed, this is partially related to sampling aspects (the discretization was the same, but the spatial band of the cost function was not the same in the case $\Delta f = 19$ MHz and in the case $\Delta f = 98$ MHz).

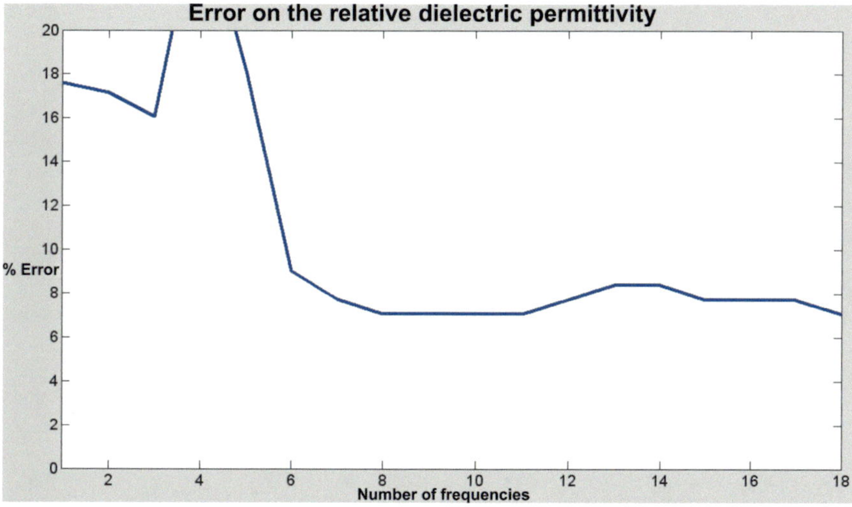

FIG. 17

Percentage error in the evaluation of the dielectric permittivity, $\Delta f = 19\,\text{MHz}$. Uniform noise on the data, SNR = 20 dB.

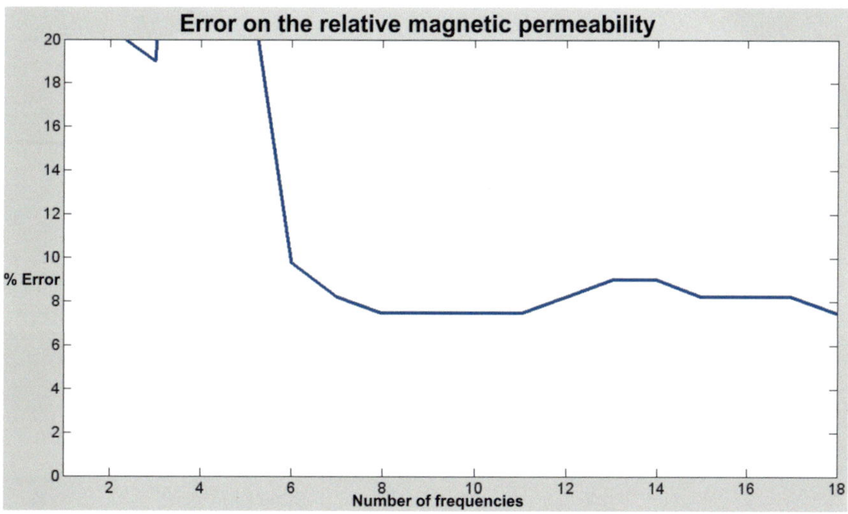

FIG. 18

Percentage error in the evaluation of the magnetic permeability, $\Delta f = 19\,\text{MHz}$. Uniform noise on the data, SNR = 20 dB.

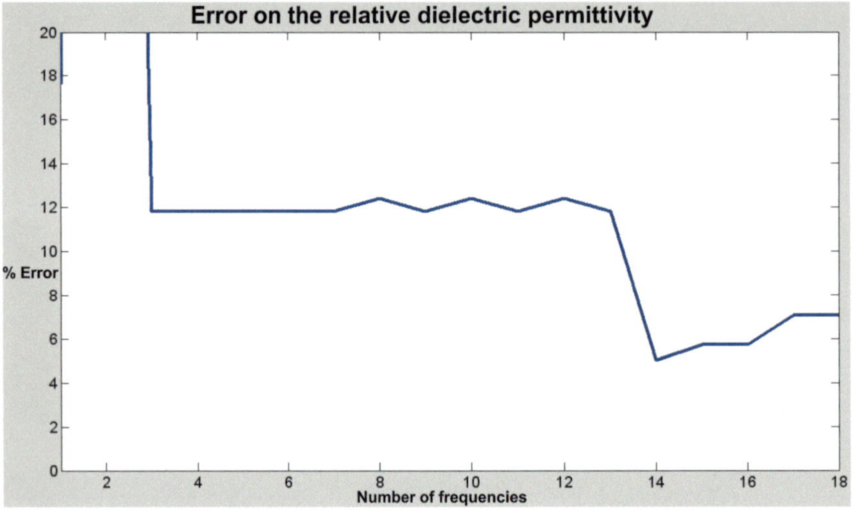

FIG. 19

Percentage error in the evaluation of the dielectric permittivity, $\Delta f = 98\,\text{MHz}$. Uniform noise on the data, SNR = 20 dB.

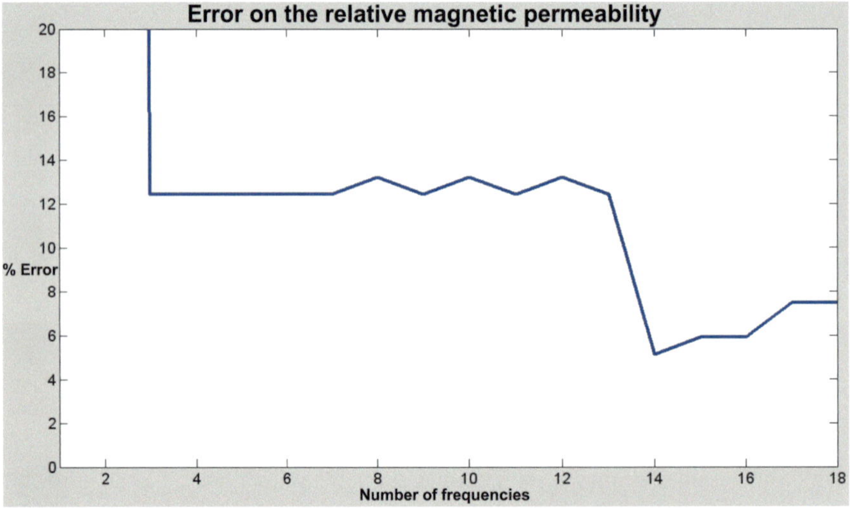

FIG. 20

Percentage error in the evaluation of the magnetic permeability, $\Delta f = 98\,\text{MHz}$. Uniform noise on the data, SNR = 20 dB.

It is not easy to establish in a general way what the best frequency step is in an absolute sense and in particular independently on the prechosen range of values R along the y-axis (see Eq. 47). So, we limit to recommend that the frequency step respects condition (47). In particular, if a comprehensive frequency band has been prechosen, so that the comprehensive number of frequency depends on Δf, making progressively narrower Δf, one makes the data redundant. Instead, if the total number of frequencies has been prechosen, so that the comprehensive band of the data is proportional to the frequency step Δf, then it is obvious that narrowing the frequency step makes the band eventually vanish so making the result unreliable. In relationship with the presented case, condition (47) means $\Delta f < 66.67$ MHz, so that the choice $\Delta f = 19$ MHz respected it and the choice $\Delta f = 98$ MHz did not respect it.

The reasoning can be repeated in the same way keeping fixed the frequency at 200 MHz (the minimum considered value) and considering a probe with extensible rods. In fact, as said above, a multifrequency and a multilength approach are equivalent (for a nondispersive material) if $\frac{\Delta D}{D_o} = \frac{\Delta f}{f_o}$. In the case at hand, it means that we can repeat the same considerations (and substantially the same graphs but for the random nature of the noise) if the frequency is kept fixed at 200 MHz and the length of the rods is prolonged from the initial value of 25 cm with a length step $\Delta D = 2.375$ cm (equivalent to $\Delta f = 19$ MHz because $\frac{\Delta f}{f_o} D_o = 2.375$cm in this case) or $\Delta D = 12.25$ cm (equivalent to $\Delta f = 98$ MHz because $\frac{\Delta f}{f_o} D_o = 12.25$cm in this case). It is immediate that if a multilength approach is used, then the equivalent of condition (47) in order to avoid the presence of a deep local minimum in the range spanned for the exhaustive minimization is

$$\Delta D < \frac{c_o}{2 f_o R} \tag{49}$$

With regard to the presented examples, in particular, condition (48) means $\Delta D < 8.33$ cm, so that the choice $\Delta D = 2.375$ cm respects condition (49), whereas $\Delta D = 12.25$ cm does not (further to being much more unpractical).

The frequency-length equivalence allows also to easily identify a maximum limit for the frequency band of the data, beyond which further data would not add further independent information. In particular, from Eq. (35), we see that the impedance transported along a lossless line is periodic with period equal to $\frac{\lambda}{2}$, where λ is the wavelength in the medium. Indeed, we do not know this quantity, but in any case, it is not larger than $\frac{\lambda_o}{2}$, namely, one-half of the wavelength in air. So, in a lossless case, it is pointless to prolong the rods beyond the limit:

$$D_{\max} = D_o + \frac{\lambda_o}{2} \tag{50}$$

Consequently, due to the already outlined frequency-length duality, it is useless to add further frequencies beyond a maximum $f_{\max}$ such as

$$\frac{f_{\max} - f_o}{f_o} = \frac{D_{\max} - D_o}{D_o} = \frac{\lambda_o}{2D_o} \Rightarrow f_{\max} = f_o + \frac{f_o \lambda_o}{2D_o} = f_o + \frac{c_o}{2D_o} \tag{51}$$

Let us remember, however, that such a limit holds only for lossless media. Paradoxically, there might be more available information in the data if the medium is lossy. Indeed, this is partially true because the reflection coefficient in the case of a lossy medium varies in both amplitude and phase, whereas it is a pure phase term for lossless media. The apparent paradox is explained because the lossless model contains itself a strong piece of information already assumed or known a priori. Of course, if the losses are too strong, the signal will be suddenly too weak and noisy, and this increase of available information becomes only theoretical. In any case, the modulus of the measured reflection coefficient (equal to 1 if and only if the medium is lossless) reveals if a lossless model is or is not suitable.

10 INVESTIGATION OF LOSSY AND/OR DISPERSIVE MATERIALS

It has been shown in the scientific literature that any lossy material is dispersive and any dispersive material is necessarily lossy [15], but the dispersion and the losses can be different for different bands. So, approximately, it is licit to consider lossy nondispersive materials, dispersive lossless materials, and dispersive lossy materials in some fixed band. We have shown in the previous section that a multilength approach is equivalent to a multifrequency approach in the case of a lossless dispersionless medium. This is still valid if we consider a dispersionless but lossy medium, but it is not only longer valid for dispersive media. In these cases, however, the multilength approach can provide the diversity of information needed for retrieving the characteristics of the medium at fixed frequency. Moreover, the multilength approach does not alter the electric cross section of the transmission line, that is, the cross section of the line with respect to the wavelength, and therefore, the multilength approach does not give rise to higher-order propagating modes. That said, in the general case of a lossy and dispersive media, our unknowns are four at each frequency, namely, the real and imaginary part of both the complex equivalent relative dielectric permittivity and the complex equivalent relative magnetic permeability. As a result, we have

$$\begin{cases} \varepsilon_r = \dot{\varepsilon}'_r - j\dot{\varepsilon}''_r \\ \mu_r = \mu'_r - j\mu''_r \end{cases} \tag{52}$$

In Eq. (52), the real parts are positive, and the imaginary parts are negative, at least for natural materials.[3] Recasting the problem according to Eq. (48) (or Eq. 44 if the dispersion is negligible), we have two complex unknowns, namely,

[3]At least for natural materials. Indeed, artificial materials exist for which, for example, the dielectric permittivity might be negative, but they are beyond our purposes.

$$\begin{cases} \sqrt{\dfrac{\mu_r}{\varepsilon_r}} = x + jz \\ \sqrt{\varepsilon_r \mu_r} = y + jw \end{cases} \tag{53}$$

which correspond to the four real unknowns x, y, z, w, with the physical constraint that x and y are positive within particular $y \geq 1$[4] and w is nonpositive (which drives also univocally the choice of the determination of the complex roots). Instead, z can be either positive or negative. The given constraints are easily imposed in the framework of an exhaustive minimization, just through the choice of the range of variability for each unknown.

After solving for the four real auxiliary unknowns x, y, z, w, the two complex unknowns of interest are given by

$$\begin{cases} \varepsilon_r = \dfrac{y + jw}{x + jz} \\ \mu_r = (x + jz)(y + jw) \end{cases} \tag{54}$$

From now on, we will focus on the theoretically more general multilength case (but the multifrequency case is analogous). So, the cost function to be minimized is the natural extension of that given in Eq. (48), namely,

$$F_{ML}(x, y, z, w) = \sum_{k=1}^{M} \left| \Gamma_{mk} - \frac{(Z_{lo}(x + jz) - Z_g) + (Z_{lo}(x + jz) + Z_g)\exp\left\{-j\dfrac{4\pi f_o [D_o + (k-1)\Delta D](y + jw)}{c_o}\right\}}{(Z_{lo}(x + jz) + Z_g) + (Z_{lo}(x + jz) - Z_g)\exp\left\{-j\dfrac{4\pi f_o [D_o + (k-1)\Delta D](y + jw)}{c_o}\right\}} \right|^2 \tag{55}$$

In order to identify the quadruple range for the minimization, with regard to x, we will exploit the range (0.1,4), that is, the same range exploited in the lossless. With regard to y, we will identify an approximate first value from measurements of return time. In order to simulate the uncertainty on that value, we will exploit in the subsequent simulation the exact value of y plus an error of 20%. The range around this point will range from this value minus 2 (or from 1 if this value minus 2 is smaller than 1) to this value plus 2. With regard to the range along the z, we have chosen the interval [−2.2]. Finally, for the range along w, we have chosen the interval [−4.0]. The quadruple range has been divided into 40 cells along each axis for computational reasons. In a first example, we have considered a medium with $\varepsilon_r = 5 - 0.3j$ and $\mu_r = 2 - 0.25j$. The size of the probe is the same of the previous lossless cases, the frequency is kept fixed at 200 MHz, and the probe is extended with a spatial step of $\Delta D = 2.375$ cm. In particular, starting from its initial length of 25 cm, 18 values are exploited for the extension of the probe, which corresponds to a maximum length of the probe equal to 65.375 cm.

After a first inversion, we have retrieved the results $\varepsilon_r = 5.0117 - 0.0810j$ and $\mu_r = 1.9417 - 0.3529j$. These results provide an acceptable estimation of the real part

[4]This condition guarantees that the propagation velocity of the waves in the considered medium is not faster than that in free space, which is a general physical condition.

of ε_r and μ_r but fail in retrieving the imaginary parts. Indeed, the coarseness of the grid has a meaningful effect on this, so the problem is also partially (even if not merely) computational. Consequently, the result can be improved with a second minimization about the solution found at the first step. In particular, we have considered four ranges of amplitude equal to 2 (instead of 4 as in the first minimization), in any case intersected with the physical conditions $x>0$, $y\geq 1$, and $w\leq 0$. In the case at hand, this amounted in the Cartesian product $[0.01,1.39]\times[2.13\text{–}4.13]\times[-1.06,0.94]\times[1.30,0]$ with respect to the variables x, y, z, and w, respectively. The second minimization has provided the results $\varepsilon_r=4.9853-0.2838j$ and $\mu_r=1.9622-0.2650j$. The final error is 0.29% on the real part of the relative dielectric permittivity, 1.9% on the real part of the relative magnetic permeability, 5.4% with regard to the imaginary part of the relative dielectric permittivity, and 6% with regard to the imaginary part of the relative magnetic permeability. The same procedure in two steps has been repeated with the addition of Gaussian white noise with SNR=40dB. The results have been $\varepsilon_r=5.0239-0.0802j$ and $\mu_r=1.9369-0.3525j$ after the first step and $\varepsilon_r=4.7163-0.2761j$ and $\mu_r=2.1086-0.2780j$ after the second step. Again, the first step provided an acceptable estimation of the real parts but not of the imaginary parts. The final error after the second minimization is 5.7% for the real part of the relative dielectric permittivity, 5.4% for the real part of the relative magnetic permeability, 7.97% for the imaginary part of the relative dielectric permittivity, and 11.2% with regard to the imaginary part of the relative magnetic permeability. As easily suspected, in lossy cases, it is important that the data are particularly clean. In particular, the results achieved with SNR=20dB were not satisfying and are not reported.

11 CONCLUSIONS

In this chapter, we have discussed the use of TDR probes in time and frequency domain, proposing a brief overview of the state of art and inserting the reasoning in the electromagnetic framework of the theory of the transmission lines. This has allowed to describe in the second part of the chapter some proposals about a possible extension of the use of the TDR probes to advanced measurements at single frequency in possibly lossy and possibly magnetic media. The first results show to be promising, even if the dealing has not been full wave but based on the only fundamental TEM propagation mode. However, also conceptually similar (but practically quite different) measurements in waveguide are based on the fundamental (TE10 in that case) propagation mode [9,10].

Essentially, this work proposes to export in the field of TDR applications the principles already exploited in laboratory measurements with regard to waveguide evaluations of the electromagnetic characteristics of a medium, with the drawback of not having at disposal the transmitted signal but with the pros of not having to pick up a sample of material and to transport it in laboratory. This can be relevant for many applications, in particular related to properties of soil not easy to retrieve with fully noninvasive methods [21–23]. Future works will be devoted to implement better gradient-based minimization algorithms (this is not very important in lossless cases, but it becomes more important in lossy cases, where the number of trials for the exhaustive minimization increases meaningfully). Moreover, full-wave numerical experiments have to be performed with the aid of simulation codes based either on the finite-difference or on the finite-element method. Finally, the design and fabrication of an extensible TDR probe would allow to perform experimental validations [24].

REFERENCES

[1] A. Cataldo, E. De Benedetto, Broadband reflectometry for diagnostics and monitoring applications, IEEE Sensors J. 11 (2) (2011) 451–459.

[2] A.M.O. Mohamed, Principles and Applications of Time Domain Electrometry in Geoenvironmental Engineering, Taylor & Francis Group, Abingdon, 2006.

[3] Time domain reflectometry theory, Agilent Application Note 1304-2, Palo Alto, CA, 2006.

[4] D.A. Robinson, S.B. Jones, J.M. Wraith, D. Or, S.P. Friedman, A review of advances in dielectric and electrical conductivity measurement in soils using time domain reflectometry, Vadose Zone J. 2 (4) (2003) 444–475.

[5] T.J. Heimovaara, Frequency domain analysis of time domain reflectometry waveforms: 1. Measurement of the complex dielectric permittivity of soils, Water Resour. Res. 30 (2) (1994) 189–199.

[6] C. Lin, Frequency domain versus travel time analyses of TDR waveforms for soil moisture measurements, Soil Sci. Soc. Am. J. 67 (2003) 720–729.

[7] M.G. Pelletier, R.C. Schwartz, G.A. Holt, J.D. Wanjura, T.R. Green, Frequency domain probe design for high frequency sensing of soil moisture, Agriculture 6 (2016) 60, https://doi.org/10.3390/agriculture6040060.

[8] K. Higasi, Y. Koga, M. Nakamura, Dielectric relaxation and molecular structure. V. Application of the single frequency method to systems with two Debye dispersions, Bull. Chem. Soc. Jpn. 44 (1971) 988–992.

[9] A.M. Nicholson, G.F. Ross, Measurement of the intrinsic properties of materials by time domain techniques, IEEE Trans. Instrum. Meas. 19 (4) (1970) 377–382.

[10] W.B. Weir, Automatic measurement of complex dielectric constant and permeability at microwave frequencies, Proc. IEEE 62 (1974) 33–36.

[11] O. Luukkonen, S.I. Maslovski, S.A. Tretyakov, A stepwise Nicolson–Ross–Weir-based material parameter extraction method, IEEE Antennas Wirel. Propag. Lett. 10 (2011) 1295–1298.

[12] D. Pozar, Microwave Engineering, fourth ed., Wiley, Hoboken, 2012.

[13] E. Pettinelli, G. Vannaroni, A. Cereti, A.R. Pisani, F. Paolucci, D. Del Vento, D. Dolfi, S. Riccioli, F. Bella, Laboratory investigations into the electromagnetic properties of magnetite//silica mixtures as Martian soil stimulants, J. Geophys. Res. 11 (2005) E04013, https://doi.org/10.1029/2004JE002375.

[14] N.E. Hill, A possible second dielectric dispersion region in polar liquids, Proc. Phys. Soc. 82 (5) (1963) 723–727.

[15] G. Franceschetti, Electromagnetics: Theory, Techniques and Engineering Paradigms, Plenu Press, New York and London, 1997.

[16] T.J. Heimovaara, J.A. Huisman, J.A. Vrugt, W. Bouten, Obtaining the spatial distribution of water content along a TDR probe using the SCEM-UA Bayesian inverse modeling scheme, Vadose Zone J. 3 (2004) 1128–1145.
[17] R. Persico, Introduction to Ground Penetrating Radar: Inverse Scattering and Data Processing, Wiley, Hoboken, 2014.
[18] J.A.R. Ball, Characteristic impedance of unbalanced TDR probes, IEEE Trans. Instrum. Meas. 51 (3) (2002) 532–536.
[19] J.A. Huisman, C.P. Lin, L. Weihermuller, H. Vereecken, Accuracy of bulk electrical conductivity measurements with time domain reflectometry, Vadose Zone J. 7 (2) (2008) 426–433.
[20] R. Persico, F. Soldovieri, R. Pierri, Convergence properties of a quadratic approach to the inverse scattering problem, J. Opt. Soc. Am. A 19 (12) (2002) 2424–2428.
[21] L. Mertens, R. Persico, L. Matera, S. Lambot, Smart automated detection of reflection hyperbolas in complex GPR images with no a priori knowledge on the medium, IEEE Trans. Geosci. Remote Sens. 54 (1) (2016) 580–596.
[22] R. Persico, G. Leucci, L. Matera, L. De Giorgi, F. Soldovieri, A. Cataldo, G. Cannazza, E. De Benedetto, Effect of the height of the observation line on the diffraction curve in GPR prospecting, Near Surv. Geophys. 13 (3) (2015) 243–252.
[23] A. Cataldo, R. Persico, G. Leucci, E. De Benedetto, G. Cannazza, L. Matera, L. De Giorgi, Time domain reflectometry, ground penetrating radar and electrical resistivity tomography: a comparative analysis of alternative approaches for leak detection in underground pipes, NDT&E Int. 62 (2014) 14–28.
[24] A.G. Gorriti, E. Slob, A new tool for accurate S-parameters measurements and permittivity reconstruction, IEEE Trans. Geosci. Remote Sens. 43 (8) (2005) 1727–1735.

CHAPTER

4

Geochemical constraints in near-surface geophysical surveying from in situ XRF spectrometry: Field trials at two aviation archaeology sites

Adam D. Booth*, Veerle Vandeginste†, Dominic Pike‡, Russell Abbey‡, Roger A. Clark*, Chris M. Green*, Nathan Howland‡

School of Earth and Environment, University of Leeds, Leeds, United Kingdom School of Chemistry, University of Nottingham, University Park, Nottingham, United Kingdom† Nuthampstead Airfield Museum, Bee Farm, Hertfordshire, United Kingdom‡*

CHAPTER OUTLINE

Innovation in Near-Surface Geophysics. https://doi.org/10.1016/B978-0-12-812429-1.00004-0

1 INTRODUCTION

Geophysical methods are widely applied to the detection and delineation of diverse near-surface targets, typically exploiting contrasts in the subsurface distribution of a number of physical properties (e.g., magnetic susceptibility, density, and electric conductivity). Constraining the target's chemical composition is often of secondary importance. For example, the remit of a typical electromagnetic survey may be to identify a response that is consistent with a buried metallic target, rather than identifying the specific metallic elements involved. However, an understanding of geochemistry could represent significant added value in a range of near-surface geophysical projects. Any process that leaves a diagnostic abundance of "marker" elements may be suitable for combined geochemical-geophysical survey; examples include industrial archaeology (e.g., Refs. [1,2]), forensic investigation [3–5], and environmental tracer testing [6].

Of course, geochemical analyses are standard in a number of fields. In quantitative archaeometric studies, they are applied to classify the composition of, for example, glass [7,8], paint [9], and ceramic remains [7,10]. They are also central to numerous forensic studies, to link samples of diverse materials to specific crime scenes [11]. However, the use of geochemical methods specifically as a field *exploration* tool is not widely reported. Instead, they are applied as laboratory methods, with measurements usually conducted on prepared (often destroyed) samples of material extracted from a site (e.g., Refs. [12–17]). This is rather different to the often quoted advantages of geophysical methods, which include the noninvasive and nondestructive nature of the survey. The recent development of field-portable geochemical analyzers represents a promising means of emulating such advantages. For example, in situ geochemical analysis using X-ray fluorescence (XRF) spectrometry has become possible with the development of handheld XRF systems (e.g., Refs. [18,19]). Such in-field equipment is particularly valuable in forensic cases (e.g., Ref. [3]) since the material under inquiry must be uncontaminated by any sampling process. The use of field-portable XRF spectrometers is reported for archaeological applications (e.g., Refs. [20–23]) but seldom using a sampling strategy consistent with standard geophysical survey (i.e., spot samples are considered, rather than systematically defined areas and/or transects).

We consider the scope of in situ geochemical exploration by deploying a field-portable XRF system alongside geophysical methods, for archaeological characterization of the crash sites of two Second World War aircraft. The in situ data compare favorably with XRF and mass spectrometry applied under laboratory conditions, but seasonally repeated surveys show variation given the changing supply of chemical elements to the ground surface. Under favorable ground conditions, the geochemical data significantly enhance the interpretation of the geophysical responses. We also suggest mechanisms by which geochemical anomalies arise and consider the circumstances in which in situ XRF surveying may be successful.

2 X-RAY FLUORESCENCE SPECTROSCOPY—THEORY AND METHODOLOGY

X-ray fluorescence spectroscopy determines the elemental composition of a sample material using high-energy, short-wavelength (X-ray) radiation (note that *spectroscopy* and *spectrometry* are distinct; the former is a technique, whereas the latter is the quantitative analysis of data). When bombarded with X-ray radiation, different elements can be identified by the characteristic *fluorescent* energy that they emit [24].

XRF responses are adversely affected by several factors, including matrix composition [25,26], surface morphology [22,27,28], and instrumental sensitivity [24]. Matrix composition effects are mitigated using manufacturers' calibrations for representative materials (e.g., mudrock, glass, and alloys). Although challenging to define, bespoke calibrations can be made [17,26] and allow the XRF data to be used as an absolute rather than relative indicator of composition [29]. Laboratory XRF practice mitigates the effects of surface morphology by (destructively) grinding samples into a fine powder. Equivalent sample preparation is impractical for in situ XRF spectrometry; hence, field-portable XRF instruments have faced skepticism in the geochemical community [30]. However, recent research (e.g., Ref. [19]) has reported similar accuracy and precision between field- and laboratory-based observations.

The instrument deployed here is a handheld *Bruker* Tracer IV-SD spectrometer (Fig. 1), an energy-dispersive instrument with a rhodium target. The spectrometer is used at different voltages, depending on whether major or trace (concentrations >1 ppm) elements are under consideration. While the detection of elements lighter than calcium can be challenging since these have a low "fluorescence yield" (i.e., their energy emissions are weak [31,32]), our system compensates this with the use of a *silicon drift detector* [33]. Sensitivity is further improved by including a *Bruker* 3V vacuum pump (Fig. 1) to inhibit the attenuation of fluorescent energy by air in the spectrometer's analysis chamber. The presence of water also impedes XRF analysis, since water scatters the X-ray radiation; therefore, in situ XRF surveys may always be vulnerable to the presence of groundwater (e.g., Ref. [34]), especially for low-yield elements.

The sample area (*spot size*) of an XRF measurement is typically 1 cm in diameter. However, the depth penetration of XRF energy in soil is on the millimeter-to-centimeter scale; hence, in situ XRF measures only the surface chemistry of host soil. While it may be detectable with geophysical methods, a target would therefore be invisible to XRF sampling unless the ground surface is enriched in relevant marker elements via some source-to-surface transport mechanism (e.g., plowing and groundwater circulation [35,36]). Even then, such transport may not only be in a vertical direction; hence, the strongest concentrations of ions may not be observed directly above the source. As such, in situ XRF prospection will likely always benefit from the a priori constraint provided by conventional geophysical survey.

FIG. 1

A *Bruker* Tracer IV-SD handheld XRF spectrometer, deployed at Nuthampstead Airfield (August 2015). Here, the Bruker spectrometer is held in the operator's right hand and the 3V vacuum pump in their left.

3 SURVEY SITES

The two crash sites (sites A and B) in this study are located within Nuthampstead Airfield (Hertfordshire, United Kingdom; National Grid Reference TL419347; Fig. 2). Nuthampstead Airfield was built in 1943, during the Second World War (1939–45). It was operated by the US Army Air Forces and was the base for the 398th Bombardment Group (Heavy). The airfield returned to British control in 1945 and was decommissioned and returned to agricultural use in 1959. Since then, the site has been extensively plowed, but runways still remain (providing access across the site), and evidence of military infrastructure remains present as crop marks (Fig. 2A). Both crash sites are hosted within clay soil: local borehole BGS TL43NW39 (UK National Grid Reference TL 41404 35166) shows that surface soils are clay-rich, beneath which is stiff clay to a depth of at least 40 m with occasional flint and chalk cobbles [37]. The underlying solid geology, the Lewes and Seaford Chalk formations, is observed at ~70 m depth.

The crash sites under investigation feature different styles of aircraft impact, with potential implications for the relative success of geophysical and geochemical survey. Additionally, ground conditions during two survey periods differed significantly, aspects we also consider when reviewing the performance of the in situ XRF survey.

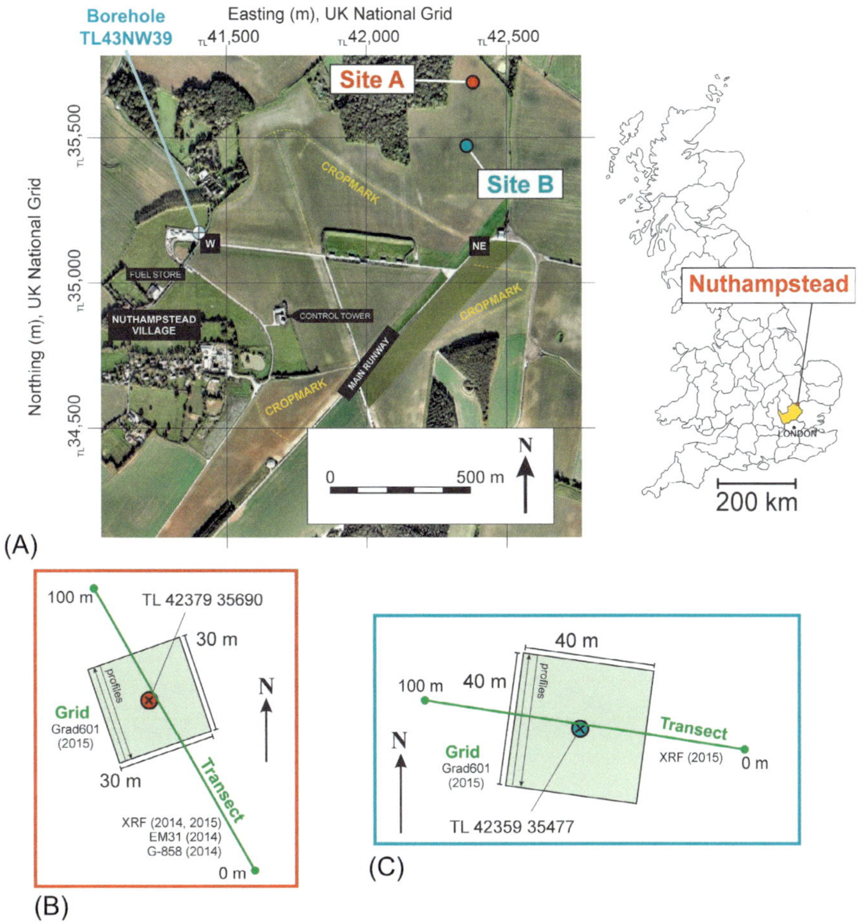

FIG. 2

Survey sites. (A) Image of Nuthampstead Airfield, © Google Earth. Runways and buildings from the original airfield remain present today, with other infrastructure visible as crop marks. The location of BGS borehole TL43NW39 is indicated, together with the aircraft crash sites. (B) Detail of surveys performed at site A. (C) Detail of surveys performed at site B.

3.1 SITE A

Site A, the main survey location in this analysis, is the crash site of de Havilland DH.98 Mosquito FB Mk.VI LR343. This aircraft crashed in the grounds of Nuthampstead, in February 1944, shortly after its takeoff from RAF Hunsdon (an airfield also in Hertfordshire). Records suggest that the port engine detached from the aircraft, causing it to invert and impact the ground at near-vertical angle. The speed and steep angle of impact into soft clay soil suggest that any remaining components of the

Mosquito could be buried at several meters depth, although records of the potential depth are seldom reported. The crash caused an intense fire and claimed the lives of the two crewmen (members of 487 Squadron Royal New Zealand Air Force). Their bodies were recovered from the site, along with some wreckage, but it is doubtful that all debris was cleared from the site and some components (including armaments and starboard engine) may remain present today.

Site A has been identified by Nuthampstead Airfield Museum using contemporary photographs of the impact (e.g., Fig. 3A) plus a local concentration of surface

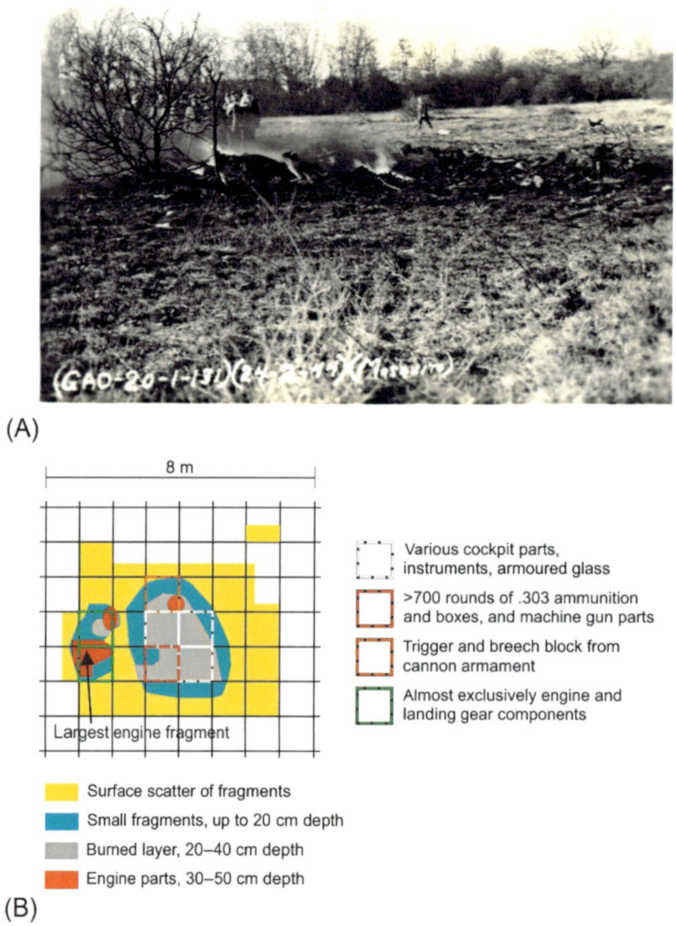

FIG. 3

(A) Contemporary photograph of the 1944 crash site at site A, in Nuthampstead Airfield. (B) Distribution of fragments identifiable as wreckage of a de Havilland Mosquito aircraft, both at surface and following shallow excavation. The map is centered on the crash site marked in Fig. 2B.

finds identifiable as Mosquito wreckage (Fig. 3B). Site A was first surveyed in November 2014, involving an exploratory transect of XRF measurements (including soil samples for laboratory validation) alongside magnetic gradiometer and electromagnetic (EM) data. Promising initial results from the XRF survey motivated a second acquisition in August 2015, in which a repeat set of XRF measurements and a larger grid of magnetic data were acquired (Fig. 2B).

3.2 SITE B

Site B is understood to be the crash site of Boeing B-17G Flying Fortress 42-102536 *Here's Hopin'*, of the 600th Bombardment Squadron, located approximately 1 mile from the end of the main runway (Fig. 2A). The aircraft exploded and burned on the ground surface (in contrast to the vertical impact at site A) on the morning of 24 December 1944, after attempting to take off from Nuthampstead during adverse weather conditions. Being prepared for an operational bombing mission, the aircraft was fully fueled and laden with 4000 lb of bombs. Two of the 10 crewmen were killed in the accident: One was trapped in the wreckage (his body was recovered later), and the other died later from the injuries he sustained. The identity of the aircraft at site B is somewhat ambiguous given that reporting could be confused with a second Flying Fortress crash at Nuthampstead Airfield, within 20 min of that of *Here's Hopin'*. There is some possibility at site B that the aircraft could be B-17G Flying Fortress 42-97374 *Madame X*, but since both aircraft exploded under similar circumstances, the geophysical/geochemical response of either of them would be similar.

It is worth noting that the Flying Fortress is a much larger aircraft than the Mosquito (unladen weights, respectively, of ~16,000 and ~6000 kg); therefore, the potential amount of wreckage is larger at site B than site A. However, identifiable debris from the Flying Fortress is recoverable across a large area of Nuthampstead Airfield, such was the force of the explosion. It is therefore unclear how much wreckage remains in the immediate locality, particularly as remaining surface debris was removed soon after the crash.

A single data archive was acquired over site B in August 2015, involving a magnetic survey grid and an XRF transect (Fig. 2C). Soil samples were again taken for laboratory validation.

4 SURVEY DESIGN AND IMPLEMENTATION

Prior to undertaking any surveys, we considered the likely expression of the crash sites in geophysical and geochemical data and used this to inform survey design.

4.1 GEOPHYSICAL CONSIDERATIONS

Most aircrafts from the Second World War are made from aluminum and steel; therefore, geophysical surveys can exploit an abundance of these metallic components in the ground with magnetic and/or electromagnetic methods (e.g., Ref. [38]).

However, this may be problematic for the sites investigated in this study. For the case of site A, the *Mosquito* aircraft was one of the few Second World War aircraft built chiefly from wood, and aluminum is only present in certain components. Consistent with site B, it is also unclear whether the larger steel components (e.g., the engine and armaments) are still present in the subsurface.

Both crashes, however, caused large fires. We therefore assumed that a magnetic survey may respond to any thermoremanent magnetic signature of the impact fire. Despite ~55 years of plowing at the site, trial excavations at site A suggest that a layer of intense burning remains present at 0.2–0.4 m depth (Fig. 3B). Aluminum fragments observed among the surface debris were melted in the impact fire (e.g., Fig. 4A(ii)); hence, the temperature of the fire must have exceeded 660°C (the melting point of aluminum). Critically, this exceeds the temperature at which iron minerals are transformed into ferrimagnetic iron oxides (200–400°C [39,40]) and approaches the Curie temperature for hematite (675°C [41]). Magnetic survey methods were therefore prioritized, with additional electromagnetic surveys in case of remaining metallic debris.

4.2 GEOCHEMICAL CONSIDERATIONS

As with geophysical considerations, a geochemical survey for aircraft wreckage might initially be designed to exploit an abundance of aluminum (Al). However, in addition to the challenges of low fluorescence yield, any small Al anomaly may be masked by the high background aluminum content in Nuthampstead's clay soil; furthermore, Al is a light element, and its XRF signature is vulnerable to groundwater masking. Alternative geochemical targets were identified by considering the XRF characteristics of surface debris samples recovered in the vicinity of site A (e.g., representative samples in Fig. 4A). These included the following:

(1) Brass ammunition cartridges (e.g., Fig. 4A(iii)): typical cartridge brass from the Second World War period is an alloy of 70% copper (Cu) and 30% zinc (Zn), occasionally containing small quantities of lead (Pb). Cartridges may also have jacket of cupronickel (Cu-Ni) alloy.

(2) Cannon rounds (e.g., Fig. 4A(iv)): this ammunition is made principally from steel, possibly alloyed with a nickel-chromium-molybdenum (Ni-Cr-Mo) blend. Other elements may be present in variants of this ammunition (e.g., phosphorous in tracer rounds).

(3) Burnt wood: although dominated by light elements (e.g., carbon and oxygen), traces of heavier elements, such as Pb, could be present in any paint residue.

In addition to these fragments, a sample of burnt soil was tested to monitor any chemical alteration (however unlikely) caused by the impact fires.

The elemental concentrations in these samples are shown in Fig. 4B. All measurements use a mudrock calibration, with the spectrometer set to 40 kV for measurement of trace elements. The manufacturer-defined mudrock setting was the most appropriate for Nuthampstead's clay-rich soil, though the lack of a bespoke calibration

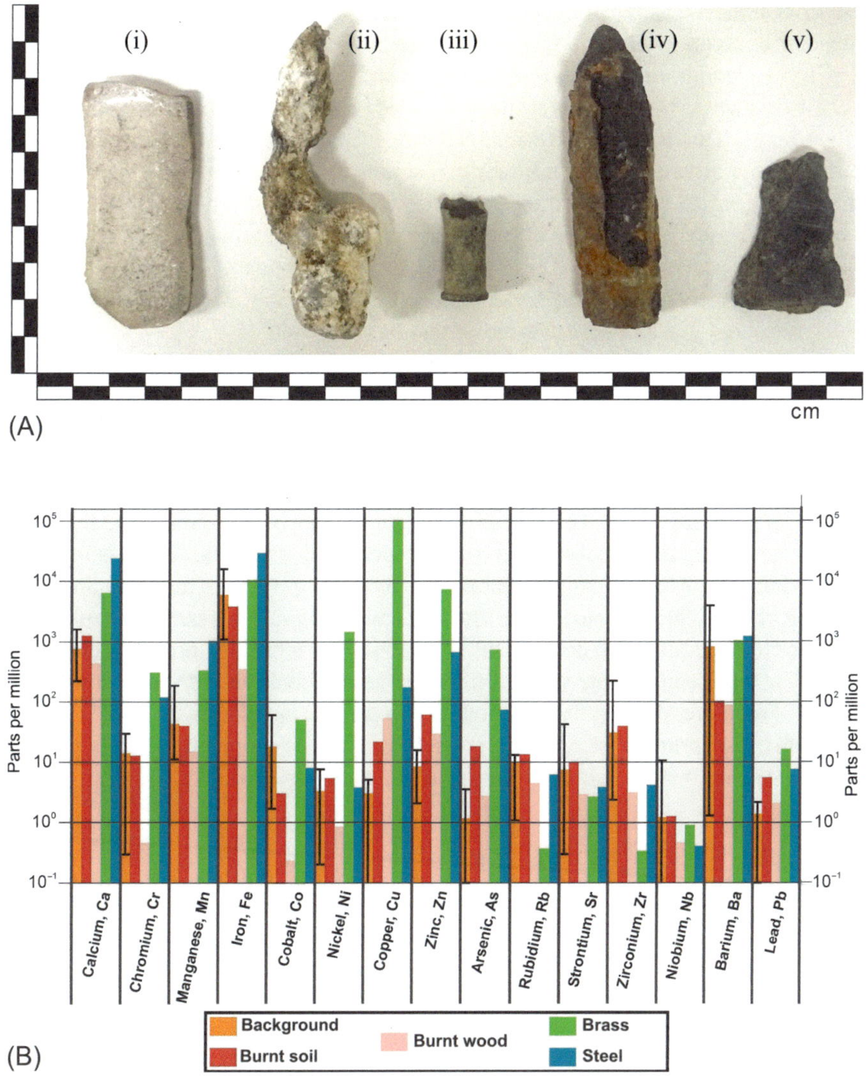

FIG. 4

(A) Representative samples of aircraft fragments recovered at the ground surface from site A: (i) heat-crazed Perspex from the cockpit; (ii) corroded aluminum, melted in the impact fire; (iii) fragment of brass ammunition cartridge; (iv) corroded steel tip of cannon round; and (v) Bakelite plastic from the cockpit. (B) Measured concentration of elements in XRF analysis of samples of burnt soil, burnt wood, brass, and steel from the Nuthampstead site, compared with background geochemistry.

implies that the measured concentrations are relative rather than absolute indicators. Elements lighter than calcium and those too scarce for even the trace element setting (e.g., molybdenum, tin, and antimony) are absent from this plot. Each concentration is compared with a background value, with error bars spanning the observed concentration range. This background value is defined for each element as the mean concentration recorded along the transect in a subset of locations outside the magnetic survey grid (Fig. 2C; omitting transect positions at 45–75 m). While it is possible that background concentrations are influenced both by any unknown land use at the site and the crash itself, this subset of samples is at least outside of the area of surface-scattered fragments. As such, these concentrations define a local background to which observed geochemical anomalies can be compared.

The brass sample is dominated by Cu, with concentration exceeding 10^5 ppm. A high Zn fraction is also recorded (~80,000 ppm), with arsenic (As) and nickel (Ni) also increased in abundance. The steel sample is Fe-enriched, although with a surprisingly low concentration of ~250,000 ppm. The low value could again indicate a calibration issue or nonideal conditions of the sample surface caused by corrosion [17]. In any case, Fe is also abundant in the background soil, therefore damaging the signal-to-noise ratio of any potential anomaly. Pb is somewhat enriched in both metallic samples but in very low concentrations that may approach the limit of instrumental sensitivity. The burnt wood sample is generally depleted in metallic elements although no element is obviously enriched against the background trend. The burnt soil sample shows little significant alteration with respect to background. Therefore, despite the vulnerability to calibration effects, elements associated with brass (primarily Cu and Zn) could therefore be promising XRF targets for either crash site, with some potential for an Fe anomaly at site B.

4.3 SURVEY IMPLEMENTATION

Surveys in November 2014 focused exclusively on site A. Acquisitions were conducted along a transect, 100 m in length, which intersected the suspected crash site at a distance of 50–60 m (Fig. 2B). Geophysical surveys were conducted with a *Geonics* G-858 caesium vapor magnetic gradiometer and a *Geonics* EM31 electric conductivity meter. Both instruments were used in a "continuous" recording mode, resulting in an along-transect sampling interval of ~0.1 m. The two sensors of the G-858 were mounted at 0.4 and 1.2 m above the ground. The EM31 was carried at a height of 1.1 m, with the antenna boom oriented parallel to the transect; electric conductivity and in-phase components of the EM31 response were recorded, since in-phase anomalies are particularly indicative of buried metal [42]. XRF measurements were conducted with the *Bruker* spectrometer along the transect at intervals ≤2.5 m, with each position irradiated for ~1 min.

During August 2015, grids were acquired over both site A and site B using a *Bartington* Grad601 vertical component fluxgate gradiometer. In each grid, profiles were separated by 1 m, with an along-profile sampling interval of ~0.1 m. Grids have dimensions 30 × 30 m over site A and 40 × 40 m over site B. The site A transect was resurveyed with the XRF spectrometer between distances of 40 and 80 m (i.e., the span of the grid along the transect) at a sample resolution of 1 m. The site B transect was sampled at 1 m intervals. Here, we show only representative data from the surveys conducted with the Grad601 instrument; the full data archive is presented by Booth et al. [43].

Given the skepticism concerning the validity of in situ XRF sampling, we also took soil samples from positions along each XRF transect for validation against standard laboratory measurements. Laboratory XRF analysis was conducted with the *Bruker* spectrometer on each soil sample, after kiln-drying them for several days (at 60°C) and then grinding them with a pestle and mortar. Seventeen samples from site A were also analyzed using an inductively coupled plasma mass spectrometer (ICP-MS), and 21 from site B were analyzed using an inductively coupled plasma optical emission spectrometer (ICP-OES) (two different spectrometers were used because different apparatuses were available at the time of analyses). These methods deliver a precise quantitative estimate of concentration [44] and are less vulnerable to calibration issues but require more extensive preparation of samples. Identical preparation was undertaken for analyses in both instruments and involved the dissolution of aliquots of 100 mg of dried and ground soil in 5 mL of hot aqua regia (37% hydrochloric acid and 68% nitric acid, in a molar ratio of 3:1) at 140°C, for 1 h. A dilution series of 1:100 was made in 2% nitric acid and analyzed for elemental concentrations. The soil samples appeared to dissolve completely in the aqua regia, although a source of error is the resistant to dissolution of any small quartz components (a potential mismatch to the XRF readings could be expected since XRF spectrometry measures the geochemistry of the whole sample, but mass and optical emission spectrometry measures only the fraction that is dissolved). For the elements of consideration in this study, comparisons were made of reported XRF versus aqua regia digestion measurements of standard soil samples to confirm that the elements considered showed little significant difference between the two methodological approaches.

Ground conditions between these two survey periods were markedly different. In November 2014, the whole airfield site had been plowed and sown with a crop of winter barley; furthermore, the survey followed a period of heavy rain, and the soil was waterlogged. By contrast, the survey in August 2015 took place after prolonged warm weather: the field was dry and dusty and covered with the stubble of a barley crop (Fig. 1).

5 RESULTS

We first review magnetic anomalies in the survey grids, before focusing on the observations from the suite of geochemical data.

5.1 GEOPHYSICAL SURVEYS

Data from the *Bartington* Grad601 surveys are shown in Fig. 5. The complete archive of geophysical data for site A is presented by Booth et al. [43].

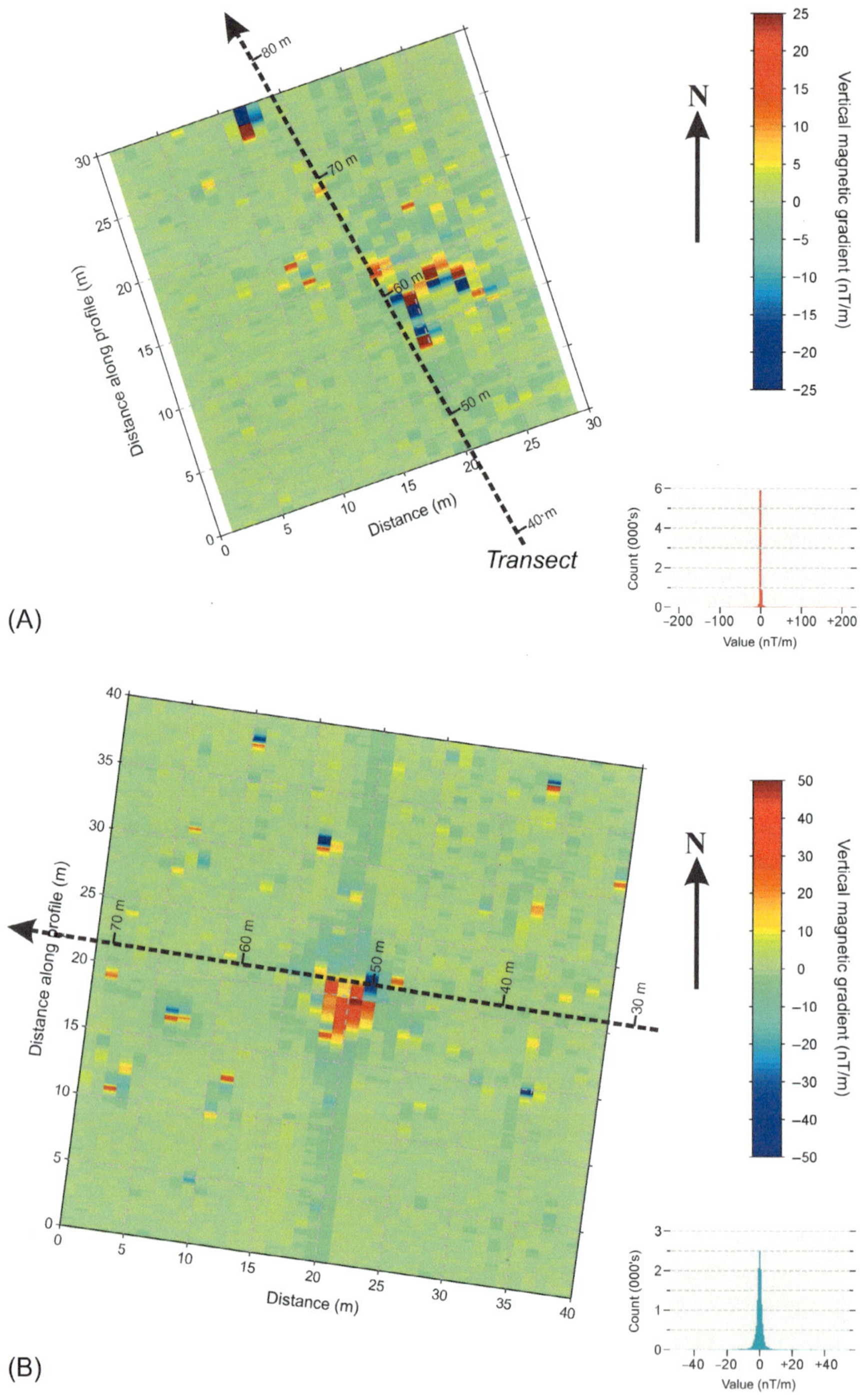

FIG. 5

Vertical magnetic gradient in (A) site A and (B) site B, acquired with the Bartington Grad601. Striping between survey lines (~4 nT/m variation) is suppressed by removing the mean value in each profile. *Inset*: histograms of observations.

The site A grid (Fig. 5A) shows a magnetic anomaly in the eastern half of the survey grid, close to the location hypothesized based on contemporary photography. The anomaly appears to elongate in an east-west orientation, extending 16 m across the profiles of the grid. The typical magnetic gradient anomaly in this region has a magnitude of ± 10 nT/m (against a mean background of approximately -1 nT/m) but exceeds ± 100 nT/m in certain profiles (e.g., at 19 m).

Site B also shows a clear magnetic anomaly at the center of the grid (Fig. 5B). However, the anomaly is approximately circular in plain view, with an approximate diameter of 5 m. The magnitude of the anomaly is generally larger than in site A (± 30 nT/m), but the distribution of responses is less prone to outliers; indeed, the extreme values in the distribution are associated with short-wavelength (presumably shallow) responses that appear separate from the main anomaly.

5.2 IN SITU XRF SPECTROMETRY

Fig. 6A shows the variation in selected XRF responses (including a three-point moving median average) observed along the site A transect in the November 2014 survey. The light and dark shading corresponds, respectively, to the span of the Grad601 grid and the extent of the magnetic anomaly (56–62 m along the transect). An enrichment anomaly is observed for Cu ~60 m along the transect, with peak values of ~110 ppm (double the 50 ppm background concentration). A moderate Zn anomaly is perceived at ~60 m with a peak approaching 150 ppm (50 ppm above the ~100 ppm background). Weaker evidence of an anomaly in As concentrations is also observed. For the repeat acquisition at site A, in August 2015 (Fig. 6B), background concentrations show a comparable magnitude with the earlier archive, but Cu, Zn, and As in particular show greater scatter about the median trend. Nonetheless, Cu appears to be consistently enriched, ~60 m along the profile. However, changes in the observed concentrations suggest a fundamental control of ground conditions, which we consider in further discussion below.

XRF data for the site B transect are shown in Fig. 6C. Background values (and the scatter about those values) are similar to those from site A, but no coherent geochemical anomalies appear to be colocated with the magnetic response. Since only one archive of data exists for this site, we cannot exclude a component of temporal variation in these records equivalent to those from site A. However, as we consider in our discussion, the absence of a geochemical anomaly could be associated with the different styles of crash.

The correlation between pairs of elemental concentrations for the site A archive is considered in Fig. 7. If the geochemical anomalies share a common origin, a correlation between their concentration would be anticipated with position along the transect. Following Bergslien [3], we define Spearman's rank correlation coefficient (r_s) for elemental concentrations of Cu, Zn, and As. Each plot lists the value of r_s, with strong correlation implied by $r_s > 0.65$, and a weak correlation by $r_s < 0.45$, and plot symbols are colored according to their position along the transect. Of particular note are the magenta and *red symbols*, which correspond to the interval 50–70 m

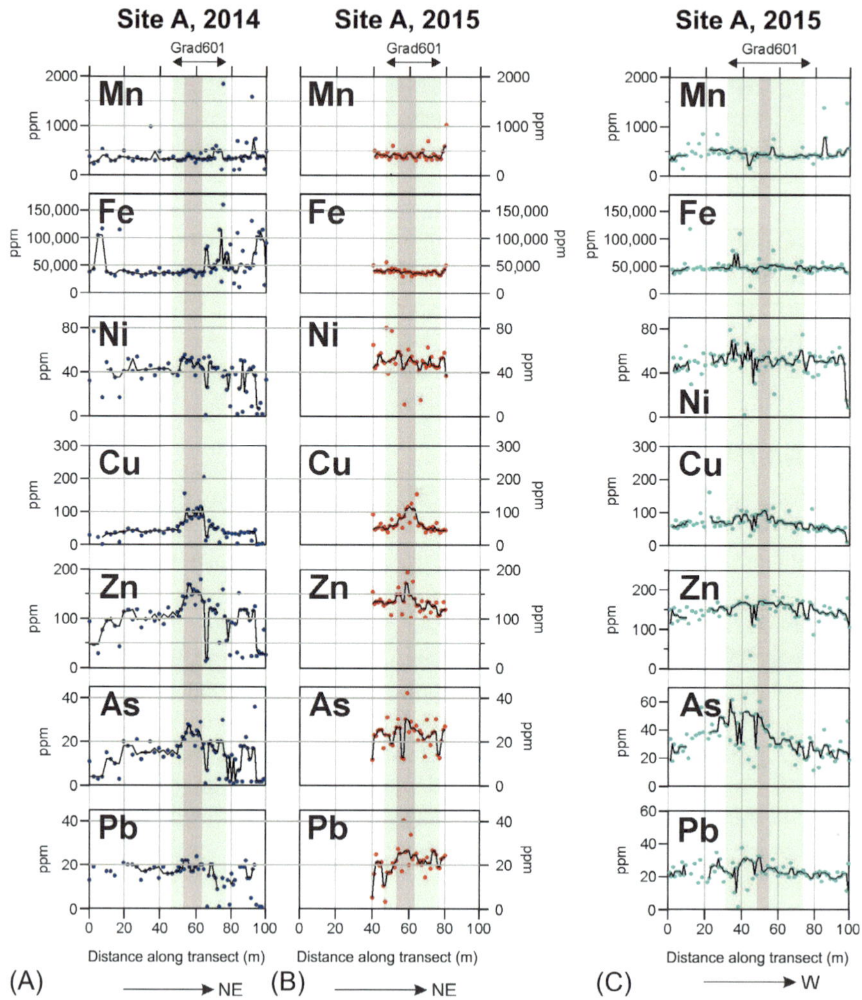

FIG. 6

Selected XRF responses along the Nuthampstead transects from (A) site A in November 2014, (B) site A in August 2015, and (C) site B in August 2015. The *black line* through each plot is a three-point moving median trend. Shaded sections show (light) the intersection with the Grad601 grids (Fig. 5) and (dark) the respective magnetic anomalies.

(consistent with the position of the magnetic anomaly). Cu and Zn are strongly correlated in the November 2014 archive (upper-right quadrant), and furthermore, their strongest correlations are associated with the 50–70m interval. Correlations are all reduced in the August 2015 archive (lower-left quadrant), with a higher degree of scatter. Nonetheless, the correlation of Cu and Zn is still classified as moderately strong in the 50–70m interval.

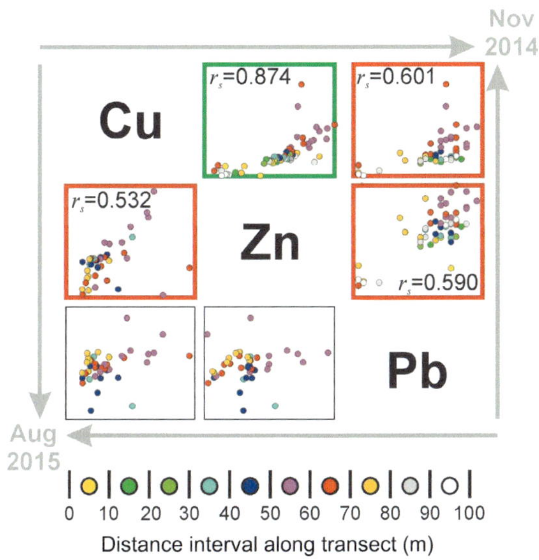

FIG. 7

Scatterplots of paired elemental data from in situ XRF surveys at site A in (upper-right quadrant) November 2014 and (lower-left quadrant) August 2015. Scatterplot frames include the value of Spearman's rank correlation coefficient, r_S.

5.3 LABORATORY COMPARISON

Results of laboratory XRF and ICP-MS analysis on site A soil samples are shown in Fig. 8A and B for selected elements; similarly, results of laboratory XRF and ICP-OES for site B samples are shown in Fig. 8C and D.

Laboratory XRF data (Fig. 8A and C) are consistent with their in situ equivalents (Fig. 6A and C), although the suppression of scatter about the median trend is notable. As an example, the root-mean-square variability about the median trend of site A's in situ Cu observations is 25 ppm, which reduces to 7 ppm for the laboratory analysis. While the underlying accuracy of in situ XRF therefore appears similar to conventional laboratory practice, in situ analysis may be insensitive to subtle geochemical anomalies that may be comparable with the background signal-to-noise ratio. However, the laboratory analysis supports the presence of a Cu and Zn anomalies in the site A transect and also the *absence* of equivalent anomalies through site B.

Concentrations determined through both ICP analyses (Fig. 8B and D) are of the same order of magnitude as the equivalent laboratory XRF data, but differences in base levels are evident. This is especially apparent for Zn, where the two trends are offset by ~70 ppm. These are attributed to the inappropriate calibration of the XRF survey, implying that these in situ surveys should be considered relative rather than absolute indicators of concentration. However, the trends between As and Pb are inconsistent; indeed, the ICP-MS/ICP-OES curves defined for Pb appear very similar

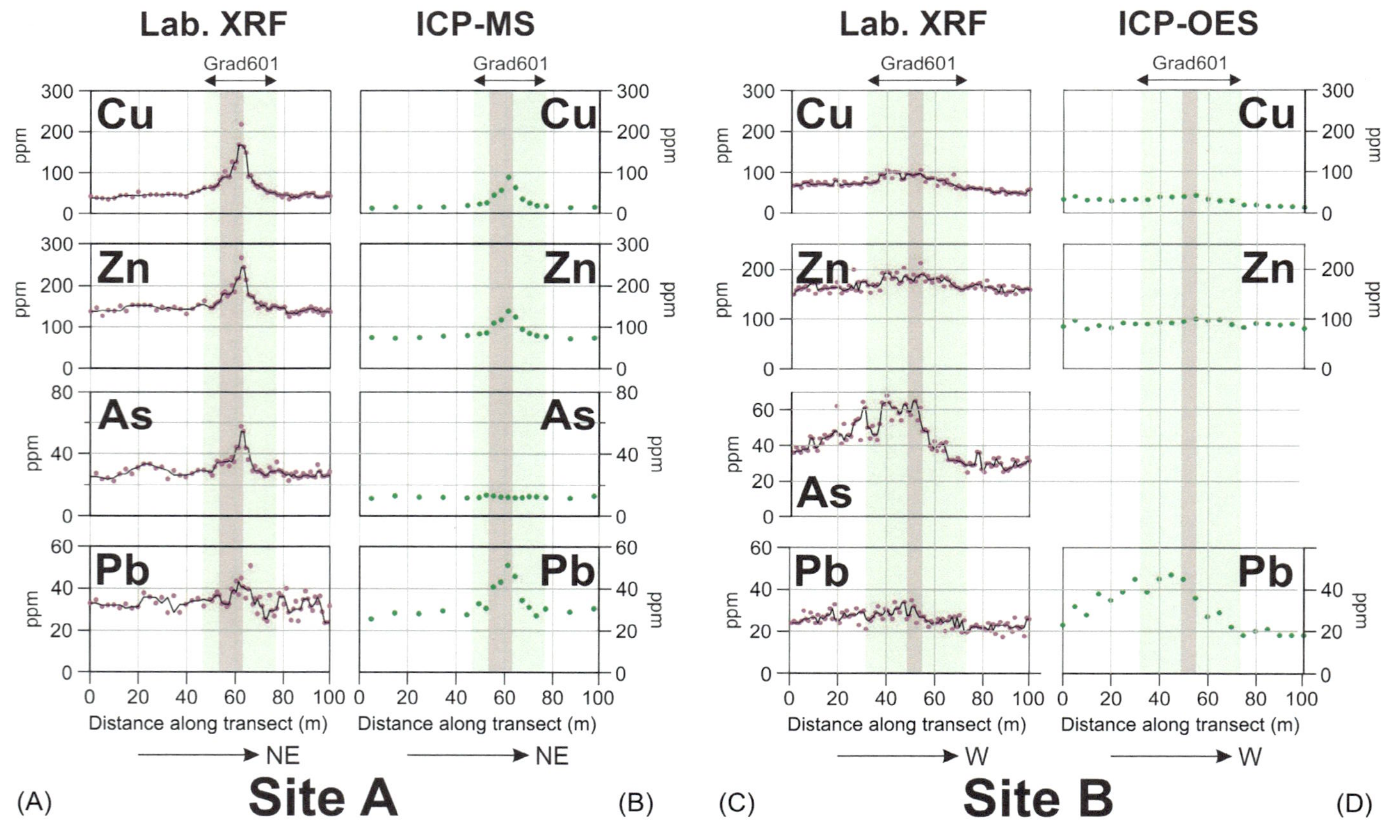

FIG. 8

Selected results from laboratory analysis of soil samples from (A and B) site A and (C and D) site B. In all cases, the black line through the plots is the three-point median trend. (A) Concentrations measured using portable XRF spectrometer on laboratory-prepared soil samples. (B) Concentrations measured from ICP-MS. (C) As. (D) Concentrations measured from ICP-OES (note that As is unavailable for this dataset).

to the XRF-derived As curves. This is attributed to a lack of resolution in XRF analysis between As and Pb, since their XRF energies are very similar (10.543 keV for Kα for As and 10.551 keV for Lα for Pb). Therefore, the spectral interference between these elements makes it challenging for XRF to distinguish between As and Pb, particularly at low concentrations. As such, the XRF anomaly in As is likely a false positive. If As and/or Pb anomalies are the specific target of an in situ XRF survey, it would clearly be prudent to consider the potential that both responses could be mutually indicative.

6 INTERPRETATION AND DISCUSSION

6.1 INTEGRATED GEOPHYSICAL AND GEOCHEMICAL INTERPRETATION

6.1.1 Site A

Colocated geophysical and geochemical anomalies are observed at site A, the reported crash site of Mosquito LR343. Specifically, a low-amplitude magnetic anomaly observed in the Grad601 data (and also in the G-858 data, reported in Booth et al. [43]) is interpreted as the response to the thermoremanence in burnt clay. Assuming a near-vertical impact, the area of this response is comparable with the facing profile of the Mosquito ($\sim$16 × 4 m) (a similar observation is made by Osgood [38], for the vertical impact of a Spitfire aircraft). Furthermore, spectral analysis [45] of the G-858 data [43] places a concentration of magnetic sources at a depth of $\sim$0.8 ± 0.1 m, comparable with the depth extent of the burnt material observed during small excavations at the site (Fig. 3B). The higher amplitude magnetic anomalies (>±100 nT/m) observed in the Grad601 grid could be responses from larger fragments of ferrous wreckage, but further survey would be required to evaluate the size and/or depth of these potential targets.

This interpretation is greatly strengthened by in situ XRF spectrometry. Colocated with the magnetic anomalies are local geochemical anomalies, particularly evident for elements (Cu and Zn) associated with brass. Besides iron and aluminum, brass is the most significant metallic component of the fully armed Mosquito aircraft. The geochemical evidence is particularly compelling since in the absence of other information, the air crash is the most plausible means of introducing these elements into the ground at this location; by contrast, the burnt layer alone could be more simply explained by, for example, disposal at some point in the recent history of the site. The full suite of geophysical and geochemical observations is therefore consistent with an air crash at the site identified within Nuthampstead Airfield; furthermore, the in situ geochemical analysis is validated with standard laboratory methods.

6.1.2 Site B

A discrete magnetic anomaly is observed in the Grad601 data from site B. This almost circular response is highly localized at the position of the Flying Fortress explosion, with a diameter of $\sim$5 m and a magnitude of ±30 nT/m. We again interpret this anomaly as the thermoremanent magnetic signature of burnt clay. Elsewhere within the grid, there are small-wavelength anomalies that could be associated with any small metallic debris (not necessarily from the aircraft), but there is no evidence that significant debris remains at the site of the explosion. Indeed, fragments of this aircraft can be found across a wide area of the airfield.

As will be considered in the following discussion, the absence of debris in the subsurface is a possible reason why no significant Cu and/or Zn anomaly was observed at site B. With no "reservoir" of metallic elements within the subsurface, any residual geochemical response at the ground surface may be rapidly attenuated. Only a small and dispersed enrichment of Pb (~20 ppm over 60 m) is observed.

6.2 CONTROLS ON THE APPLICABILITY OF IN SITU XRF SPECTROMETRY

Our suite of geochemical analyses allows us to suggest a set of requirements for effective in situ XRF surveying at a given site. Aside from purely instrumental effects (e.g., calibration and sensitivity), the detectability of a given geochemical target depends on

(i) the abundance of a diagnostic element in the source material,
(ii) the groundwater solubility and adsorption potential of that given element,
(iii) the efficiency of any source-to-surface transport mechanism.

Although having an abundance of an element seems obvious, our test sites show the importance of considering the detailed origin of a particular target. While both of our test cases are sites of aviation archaeology, they originate from different styles of crash. At site A, the Mosquito aircraft impacted the ground at a near-vertical angle, whereas the Flying Fortress at site B exploded at surface, distributing wreckage over a very large area, potentially leaving little material remaining within the subsurface. We also emphasize the need for a *diagnostic* chemical element, which is abundant in sufficient quantity in the subsurface and anomalous with respect to background (i.e., soil) geochemistry. At site A, elements associated with brass (Cu and Zn) were abundant at the ground surface, as was anticipated from initial geochemical surveying of recovered ammunition fragments.

Regarding the solubility of elements, we consider that XRF surveys are essentially mapping geochemical contaminants in surface water and soil. The limited depth of penetration of the XRF instrument means that all observed geochemistry pertains only to the uppermost "skin" of the ground. While it is possible that small metallic fragments exist at the ground surface, no such fragments were observed when samples were ground for laboratory analysis. Instead, elements are likely present in surface water or have adsorbed onto soil particles. Any target geochemical element must therefore be dissolvable in groundwater but persist in detectable concentration when adsorbed onto soil.

Regarding a source-to-surface transport mechanism, plowing appears to be effective at Nuthampstead. Furthermore, the time since plowing could be a key control on the signal-to-noise ratio of the XRF anomalies. The site A survey in November 2014 was conducted soon after plowing, thereby providing the ground surface with a "fresh charge" of metal-rich groundwater. By August 2015, several months since plowing, the magnitude and correlation coefficients of anomalies had been suppressed. Ordinarily, it might be expected that the drier ground conditions in summer would yield higher geochemical concentrations (e.g., Ref. [19]), but at the time of this acquisition, the ground had been undisturbed for several months. Metal ions could therefore have been flushed from the site by, for example, rainfall or transported back into the subsurface. Given that the sample size of the XRF instrument is $\sim 1\,cm^2$, it is unlikely that analyses are conducted at precisely the same location between different time periods; however, the changes in the XRF responses are not a shift in the position of the geochemical anomalies, but in the scatter, correlation, and absolute values of geochemical concentrations.

Of course, calibration issues are unavoidable in in situ XRF surveying (e.g., Ref. [17]). Samples lack the rigorous preparation undertaken in laboratory analysis preparation, and more importantly, it is difficult to define bespoke calibrations for all survey environments and soil types. A nonspecialist should therefore consider XRF spectrometry as a qualitative tool for "anomaly definition," rather than interpreting the absolute values of the recorded concentrations. Bespoke calibrations are recommended if absolute concentrations are required, for example, for comparative archaeometric purposes [17], or where forensic analysis may lead to litigation [3–5]. Validation with laboratory analysis is also advocated since XRF scattering effects are minimized in powdered samples. Whereas in situ XRF sampling considers only the geochemistry of a surface "skin" of the soil (itself a nonideal surface to be sampled), the concentrations measured in the laboratory represent the homogenized geochemistry of a finite soil volume.

6.3 PRACTICALITY OF XRF SURVEYING

With respect to the efficiency of acquisition, in situ XRF spectrometry compares favorably with established geophysical methods. Not only is the cost of equipment similar to many geophysical systems, but also the rate of data return (40 samples/h, here distributed across a 100 m transect) is comparable with, for example, surveying with electric resistivity tomography. While XRF spectrometry would likely be impractical as an initial reconnaissance tool, it can contribute valuable insight to the understanding of a target once that target has been identified. Geophysical surveys are generally limited to identifying anomalies, rather than characterizing diagnostic elements; the practice we report here offers greater diagnostic potential than currently provided by conventional field geophysics.

7 CONCLUSIONS

In situ XRF spectrometry shows potential as a geochemical complement to a variety of near-surface geophysical surveys. Here, we investigated both the advantages and limitations of XRF spectrometry for the characterization of two aircraft crash sites.

At one site, increases in the local abundance of Cu and Zn were considered to originate with brass ammunition cartridges among the aircraft wreckage. However, despite a clear geophysical anomaly, equivalent results were not observed at a second site likely because the nature of the crash resulted in little subsurface debris. These different responses suggest that a detectable and significant XRF anomaly requires not only that a source of geochemical contamination is present but also that this contamination can enter subsurface groundwater and that a source-to-surface transport mechanism is also active. Under such favorable conditions, in situ XRF surveying offers significant diagnostic benefits for the interpretation of geophysical anomaly, allowing constraint on the geochemistry that underlies a given geophysical response.

ACKNOWLEDGMENTS

This work is dedicated to the memory of Plt Off Raymond Payne and Fg Off John "Dunch" McMillan (487 Squadron Royal New Zealand Air Force) and Second Lt Francis Harrod and Cpl David Flores (398th Bombardment Group (Heavy) US Army Air Forces). The late Farmer Martin Barker allowed access to the Nuthampstead site, and Jayne Rollo assisted with field acquisition during 2015. Andrew Kyprianou and Stephanie Trayhurn assisted with logistics.

REFERENCES

[1] A.R. Millard, Geochemistry and the early alum industry, Geoarchaeology: Exploration, Environments, Resources, vol. 165, Geological Society, London, 1999, pp. 139–146 (Special Publications).

[2] A.B. MacKenzie, I.D. Pulford, Investigation of contaminant metal dispersal from a disused mine site at Tyndrum, Scotland, using concentration gradients and stable Pb isotope ratios, Appl. Geochem. 17 (8) (2002) 1093–1103.

[3] E.T. Bergslien, X-ray diffraction and field portable X-ray fluorescence analysis and screening of soils: project design, in: D. Pirrie, A. Ruffell, L.A. Dawson (Eds.), Environmental and Criminal Geoforensics, The Geological Society, London, 2013 (Geological Society Special Publication No. 394).

[4] A. Ruffell, P. Wiltshire, Conjunctive use of quantitative and qualitative X-ray diffraction analysis of soils and rocks for forensic analysis, Forensic Sci. Int. 145 (1) (2004) 13–23.

[5] V.M. Sbarato, H.J. Sánchez, Analysis of arsenic pollution in groundwater aquifers by X-ray fluorescence, Appl. Radiat. Isot. 54 (5) (2001) 737–740, https://doi.org/10.1016/S0969-8043(00)00349-3.

[6] C. Soulsby, J. Bradford, J. Dick, J.P. McNamara, J. Geris, J. Lessels, M. Blumstock, D. Tetzlapp, Using geophysical surveys to test tracer-based storage estimates in headwater catchments, Hydrol. Process. 30 (2016) 4434–4445, https://doi.org/10.1002/hyp.10889.

[7] E.V. Aidona, A. Sarris, D. Kondopopulou, I. Sanakis, Application of magnetic and spectrometry methods in the detection of human activity in soils: a case study at the archaeological site of Kitros (Northern Greece). Archaeol. Prospect. 8 (2001) 187–198, https://doi.org/10.1002/arp.168.

[8] R. Falcone, W. Renier, M. Vertià, Wavelength-dispersive X-ray fluorescence analysis of ancient glasses, Archaeometry 44 (4) (2008) 531–542.

[9] L. Bonizzoni, S. Bruni, V. Guglielmi, M. Milazzo, O. Neri, Field and laboratory multi-technique analysis of pigments and organic painting media from an Egyptian coffin (26th Dynasty), Archaeometry 53 (6) (2011) 1212–1230.

[10] G. Cultrone, E. Molina, C. Grifa, E. Sebastián, Iberian ceramic production from Basti (Baza, Spain): first geochemical, mineralogical and textural characterization, Archaeometry 53 (2) (2011) 340–363.

[11] C. Roux, C. Lennard, X-ray fluorescence in forensic science. Encyclopedia of Analytical Chemistry, 2006, https://doi.org/10.1002/9780470027318.a1124.

[12] C.J. Carey, H.J. Wickstead, G. Juleff, J.C. Anderson, M.J. Barber, Geochemical survey and metalworking: analysis of chemical residues derived from experimental non-ferrous metallurgical processes in a reconstructed roundhouse, J. Archaeol. Sci. 49 (2014) 383–397.

[13] S.R. Cook, R.Y. Banerjea, L.-J. Marshall, M. Fulford, A. Clarke, C. van Zwieten, Concentrations of copper, zinc and lead as indicators of hearth usage at the Roman town of Calleva Atrebatum (Silchester, Hampshire, UK), J. Archaeol. Sci. 37 (4) (2010) 871–879.

[14] K. Dirix, P. Muchez, P. Degryse, B. Mušič, J. Poblome, Integrating multi-element geochemical and magnetic survey at Ancient Sagalassos (Southwest Turkey): anthropogenic versus natural anomalies, Archaeol. Prospect. 20 (2013) 233–247.

[15] L. Vittori Antasari, S. Cremonini, P. Desantis, C. Calastri, G. Vianello, Chemical characterisation of *anthro-technosols* from Bronze to Middle Age in Bologna (Italy), J. Archaeol. Sci. 40 (1) (2013) 3660–3671.

[16] C.A. Wilson, D.A. Davidson, M.S. Cresser, Multi-element soil analysis: an assessment of its potential as an aid to archaeological interpretation, J. Archaeol. Sci. 35 (2) (2008) 412–414.

[17] R.B. Scott, K. Eekelers, P. Degryse, Quantitative chemical analysis of archaeological slag material using handheld X-ray fluorescence spectrometry, Appl. Spectrosc. 70 (1) (2016) 94–109, https://doi.org/10.1177/0003702815616741.

[18] A.X. Lu, J.H. Wang, L.G. Pan, P. Han, Y. Han, Determination of Cr, Cu, Zn, Pb and As in soil by field portable X-ray fluorescence spectrometry, Spectrosc. Spectr. Anal. 30 (10) (2010) 2848–2852, https://doi.org/10.3964/j.issn.1000-0593(1010)10-2848-05.

[19] A.R. Schneider, B. Cancès, C. Breton, M. Ponthieu, X. Morvan, A. Conreux, B. Marin, Comparison of field portable XRF and aqua regia/ICPAES soil analysis and evaluation of soil moisture influence on FPXRF results, J. Soil. Sediment. 16 (2016) 438–448, https://doi.org/10.1007/s11368-015-1252-x.

[20] N. Del Solar Valarde, S. Kinis, R. Chapoulie, R. Joannes-Boyau, L.J. Castillo, Characterization of pre-Columbian artefacts in situ through handheld portable X-ray fluorescence spectrometry: the case of ceramics from the Mochica site of San José de Moro (Peru), Herit. Sci. 4 (37) (2016), https://doi.org/10.1186/s40494-016-0109-y.

[21] R. Fernandes, B.J.H. van Os, H.D.J. Huisman, The use of hand-held XRF for investigating the composition and corrosion of Roman copper-alloyed artefacts, Herit. Sci. 1 (2013) 30, https://doi.org/10.1186/2050-7445-1-30.

[22] A. Shugar, R.A. Armitage, J.H. Burton (Eds.), Portable X-ray fluorescence and archaeology: limitations of the instrument and suggested methods to achieve desired results, Symposium on Archaeological Chemistry: American Chemical Society, Division of the History of Chemistry, American Chemical Society, Washington, DC, 2013.

[23] M. Sepúlveda, S. Gutierrez, J. Carcamo, A. Oyaneder, D. Valenzuela, I. Montt, C. M. Santoro, *In situ* X-ray fluorescence analysis of rock art paintings along the coast and valleys of the Atacama Desert, Northern Chile, J. Chil. Chem. Soc. 60 (1) (2015) 2822–2826.

[24] G.J. Weltje, R. Tjallingii, Calibration of XRF core scanners for quantitative geochemical logging of sediment cores, Earth Planet. Sci. Lett. 274 (2008) 423–438.

[25] G.E.M. Hall, G.F. Bonham-Carter, A. Buchar, Evaluation of portable X-ray fluorescence (pXRF) in exploration and mining: phase 1, control reference materials, Geochem. Explor. Environ. Anal. 14 (2014) 99–123.

[26] J. Quye-Sawyer, V. Vandeginste, K.J. Johnston, Application of handheld energy-dispersive X-ray fluorescence spectrometry to carbonate studies: opportunities and challenges, J. Anal. At. Spectrom. 30 (2015) 1490–1499.

[27] N. Forster, P. Grave, N. Vickery, L. Kealhofer, Non-destructive analysis using pXRF: methodology and application to archaeological ceramics, X-Ray Spectrom. 40 (2011) 389–398.

[28] P.J. Potts, P.C. Webb, O. Williams Thorpe, Investigation of a correction procedure for surface irregularity effects based on scatter peak intensities in the field analysis of geological and archaeological rock samples by portable X-ray fluorescence spectrometry, J. Anal. At. Spectrom. 12 (1997) 769–776.

[29] J. Środoń, V. Drits, D.K. McCarty, J.C.C. Hsieh, D.D. Eberi, Quantitative X-ray diffraction analysis of clay-bearing rocks from random preparations, Clay Clay Miner. 49 (2001) 514–528.

[30] E. Frahm, Validity of "off-the-shelf" handheld portable XRF for sourcing Near Eastern obsidian chip debris, J. Archaeol. Sci. 40 (2013) 1080–1092.

[31] M.O. Krause, Atomic radiative and radiationless yields for K and L shells, J. Phys. Chem. Ref. Data 8 (2) (1979) 307–327.

[32] J. Berlin, Analysis of boron with energy-dispersive X-ray spectrometry: advances in light element analysis with SDD technology, Imaging Micrometry 13 (2011) 19–21.

[33] R.J. Speakman, N.C. Little, D. Creel, M.R. Miller, J.G. Inanez, Sourcing ceramics with portable XRF spectrometers? A comparison with INAA using Mimbres pottery from the American Southwest, J. Archaeol. Sci. 38 (2011) 3483–3496.

[34] R. Tjallingil, U. Röhl, M. Kölling, T. Bickert, Influence of the water content on X-ray fluorescence core-scanning measurements in soft marine sediments, Geochem. Geophys. Geosyst. 8 (2) (2007) Q02004, https://doi.org/10.1029/2006GC001393.

[35] R.E.M. Hedges, A.R. Millard, Bones and groundwater: towards the modelling of diagenetic processes, J. Archaeol. Sci. 22 (1995) 155–164.

[36] S. Campana, S. Campana, S. Piro, Archaeological site detection and mapping: some thoughts on differing scales of detail and archaeological 'non-visibility', Seeing the Unseen: Geophysics and Landscape Archaeology, Taylor and Francis, Boca Raton, FL, 2009 (Chapter 2).

[37] BGS, Borehole Record TL43NW39 (BGS ID 19536802), Completed 12/06/2014 at Morrice Green Farm, Nuthampstead, http://scans.bgs.ac.uk/sobi_scans/boreholes/19536802/images/19536797.html, 2014. Accessed 30 March 2016.

[38] R. Osgood, Exercise Tally Ho!—Archaeological Project Report for the Recovery of Spitfire P9503 at Lidbury, near Upavon, Wiltshire, Under the Protection of Military Remains Act (1986), Historic England, 2014.

[39] U. Schwertmann, R.M. Taylor, Iron oxides, Minerals in Soil Environments, second ed., SSSA Book Series, vol. 1, Soil Science Society of America, 1989, pp. 379–438. ISBN: 13 978-0891187875.

[40] M. Hanesch, H. Stanjek, N. Petersen, Thermomagnetic measurements of soil iron minerals: the role of organic carbon. Geophys. J. Int. 165 (2006) 53–61, https://doi.org/10.111/j.1365-246X.2006.02933.x.

[41] N. Herz, E.G. Garrison, Geological Methods for Archaeology; Part II—Dating Techniques, first ed., Oxford University Press, 1997. ISBN: 13 978-0195090246.

[42] J.D. McNeill, Technical Note TN-11: Use of EM31 Inphase Information, Geonics Limited, Mississauga, Ontario, Canada, 1983.

[43] A.D. Booth, V. Vandeginste, D. Pike, R. Abbey, R.A. Clark, C.M. Green, N. Howland, Geochemical insight during archaeological geophysical exploration through *in situ* X-ray fluorescence spectrometry, Archaeol. Prospect. 24 (2017) 361–372.

[44] K. Pye, D. Croft, Forensic analysis of soil and sediment traces by scanning electron micrometry and energy-dispersive X-ray analysis: an experimental investigation, Forensic Sci. Int. 165 (1) (2007) 52–63.

[45] A. Spector, F.S. Grant, Statistical models for interpreting aeromagnetic data, Geophysics 35 (1970) 293–302.

FURTHER READING

[46] D.B. Dungworth, Roman copper alloys: analysis of artefacts from northern Britain, J. Archaeol. Sci. 24 (1997) 901–910.

[47] G. Jones, Geophysical investigation at the Falling Creek Ironworks, an early industrial site in Virginia, Archaeol. Prospect. 8 (2001) 247–256, https://doi.org/10.1002/arp.173.

[48] N. Masini, R. Lasaponara, Satellite-based recognition of landscape archaeological features related to ancient human transformation, J. Geophys. Eng. 3 (3) (2006) 230–235, https://doi.org/10.1088/1742-2132/3/3/004.

[49] H. White, D. Dungworth, Warmley brassworks, siston, bristol: analysis of some eighteenth-century brassworking debris, Historic England Report, Heritage Protection Department, Report 54/2007, 2007.

CHAPTER

Advanced magnetic prospecting for archaeology with a vehicle-towed array of cesium magnetometers

5

Neil Linford, Paul Linford, Andrew Payne
Geophysics Team, Historic England, Portsmouth, United Kingdom

CHAPTER OUTLINE

1 INTRODUCTION

Aerial- or satellite-based photography and, increasingly, airborne lidar provide an important resource for the identification of archaeological activity at a landscape scale. However, to be successful, site conditions must be favorable for the production of crop marks or the preservation of some topographic variation. This can often vary with changing land use over recent historic time since widespread aerial coverage has been available and can lead to varying degrees of representation or survival within the archaeological record. The introduction of modern mechanical agriculture has, undoubtedly, led to the loss of some topographic expression for unprotected or unrecognized monuments through repeated episodes of plowing, although the crop or parch marks produced under suitable conditions within these arable fields may

Innovation in Near-Surface Geophysics. https://doi.org/10.1016/B978-0-12-812429-1.00005-2

partially secure the fidelity of the record. Other local conditions may further influence the success of aerial photography through masking deposits of alluvium in low-lying river valleys or colluvium at the bottom of actively eroding hill slopes. Of course, this must be balanced against the influence of more recent airborne techniques, such as the use of large area lidar coverage, which has produced impressive results in many areas of woodland where the topographic survival of monuments and other features has been able to resist the attentions of the plow.

While ground-based geophysical techniques have traditionally been applied to the investigation of known archaeological sites, often originally identified from the air, the ability to expand realistic areas of coverage allows a much greater scrutiny of the landscape. Ground-based techniques can, increasingly, be applied at a scale allowing the strength of the different techniques to be more fully complemented, corroborated, and interpreted in combination. Vehicle-towed magnetic arrays allow levels of coverage that can provide the important intrasite scale to investigate potential variations in response between aerial photography and geophysics, although of course the success of the technique may also vary with geology, soil, and land use conditions (e.g., Refs. [1,2]). Certainly, within much of England, conditions are generally well suited to magnetic surveys, and the application of high sample density coverage over larger areas with modern instrumentation has led to the partial reassessment of some geologies that may have been historically underrepresented. Large-scale magnetic survey is also of considerable use in a number of other application areas, for example, near-surface mineral mapping and agricultural and soil mapping, together with the detection of utilities and unexploded ordinance [3].

2 METHODOLOGY

Attempts to develop vehicle-towed systems for magnetic acquisition have been sought for some period of time and are, perhaps, not fully represented in the literature (e.g., Ref. [4]). In the context of this chapter, a distinction is made between truck-mounted magnetometer systems that produce very large heading errors (e.g., Refs. [5,6]) and those towed by light vehicles likely to be better suited for archaeological prospecting. An early example of the latter known to the authors was developed using a single analogue Plessey fluxgate magnetometer towed behind a camper van that was deployed on sites being investigated in Italy by the British School of Archaeology in Rome in the 1980s (Basil Turton, pers. comm.). It is unknown whether any published comparison of this system against contemporary handheld instruments exists, but such pioneering attempts were certainly in the vanguard of more recently developed technology able to utilize both digital data capture and satellite-based geopositioning [7–11]. While many array-based towed systems have used fluxgate magnetometer sensors, due to the relatively lower cost per sensor, small size, and low power consumption, in this case, it seemed logical to build on the high-quality magnetic data produced by the original English Heritage hand-operated cesium instrument array [12].

The initial vehicle-towed array was based around four modified Scintrex SM-4 SmartMag sensors with extended cabling to allow the total field sensors to be separated by up to 2 m from the control electronics. A series of prototype platforms for mounting the magnetometer sensors and GPS receiver were trialed during the initial development and towed behind a lightweight Kawasaki Mule All-Terrain Vehicle (ATV) using a 6 m umbilical towrope. A high-capacity traction battery mounted on the ATV fed through an inverter provided a 24V power supply to the sensor electronics (Fig. 1) and, with a second backup battery being partially recharged from the engine-driven alternator while the survey progressed, the system could be continuously operated throughout a full day of acquisition. Standard RS232 cables were also mounted on the umbilical to allow data transfer from all of the individual sensors and Trimble GNSS receiver.

This system was generally operated with the four total field sensors in a horizontal array, separated by 0.5 m, without any direct gradient measurement or local base station sensor. While this configuration worked well providing high-quality data over sites without immediate sources of ambient magnetic noise such as passing vehicles, using one sensor to record the vertical gradient can be highly beneficial for towed array systems. In addition practical limitations on the length of the umbilical towrope result in the influence of the magnetic field due to the ATV, both from the steel components of the vehicle and the variable response generated by the engine-driven alternator [13].

These considerations led to a further development of the system through the replacement of the original Scintrex instrumentation with six Geometrics G862 cesium vapor sensors, offering both an increased sensitivity ($<0.004\,nT/\sqrt{Hz}$ rms) and greater bandwidth allowing acquisition rates of up to 100 Hz if required. This revised configuration allows a linear array of five total field sensors mounted 0.5 m apart, approximately 0.36 m above the ground surface with the sixth magnetometer mounted 1.0 m above the central of the horizontal array to act as a gradient sensor (Fig. 1D). The towrope was also lengthened to 15 m to increase the separation between magnetic components on the ATV and the magnetometer sensors. Over typical sites, where the terrain may not be entirely even, the sensors are operated at 25 Hz and a ground speed of ~4.0 m/s resulting in an along line sample density of ~0.15 m and a crossline spacing of 0.5 m over a 2.5 m wide instrument swath.

The towed instrument platform on which the cesium sensors are mounted has evolved through a series of prototypes beginning with a hybrid lightweight sledge with a central wheeled subchassis (Fig. 1C) to the current adoption of the commercially available Geomatrix Geophysical Equipment Exploration Platform (GEEP) twin-hulled sledge (Fig. 1E). The use and development of each system have been determined by a combination of field conditions, including access constraints to more remote upland sites, where the lightweight system proved highly beneficial, and the requirement for a more robust system to cope with heavy soils over arable land with replaceable nylon sacrificial skid plates to cope with sharp, abrasive flint nodules. The GEEP platform offers great flexibility in the mounting position of individual sensors, maximizing their separation from the control electronics and other immediate sources of magnetic interference, such as the Trimble R8 GNSS receiver,

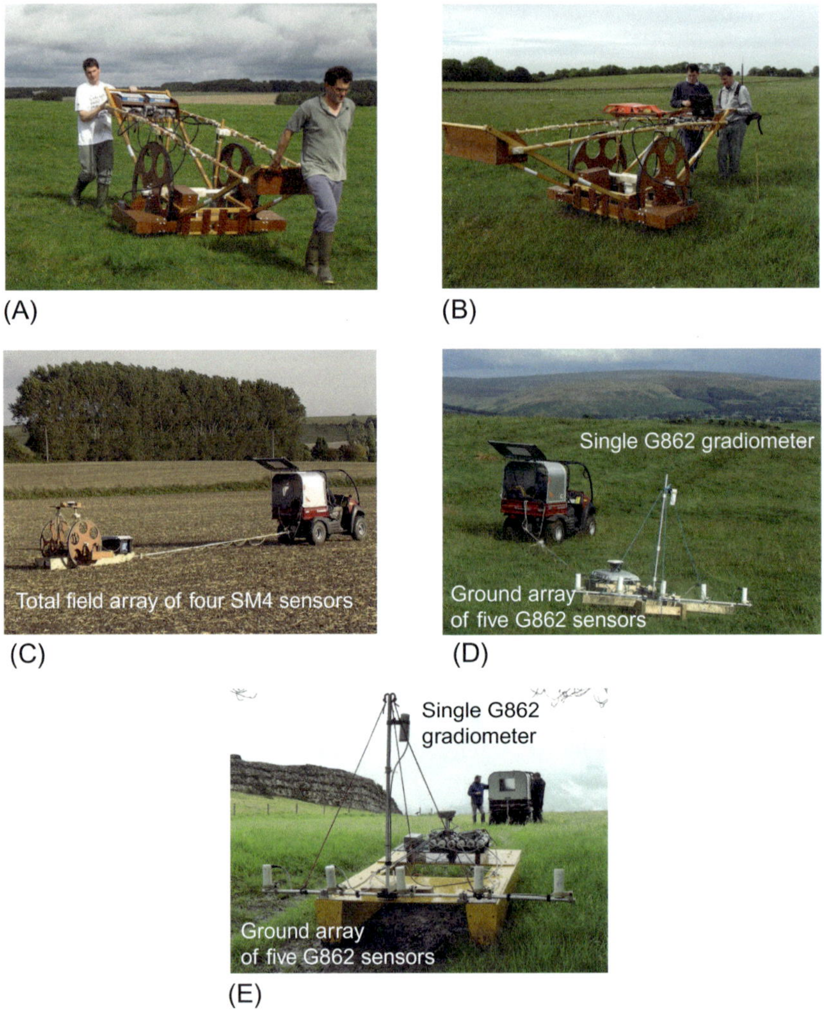

FIG. 1

Development of the cesium magnetometer array from (A) a hand-pushed nonmagnetic cart with four Scintrex SM4 sensors and no positional control; (B) the same hand-pushed cart and sensors with a comounted GPS receiver for grid-less surveying; (C) prototype vehicle-towed system utilizing the SM4 sensors as a horizontal array of total field sensors mounted on a hybrid sledge-cart platform; (D) ground array of five Geometrics G862 sensors with an additional G862 mounted as a central gradiometer, on a lightweight prototype platform together with a centrally mounted GPS receiver; and (E) G862 sensors mounted on the GEEP sledge platform to maximize the separation from the control electronics and Trimble R8 GNSS receiver. Power supply and data logging for the vehicle-towed systems are provided through the umbilical link between the platform and vehicle.

which is mounted 1.65 m in front of the central magnetometer in the array. In addition, a bespoke concatenation and power distribution board was commissioned to combine the output data strings from each of the six sensors together with the NMEA string from the GNSS receiver. A point-per-second (PPS) pulse from the GNNS receiver can also be used to synchronize the timing clock used to trigger the measurement sequence, and the board will also distribute control messages to all of the magnetometers. Operation with a second Trimble R8 receiver established as an on-site base station using the Ordnance Survey VRS Now correction service results in the accurate geopositioning of every magnetic measurement at the time of acquisition.

Acquisition software developed by the geophysics team is used to control the initial setup parameters of the instruments and to receive and log the data stream from the array to a ruggedized laptop PC mounted on the ATV (Fig. 2A). This software

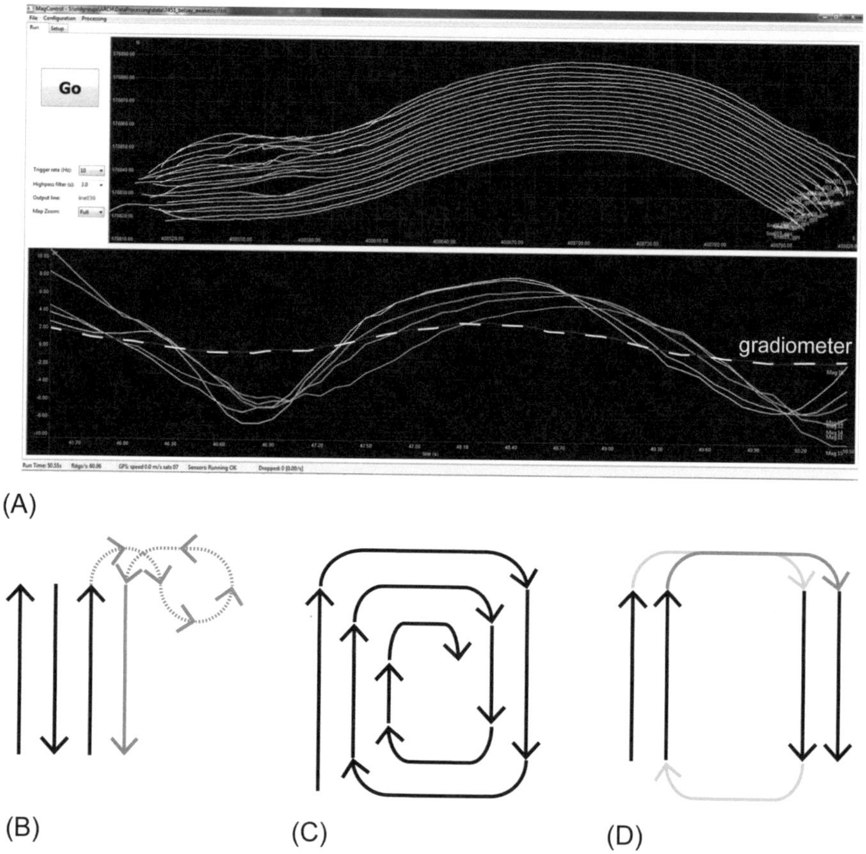

FIG. 2

(A) Screenshot of the field acquisition software in use showing the *X-Y* trace plot (lower panel) for each of the six G862 sensors; the central *dashed line* shows the output from the gradiometer sensor. The upper panel shows a "cookie trail" of the positional data from the comounted GNSS receiver. Alternative modes of field acquisition are illustrated by cartoons showing (B) parallel "zig-zag" lines, (C) spiral pattern, and (D) adjacent blocks of parallel lines.

provides a continuous plot of the output from each sensor, for real-time monitoring of data quality, together with a live "cookie trail" showing the GNSS positional data as they are collected. The immediate display of each GNSS coordinate aids navigation, allowing grid-less data acquisition, and indicates when the current line of readings has either veered away from the intended course or is overlapping more than necessary with an adjacent previously collected instrument swath.

3 FIELD ACQUISITION

Consideration has also been given to the most appropriate field acquisition methodology to allow rapid coverage without compromising data quality. Given the length of the towrope, a conventional pattern of parallel, "zig-zag" data swaths can slow acquisition, due to the wide turning circle required at the end of each line (Fig. 2B). One potential solution is to collect data over a continuous spiral or pivot point pattern, working from the outside circumference of the field into the center (Fig. 2C). However, this mode of acquisition can produce an inconsistent transient magnetic anomaly influenced by the towing ATV as the system is maneuvered around corners (e.g., Ref. [14]). A favorable compromise is to collect data from two adjacent blocks of parallel swaths, separated by approximately 50 m (Fig. 2D). Once sufficient lines have been collected to conjoin the two initial blocks, a new block of data is established, and coverage progresses block by block across the available area. This minimizes the time required to turn the array at the end of each line and maintains a consistent swath orientation to assist with the subsequent data processing.

4 DATA PROCESSING

Raw data from each sensor are processed using a 1D high-pass median filter along each sensor traverse to suppress the magnetic response to the ATV and heading error across the sensor channels (Fig. 3A). Filter window length varies between 30 and 60 m depending on the characteristics of the data set in question.

However, it was found that archaeological anomalies of similar dimension to the filter window length running parallel to the survey traverse direction could be degraded by this process (Fig. 3B). In earlier surveys with the Scintrex system, where this was identified as a problem, striping effects were instead removed using a directional cosine filter oriented at right angles to the traverse direction. This was followed by application of an 8 m high-pass filter (four times the traverse swath width of 2 m) to emphasize along traverse striping while suppressing anomalies of interest. While both the directional and high-pass filters were conducted in the frequency domain, the resulting space domain data set emphasizes the linear artifacts after the application of an inverse Fourier transform (Fig. 3C). Subtracting this data set with the linear artifacts from the original data removes the striping while preserving anomalies of interest (Fig. 3D).

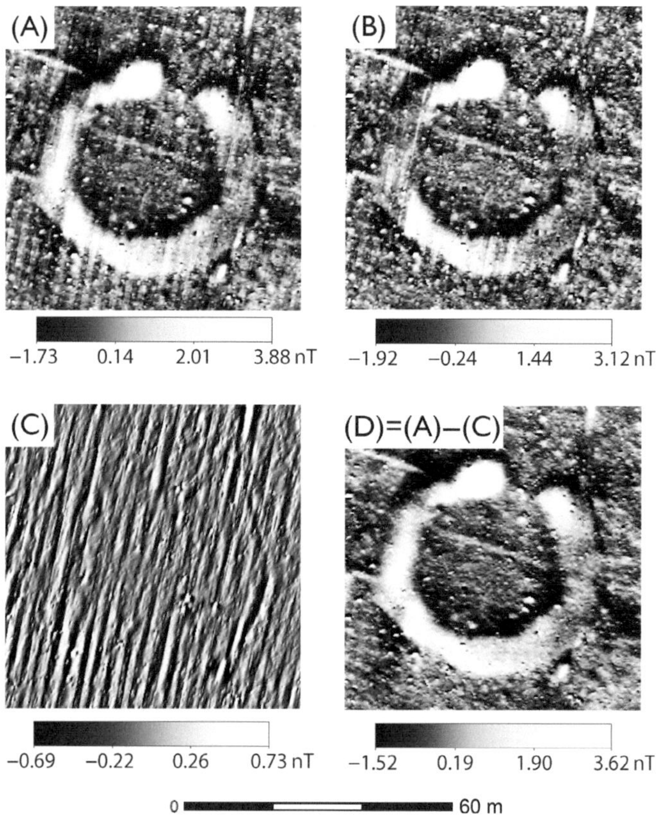

FIG. 3

Initial data processing scheme applied to vehicle-towed data collected over the Wilsford Henge monument (cf Fig. 6) showing (A) unprocessed data with striping due to the heading error created by the towing vehicle, (B) following standard processing with a 20 m 1D median filter destriping algorithm that partially degrades the anomaly over the henge ditch, (C) the unprocessed data treated with a fast Fourier transform (FFT) directional cosine filter to better isolate the striping, and (D) final processed version of the data following the subtraction of the striping avoiding image degradation.

A drawback to this approach is that data must be collected in straight, parallel traverses so that all striping artifacts can be emphasized by applying the directional cosine filter in a single direction obviating to some extent the advantages of direct GPS positioning. Hence, more recent surveys using the G862 system have employed one of the six sensors as a gradiometer sensor positioned 1 m above the center of the array of five near-ground sensors. Subtracting its output from corresponding measurements made with each of the near-ground sensors creates a pseudovertical gradient total field data set that largely eliminates striping effects, greatly reducing the need for high-pass filtering.

Processing now consists of subtracting the measurement made by the single gradiometer sensor from the simultaneous measurements being made by the ground-array sensors then any slight heading errors that remain can be eliminated using the 1D median filter previously described. Determining the best way to subtract the gradiometer measurement has required some experimentation. Since there is only one such sensor above the center of the array, its measurement is strictly only accurate for the sensor immediately below it. If there is any variation in the local magnetic field along the line of the sensors (i.e., transverse to the direction of travel), the true gradiometer measurement above sensors to the left and right of the center will be slightly different. Directly subtracting the central unmodified gradiometer measurement from the simultaneous measurements being made by each ground-array sensor eliminates transient magnetic anomalies, such as those from passing cars, very effectively but introduces slight "streaking" discontinuities between adjacent swaths in the vicinity of strong magnetic anomalies (Fig. 4A).

Alternatively, the gradiometer measurements can be processed and interpolated onto a regular grid as an entirely separate survey and then subtracted from the corresponding grid created from the ground-array sensors afterward. This allows the magnetic field at the height of the gradiometer sensor to be more accurately modeled eliminating the "streaking" discontinuities but introducing strong spurious anomalies in places where any transient magnetic anomalies occurred (Fig. 4B).

A compromise is to subtract a modified gradiometer measurement from the ground-array sensors to the left and right of the center. The slope (or horizontal gradient) of the local magnetic field at the gradiometer sensor's height is estimated by time-domain upward continuation of all ground-array-based measurements within 2m of the position of interest in the same swath [15] (p. 316, eqn 12.4). As Blakely points out, for accurate upward continuation, a much wider set of measurements are needed, but in this case, only an estimate of the local horizontal gradient is required as the true value of the field at gradiometer height is known in the center of the array. This estimate of the horizontal field gradient along the array is used to modify the gradiometer measurement subtracted from the ground measuring sensors to the left and right of the center. This method takes account of transient magnetic anomalies that affect all the sensors simultaneously and to the same degree eliminates "streaking" discontinuities (Fig. 4C).

5 CASE STUDIES

A series of case studies are presented illustrating the application of the vehicle-towed cesium magnetometer array to a range of archaeological sites and landscapes within England. These sites have been chosen to represent a range of geologies and types of site and also indicate the efficient use of limited resources that can result from the use of such technology.

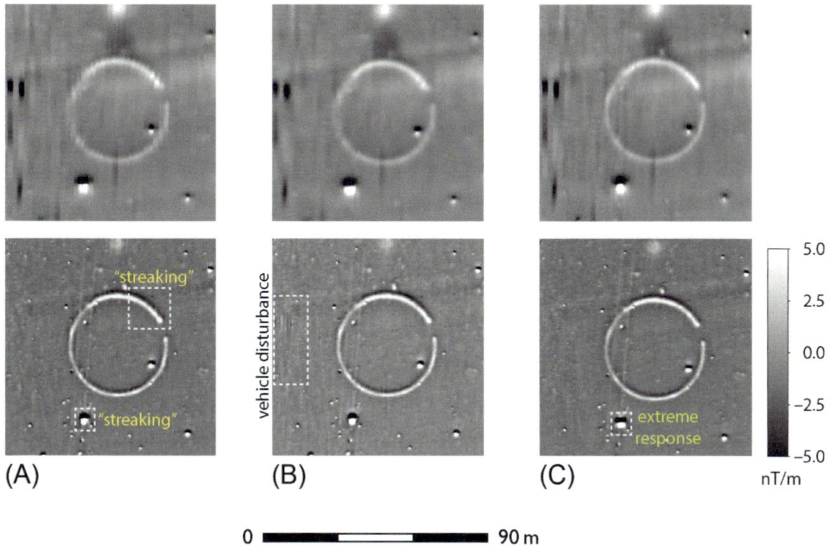

FIG. 4

Data collected over a henge monument using the most recent iteration of the system to illustrate the effects of different ways of subtracting the gradiometer sensor. In each column, the upper plot shows the processed gradiometer data on its own, while the lower plot shows the results of subtracting this from the ground-array sensors. (A) Direct subtraction of the gradiometer measurement showing slight "streaking" discontinuities near strong magnetic anomalies as indicated by the *annotated dashed white boxes inset on the lower plot.* (B) Gridding and interpolating the gradiometer measurements separately avoids "streaking" discontinuities but introduces spurious anomalies where vehicles passed by on a nearby road *(annotated white dashed box inset on the lower plot).* (C) Using upward continuation to modify the gradiometer measurement value subtracted from the sensors not directly beneath the gradient sensor both removes "streaking" and handles the vehicle anomalies at the expense of slight distortion of some extreme magnitude ferrous anomalies *(annotated white dashed box inset on the lower plot).*

5.1 HAM HILL, SOMERSET

Ham Hill is an unusually large Iron Age hillfort extending over 80ha rich in the remains of later prehistoric and Roman activity. Parts of the site have been quarried for Ham Hill Stone since the Roman period, and two active quarries continue in operation, with the remainder utilized as a country park managed by South Somerset District Council. Aerial photography for the site has been of more limited use, perhaps due to the relatively thin soil cover over the Upper Lias Shelly Limestone and predominance of permanent pasture. However, magnetic surveys have provided detailed evidence for activity within the hillfort interior from the Bronze Age to Roman periods, together with later quarrying sites that have obscured traces of earlier occupation in some areas.

The site has provided a useful test for the development of cesium magnetometer systems, as the survey of the first of the three fields was conducted with the original Scintrex sensors mounted on a hand-operated, nonmagnetic cart array. A second survey, undertaken with the same magnetometers and hand cart, was conducted in the adjacent field but utilized a comounted GPS receiver to simultaneously record positional information together with the signal from the Scintrex sensors. This allowed initial experiments to be conducted for collecting data using the grid-less system, resulting in a much improved rate of data acquisition, approximately twice the equivalent speed of the hand-pushed grid-based survey, at an identical 0.125×0.5 m sample density with no degradation in data quality. The total survey area of 20 ha took 10 days in the field to complete (Fig. 5) [16].

The geophysical surveys identified multiperiod settlement evidence that could not be readily discerned from the aerial photography including an early phase of probable Bronze Age field systems and linear boundaries, preceding an intense phase of Iron Age activity represented by a major arterial roadway running between the main entrances to the hillfort, together with numerous ditched enclosures, associated pit clusters, and round houses, indicated by the annotation on Fig. 6. Subsequent Roman activity in the form of a small corridor villa with further associated enclosures and quarries is found to the east of the site. It is of interest to note that the magnitude of responses from the enclosures sharing a similar orientation to the villa is enhanced compared with those related to the earlier phases, perhaps due to increased occupation or semi-industrial activity incorporating more thermal magnetic enhancement of soils associated with the Roman settlement. The importance of quarrying for building stone is indicated through several early phases of quarrying activity, possibly dating as far back as the Roman period. These appear to have disturbed earlier traces of occupation where they occur though fortunately their extent is limited and, in general, the quarries occur to the north of a change in the underlying geology marked by a linear anomaly and an apparent increase in the background magnetic enhancement of the soils reflected by an increase in the magnitude of the response to the enclosure ditch anomalies.

Although fluxgate gradiometer survey data already existed for an additional 9 ha field to the south [17–20], this area was used for repeat survey to test the new Geometrics sensors mounted on a vehicle-towed sledge, where the clarity of response could be readily compared with the previous data collected with the Scintrex instruments. Taking into account the variation in magnetization due to changes in the underlying geology and nature of the archaeology, the vehicle-towed Geometrics data correspond well with the previous Scintrex cesium vapor results, particularly when comparing the magnitude of response from the enclosure ditches found to the south west of the site (Fig. 5, inset). To enable a more direct comparison between the two instrument arrays, the Scintrex total field data were converted to the equivalent pseudogradiometer values following Blakely [15]. There are certainly no apparent artifacts in the data due to the influence of the towing vehicle, and the use of a central gradiometer sensor has allowed coverage closer to field boundaries where the degrading effect of both ferrous fencing and passing vehicles on the road

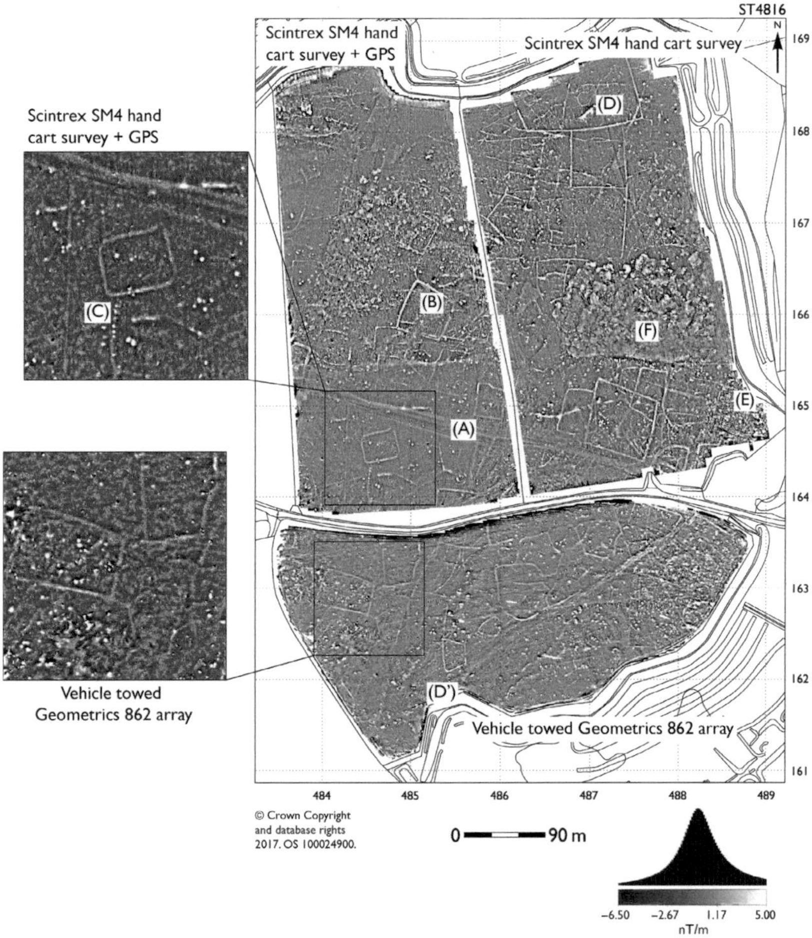

FIG. 5

Results from the cesium magnetometer survey conducted at Ham Hill, Somerset, showing the wealth of anomalies related to the multiperiod settlement of the site. The figure combines three data sets collected using the original Scintrex SM4 sensor array on a hand-operated cart and preestablished survey grid, Scintrex sensors on a hand-operated cart using GPS for grid-less positional control and the results from the current Geometrics G862 array collected using a grid-less, vehicle-towed array. There appears to be no degradation in the data quality with respect to the acquisition methodology (see inset enlarged portions of the data showing weakly magnetized enclosure ditches from adjacent areas of the survey collected with the two different instrument types). The recorded data are shown as a linear greyscale image plotted between −6.5 and +5 nT/m as indicated by the inset grayscale map key. The Scintrex total field data were converted to the equivalent pseudogradiometer values following Blakely [15], and the geometric measurements were determined by subtracting the central gradiometer sensor from the array. Annotation on the figure indicates (A) the arterial roadway, ditched enclosures (B), examples of pit alignments (C), round houses (D) and (D′), and the Roman villa (E) and quarry (F).

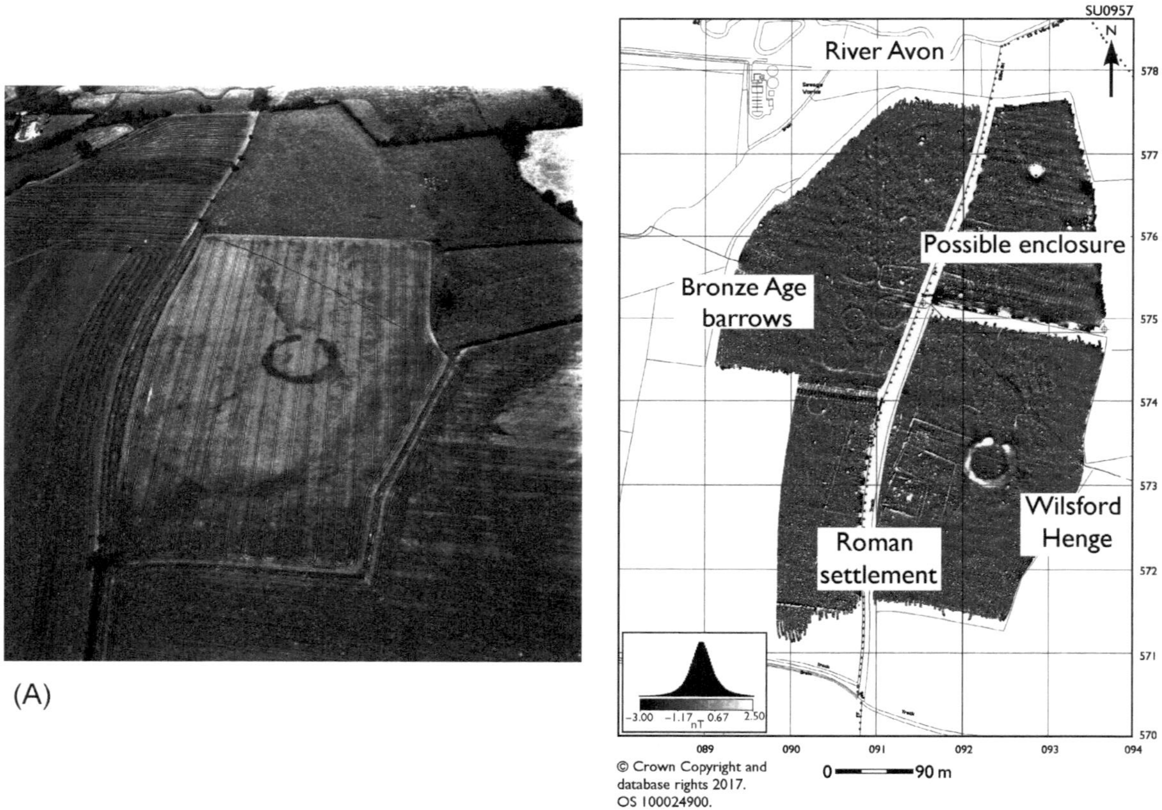

(A) (B)

FIG. 6

(A) Aerial photograph showing general view of Wilsford Henge NMR SU0957/16 (4653/39) 17-JUL-1990 ©Crown copyright, together with (B) vehicle-towed cesium magnetometer array results using the Scintrex sensors from the Marden barrow cemetery and Wilsford henge sites. A previously unrecognized Roman settlement has been revealed by the geophysical survey data, together with the known location of the Wilsford henge, a series of circular ditched Bronze Age barrows and a rectilinear enclosure extending across the central raised track.

could be minimized. Utilizing a vehicle-towed array allowed the data to be acquired at approximately twice the rate of the hand-operated cart-based, grid-less cesium surveys using the Scintrex sensors to the north with no significant artifacts due to the vehicle-towed mode of acquisition.

As the archaeology in this area shows a very variable magnetic contrast, it is difficult to determine how much the sensitivity of the instruments influences the results from a purely visual comparison, especially in a $<0.004\,\mathrm{nT}$ range. While beyond the scope of the study at Ham Hill, due to time restrictions for the field survey, comparable data from the same area should ideally be collected to allow the processed data recorded with the vehicle-towed system to be subtracted from the data recorded by a hand-operated system. It is hoped that this might be achieved at Ham Hill or another suitable site in the future.

The new cesium vapor data combined with the earlier geophysical surveys have provided colleagues from Cardiff University detailed information to build a near complete archaeological map of the internal character of the hillfort [21,22]. This will be used to inform future research and the management of the monument and balance the protection of an important archaeological resource against the need to allow some limited quarrying of Ham Hill Stone for the repair of significant historic buildings in the locality. While a direct comparison between the fluxgate and cesium vapor data sets was not possible at this site, due to the unavailability of the earlier commercial survey results, some minor elaboration of the existing interpretation has been possible, for example, the identification of hut circles that were not previously visible.

5.2 MARDEN HENGE ENVIRONS, WILTSHIRE

This project builds on earlier work using noninvasive techniques including aerial photographic analysis, earthwork and geophysical survey [23], and a small-scale excavation focused on the Neolithic henge at Marden, Wiltshire [24]. The aerial photographic analysis extended beyond the main henge and investigated the distribution of associated monuments throughout the Vale of Pewsey, identifying a number of relatively extensive sites where follow-up work was recommended to contribute to the enhanced understanding of the monuments involved.

While magnetic survey conducted with the original Scintrex sensors and hand cart system over the Marden henge produced relatively poor results, due to the generally less favorable Upper Greensand geology, two further sites found to the south of the river Avon were chosen to test the vehicle-towed array. Both of these sites are recognized from the aerial photographic record: the barrow cemetery and Roman villa known at Charlton St. Peter [25] and a site immediately south of the Marden henge containing a further barrow cemetery and the smaller scale, Wilsford henge (e.g., Fig. 6). Ground conditions varied from permanent pasture at Charlton St. Peter, where some comparison could be made with the more limited fluxgate gradiometer coverage [25], and then over the arable sites at Wilsford to test the suitability of the system on more rugged terrain.

Surveys at both sites proved successful, although the transitional geology from sand to chalk produced an extremely subtle response to certain buried remains that show clearly on the aerial photography but certainly demonstrated an improvement over the results from the less responsive Greensand at Marden Henge. For example, some of the weaker barrow ditches at Charlton St. Peter (Fig. 7) recorded $<0.5\,nT$ magnitude anomalies, demonstrating the importance of ensuring that the full sensitivity of the Scintrex cesium sensors was not significantly compromised when deployed in a vehicle-towed system [7]. Additional modification of the instrument platform, through isolating the sensors from both the GPS receiver and the instrument electronics, was also envisaged during the development of the prototype system.

At Charlton St. Peter, the new Scintrex cesium survey data expanded the previous area of coverage, focused on the Roman villa site, to include the full extent of the barrow cemetery found in the same field to the north [26]. While the new survey data have not revealed any additional barrows other than those identified by the initial aerial photography, some tentative evidence has been provided to support a possible additional monument indicated by more recent lidar coverage that has not produced a crop mark. Some greater clarity has been provided over the Roman site too, with an indication of the presence of additional building remains that were not covered by the 1994 geophysical survey or recognized through aerial photography (Fig. 7).

Results from the 16 ha survey at Wilsford demonstrated a slightly greater magnitude of response over the barrows ($\sim 1.0\,nT$) and also revealed a previously unrecognized Roman settlement abutting the Neolithic henge [27] (Fig. 6). This represents a significant enhancement of the existing record of the site established through aerial photography and, given the extremely clear crop marks produced by the Wilsford Henge, raises important questions regarding the visibility of certain phases of archaeological remains under particular conditions. Perhaps of equal interest is the response to a rectilinear crop mark that has only appeared on a single aerial photograph to date and was initially interpreted as some form of prehistoric enclosure. These crop marks are all replicated by magnetic anomalies including the extension of the enclosure into the paddock to the east, where it had never produced a visible crop mark. Corroborating such ephemeral crop marks provides an important role for ground-based geophysics, together with enhancing discernible detail such as the prominent terminal pits forming the entrance to the Wilsford henge and the suggestion of an internal ring of post holes visible in the magnetic data that were confirmed through subsequent excavation (Fig. 3).

5.3 NATIONAL ARCHAEOLOGICAL IDENTIFICATION SURVEYS (NAIS)

This group of projects utilize a landscape-based approach to identify and characterize archaeological activity, combining aerial mapping supported by the more targeted analytic earthwork survey, geophysical survey, and small-scale excavation. Given the expansive coverage offered by aerial mapping, vehicle-towed magnetic

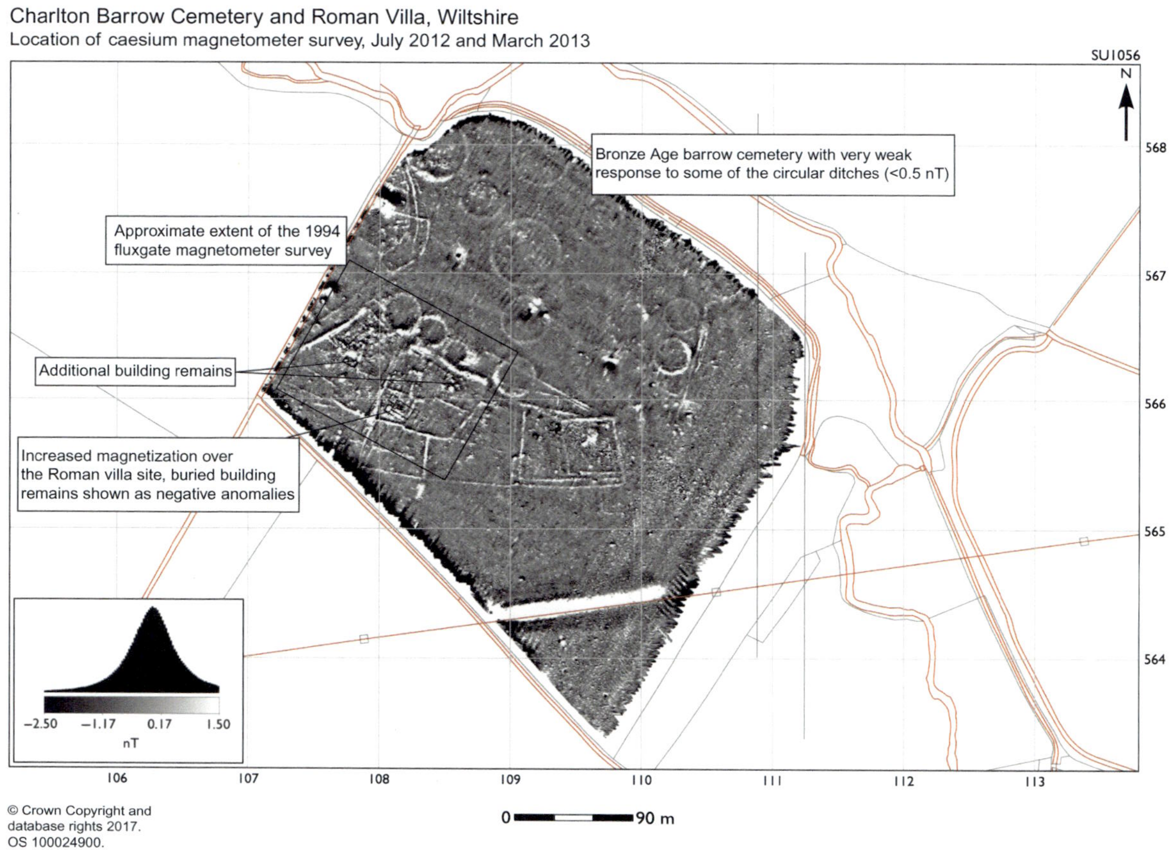

FIG. 7

Vehicle-towed cesium magnetometer array results using the Scintrex sensors from the survey over the Bronze Age barrow cemetery and later Roman villa complex at Charlton St. Peter, Wiltshire. The extent of the previous fluxgate magnetometer survey [25] is shown by the inset box over the very well-magnetized Roman remains, compared with the weaker response exhibited by the circular barrow ditches. A high-voltage power transmission line crosses the site to the south.

survey was deemed essential to provide an appropriate level of coverage over a sample of the many identified sites. Two pilot project areas were selected that contrasted an upland environment in the Lakes and Dales in the north of England, with a lowland study area in West Wiltshire. The challenge for the upland project was the accessibility of sites and the often rugged nature of the terrain, a factor that had no direct impact on the aerial mapping phase. To cope with these conditions, the lightweight prototype sledge using the Geometrics sensors tested at Ham Hill was deployed that allowed the system to set up at some distance from the nearest road access point, even where this required fording a shallow stream to reach the survey site. Numerous surface obstacles in the field, such as extant boulders or impenetrable vegetation, could also be negotiated by repositioning the lightweight platform to ensure even coverage with minimal interruption from blank areas. It was important here to demonstrate that appropriate vehicle-towed systems can be deployed in areas that may otherwise only be accessible to handheld instrumentation.

For example, the curvilinear "teardrop"-shape earthwork features, identified from the aerial photography site on the eastern side of Kitridding Hill, suggested a defended enclosure of probable prehistoric origin defined by a ditch with both inner and outer banks for much of its perimeter (Fig. 8A and B). It was hoped that geophysical survey might help define a series of internal earthwork scoops or platforms potentially identified as hut sites or subenclosures and external linear banks extending from the monument that appear to have been degraded through agricultural land improvement.

Despite challenging surface conditions and relatively weak magnetic response, the cesium magnetometer survey with the Geometrics sensors (Fig. 8C) [28] has successfully provided useful results to elucidate the aerial mapping. In particular, the course of the enclosure ditch and bank, although less complete in the geophysical coverage, confirms an apparently more geometric "shield"-shaped form, only partly described by the earthworks. A possible interruption in the northeast corner of the enclosure is suggested by the earthworks, although the bank and ditch are poorly preserved at this point, and appears as a more continuous ditch-type magnetic anomaly, forming a sharp angle on the northwest corner, so a deliberate entrance here seems unlikely; whether there was an entrance gap on the west side is, however, less clear. The magnetic survey also provided additional evidence for internal subdivision and possible occupation activity within the enclosure. While some of this activity is associated with visible platforms or other topographic variation, many significant geophysical anomalies have no apparent surface expression. Wider survey of the area immediately surrounding the enclosure was limited due to the severe topography of the hilltop location although some additional, perhaps geological, anomalies were recorded.

Access to the terrain in the lowland project area in Wiltshire was, in some regards, less challenging and consisted mainly of large, arable fields that produced an extensive crop mark record. To provide complementary geophysical survey results required the rapid acquisition of data over very large areas aiming both to enhance

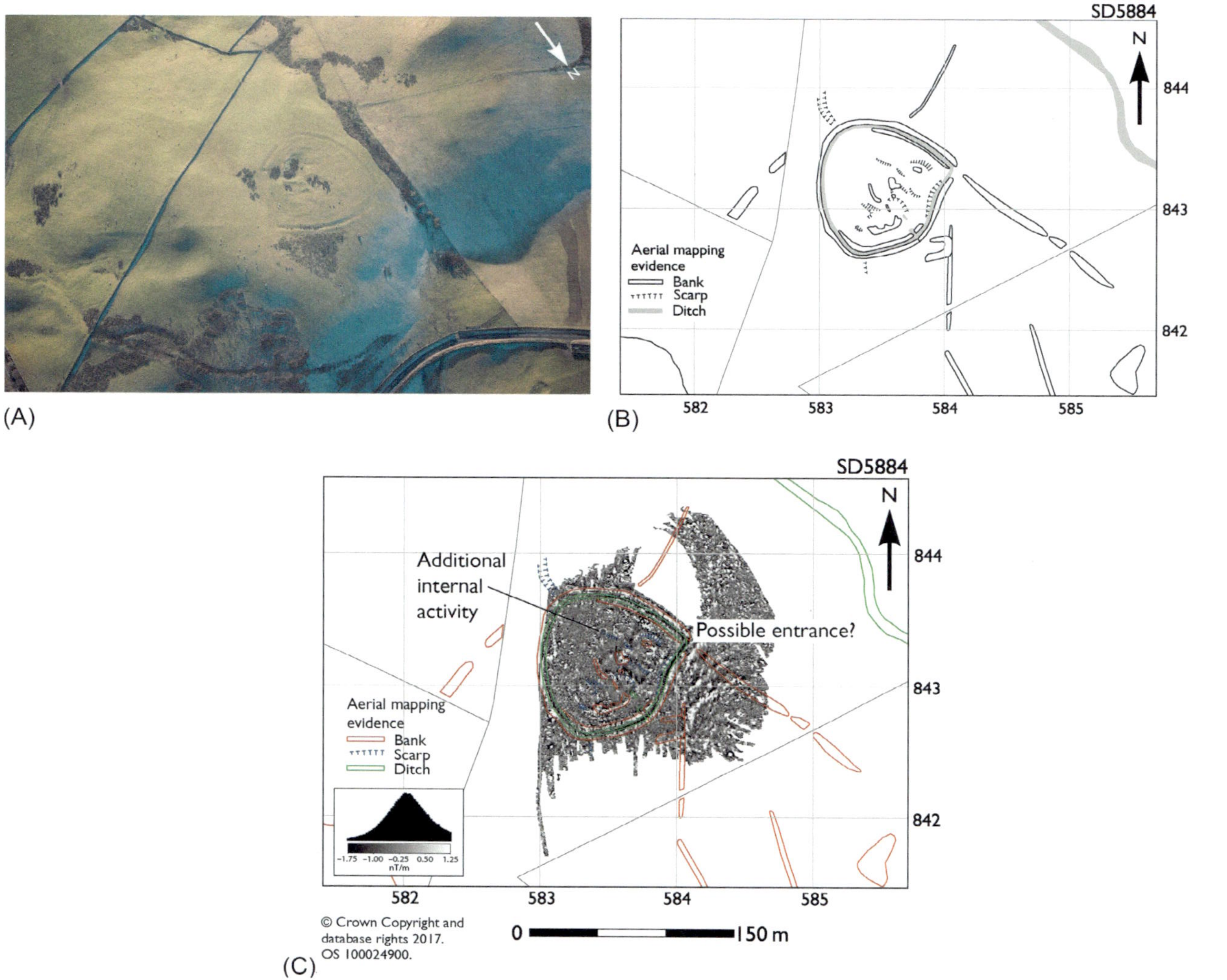

FIG. 8

(A) Oblique aerial photograph of the Kitridding curvilinear enclosure (NMR 28366/17 11-DEC-2012) together with cesium magnetometer survey using the Geometrics sensors over the weakly magnetized enclosure at the site. The data are shown as a linear grayscale image plotted between −1.75 and +1.25 nT/m as indicated by the inset grayscale map key. The aerial photographic evidence is summarized in (B) and (C) overlying the geophysical data, and shows the potential entrance gap indicated by the earthworks, potentially discounted by the presence of a continuous ditch in the magnetic data that also shows more evidence for internal occupation within the enclosure. Color overlay of the aerial photographic evidence shown in (C) is reproduced in the electronic version of this publication.

the aerial photographic record and provide some confirmation of survival of the remains. Limited time windows for access were available due to the cropping regimes over the sites, so the ability to acquire large areas of coverage was of particular benefit.

Fig. 9 shows the results from the site of a rectilinear ditched enclosure at Little Chalfield, connected to an extended double-ditched linear "driveway" approach feature with an adjoining field system of probable Iron Age to Roman date [29]. The cesium magnetometer array using the Geometrics sensors was used to cover a 16 ha area and confirmed the presence of the enclosure, providing additional internal definition and evidence for occupation activity including a possible hut circle. While a combination of geology, soils and modern interference from ferrous pipes have degraded the magnetic response in parts of the survey area, the geophysical results have also enhanced the wider mapping of plowed-out field systems surrounding the enclosure.

Similar areas of coverage were achieved over an area of 27.5 ha encompassing a range of probable Iron Age to Roman enclosed settlements at Paxcroft Farm, Hilperton, Wiltshire, as part of the same lowland project (Fig. 10) [30], where the vehicle-towed magnetic survey successfully clarified the extent and layout of the plow-leveled enclosure remains, resolving some confusion between geologic and archaeological responses that had hindered the interpretation of the aerial photographic evidence. This refined both the morphology of the perimeter ditches together with additional detail, such as possible entrance gaps, the presence of more complex outworks, and an internal hut circle within the southern oval enclosure.

While more limited coverage was possible due to the presence of a standing crop at Kellaways Farm, Langley Burrell Without, Wiltshire, rapid vehicle-towed survey of 0.7 ha still proved useful to confirm both the location of the Roman enclosure and provide some additional, internal detail (Fig. 11) [31]. The limited ground impact of the all-terrain vehicle combined with the lightweight sledge allowed the survey to be conducted with minimal damage to the crop. The local geology has, again, partially confused the full interpretation of the magnetic data, with a number of amorphous anomalies apparent that do not correlate with the aerial photographic record. However, the survey did provide evidence for localized semi-industrial activity within the eastern extent of the enclosure through an area of enhanced magnitude of response, partially borne out through subsequent excavation.

Following the important developmental work conducted using lightweight prototype instrument platforms for mounting the sensors, successful trials were made using the commercially available Geophysical Equipment Exploration Platform (GEEP) sledge system [32]. While this system is designed to mount multiple instrument types, initial comparative tests using cesium magnetometer sensors have demonstrated the applicability of this minimally magnetic platform [10]. In this case, the GEEP has been configured from two robust fiberglass hulls interconnected by a fore deck, comprising all the electronic components including the GNSS receiver, and a separate aft deck where the six cesium sensors are mounted away from immediate

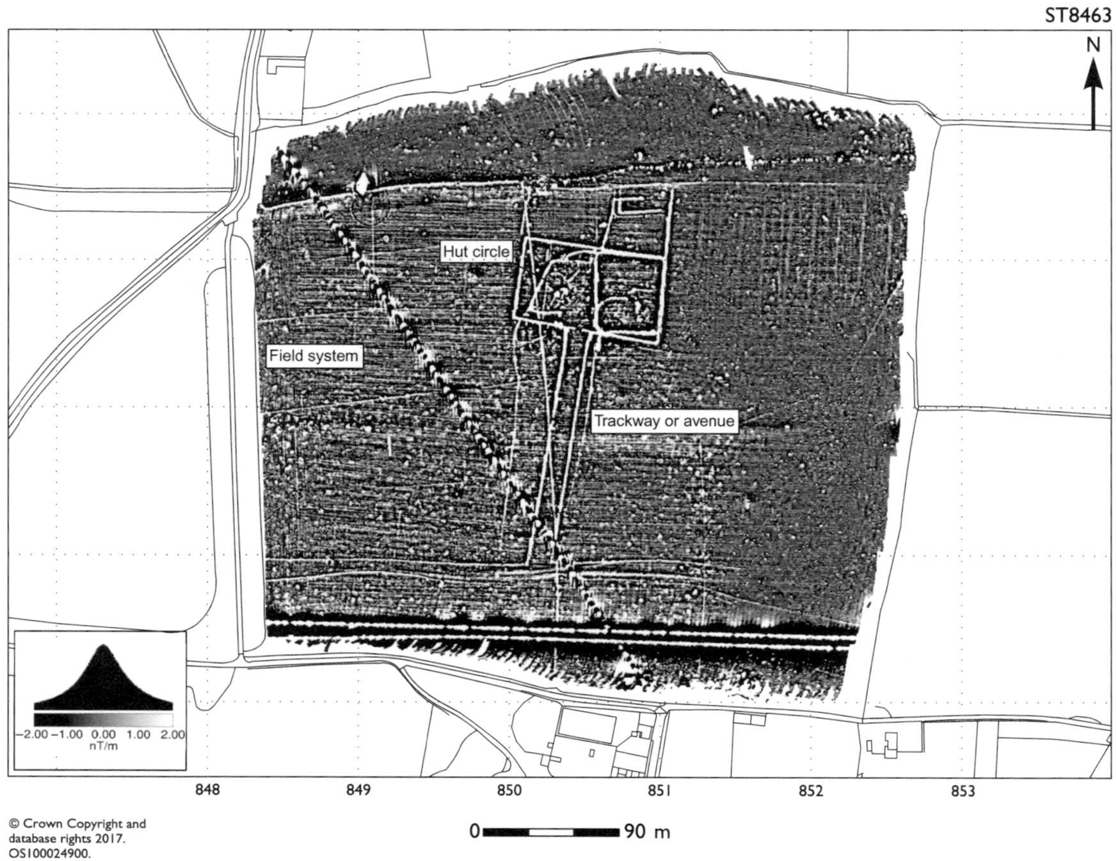

FIG. 9

Cesium magnetometer survey using the Geometrics sensors over the Iron Age and Roman multiphase settlement at Little Chalfield, consisting of a number of intercutting rectilinear enclosures with an adjoining trackway or avenue approach, surrounded by an extensive field system. A hut circle anomaly within the main enclosure is also indicated.

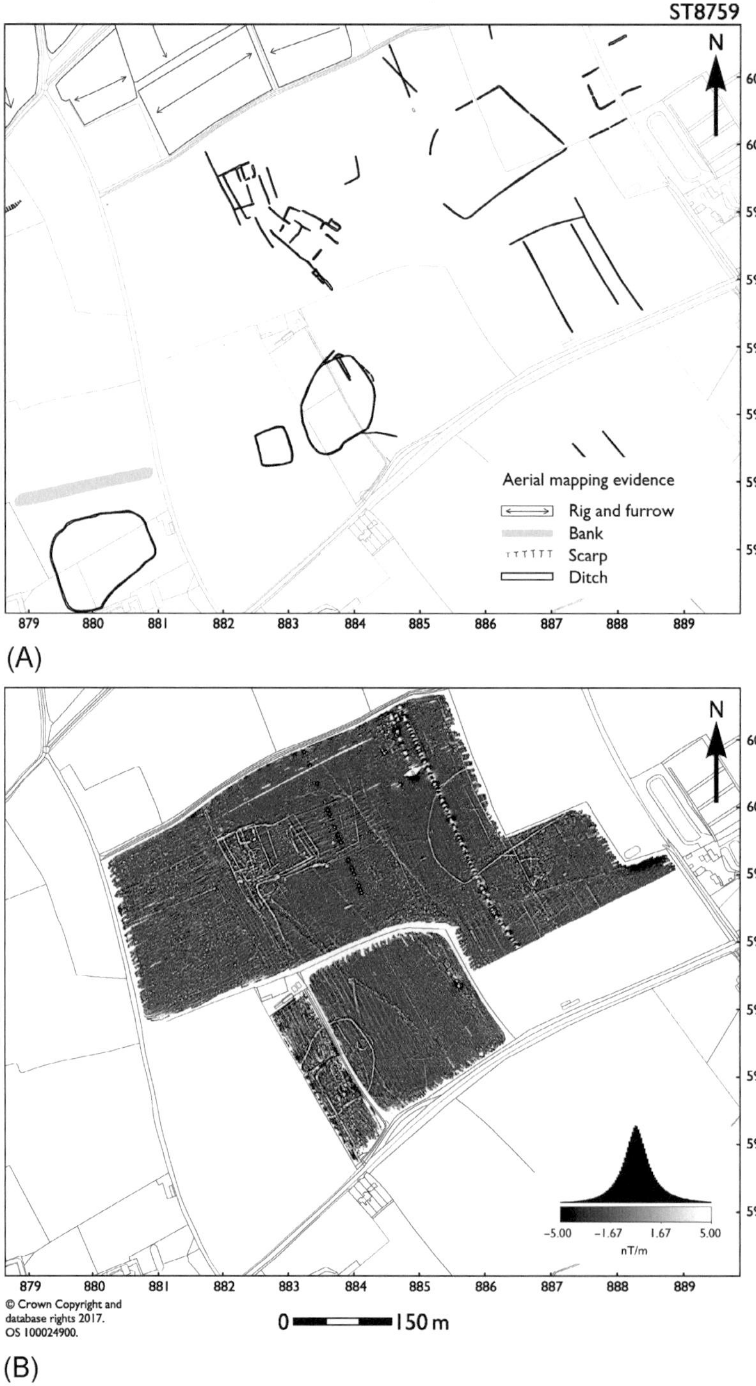

FIG. 10

Vehicle-towed magnetometer survey results from Paxcroft Farm with (A) the transcription of the aerial photographic record and (B) the grayscale image of the magnetic data.

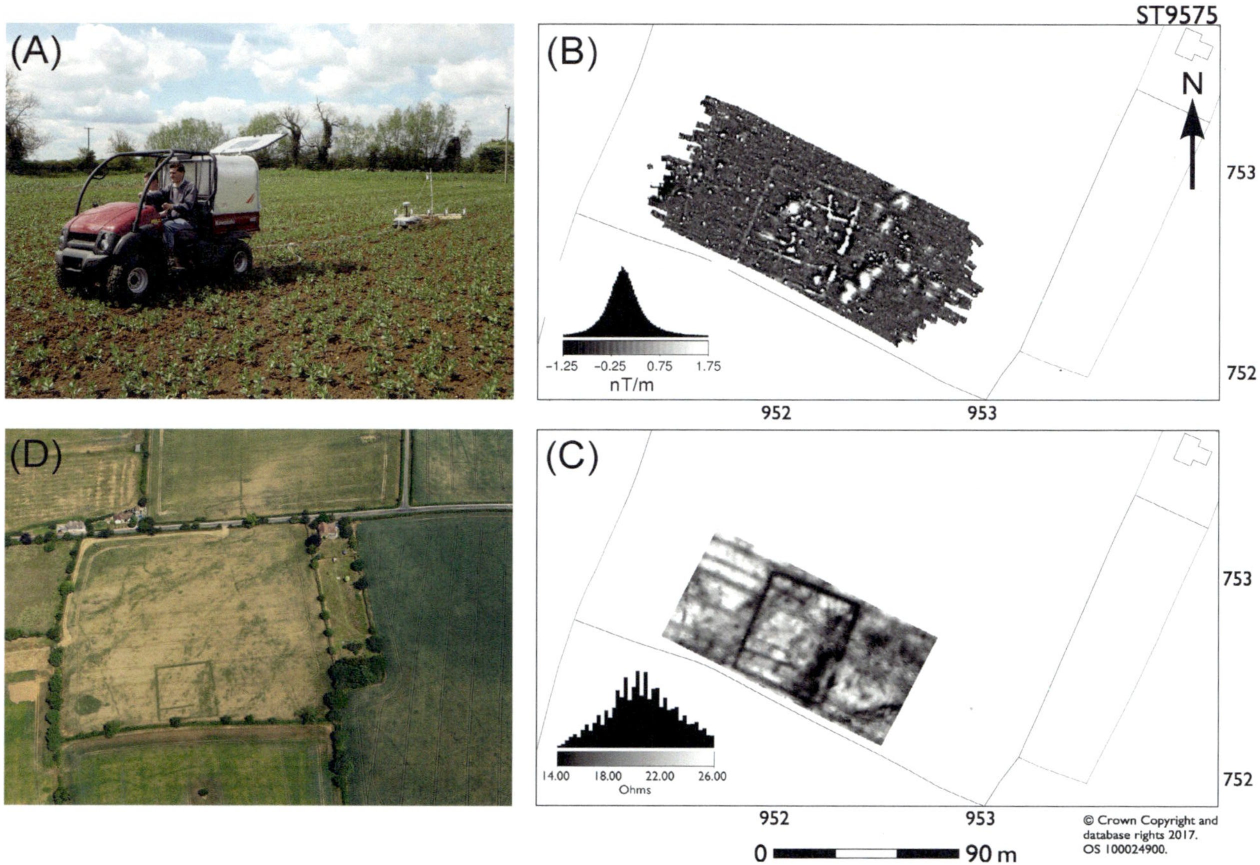

FIG. 11

(A) View of the survey in progress at Kellaways Farm where the combination of a low-impact ATV and lightweight instrument platform minimized damage to the extant crop in the field at the time of the survey. The results of the magnetic survey (B) identified anomalies related to the ditches of the Roman enclosure and provided evidence for internal, semi-industrial activity and allowed accurate placement of (C) an earth resistance survey, which together enhanced (D) the aerial photographic record.

sources of magnetic interference. The platform can be quickly assembled on site using nonmagnetic fasteners and, through the use of sacrificial plastic skid plates on the base of each hull, is able to cope with the high levels of abrasion that occur over many ground surfaces during large-scale acquisition.

5.4 STONEHENGE SOUTHERN WORLD HERITAGE SITE

While a considerable degree of geophysical coverage has been acquired throughout the Stonehenge World Heritage Site (WHS), much of this has occurred to the north of the A303 trunk road in the area containing the monument itself and large earthwork features such as the Stonehenge cursus and the majority of the avenue. Recent collaborative projects have led to an impressive degree of coverage using a variety of geophysical techniques, including extensive acquisition with vehicle-towed fluxgate gradiometer arrays [8,9]. Access to the Southern World Heritage Site has been more restricted due to a transition from predominantly pastoral land use, surrounding the monument itself, to a more varied regime of arable, pig farming and an RSPB stone curlew reserve to the south. While some specific monuments and areas have been targeted in the past, there is clearly much to be obtained from extending detailed geophysical coverage throughout the WHS and this led to a project that required 137 ha of magnetic and 48 ha of multichannel ground-penetrating radar (GPR) acquisition by the authors within a four-week program of field work timed to make best use of limited access windows for part of the pig farm and the RSPB bird reserve over the winter. The use of the vehicle-towed cesium magnetometer array proved essential, particularly with the GEEP instrument platform that minimized equipment down time despite the highly abrasive flint nodules scattered throughout the extensive arable fields.

Many of the anomalies detected by the survey relate to extant monuments or those previously identified through aerial photography, particularly in the fields under arable agriculture where plowing may have leveled any surface expression of the archaeology. In this case, the geophysical survey is able to provide confirmation of the location and survival of the archaeology when it may not, due to different crop rotations and climatic conditions, appear regularly from the air. For example, to the west of the WHS, a brief window allowed access to one of the fields used by the large pig farm where the magnetic survey confirmed the survival of a series of closely spaced or interlocking, more weakly magnetized subcircular ring ditches partially known from aerial photography within a group of three scheduled bowl barrows (Fig. 12) [33,34]. Such subtle archaeology may well struggle to survive, but the geophysics has been able to both confirm and enhance the number of ring ditches present together with a more strongly magnetized, linear ditch section crossing the field boundary where it appears to terminate, suggesting an intriguing target for subsequent excavation.

The more recent history of the WHS is reflected in the course of the military light railway seen in Fig. 12 running parallel with the A360 road to the west and passing through the circular ditch of a henge monument (Winterbourne Stoke 74) identified by an earlier geophysical survey in the 1990s targeting a ring ditch seen on aerial

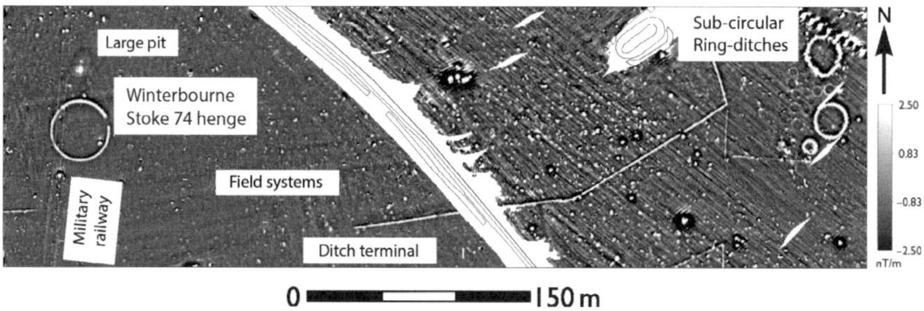

FIG. 12

Extract from the vehicle-towed cesium magnetometer survey of the Stonehenge southern World Heritage Site, Diamonds Field, showing magnetic anomalies related to both prehistoric and more recent activity at the site.

photography [35]. Despite the modern interference from the railway, there is good evidence for a possible entrance gap to the northeast of the henge, but no indication of internal activity. However, a group of pits immediately to the north was considered to be possibly associated with the henge. The survey also covered the site of Coneybury henge to the west of the WHS that provides an interesting comparison between an earlier fluxgate gradiometer survey, conducted in 1980 prior to the partial excavation of the monument, and modern high-sensitivity cesium coverage [36,37]. Fig. 13 shows that the new data replicate the previously recorded curvilinear anomalies over the henge ditch, with an equal magnitude of response and a similar entrance gap to the northeast. A less pronounced break in the ditch to the south marks the location of an excavation trench as it is absent from the original survey data. The presence of internal intense magnetic anomalies, only mapped by the current survey, seems most likely to be due to a more recent origin, possibly even the excavation itself.

The western extent of the WHS also produced a wealth of linear anomalies in both the magnetic and GPR data sets that appear to indicate a network of ditches not fully represented in aerial photography. One of these, shown in Fig. 14, demonstrated an interesting bifurcation in the magnetic data suggesting some greater significance, particularly given the location on high ground overlooking Stonehenge about 1 km to the northwest. Subsequent excavation of this ditch proved particularly fruitful, revealing two superimposed middle Bronze Age burials suggesting these linear anomalies, some only represented as quite subtle responses in the GPR data, define a network of linear boundaries demarking the prehistoric landscape [37].

5.5 SILCHESTER ENVIRONS PROJECT

The ability to acquire large-scale magnetic data sets with vehicle-towed systems can also be used to test the suitability and variation of different geologies and soils. The Silchester Environs project is based around the high-quality, handheld fluxgate survey data collected over the entire walled Roman town and immediate vicinity of

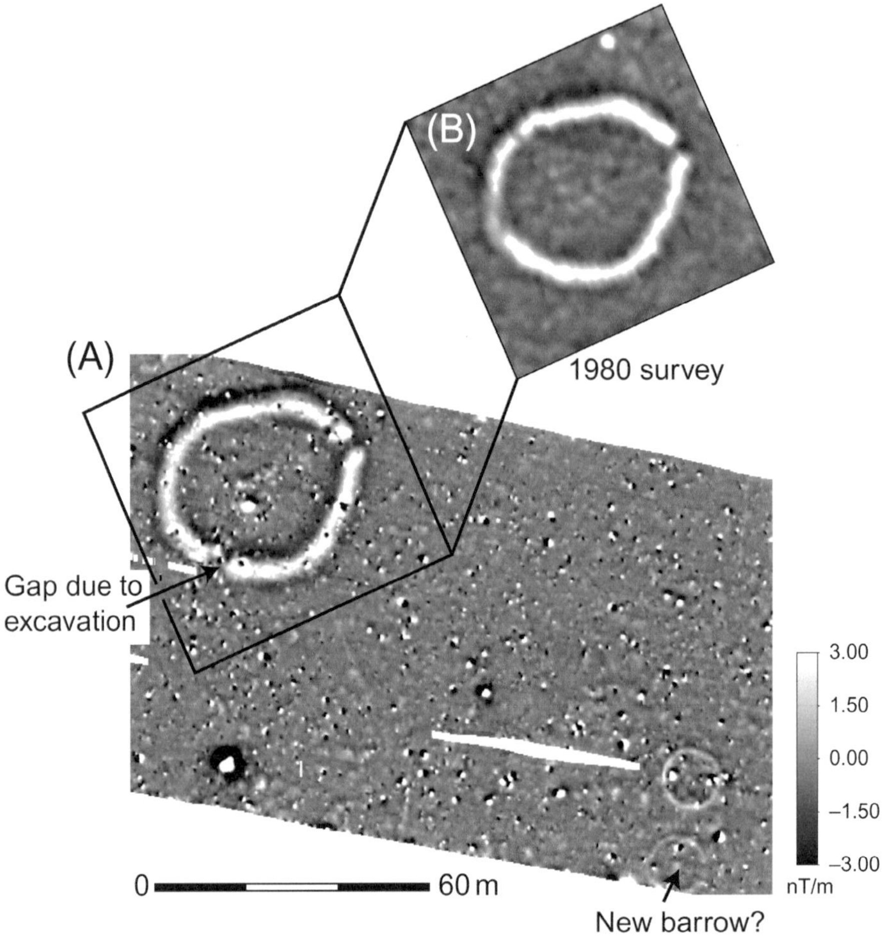

FIG. 13

(A) Extract from the vehicle-towed cesium magnetometer survey over Coneybury henge, together with (B) the original fluxgate data for comparison. The wider extent of the new survey data has revealed a small, circular anomaly possibly due to a previously unrecognized barrow partially obscured by the field boundary.

Calleva Atrebatum, conducted by the University of Reading [38]. The surrounding landscape contains considerable evidence for the late prehistoric precursors to the Roman settlement, but many of these monuments are being lost to either the plow or development pressure between the two conurbations of Basingstoke and Reading. Aerial photography and more recent lidar coverage have done much to enhance the identification of archaeological activity, particularly the network of long linear earthworks that run throughout this landscape. However, successful magnetic survey has

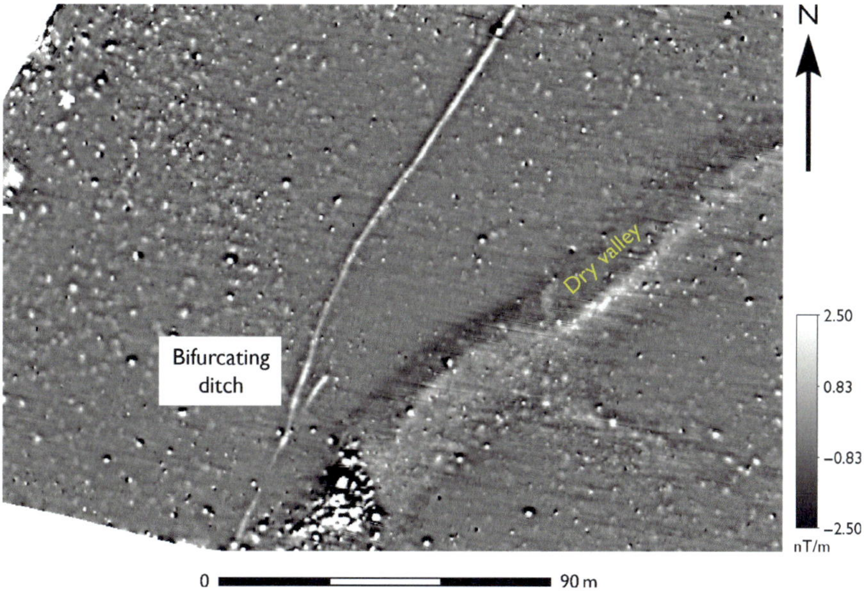

FIG. 14

Extract from the vehicle-towed cesium magnetometer survey of the Stonehenge southern World Heritage Site, at West Amesbury, showing the response to the bifurcating ditch where subsequent excavation revealed the presence of two burials. Other magnetic anomalies are due to the sediments accumulated within the course of dry valley crossing the site and numerous, discrete, responses due to individual pits or tree throws can also be discerned.

proved to be dependent upon the local geology, and the response to known archaeological features has varied dramatically between the sand and gravel deposits and less favorable clays.

Rapid data acquisition with a vehicle-towed system has allowed greater coverage over areas of more challenging geology, examining the variation of response at a scale and commitment of field resource more able to cope with the size of the geologic units. For example, extensive survey surrounding the village of Little London was able to successfully identify the location of highly magnetic, Roman tile kilns suspected from the previous recovery of surface artifacts and the presence of clay deposits (Fig. 15). Extension of the survey area to cover a large, ditched enclosure known from limited aerial photography over the same geology immediately to the west failed to produce any discernible magnetic anomaly. GPR survey over the same enclosure did reveal a very subtle response to the ditches, but at present, it is unclear how well the features survive at the site. Further survey of similar enclosures within the project area over the same geology at Simms Copse produced more successful results with discernible magnetic anomalies, although the GPR survey again produced a more definitive response.

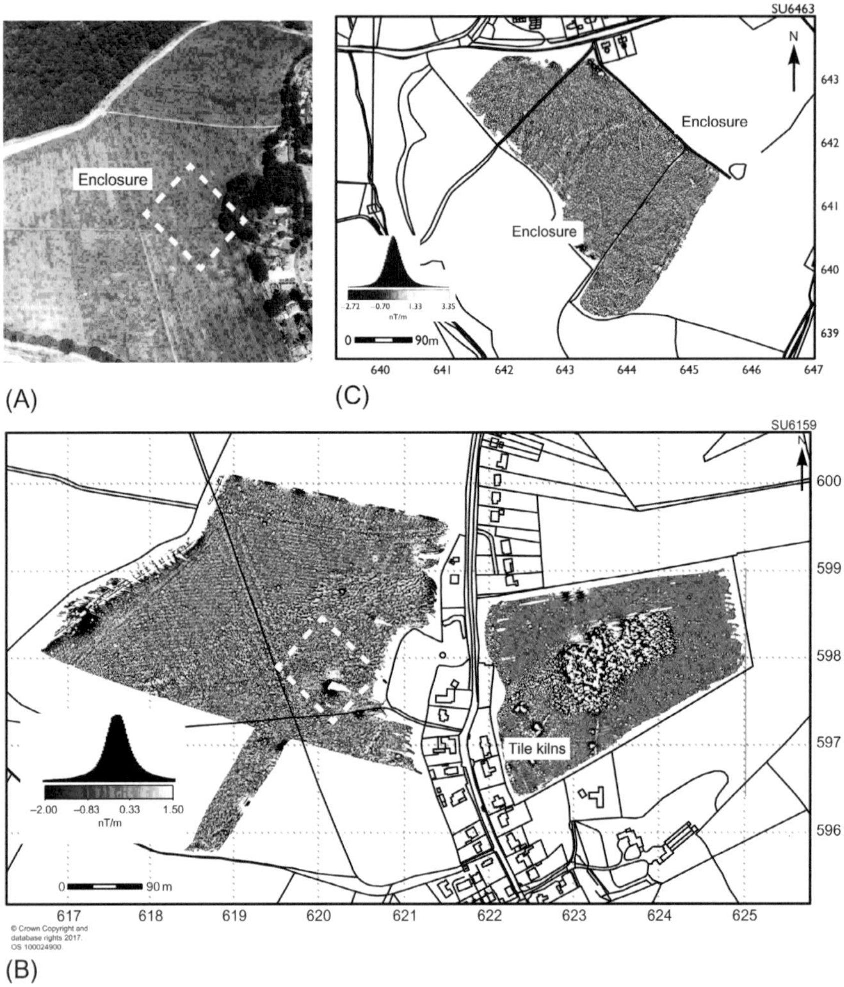

FIG. 15

(A) Aerial photograph showing the enclosure *(annotated by a dashed white box inset)* at the Little London Roman tile kiln site. The magnetic data from the same site (B) detected the well-magnetized tile kilns but (C) failed to identify the enclosure ditch. Similar enclosures at the nearby site of Simms's Copse produce just discernible magnetic anomalies.

6 CONCLUSION

The increased availability of vehicle-towed magnetometer arrays provides the possibility to extend the investigation of individual archaeological sites into a much wider archaeological context, maximizing the information that can be recovered in a relatively short period of field work. This allows the most efficient use of resources and can often overcome the difficulty of limited windows to access certain areas due to the specific agricultural or ecological regime in place. Multiple sensor arrays can also be deployed with support for geopositional control, data logging and power supply for the instruments provided within the towed system. While multiple fluxgate magnetometer sensors can be operated from hand or hand-pushed cart systems, the power requirements of very large arrays or to support the current generation of cesium vapor instrumentation is better suited to a towed system. However, it is important to recognize that vehicle-towed systems are not necessarily able to access all types of terrain, for example, where either the presence of standing stones or vegetation prohibits the use of a vehicle-towed system.

Even under partially unfavorable field conditions, the vehicle-towed array described here allowed for data acquisition rates exceeding 2.0 ha per hour, using a five-sensor array with a 0.5 m crossline spacing. Higher rates of acquisition are certainly possible with larger arrays of instruments and seem likely to be increased further with the development of new sensor technology, such as the next-generation solid-state cesium vapor sensors. Data quality remains paramount to any vehicle-towed system, and despite the relatively short distance between the towing vehicle and the magnetometer array (often <5 m) required to maintain maneuverability in the field, with appropriate design and data processing considerations, good quality data can be recorded free of any apparent artifacts, even over quite weakly magnetized sites. Certainly, the use of a minimally magnetic all-terrain vehicle and a central gradient sensor on the magnetometer array together with appropriate data processing taking consideration of the full influence of the magnetic anomaly due to the towing vehicle is required.

REFERENCES

[1] English Heritage, Geophysical Survey in Archaeological Field Evaluation, second ed., English Heritage, Swindon, 2008.

[2] J. Gerrard, L. Caldwell, A. Kennedy, Short report: green waste and archaeological geophysics, Archaeol. Prospect. 22 (2) (2015) 139–142.

[3] J.M. Reynolds, An Introduction to Applied and Environmental Geophysics, Wiley, Chichester, 1997.

[4] T. Herbich, Magnetic prospecting in archaeological research: a historical outline, Archaeol. Polona 53 (2015) 21–68.

[5] P.E. Gettings, M.E. Gettings, M.W. Bultman, Data Collection and Reduction Procedures for 1900 km of Total Intensity Magnetic Field Data Collected With a Truck-Mounted System in Southeastern Arizona, Southwestern Colorado, and Northeastern Wyoming, (1995). (U.S. Geological Survey Open-File Report 95-32).

[6] J.E. Tilden, D.A. Ponce, J.M.G. Glen, B.A. Chuchel, K. Tushman, A. Duvall, Gravity, Magnetic, and Physical Property Data in the Smoke Creek Desert Area, northwest Nevada, (2006). (U.S. Geological Survey Open-File Report 2006-1197).

[7] N. Athens, J. Glen, R. Morin, S. Klemperer, ATV magnetometer systems for efficient ground magnetic surveying, Lead. Edge 4 (2011) (2011) 394–398.

[8] T. Darvill, F. Leuth, K. Rassmann, A. Fischer, K. Winkelmann, Stonehenge, Wiltshire, UK: high resolution geophysical surveys in the surrounding landscape, 2011, Eur. J. Archaeol. 16 (1) (2013) 63–93.
[9] C. Gaffney, V. Gaffney, W. Neubauer, E. Baldwin, H. Chapman, P. Garwood, H. Moulden, T. Sparrow, R. Bates, K. Löcker, A. Hinterleitner, I. Trinks, E. Nau, T. Zitz, S. Floery, G. Verhoeven, M. Doneus, The Stonehenge Hidden Landscapes Project, Archaeol. Prospect. 19 (2012) 147–155.
[10] I.A. Hill, N. Linford, Rapid and precise archaeogeophysical survey using a multi-sensor platform (MSP), Proceedings of the EAGE-NSG Meeting, Utrecht, September 2004.
[11] N. Linford, P. Linford, A. Payne, Chasing aeroplanes: developing a vehicle-towed caesium magnetometer array to complement aerial photography over three recently surveyed sites in the UK, Near Surv. Geophys. 13 (6) (2015) 623–631.
[12] N. Linford, P. Linford, L. Martin, A. Payne, Recent results from the English Heritage caesium magnetometer system in comparison to recent fluxgate gradiometers, Archaeol. Prospect. 14 (3) (2007) 151–166.
[13] R. Siegel, Use of Commercial-Off-the-Shelf Vehicles for Towed Array Magnetometry: Geophysical Procedures for Vehicle Signature Measurement ESTCP Project MM-0605 (Technical Report), SAIC, 2007.
[14] I.A. Hill, N. Linford, Comparison of rapid and precise magnetometer survey techniques over Wroxeter roman city site, Shropshire, American Geophysical Union, Spring Meeting 2005 #NS32A-05, 2005.
[15] R.J. Blakely, Potential Theory in Gravity and Magnetic Applications, Cambridge University Press, Cambridge, 1995.
[16] A. Payne, N. Linford, P. Linford, Ham Hill, Stoke sub Hamdon, Somerset. Report on Geophysical Surveys, March, May and December 2011, (2012). (English Heritage Research Reports Series 22/2012).
[17] Geophysical Surveys of Bradford, Ham Hill: Report on Geophysical Survey, (1992) (Unpublished 92/101).
[18] Geophysical Surveys of Bradford, Ham Hill: Report on Geophysical Survey II, (1993) (Unpublished 93/134).
[19] Geophysical Surveys of Bradford, Ham Hill: Report on Geophysical Survey III, (1994) (Unpublished 94/123).
[20] GSB Prospection, Hamdon Hill, Somerset, (2001)(Unpublished 2001/35).
[21] N. Linford, P. Linford, A. Payne, I. Hardwick, Z. Edwards, Ham Hill, Stoke Sub Hamdon, Somerset. Report on Geophysical Surveys, November 2013, (2014) (English Heritage Research Reports Series 67/2014).
[22] N. Sharples, C. Evans, A. Slater, A. Payne, P. Linford, N. Linford, Ham and mustard, the prehistoric hillfort on Ham Hill, Somerset, Br. Archaeol. 123 (2012) 34–39.
[23] D. Field, L. Martin, H. Winton, The Hatfield Earthworks, Marden, Wiltshire. Archaeological Survey Report, (2009). (English Heritage Research Department Reports 96/2009).
[24] D. Field, J. Leary, Recent work at Marden Henge, Wiltshire, in: Research News—Newsletter of the English Heritage Research Department, vol. 15, English Heritage, London, 2010, pp. 10–13.
[25] M. Corney, C. Gaffney, J. Gater, Geophysical investigations at the Charlton Villa, Wiltshire (England), Archaeol. Prospect. 1 (2) (1994) 121–128.

[26] N. Linford, P. Linford, A. Payne, Charlton Barrow Cemetery and Roman Villa, Wiltshire, Report on Geophysical Surveys, July 2012 and March 2013, (2013) (English Heritage Research Reports Series 9/2013).
[27] N. Linford, P. Linford, A. Payne, Marden Barrow Cemetery and Wilsford Henge, Wiltshire, Report on Geophysical Surveys, September 2012 and March 2013, (2013) (English Heritage Research Reports Series 8/2013).
[28] N. Linford, P. Linford, A. Payne, I. Hardwick, Lakes and Dales Nais, Kitridding Hill, Lupton, Cumbria Report on Geophysical Surveys, July 2013, (2013) (English Heritage Research Reports Series 56/2013).
[29] N. Linford, P. Linford, A. Payne, Little Chalfield, Atworth, Wiltshire, Report on Geophysical Survey, August 2014, (2014) (English Heritage Research Reports Series 78/2014).
[30] N. Linford, P. Linford, A. Payne, Z. Edwards, Paxcroft, Hilperton, Wiltshire, Report on Geophysical Surveys, May and September 2014, (2015) (English Heritage Research Reports Series 9/2015).
[31] N. Linford, P. Linford, A. Payne, Z. Edwards, Kellaways Farm, Langley Burrell Without, Wiltshire, Report on Geophysical Survey, May 2014, (2015) (English Heritage Research Reports Series 79/2015).
[32] I.A. Hill, T. Grossey, C. Leech, High-resolution multi-sensor geophysical surveys for near surface applications can be rapid and cost-effective, Lead. Edge 23 (7) (2004) 684–689.
[33] N. Linford, P. Linford, A. Payne, Southern WHS Survey, Diamonds Field, Boreland Farm, Wiltshire, Report on Geophysical Surveys, August 2015, (2015) (Historic England Research Reports Series 93/2015).
[34] N. Linford, P. Linford, A. Payne, Southern WHS Survey, Diamonds Field, Druid's Lodge, Wiltshire, Report on Geophysical Surveys, August 2015, (2015) (Historic England Research Reports Series 94/2015).
[35] Geophysical Surveys of Bradford, Report on Geophysical Survey: A303 Amesbury to Berwick Down, (1992) (Report No. 92/03).
[36] A.D.H. Bartlett, Geophysical Survey at Coneybury Henge, Amesbury, Wiltshire, 1980, (1988). (Ancient Monuments Laboratory Reports Series 169/88).
[37] N. Linford, P. Linford, A. Payne, Southern WHS Survey, West Amesbury Farm, Wiltshire, Report on Geophysical Surveys, August 2015, (2015) (Historic England Research Reports Series 95/2015).
[38] J. Creighton, R. Fry, Silchester: Changing Visions of a Roman Town. Integrating Geophysics and Archaeology-the Results of the Silchester Mapping Project 2005-10, Society for the Promotion of Roman Studies, London, 2016.

CHAPTER

6 Making sense of anomalies: Practices and challenges in the archaeological interpretation of geophysical data

Lieven Verdonck*, Philippe De Smedt[†], Jeroen Verhegge*,[‡]

Department of Archaeology, Ghent University, Ghent, Belgium Department of Environment, Ghent University, Ghent, Belgium[†] Geosonda Environment NV, Ghent, Belgium[‡]*

CHAPTER OUTLINE

Innovation in Near-Surface Geophysics. https://doi.org/10.1016/B978-0-12-812429-1.00006-4

1 INTRODUCTION

If we describe archaeology as "the study of past human cultures through the analysis and interpretation of artefacts and material remains" [1], the ability to interpret and extract archaeological meaning from those remains is as important as the processes of acquiring and processing the measurements. This chapter deals with the interpretation of material remains detected through geophysical methods. These record variations in subsurface physical properties (in particular, magnetic, electrical, and electromagnetic properties), by means of nondestructive measurements on or close to the surface. They result in mostly digital data that may contain anomalies (measurements that deviate from the normal, uniform background; [2]) approximately representing the investigated archaeological targets.

In the past decades, archaeological geophysics has gained wider acceptance within the archaeological community and has transformed from a preliminary step leading to excavations to an important independent source from which to study the past [3]. The emphasis is no longer only on settlements and known sites but also on the differentiation of zones within larger sites, on the space between sites, and on the prospection for unknown sites [4–8]. This is in line with landscape archaeology and the debate in aerial archaeology on observer-directed versus landscape-oriented methodology [9].

The scale of archaeological geophysics has continued to grow. Vehicle-towed platforms carry large arrays of magnetometer and ground-penetrating radar (GPR) sensors [10,11], as well as electromagnetic induction (EMI) systems and earth resistance meters [12,13]. More and more case studies illustrate the increased potential of geophysical techniques as a means to study the human past and not just to locate buried remains [14]. However, archaeological interpretation has not kept up with this rapid development, and the interpretative potential of the geophysical data remains underutilized. This is caused partially by the fact that in many publications, survey results are presented mainly in geophysical and less in archaeological terms (perhaps due to the fact that the majority of the practitioners are geophysicists rather than archaeologists), while geophysical techniques are still seen by many archaeologists merely as an exploratory tool that can direct later excavations. More fundamental problems that complicate archaeological interpretation are related to the complex relationship between the measured anomalies and the causative buried features and to the subjective character of the interpretation process.

1.1 COMPLEX RELATIONSHIP BETWEEN MEASUREMENTS AND BURIED FEATURES

1.1.1 Contrast

The geophysical response and its interpretability strongly depend on the presence of contrasts in physical properties between the soil and the archaeological features embedded in it (in Table 1, an overview is given of the primarily relevant physical properties detected by the geophysical techniques discussed in this chapter). In their

Table 1 Physical Properties Upon Which the Geophysical Prospection Techniques, Discussed in This Chapter, Are Based

Geophysical Technique	Physical Property of the Soil
Magnetometer survey	Magnetic susceptibility Remanent magnetization
Topsoil magnetic susceptibility survey	Magnetic susceptibility
Earth resistance survey	Electrical resistivity
Electrical resistivity imaging	Electrical resistivity
Low-frequency electromagnetic methods (electromagnetic induction)	Magnetic susceptibility Electrical conductivity
Ground-penetrating radar	Dielectric permittivity

An archaeological feature can only be detected if a contrast exists in one or more of these properties.

turn, these contrasts depend not only on the type of archaeology but also equally on geology, soil, hydrology, climate, agricultural practices, etc. [15]. If a contrast is very weak, this may severely hamper data interpretation. An example is the influence of varying soil moisture on electrical conductivity survey data. Fig. 1 shows apparent electrical conductivity (ECa) data resulting from a frequency-domain EMI survey on a rendzina soil in Wiltshire (the United Kingdom), conducted in dry and wet conditions. The moisture variations influence the discrimination of archaeological features. While in dry conditions (Fig. 1A) rectangular features relating to prehistoric field systems retain more moisture than the surrounding soil and can clearly be discerned in the western part of the area, this contrast is obscured by an overall higher ECa when the soil is nearly water saturated (Fig. 1B). An unexpected contrast may equally complicate the interpretation. For instance, a ditch refilled with magnetically enriched topsoil normally generates a positive magnetic anomaly (it is more magnetic than the surrounding soil). However, dissolution of ferromagnetic particles and precipitation as goethite or lepidocrocite can transform the positive anomaly into a negative one (less magnetic than the background value) in soils with stagnant moisture and a changing groundwater table [16].

1.1.2 Nonuniqueness

After basic processing, a geophysical area survey mostly produces a two-dimensional (2-D) map from which we can derive the lateral position of the buried structures. These maps are complex functions of the true 3-D distribution of the measured physical soil properties (e.g., magnetic susceptibility or electrical resistivity). In order to obtain more information on the soil property distribution, inversion is commonly applied to earth resistance and EMI measurements carried out with different electrode or coil separations or configurations [17,18], and less frequently to magnetometer data [19,20]. By means of inversion, a subsurface model can be found that results in the best agreement between the sensor observations and the response computed by applying a forward operator to the model [21]. The forward operator is

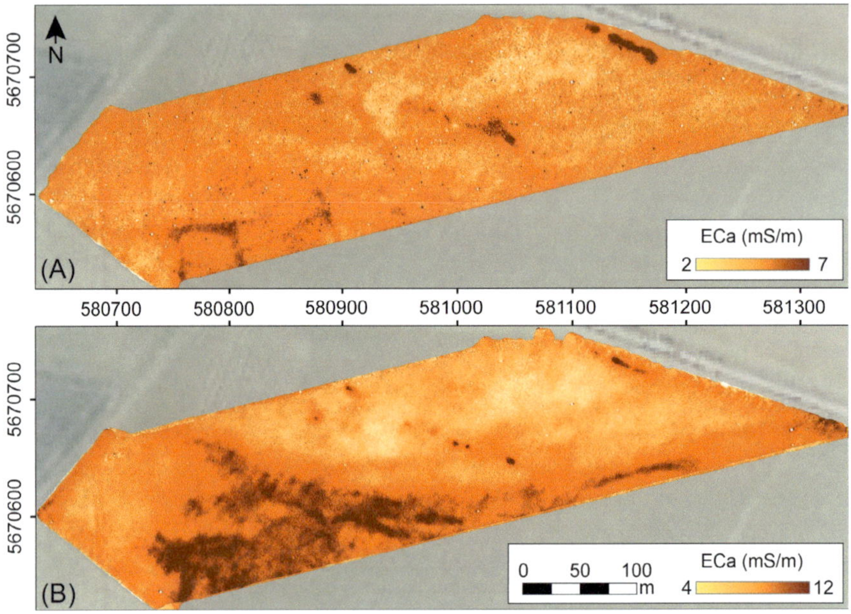

FIG. 1

Apparent electrical conductivity (ECa) data recorded with a Dualem-21S EMI instrument (2.1 m perpendicular coil configuration) in the same survey area in July 2015 (A) and November 2015 (B). The July measurements, with soil moisture content near the wilting point, render ECa variations that differ strongly from the data obtained when the soil was nearly water saturated in November.

a mathematical description of physical relationships, which allows calculating the response when the earth model and the energy source are known.

However, this inversion process is nonunique, since a multitude of models could result in a computed response that fits the observed values. Additional information is needed to reduce this uncertainty [22] and to constrain the models that we generate from our geophysical data. By constraining, we reduce the number of models that could possibly fit our observations, through the addition of prior knowledge (from geophysical measurements, other prospection data, or excavations).

Earth resistance survey, magnetometry, and EMI are potential field methods, yielding information on the bulk properties of the earth. Therefore, a measurement is always influenced by a certain soil volume and cannot be attributed precisely to a point in 3-D space unless an inversion is carried out. The situation is different for GPR, which is an example of a wavefield method, as it is based on the propagation of energy in the form of waves, which can be reflected. Each measurement on the surface yields a series of samples, corresponding to GPR reflections whose travel times are known precisely [23]. This 3-D nature of the observed data matches the dimensionality of the physical property distribution that needs to be recovered.

Nevertheless, the conversion of travel times to depth is often difficult since the radar wave velocity has to be determined precisely. Moreover, also for GPR, further processing (migration, inversion) is needed to estimate the geometry of the features and obtain quantitative information on the distribution of the physical properties in the soil [24].

1.1.3 Noise

Besides nonuniqueness and the difficulty to obtain quantitative, 3-D information through inversion, which in archaeological geophysics is in many cases unnecessary since important information about the shape and location of the anomalies can be determined from the 2-D data plot, other parameters such as noise and limited resolution and sensitivity result in uncertainty in the interpretation. Noise can take many forms, but random noise usually plays an important role [25]. If the useful signal is weak compared with the noise, there can be uncertainty related to the nature and even the presence of archaeological features. As an example, the results of a GPR survey conducted at the Roman town Interamna Lirenas (Italy) show building foundations in the largest part of the survey area, although most are very shallow (with a maximum two-way travel time of ~13 ns, which corresponds to an estimated depth of 0.65 m). In the northern part of the town, the built-up area seems to end abruptly (Fig. 2A). Since also the magnetic contrasts produced by the foundations are weak (Fig. 2B), information from other geophysical techniques (such as electrical resistivity imaging; see, e.g., [27]) or invasive methods (e.g., augering; see Section 4.3.2) would

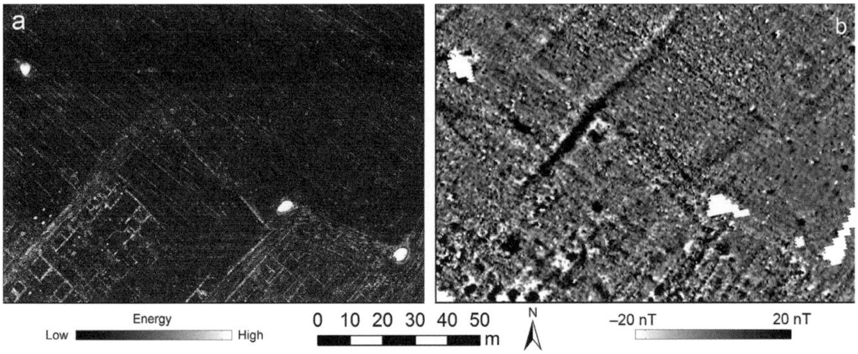

FIG. 2

(A) GPR energy slice at 11–12 ns from the Roman town Interamna Lirenas (Italy) showing how the building foundations disappear as one moves in northern direction, perhaps because of signal attenuation in combination with the deposition of colluvium and more deeply buried structures (the terrain slopes down in northern direction). (B) The fluxgate gradiometer data from the same area do not provide more information on the foundations, although the roads show more clearly (see Ref. [26]).

Fluxgate gradiometer data courtesy of the Faculty of Classics, University of Cambridge, and the British School at Rome.

be needed to answer the question whether the absence of foundations in parts of the GPR data is caused by the attenuation (i.e., loss of energy) of the signal or by a real absence of structures in that area.

1.2 ARCHAEOLOGICAL INTERPRETATION: A SUBJECTIVE PROCESS

Combining geophysical results with archaeological data allows to move from a geophysical interpretation, describing the strength and extent of the geophysical anomalies, to an archaeological one, allowing to determine the origin of the anomalies (archaeological, geological, or of recent origin), to answer questions on the nature and function of the archaeological structures (e.g., wall or ditch), and in some cases even to propose an approximate date. This archaeological interpretation, which is the ultimate aim why geophysical data are acquired, happens by mentally comparing observed archaeologically relevant features with known features [28]. This process is subjective, meaning that the results will vary with different interpreters. It relies on identifying what is familiar and assigning meaning based on experience and on the knowledge of parallels. Therefore, a clear distinction should be made between facts that can be deduced in a relatively secure way and inference that is more hypothetical or uncertain [15].

Deriving an archaeological interpretation from geophysical data is often linked to the morphology of the detected anomalies. When the anomalies are amorphous, integrating geophysical data into the archaeological process becomes cumbersome. Consider, for instance, Fig. 3A, showing in-phase magnetic susceptibility data collected with an EMI instrument in the Stonehenge landscape [29]. Any subsurface disturbance of the rendzina soil results in a magnetic anomaly due to the strong magnetic contrast between the diamagnetic bedrock (i.e., with a very weak, negative magnetic susceptibility) and the topsoil with a strong, positive magnetic susceptibility. The data are peppered with discrete magnetic anomalies of varying amplitude

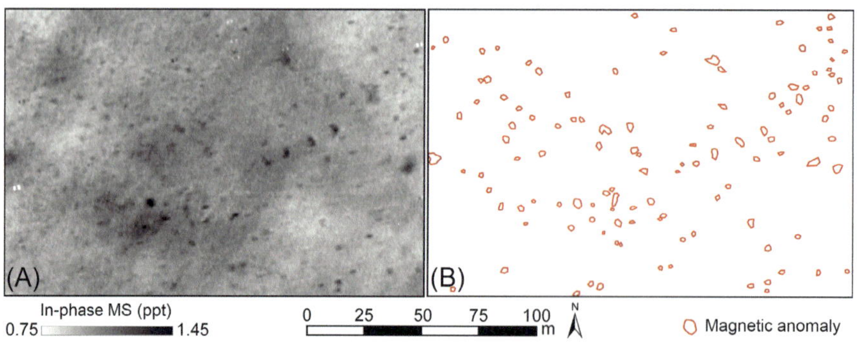

FIG. 3

(A) In-phase magnetic susceptibility EMI data from Stonehenge (the United Kingdom), collected with a 2 m horizontal coplanar coil configuration. (B) Location map of the magnetic anomalies present in the data shown in (A).

(Fig. 3B), which could be interpreted as dug features (pits). However, solely on the basis of these geophysical data, attributing archaeological relevance to the anomalies is impossible. The potential causes for their presence are manifold, ranging from recent soil intrusion to tree throws and local heterogeneities in the chalk bedrock and soil composition. As such, the inclusion of ancillary data is needed to transform these geophysical variations to archaeologically relevant information.

1.3 AIM OF THIS CHAPTER

The level of detail that can be achieved during interpretation depends on the intensity of the data collection. Gaffney and Gater [30] distinguished three levels of intensity: prospection (identifying areas of potential archaeological interest), assessment (delimiting and mapping sites, zones within sites, and individual features), and characterization (detailed investigation to reconstruct the shape of the buried features as exactly as possible and extract all possible other information). As methods of data acquisition became faster, also for large-scale projects, total coverage at a high sample density, allowing detailed characterization, has become within reach. Characterization, the most intensive level of investigation, allowing an interpretation as secure and reliable as possible, is the subject of this chapter.

Notwithstanding the problems outlined above, we believe the enormous potential offered by geophysical prospection in increasing our knowledge about the human past can be exploited more fully. In what follows, we propose some directions that can, in our view, contribute to this. Firstly, starting from the fact that geophysical results are different from excavation data, we would like to stimulate discussion about how geophysical survey can be most useful to gain a clearer understanding of archaeological problems and what are answerable questions to be asked from it. Secondly, we stress the need for a considered choice of instrumentation, a careful field practice and suitable data processing strategies, when characterizing archaeological sites or landscapes. Thirdly, we discuss how the combination of geophysical data with other geophysical and remote sensing data and with data from other invasive or noninvasive archaeological methods can result in an enhanced level of interpretation. Finally, we focus on the increasing role of computer-aided interpretation as a means of handling ever-growing data volumes, in a more objective way.

2 THE NEED TO ASK ANSWERABLE ARCHAEOLOGICAL QUESTIONS

The uncertainty that accompanies archaeological interpretation of geophysical survey data urge us to ask the question how these can be useful for enhancing our understanding of the human past. Despite the widespread adoption of geophysical techniques for the increasingly detailed investigation of large areas, there is little debate about what new archaeological evidence it can provide and what are answerable archaeological questions to ask of geophysical data. Nowadays, a critical attitude is adopted toward the methodology used to interpret data resulting from archaeological excavations. For example, in the study of Roman houses, one conclusion is that insights into the everyday life of the inhabitants cannot be based merely on architectural evidence. A careful study of artifacts and small finds plays a crucial part, although even at rich sites like Pompeii, this is complicated by the crude methods of early excavators. Artifact analysis has, for example, demonstrated the multifunctionality of the *atrium* (the central hall in the traditional aristocratic Roman town house), which served not only as a space for altars to household deities and ancestral images but also as a storage space and a workroom, given the frequently discovered loom weights [31]. This critical attitude seems justified also for prospection data, which is unable to answer all archaeological questions. Taking again the example of Roman domestic space, the resolution of geophysical methods does not allow to discern furniture or equipment, which could give clues to social behavior, to economic activities or to room functions. Neither can we obtain information on the decoration of the houses (mosaics and wall paintings), which is related to wealth and status. More answerable questions could, for instance, be situated on the interface between private space and urban planning [32]. On the basis of excavation data from Roman towns in Central Italy, there are indications for the presence of residential zones with large parcels close to the *forum*, whereas smaller plots tend to be located further away. Furthermore, large houses seem to be placed on wider streets and on more elevated ground [33]. The question arises to which extent these hypotheses, based on relatively scarce excavated examples, coincide with evidence from geophysical and other prospections.

3 INTERPRETABILITY RELIES ON CAREFUL DATA ACQUISITION AND PROCESSING

The interpretation is affected by the choice of the instrumentation and how it is configured. In magnetometer surveys, for instance, the strength of a magnetic anomaly depends not only on the sensitivity of the instrument [25] but also on the configuration, e.g., differential (the second sensor stays at a fixed position in the field) versus gradiometer configuration (the second sensor is at a fixed distance from the first one, usually above it; [34]). When using an impulse GPR system, the choice of the antenna frequency affects the vertical and horizontal resolution and therefore the interpretability of smaller features (a higher frequency allows a higher resolution; [35]).

Geophysical measurements often show a complex response, dependent on the configuration of the instrument. For example, the readings obtained with different transmitter-receiver configurations in frequency-domain EMI instruments (e.g., horizontal coplanar, vertical coplanar, or perpendicular) vary strongly, affecting the detection of the targeted archaeological features [36]. In a comparable manner, the interpretative potential of earth resistance measurements is dependent on the electrode array and separation. When a twin probe or square array is used and the electrode separation is large in comparison with the depth of the archaeological features, a single feature can even show as a double anomaly [37]. This was the case at Vieil-Evreux (France), where the shallow walls of the Gallo-Roman sanctuary are visible as double, parallel structures in the data acquired with the *automatic resistivity profiling* system (a quad-towed square array) using an electrode separation of 2 m [38].

Also the sample density is essential for a reliable archaeological interpretation. Fig. 4A shows GPR data collected at the Roman town Ammaia (Portugal) with both the transect spacing and the distance between the samples along a transect being equal to 0.05 m. The arrow indicates a horseshoe-shaped anomaly, probably to be interpreted as a bread oven (Fig. 4C). When a subset is extracted to simulate a transect spacing of 0.10 m and interpolated to 0.05 m, this shape can still be recognized. This is, however, no longer possible in the subset with a 0.25 m transect spacing (Fig. 4B), where the anomaly shows as a circle and may be interpreted wrongly, e.g., as a well. Therefore, in this example, a transect spacing of 0.25 m is too large for a detailed interpretation of the weaker features.

It is not always easy to determine the right sample density before the survey is undertaken, as the size of the relevant archaeological features is often unknown. To circumvent this problem, one can attempt to get an idea of the maximum spatial frequency of the signal (the signal is the continuous change of the physical property that is measured). The maximum spatial frequency can be described as the maximum

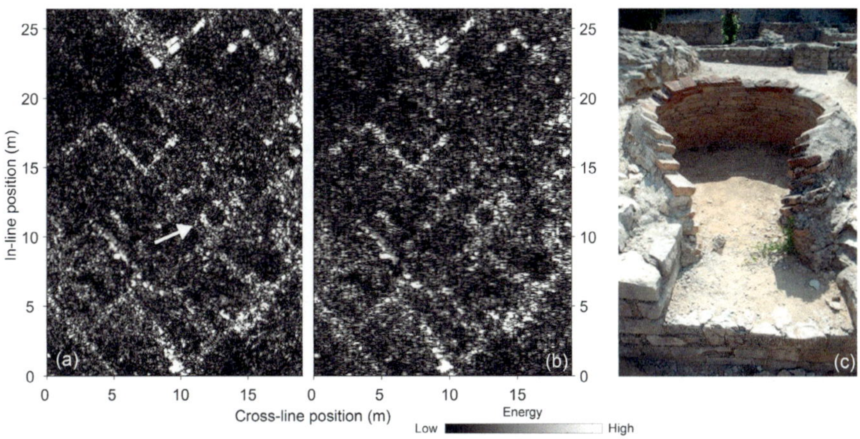

FIG. 4

(A) GPR energy slice from the Roman town Ammaia (Portugal) surveyed in vertical direction with a transect spacing of 0.05 m. (B) Data set extracted from (A) to simulate a transect spacing of 0.25 m, with subsequent interpolation to 0.05 m. The arrow in (A) shows an anomaly, probably to be interpreted as a bread oven given its horseshoe shape. It is recognizable only in the data with transect spacing of 0.05 m [39]. (C) Excavated example of a bread oven from Vaison-la-Romaine (France).

number of wave cycles per unit distance in the spectrum of the measured signal (i.e., after Fourier transform). This maximum can be found, for example, through the measurement of a test traverse that is converted to the spatial frequency domain [39]. The necessary sample density can then approximately be determined by means of the sampling theorem, which states that the sampling rate should be at least twice the maximum spatial frequency [40].

Most geophysical data need at least minimal processing so that they can be interpreted in a meaningful way. Drift in EMI data, for instance, which is caused by instrument characteristics and environmental circumstances, may make an interpretation impossible without correction, best performed using a calibration line driven across the entire survey area, intersecting all survey lines. Besides the removal of baseline drift [41], correcting the remaining errors (microdrift) can further enhance the quality of the data [12], as is shown in Fig. 5. If GPR data are collected over an area with variations in the topography, they should be adjusted for elevation. Only then can the true shape of a horizontal floor or living surface become clear [42]. Whereas conventional elevation static shifts (vertically shifting the GPR traces to compensate for differences in topography) are mostly sufficient, the use of more advanced processing methods is necessary to reconstruct the size and position of the reflectors, when the acquisition surface is very irregular (with surface gradients

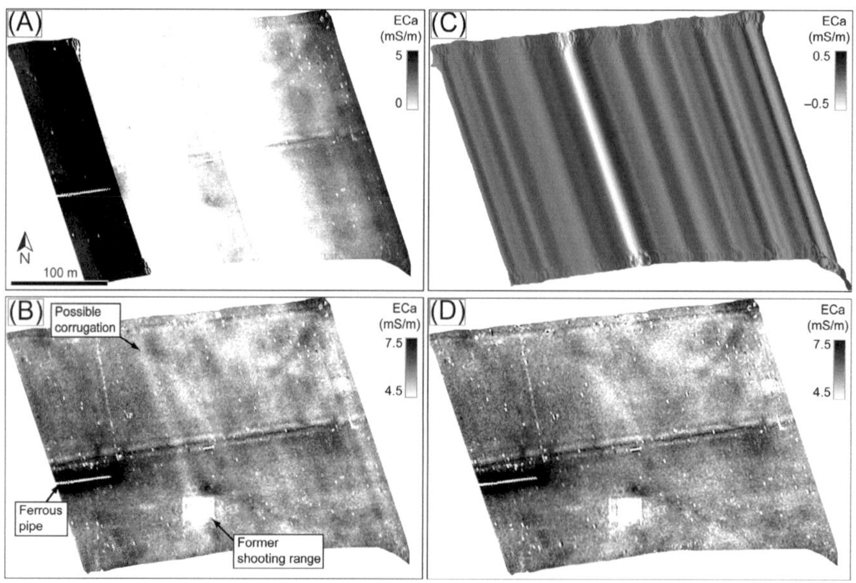

FIG. 5

Different steps in drift correction applied to ECa data recorded at Stonehenge with a Dualem-21S EMI instrument (2 m horizontal coplanar coil configuration). (A) Raw data. (B) Baseline drift-corrected data. The microdrift (C) was then isolated and removed to produce the microleveled data (D) [12].

>10%; [43,44]). It is also essential to be aware of possible artifacts generated by data processing. For example, the archaeologically relevant information can be enhanced by high-pass filtering (i.e., removing the slowly varying background; [45]), especially in earth resistance data where archaeological features are often obscured by this background response [27,30]. However, this can introduce pronounced artifacts that should be borne in mind to avoid misinterpretations [37].

4 ENHANCING THE INTERPRETATIVE POTENTIAL THROUGH DATA COMBINATION

An important means for tackling the issues described in the introduction and enhancing the interpretability of geophysical survey results is to compare and combine them with other sources of information. Two categories can be distinguished. The first one is information that is related to the area where the geophysical prospection took place, and can be mapped. This includes the calculation of attributes from the measured data (Section 4.1), the combination with other geophysical data, with air- or spaceborne remote sensing imagery (Section 4.2), or with the results of field walking (Section 4.3.1). Furthermore, it comprises the analysis of old maps, the survey of extant archaeological remains, geomorphological investigations, and invasive methods such as coring (Section 4.3.2), test pits, or excavations (Section 4.3.3). The second category consists of historical and documentary sources that cannot be mapped, and of all possible prospection or excavation data from other sites or regions, together with their interpretation (Section 4.4).

4.1 CALCULATION OF ATTRIBUTES

An attribute can be described as any measure that helps to reveal or quantify features that are not obvious in the original geophysical data [46]. For example, in GPR data processing, the energy and the instantaneous amplitude attributes are frequently used to build time slices. In the last decade, multitrace GPR attributes such as coherency and similarity, and combinations of attributes (e.g., the similarity of the energy; [47]) have also been employed. In Fig. 6, we illustrate the extraction of attributes by means of a GPR data set collected at Bourne Park (Kent, the United Kingdom), within the Canterbury Hinterland Project [48]. From the migrated data representing an estimated depth of 0.30–0.80 m, first, the energy attribute was calculated (the averaged sum of the squared sample values within this depth window; Fig. 6A and B). Given the symmetrical nature of the wall foundations, we also examined whether phase symmetry could enhance the weak anomalies in our data set. Phase symmetry obtains local frequency information by means of wavelet transform, employing Log-Gabor functions. By convolving the signal with sine and cosine wavelets based on these functions, areas of symmetry can be localized. For each pixel in a 2-D image, this procedure is repeated for different wavelet scales and different orientations, and the average of the results is taken [49]. This method, which is independent on the

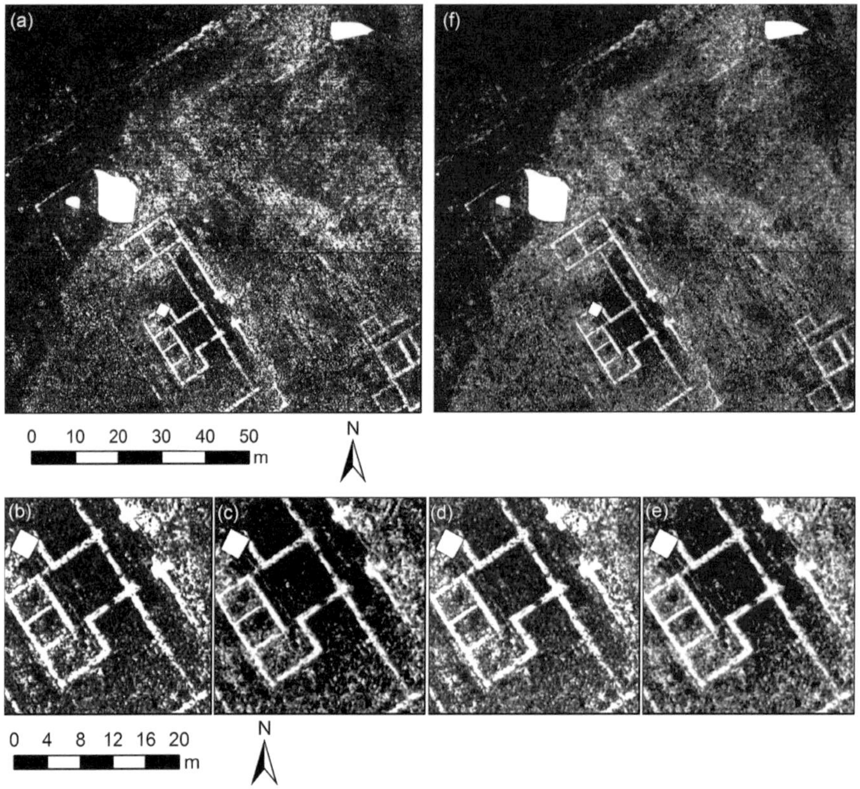

FIG. 6

Attributes extracted from GPR data collected at Bourne Park (the United Kingdom).
(A) Energy slice at a depth of 0.30–0.80 m. (B) Detail from (A). (C) Phase symmetry image, showing for each pixel the highest value of all 0.01 m-thick phase symmetry slices between 0.30 and 0.80 m depth. (D) Scale parameter of the Weibull distribution, fitted to the amplitude spectrum of the GPR traces. (E and F) First principal component computed from the three attributes shown in (A and B), (C), and (D).

local contrast, has been applied to GPR utility detection [50]. We calculated a phase symmetry image for each 0.01 m-thick amplitude depth slice in our 0.30–0.80 m depth window. A single slice was created by selecting for each pixel the highest phase symmetry value of all the resulting images (Fig. 6C). The third attribute calculated from the GPR data is based on distribution fitting, in which a probability density function (PDF) is fitted to a sequence of numbers, resulting in parameters that characterize the PDF. Distribution fitting has, for example, been used in airborne imaging spectroscopy, where the parameters of PDFs fitted to the frequency distribution of hyperspectral pixels produce grayscale or false-color images, depending on whether these images are based on one parameter or on two or three parameters [51].

Similarly, we fitted PDFs to the amplitude spectrum of each GPR trace. We used the Weibull distribution that can conveniently model the right skewed GPR amplitude spectrum (Fig. 6D). In Fig. 6E and F, the three attributes were combined into a single image, as is described in Section 4.2.2.

4.2 COMBINATION OF DIFFERENT GEOPHYSICAL AND REMOTE SENSING DATA SETS

4.2.1 Combined analysis of different data sets

As geophysical methods in most cases inform only on one physical soil property, combining different types of geophysical survey data increases the interpretative potential of the entire geophysical data set. The example shown in Fig. 7 compares three different geomagnetic data sets of the same group of anomalies detected in a low magnetic background (sandy soil) in Flanders (Belgium). Two in-phase magnetic susceptibility EMI data sets are shown (Fig. 7A and B), alongside a fluxgate gradiometer data set (Fig. 7C). Whereas all data sets reveal highly magnetic features, the EMI data only inform on the induced magnetization and thus allow determining the presence of material with a high magnetic susceptibility. The magnetometer data, on the other hand, display strong magnetic dipoles with different orientations indicating strong magnetic remanence of the detected features. From the latter, it can be deduced that the detected structural remains are likely brick structures, an interpretation that cannot be inferred directly from the in-phase EMI data. Using multiple EMI measurements, obtained with different coil configurations (the 1 and 2m

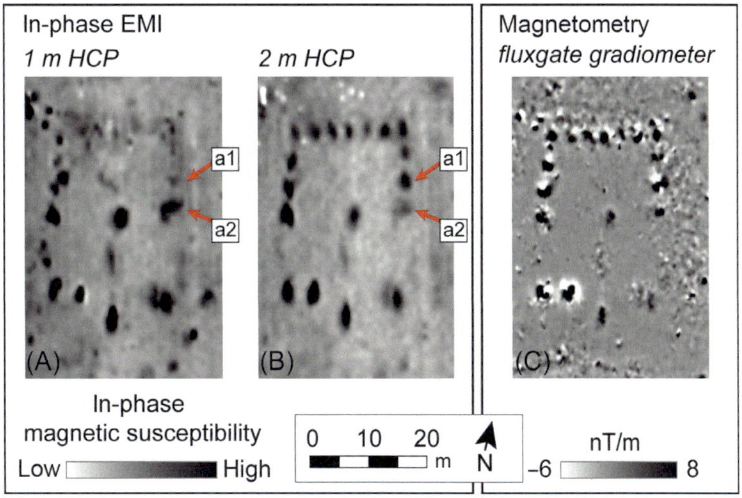

FIG. 7

Comparison between two in-phase magnetic susceptibility EMI data layers (A and B) and a fluxgate gradiometer data set (C) showing a group of brick staddle post foundations.

Magnetometer data courtesy of D. Simpson (GeoSurveys Ltd, Portugal).

horizontal coplanar (HCP) coil pairs of a Dualem-21S instrument), the in-phase magnetic susceptibility of different soil volumes can be compared. While the 2 m HCP data show at least 18 high susceptibility anomalies (comparable with the strong responses in the magnetometer data), the 1 m HCP data display varying responses at each location. The feature at (a1) results in a low-amplitude anomaly in the 1 m HCP data, whereas a strong magnetic response is obtained at (a1) with the 2 m coil pair. The inverse occurs at (a2), where a high-amplitude and larger spatial extent is recorded with the 1 m coil pair. These different responses relate to different depth extents of the features at (a1) and (a2), information that is absent in the fluxgate gradiometer data. Excavations of both features showed that this different magnetic stratigraphy is related to the preservation level of the brick foundations; while (a1) was largely intact, the foundation at (a2) was affected heavily by plowing, which had removed bricks from their original configuration, mixing them with the topsoil [52]. As such, the variation in the combined magnetic responses in the 1 and 2 m HCP data can be related to the preservation of the detected features, where those with a strong response in the 1 m HCP data are more likely to have been disturbed by recent agricultural practice.

4.2.2 Data integration

One step further than the side-by-side analysis of different geophysical data sets is the combination of the data in order to produce a single output image. Often, here, the terms "integration" and "fusion" are used interchangeably [53,54]. In this chapter, "fusion" denotes the case where many types of data are included in the visualization, such as different types of geophysical data, manually interpreted features, and objects detected in more automated ways (examples in 3-D are [55–57]). For a combination merely including images from multiple geophysical surveys (as in the examples below), we have used the term "integration." Data integration and fusion techniques have been used for decades in medical imaging, military and security applications, navigation, and robotics. In remote sensing, since the 1980s, images with varying spatial and spectral resolution have been combined. For example, a high-resolution optical panchromatic image can be combined with a lower-resolution multispectral image to add spectral information from different wavelengths (color), a technique known as pansharpening. Data integration has not been used widely in archaeology (examples in archaeological remote sensing are [58] and [59]). Also in archaeological geophysics, experiments in the integration of different geophysical techniques have been published. Below, we give a short overview and provide some examples based on the data from Bourne Park, to examine the usefulness of data integration for archaeological interpretation. An in-depth discussion of geophysical data integration is given elsewhere in this volume by Kvamme et al.

Graphical integration methods (contour lines over a color or grayscale image, pseudo-3-D surfaces, red-green-blue (RGB) compositing, and alpha compositing) can be helpful in the interpretation of multimethod geophysical surveys or different attributes of the same data set [60]. Their advantage is that the original data sets can be distinguished after merging them [61]. In RGB composites, three data sets are assigned to the red, green, and blue color channels. For example, in Fig. 8A and

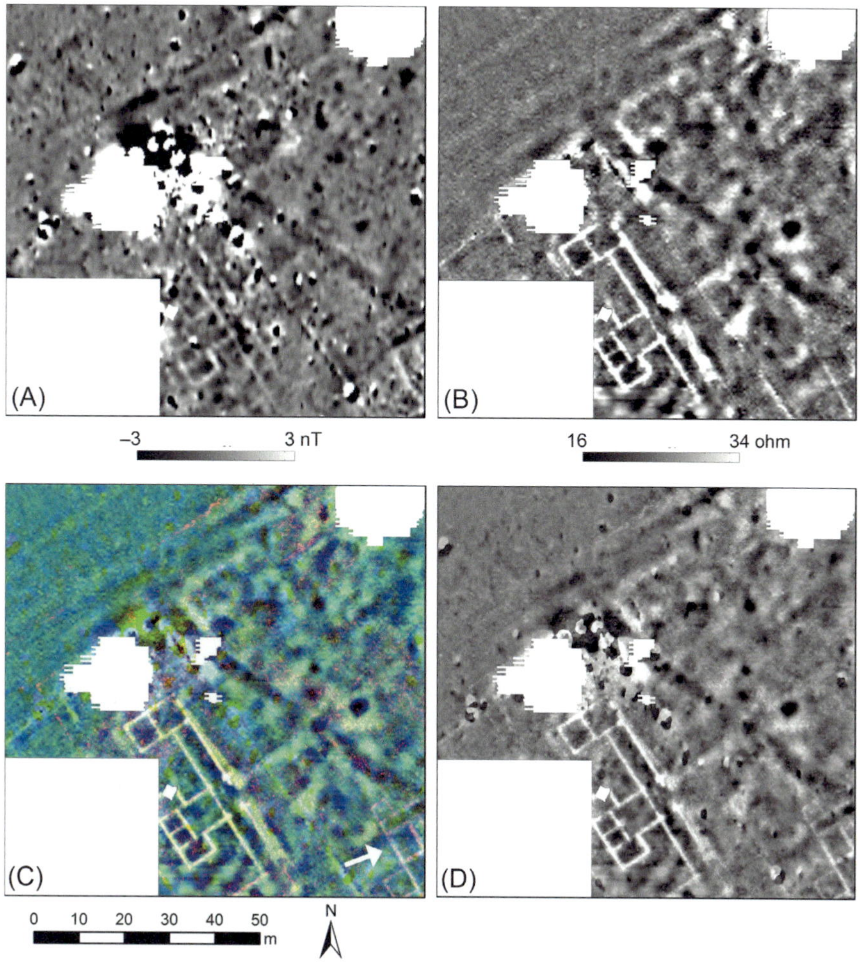

FIG. 8

(A) Extract from a fluxgate gradiometer survey conducted at Bourne Park (the United Kingdom). The polarity of the data was inverted. (B) Earth resistance survey covering the same area [48]. (C) RGB composite using the GPR energy slice shown in Fig. 6A *(red channel)*, the earth resistance data *(green channel)*, and the magnetometer data *(blue channel)*. (D) First principal component computed from the same three data sets. (For interpretation of the references to color in this figure legend, the reader is referred to the web version of this article.)

Fluxgate gradiometer and earth resistance data courtesy of the Canterbury Hinterland Project.

B, fluxgate gradiometer and earth resistance data from Bourne Park are displayed. These two data sets and the GPR energy slice with depth of 0.30–0.80 m (Fig. 6A) were resampled so that a common sample density of 0.1 m × 0.1 m was obtained, and standardized by subtracting the mean and dividing by the standard deviation [54]. The magnetometer data were inverted so that the building foundations are represented by high values, as is the case for the earth resistance and GPR data. In the RGB composite, the yellow color of some walls of the western building indicates that these were detected by the earth resistance survey (the green channel) and the GPR survey (the red channel), whereas the foundations of the eastern building were mainly present in the GPR data (see the arrow in Fig. 8C). The weakness of graphical data integration is that they only serve a clearer visualization. No new data are created that can be used for later interpretation [54].

Binary operations (Boolean union, Boolean intersection, and sum of binary data sets) are powerful but require the determination of global thresholds. This is difficult in data sets that are characterized by strong variations in intensity, where a lot of useful information can get lost before data combination occurs. This problem is circumvented when starting from the raw, continuous data. On continuous magnetometer and earth resistance measurements, the four basic arithmetic operations were applied by Neubauer and Eder-Hinterleitner [28], while Piro et al. [62] examined the mean and the product of three data sets, and more complex numerical combinations were carried out by von der Osten [63]. Data integration can also use statistical relationships between the input images, such as the data maximum, principal component analysis (PCA), logistic regression, or *k*-means clustering [54,64]. PCA is a well-established technique in remote sensing, and reduces the dimensionality of multivariate data by calculating linear combinations of the input variables (the different geophysical or remotely sensed data sets). This results in a new set of variables, the principal components (PCs), which are uncorrelated and represent decreasing amounts of variance [65]. The first few PCs often contain the bulk of the archaeological information, showing a high degree of correlation between the original data sets. PCA has been performed successfully between successive GPR amplitude time slices in order to enhance weak anomalies [66]. In Fig. 6E and F, three GPR attribute plots from Bourne Park (see Section 4.1) were integrated into a single image using PCA. Fig. 8D shows the first PC calculated from the three data sets in Figs. 6A and 8A and B.

Data integration methods developed in other fields and so far little applied in archaeology, can also prove effective. In remote sensing, a number of algorithms use a multiresolution approach, such as wavelet transform (WT). The WT decomposes a signal into a set of scaled and shifted versions of an original wavelet (the mother wavelet; [67]). This decomposition in the 2-D case results in an approximation (components with low spatial frequency) and horizontal, diagonal, and vertical details (components with high spatial frequency). It can be iterated by taking the approximation as the input of another filtering: a multiple-level wavelet transform [68]. In remote sensing, selected detail panels of multispectral images are often substituted by or added to the corresponding panels of a panchromatic image [69]. A less complex application of the WT to the data from Bourne Park consisted in performing a level-5 wavelet decomposition of the three images in Figs. 6A and 8A and B using the db2 wavelet, taking the mean of the resulting approximations and details and restoring the integrated image from these average decompositions (Fig. 9A).

Another possibility is to integrate the data in the gradient domain. The magnitude of the gradient denotes the difference between neighboring values in the direction of

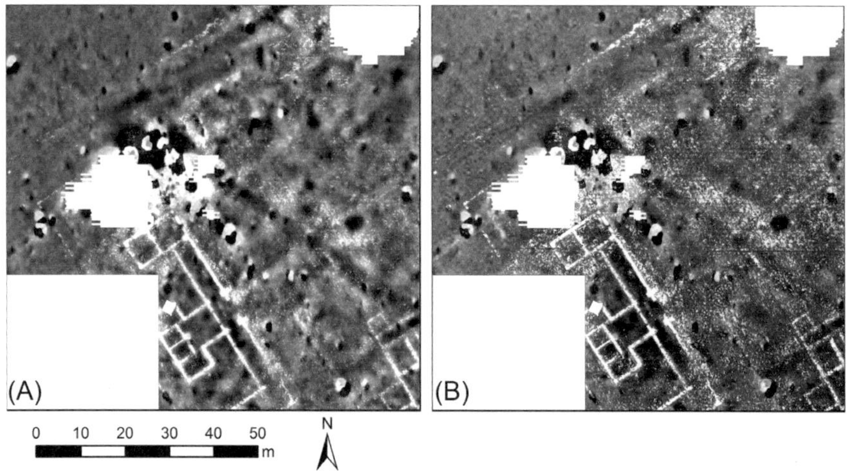

FIG. 9

(A) Data integration based on wavelet transform, computed from fluxgate gradiometer and earth resistance data (Fig. 8A and B) and from a GPR energy slice (Fig. 6A), all from Bourne Park. (B) Gradient domain integration computed from the same three data sets.

the greatest change and can be calculated from the input images, for example, by convolving them with a Sobel filter. It emphasizes the small details of the features in the geophysical data. Performing an integration based on the gradient has the advantage that for each pixel in the output image, the input data set with the largest gradient can be selected, so that the result contains more detail than any of the input data sets. However, converting the integrated gradient back to the original image domain is complex. From a mathematical point of view, the gradient can be seen as the derivative, so it needs to be integrated in order to obtain the output image. However, artifacts may be created when the image is reconstructed since the integrated gradient is a combination of the gradients of different input images. Tackling this problem requires solving Poisson's equation [70]. To do so, Paul et al. [71] use an iterative reconstruction technique based on a modified Haar wavelet transform. Their algorithm, originally developed for the integration of multiexposure and multifocus images, can be used for RGB images or grayscale images such as geophysical data sets. A gradient domain integration of the fluxgate gradiometer, earth resistance, and GPR data from Bourne Park is shown in Fig. 9B.

4.3 EVALUATION BY MEANS OF FIELD WALKING AND INVASIVE METHODS

4.3.1 Field walking

A series of increasingly invasive techniques can be applied to verify the results of a geophysical survey and compare it with independent evidence. The collection of surface finds through field walking can yield complementary information, especially about the temporal development of the study area [72]. Trachet et al. [73] studied the correlation between surface artifacts, collected while registering the exact provenance of each artifact using an RTK GNSS receiver, and the results of an EMI survey at two abandoned medieval harbor sites near Bruges (Belgium). Whereas no strong correlation exists at the level of the individual in-phase (IP) magnetic susceptibility features, much stronger correlations appear when the EMI data are classified in deciles and the amount of finds is counted for every decile: artifact clusters coincide with the delineation of the historic allotments recorded in the IP response, suggesting limited lateral displacement of plow soil artifacts. At one of the sites, a clear correlation also exists between the thirteenth-century artifacts (corresponding to the original nucleus of the settlement) and the soil variations reflected in the ECa values: less conductive (sandy and dry) soils show high artifact densities, while the more conductive (wet) zones related to a former tidal inlet remain empty of occupation. The fourteenth- and fifteenth-century artifacts indicate an expansion to the more peripheral grounds.

4.3.2 Coring, borehole logging and direct push soil sensing

A variety of manual and mechanical coring and augering techniques are available for the calibration, validation, or archaeological interpretation of geophysical survey results. Coring results in the retrieval of a continuous section of sediment, whereas in the case of auger drilling, the soil sample is disturbed [74–76]. These methods are primarily used in deeply stratified landscapes and sites [77]. Rapid macroscopic description, field data registration, and interpretation of the retrieved archaeological or environmental soil materials are possible on site, but sample collection in plastic sample liner containers permits more detailed description or data registration in the laboratory for further geophysical, sedimentological, or paleoenvironmental characterization (e.g., [78]). However, the physical properties of the soil can change as soon as the sample is retrieved. To avoid this, geophysical sensors can be lowered down a predrilled borehole, or alternatively, the subsoil is penetrated by direct-push sensors (through continuous vertical force or hammering) to measure in situ soil properties in relatively soft soils. For example, Dalan et al. [79] used a combined direct-push color system and electrical conductivity dipole. Aside from being less destructive, direct-push technology is more efficient in collecting interpretation data because no time is spent bringing a sample to the surface to measure its (geophysical) properties. As such, it may offer complementary information when surface geophysical survey only is not able to answer the research questions adequately (e.g., [80]).

The geophysical properties, registered through core, borehole, or in situ sensing, and the derived transition depths of soil or archaeological units are suitable for geophysical subsurface modeling. Alternative or additional proxy measurements (e.g., color and mechanical resistance) assist in determining calibration depths [81] or are used qualitatively for further characterization or interpretation of the geophysical survey data. As such, Verhegge et al. [82] distinguished whether the ECa variations measured at Doelpolder Noord (Belgium) through EMI survey and electrical resistivity imaging (ERI) were due to variations in the prehistoric landscape (peat depth

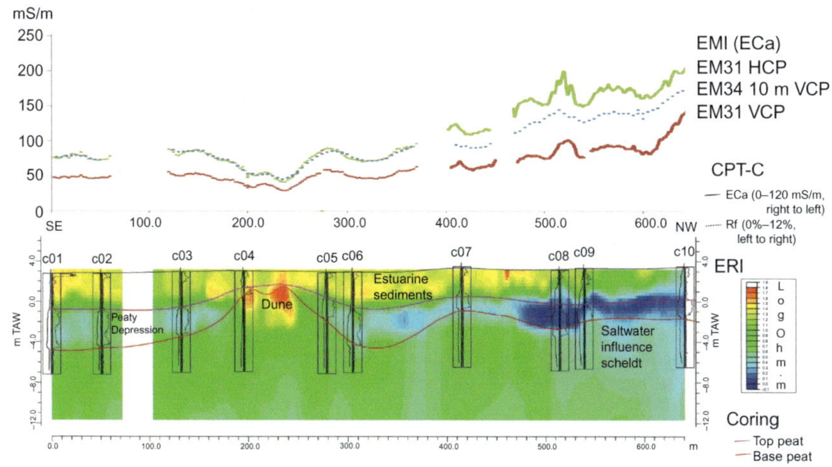

FIG. 10

(A) ECa responses resulting from surface EMI surveys along a section through Doelpolder Noord (Belgium), using different instruments in the horizontal (HCP) and vertical coplanar (VCP) coil configurations. (B) Results of an ERI profile along the same section; superimposed are the results of cone-penetration tests with an electrical conductivity sensor (CPT-C) and a geological subsurface model based on soil coring [82].

and thickness) only or also to saltwater variability. Peat layers were identified through cone-penetration testing [83], using the friction ratio (ratio of sleeve friction and tip resistance; Rf). When saltwater was present within the soil profile outside the peat layer, this was observed as an increased (in situ recorded) ECa, which corresponded with the surface EMI and ERI results (Fig. 10).

Coring, borehole logging, and direct-push soil sensing are most suitable for detecting layers with a continuous horizontal extent. The limited horizontal impact area of these methods renders them less appropriate when the archaeological or soil feature represented by the geophysical anomaly is horizontally discrete or discontinuous. Detecting the latter requires such a small spacing between measurement locations that more invasive techniques such as test pitting, trial trenching, or excavation prevail.

4.3.3 Excavation

Small, carefully placed stratigraphic excavation can be an efficient method to verify the presence of archaeological structures and to provide details about their nature (see, e.g., [84]). In most cases, it is needed to provide a refined estimate of the chronology. Nevertheless, not too much weight should be placed on the evidence of small excavation trenches when investigating large building structures or complex stratigraphy. Often, the excavation of a much larger, contiguous area is needed to elucidate problems of interpretation, and small excavations are therefore not always superior to other sources of information, for example, from field walking or typological analysis of geophysical survey results [85]. In some cases, old excavations lacking appropriate georeferencing can be located by fitting them onto the geophysical prospection map, and in their turn these excavations can give indications on how the geophysical data can be archaeologically interpreted. At Mariana (Corsica), GPR and fluxgate gradiometer data (Fig. 11A and B) and the mapping of a few wall foundations visible aboveground allowed locating a trench dug in the 1930s (Fig. 11C, 4–6). In their report, the excavators describe several rooms, visible in the gradiometer data, which they assume to be part of a building complex with a public character. Several other structures visible in the geophysical data probably belonged to the

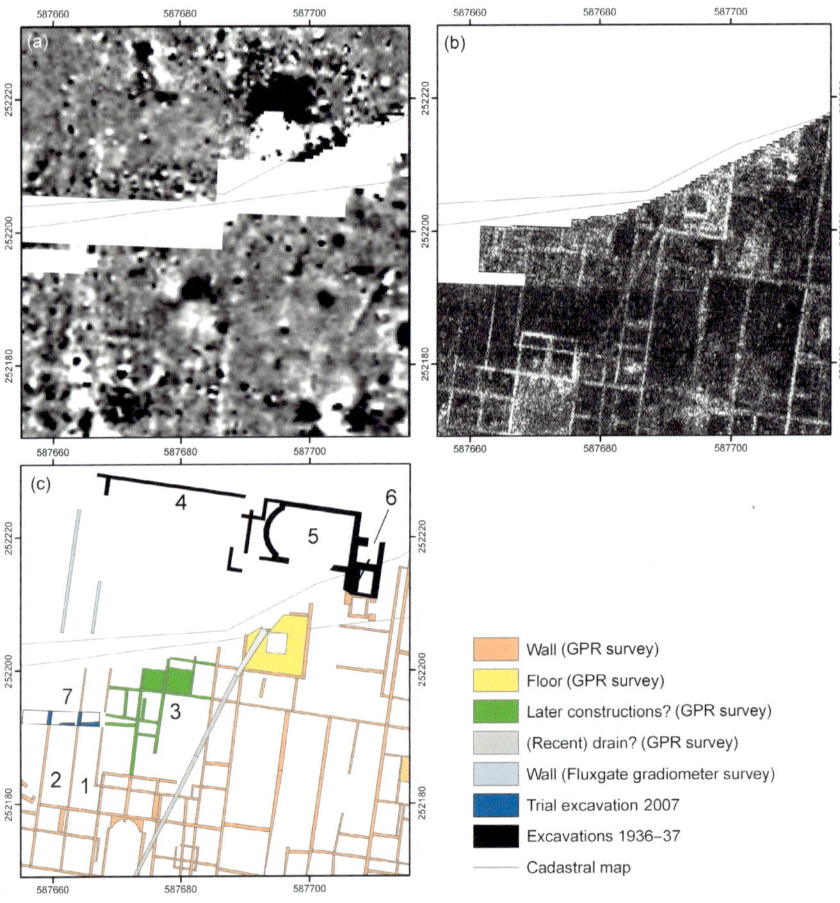

FIG. 11

(A) Fluxgate gradiometer data gathered over a building with courtyard at Mariana (Corsica), from −15nT *(white)* to +15nT *(black)*. (B) GPR energy slice at 80–85cm depth. (C) Interpretation: 1, portico; 2, corridor; 3, internal divisions within courtyard; 4–6, structures uncovered during excavations in 1936–37; and 7, 2007 excavation trench [86].

same complex. On the basis of the building techniques observed in a small excavation conducted in 2007 (Fig. 11C, 7), a chronology for the construction of this complex was proposed in the first-second century AD, but the absence of datable finds from the foundation trenches prevents a more precise chronology [86].

4.4 LEGACY DATA AND GREY LITERATURE

Feedback from excavations in the area where the geophysical anomalies were detected is available only in a limited number of cases. However, parallels from other locations offered by legacy excavation data and gray literature can also aid interpretation. For example, for Roman domestic architecture, the comparison with excavated sites often facilitates the interpretation. Such was the case at Ammaia (Portugal), where a large house with *peristylium* (i.e., a courtyard surrounded by columns) in the GPR data (Fig. 12A and B) resembles rich town houses elsewhere in Roman Hispania (Fig. 12C). The function of some rooms can be determined (*triclinium* or dining room and reception rooms), and there is evidence of the enlargement of some rooms at the expense of others (Fig.12D; [87]). For more modest town dwellings, uncertainty is usually higher, and even the delimitation of different houses in an *insula* can be problematic: they may have been combined into a single unit at a later date, or have functioned as separate houses even when linked by interconnecting doors [88].

Legacy data are useful especially if they link excavations with geophysical survey. An example is the report that resulted from the excavation of prehistoric pits, Bronze Age ditches, and an Iron Age enclosure at the Lismullin 1 site, Meath County, Ireland, on behalf of the National Roads Authority (now Transport Infrastructure Ireland). In the report, detailed comparisons are made between the anomalies detected by previous cesium magnetometer and magnetic susceptibility surveys, and excavation results, stating whether the geophysical anomalies were identified during the excavation and whether an archaeological meaning could be attributed to them [89]. A large-scale comparison between magnetometer data and excavations, and an assessment of the success of geophysical survey, encompassing tens of sites in Ireland, were conducted by Bonsall et al. [90]. In the unlocking of legacy data and gray literature, an important role is played by online databases such as the gray literature archive of the Archaeology Data Service (the United Kingdom) or the recently launched GIS-based portal Chronocarto (Ecole normale supérieure, France).

4.5 DISCUSSION

Combination of different geophysical and nongeophysical data can provide a more complete view and increase confidence in the interpretation. When inversion is performed, the addition of prior knowledge can constrain our model of the subsoil (Section 1.1.2). For example, the reconstruction of the 3-D magnetic susceptibility distribution in a Neolithic ditch by Neubauer and Eder-Hinterleitner [19] was constrained by using magnetic susceptibility measurements from other excavated sites

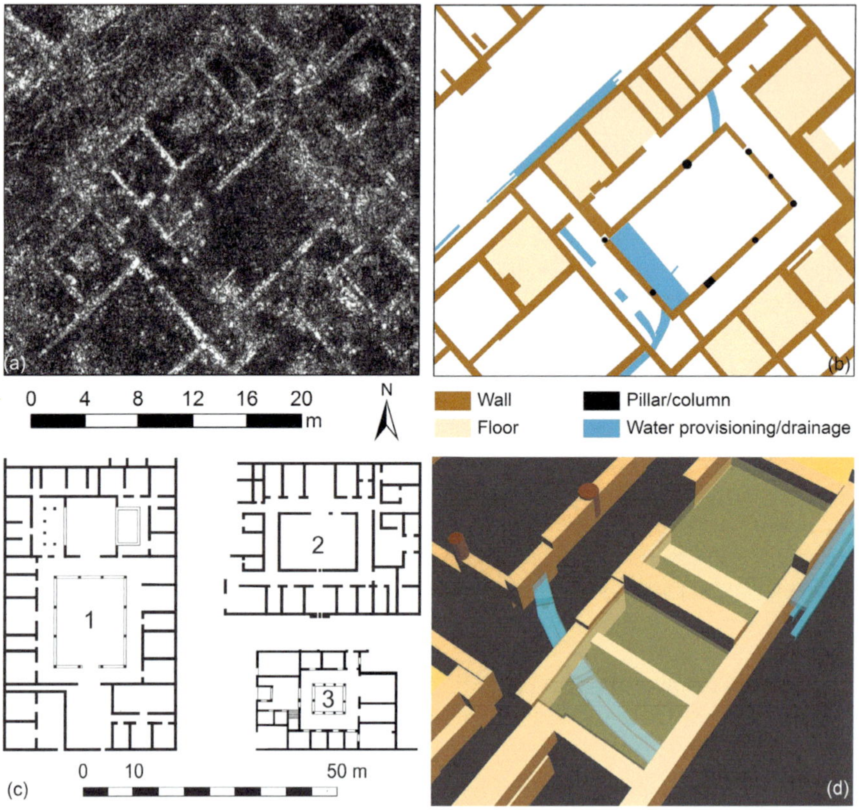

FIG. 12

(A) GPR energy slice at a depth of 0.55–0.60 m from Ammaia (Portugal), showing a courtyard house. (B) Interpretation of the GPR data (all depth slices). (C) Excavated Roman houses on the Iberian peninsula, used for comparison with the geophysical survey results: (1) Casa de los Pajaros (Italica), (2) Casa 1 (Iuliobriga), and (3) Baelo Claudia. (D) Three-dimensional interpretation of the northern corner of the house, with the floors made transparent, so that the older walls in the original arrangement of the rooms become visible. Vertical exaggeration = 2.

C: Modified after K.E. Meyer, Axial peristyle houses in the western empire, J. Roman Archaeol. 12 (1999) 101–121; F. Teichner, Romanisierung und Keltische Resistenz: Die "kleinen" Städte im Nordwesten Hispaniens, in: Die Selbstdarstellung der römischen Gesellschaft in den Provinzen im Spiegel der Steindenkmäler. Akten des IX Internationalen Kolloquiums über provinzialrömisches Kunstschaffen, Innsbruck, 2005. Innsbruck University Press, 2007, pp. 202–216; J.J. Cepeda Ocampo, J.M. Iglesias Gil, A. Ruiz Guiérrez, Territorio rural y espacio urbano en Iuliobriga (Cantabria), in: J. Mangas Manjarrés, M.A. Novillo (Eds.), El territorio de las ciudades romanas, Ediciones Sísifo, Madrid, 2008, pp. 309–331.

and information about the shape of the investigated ditch from an excavated section. For the inversion of magnetometer data from the Raganello Archaeological Project (Calabria, Italy), Schmidt et al. [20] used two approaches. In the first one, only the magnetic susceptibility, the size, and the position of polygons at different depths were adjusted (keeping the depth and thickness of the layers invariable), resulting in reasonably consistent outcomes that approximate the buried features in a better way than the initial interpretation. The second approach consisted in designing models based on known archaeological features and included inverting for the depth and thickness of the different layers. However, this resulted in very different models all fitting the measured data equally well and hence in considerable uncertainty about the features that caused the measured magnetic anomalies.

Increasing the number of prospection techniques reduces the size of the area that can be surveyed (within a given budget and available time), and this fact should be balanced against the usefulness of multimethod surveys. The use of multiple geophysical surveys is legitimate when otherwise unreliable responses can be validated by information from different sources, when there is little *a priori* knowledge about the buried features investigated and it is therefore difficult to predict which technique will be most successful, or when the features are expected to cause anomalies in different physical soil properties [91]. Each individual data set should be of sufficient quality, and its use should be justified from an archaeological point of view [92]. Furthermore, the amount of additional information decreases as the number of detection techniques becomes larger [54]. On the other hand, the additional cost of using multimethod geophysical survey has become less prohibitive given the wider availability of high-resolution geophysical technology capable of covering large areas in a short time span and more advanced data processing. Interpretation has not experienced a similar progress as it is still performed largely manually. For this reason, the cost-to-benefit ratio of acquiring and combining different data sets has gone down, and the extra effort seems justified when it holds the prospect of facilitating the interpretation.

Data integration, in particular, can be an efficient tool to analyze the growing number of different geophysical and remote sensing techniques in a meaningful way. When geophysical data sets reveal different subsurface features, integrating them creates a more complete picture. In other cases, anomalies detected through different techniques reveal the same buried features, indicating a high correlation between the data sets (although this correlation often becomes clear only when analyzed locally; [93]). This results in an improved understanding of the dimensions and sometimes the physical properties of the buried remains and allows a more detailed characterization [94–96].

Nevertheless, there is only a value in performing data integration when the composite image contains information of a greater quality than can be extracted from each of the individual input data sets [97]. In remote sensing, a variety of measures for the quantitative assessment of the gain in quality exist, but most require a reference image [69], which is not available in archaeological geophysics. Although no generally accepted quantitative index is available that can replace visual assessment, quality control seems important for the future development of data integration in archaeology. Further research is also needed into the integration of 3-D geophysical data such as GPR or ERI, which are difficult to handle using current 2-D data integration techniques.

To further the archaeological interpretation of geophysical data, invasive methods can be highly useful. Targeted borehole survey, test pitting, or stratigraphic excavations are methods that allow adding detailed archaeological and geological information needed to transform geophysical measurements to archaeological information. However, an ambivalence needs to be taken into account when bridging this gap. While direct, robust archaeological information can be obtained through invasive research, this does not guarantee a satisfactory interpretation of the geophysical data. Understanding the presence or absence of a geophysical contrast relating to archaeological features can only be addressed through in situ evaluation of the targeted physical property (e.g., by recording the magnetic properties of sections through the features of interest or by geophysical borehole logging; [98]) or by documenting other relevant subsurface variables (e.g., changes in soil texture when interpreting variations in apparent resistivity).

5 COMPUTER-AIDED OBJECT DETECTION

5.1 HUMAN INTERPRETATION AND COMPUTER VISION

Nowadays, the creation of interpretative diagrams still largely consists of the visual analysis of geophysical maps, followed by the laborious, manual delineating of likely archaeological anomalies by means of lines and polygons, mostly using a geographical information system (GIS) [54]. Growing data volumes raise the question if these traditional interpretation methods will continue to be sufficient and whether computer-aided methods of archaeological object detection should not become more important. Applications of automated and semiautomated object recognition exist in many domains such as medical imaging, number plate and face recognition, astronomy, military reconnaissance, optical character recognition, navigation, and robotics. In archaeology, aspects of automated object detection were introduced in the 1990s by satellite remote sensing specialists, and since then, in aerial and remote sensing archaeology, a number of papers have been published, and an animated debate has grown (see, e.g., [99]). In archaeological geophysics, object detection methods were applied for instance to the interpretation of magnetic data, by Sheen and Aspinall [100]. They used theoretically calculated anomalies to train an artificial neural network that recognizes anomalies in real survey data. Bescoby et al. [101] used modeled magnetic responses as training data for five identically trained neural networks, to estimate the depth of magnetic features in field data from Butrint (Albania) and create a 3-D subsurface reconstruction of the surviving wall foundations. More recently, aspects have been incorporated in papers mentioned in the following sections.

Overall, computer-aided object detection (CAOD) remains rarely used in archaeology and is often viewed with suspicion. The skepticism comes down to the idea that computers are not able to recognize the wide variety of patterns indicating past human activity and separate them from geological anomalies or disturbances of modern origin, and that computer vision cannot therefore be a substitute for manual interpretation [102,103]. In part, this suspicion is based on the fact that expertise on CAOD has come from outside archaeology and from fields where often the detection of one class of features in one type of environment is required, which is different from extracting a variety of archaeological feature types across different types of soils or landscapes [104]. Where the archaeological features belong to one class with a simple shape (e.g., circular structures), relatively simple algorithms can be used. For the detection of features with different sizes and shapes, more complex schemes are needed [105]. This can lead to highly specified algorithms that apply only to a reduced class of problems.

Detection algorithms do not have the same ability as the human eyes, especially when features are only partially visible in the prospection data or are in a bad state of preservation [106]. On the other hand, manual visual inspection is time-consuming and may not be able to analyze vast data sets. Moreover, detection algorithms can analyze the data in ways that allow to observe beyond what is familiar or readily visible. The fact that they are not as flexible as a human interpreter can be turned into an advantage, as the detection is less biased, more objective, and more consistent [107]. All features that meet certain criteria are detected, including those not expected by the archaeologist [104,108]. This could lead to a better understanding of the way in which traditional manual interpretation occurs. Defining classes may make the interpreter more aware of possible biases in his manual approach and force him to define more properly what features are to be identified, less driven by expectation. It is clear that CAOD techniques are no substitute but rather a complement to knowledge-based manual detection. For that reason, we avoid using the terms "automated" or "semiautomated." Computer detection results in a segmented image with potential archaeological structures, including a number of false positives that need to be verified. It does not provide an archaeological interpretation, which occurs in the next phase by the human interpreter, who makes the ultimate decision on the relevance and accuracy of the detection results and assigns archaeological meaning to them.

A large number of object detection approaches have been developed, and many classifications have been proposed. In one taxonomy based on remote sensing literature [109], these methods are divided into four categories, which partially overlap: template matching, knowledge-based methods, object-oriented methods, and machine-learning-based methods. Two of these techniques (template matching and object-oriented methods) are discussed in more detail in Sections 5.2 and 5.3 and applied to the data from Bourne Park.

5.2 TEMPLATE MATCHING-BASED OBJECT DETECTION

Template matching involves the creation of a template that is translated over the image in order to determine to what extent it represents certain regions in the image and thereby locate interesting objects. A similarity metric such as the sum of absolute differences, the sum of squared differences, or the normalized cross correlation (NCC) is used to find the best matches. In archaeological remote sensing, template matching has been used mainly for the detection of circular features in aerial photography or data from airborne laser scanning, such as burial mounds, dwellings, or kilns [110–112]. Circular features have the advantage that they are rotation invariant. For other shapes such as rectangles, the workflow is more complex, requiring an additional pass for each rotation of the template. The same is true if templates with different scales are used. Below, we examine the usefulness of template matching for the detection of rectilinear Roman wall foundations in two subsets of the GPR data from Bourne Park [113], using three GPR attributes combined into a single image (see Section 4.1 and Fig. 6E and F). We defined rectangular templates with different scales to capture the range of wall sizes. Matching the template models to the GPR data and calculating the NCC resulted in as many correlation matrices as there were templates. For each pixel in the GPR image, the highest NCC was used to create a final correlation image. The foundations at Bourne Park were built according to two slightly different orientations. It is more straightforward to use horizontal and vertical templates than templates rotated at an arbitrary angle, so the GPR image was rotated instead of the templates (Fig. 13A–C). In most remote sensing applications of template matching, multiple detections within a predefined distance are assumed to represent the same feature, and only the detection with the highest NCC is important. By contrast, in our implementation, all pixels with an NCC higher than a certain threshold were kept.

The resulting plots not only show the walls but also include numerous false positives (Fig. 13D and E). A simple classification, based on a size threshold equal to half the size of the smallest wall template and excluding isolated detections that do not touch any other structure, removed the majority of the false positives. For the 2-D visualization, the remaining objects were enclosed in a rectangular bounding box. This procedure resulted in a completeness (COM) of 73.6% and a correctness (COR) of 81.5%, when compared with the manual interpretation (Fig. 14). The completeness was found by calculating the area of the archaeological structures in the manual interpretation and determining what portion of that area was detected by means of the automatic identification procedure. The correctness was obtained by calculating the ratio between the area belonging to both automatically and manually mapped features, and the total area of the automatically detected features. Since from visual inspection the walls appeared too long and too narrow, correction factors for the length and width were applied. This reduced the difference between COM and COR but only marginally improved the F-measure, which is defined as 2 $(COM \cdot COR)/(COM+COR)$ when equal weight is given to COM and COR [114] and for the experiment described here was 77.5%.

To extract 3-D objects, the 2-D rectangles were extruded so that they span the entire depth of the GPR energy cube, which first was subjected to median, dilation, and erosion filtering. For each 3-D volume obtained in this way, the mean of the energy μ_E was calculated. The final upper and lower boundaries of the volumes were then defined as the depth where the energy equaled μ_E multiplied by a factor. To allow the upper and lower boundaries to vary within one 3-D volume, these upper and lower boundaries were not defined for the entire 3-D volumes, but for small segments (Fig. 15). The success rate for the 3-D extraction was slightly lower than for

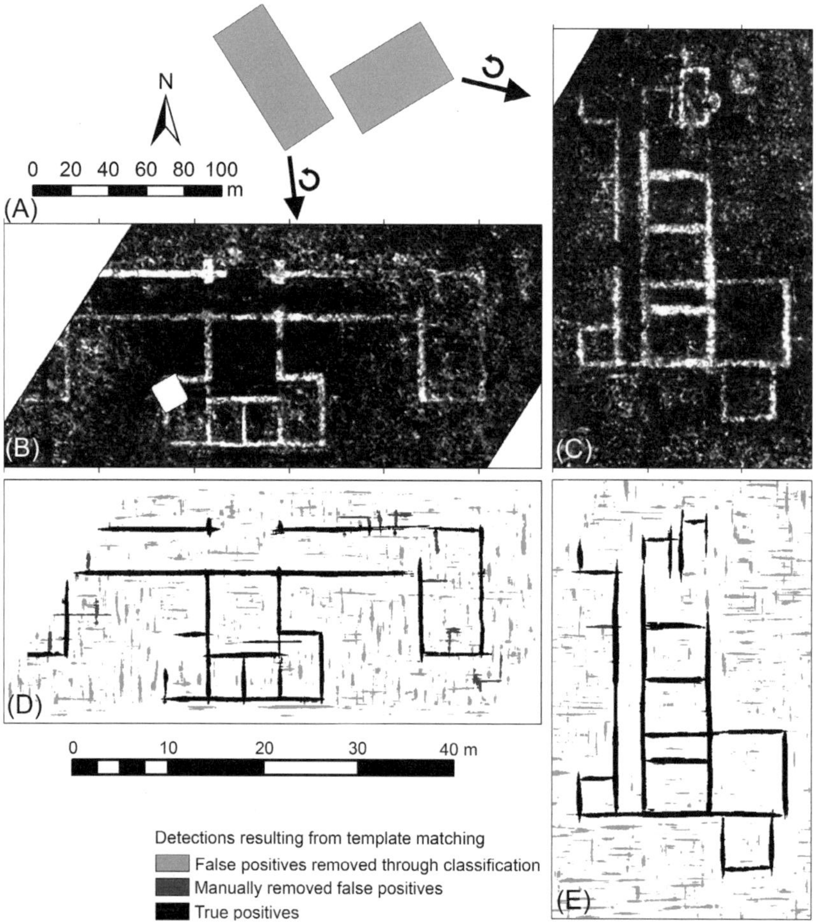

FIG. 13

(A) Original location of the two areas at Bourne Park where template matching was applied. (B and C) The two areas were rotated separately so that the walls run in the direction of the rows and columns of the GPR image. (D and E) Regions where the highest correlation coefficient (resulting from cross-correlation tests with different templates) exceeded 0.18 [113].

the 2-D case, with an F-measure of 71.3%. In this case, COM and COR were defined in terms of volumes instead of areas [113].

5.3 SEGMENTATION AND OBJECT-BASED IMAGE ANALYSIS

5.3.1 Segmentation

Segmentation can be defined as the partitioning of an image into homogeneous pixel groups [115]. A number of segmentation techniques have been developed. Segmentation can take place on the pixel level, for example, by means of a global threshold.

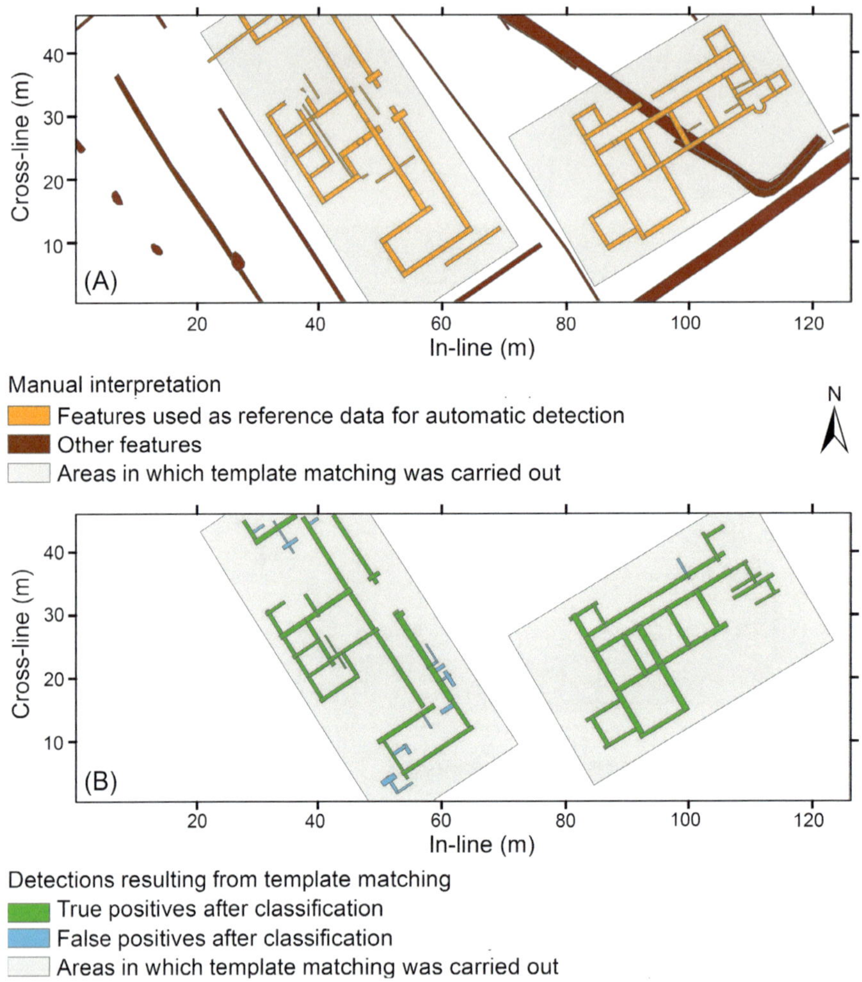

FIG. 14

Comparison of (A) the manual interpretation of the data in Fig. 13B and C and (B) the features detected through template matching [113].

In a study by Schmidt and Tsetskhladze [116], the values above the threshold were converted to polygons, which were subsequently adjusted by means of vector processing. Linford and Linford [117] equally applied a global threshold to GPR slices, to create an image with binary objects, whose boundaries were traced to obtain vector polygons. The properties of these polygons (area, perimeter size, and orientation) were then used to classify circular objects (pits) or rectilinear structures. Pixel-based segmentation and classification algorithms were also used by Ernenwein [118]: better results were achieved with supervised (Mahalanobis) classification than with unsupervised classification (*k*-means clustering).

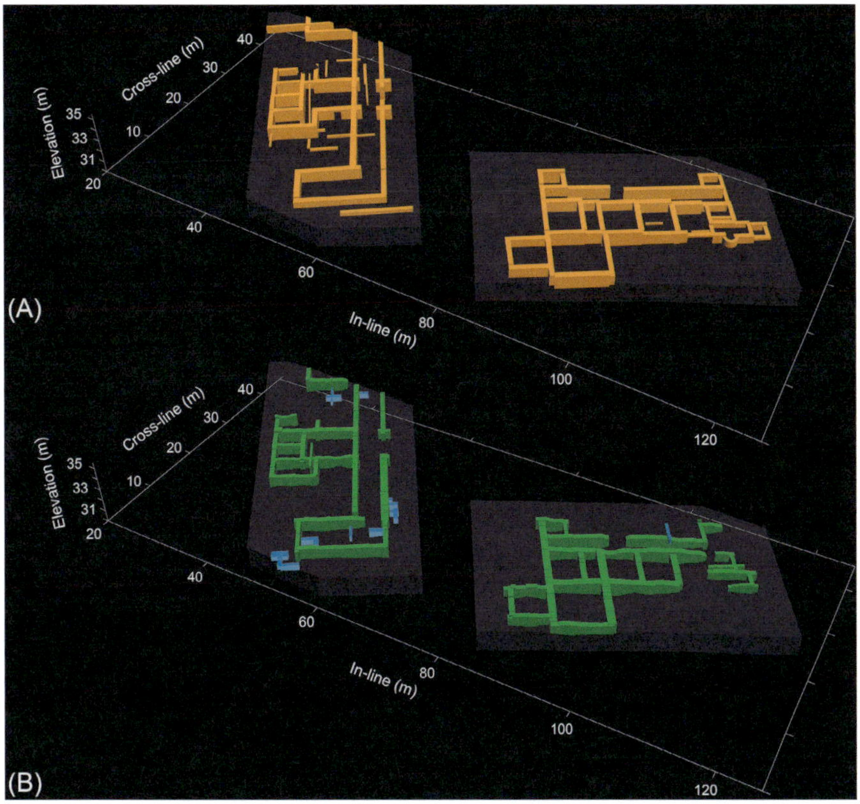

FIG. 15

Comparison of (A) the manual 3-D interpretation of the GPR data from Bourne Park and (B) the features detected through 2-D template matching and after determination of the upper and lower boundaries as described in the text. For the explanation of the colors, see Fig. 14. The transparent boxes represent the areas where template matching took place [113].

With the growing availability of high-resolution imagery in remote sensing, the situation where the objects in the landscape are much larger than the resolution of the image became more common. As a consequence, it was recognized that image analysis should no longer occur merely on the pixel level and that also the structural parameters of an image such as the color, the shape, or the context should be exploited. In this way, meaningful image objects (real-world objects on remotely sensed images such as houses or agricultural fields, or buried features in geophysical images) can be distinguished. This approach is mostly referred to as object-based image analysis (OBIA) or sometimes as geographic OBIA (GEOBIA), to distinguish it from the use of object-oriented techniques in biomedical imaging, materials sciences, or computer vision [119].

OBIA involves two steps: the segmentation of the image into objects and the classification of those objects. A frequently used segmentation technique used within

OBIA is multiresolution segmentation (MRS). The MRS procedure starts with image objects consisting of a single pixel. At each iteration, an object A is merged with its neighbor object B if A is most similar to B and *vice versa*, so that the increase in heterogeneity is minimized. The homogeneity criterion is based on the objects' mean spectral value (or mean geophysical response value), weighted by their size, and on their shape (compactness and smoothness). To make sure that the scale of all merged objects remains similar and that segments do not grow faster in regions with low variance, all existing objects are treated before any object is allowed to grow further. The merging of objects continues as long as a predefined heterogeneity threshold is not exceeded. As the scale of the objects depends on this threshold, it is called the scale parameter. By means of this parameter, the scale of the objects resulting from the segmentation is adapted to the size of the objects of interest [120]. Given the importance of the scale parameter, tools have been developed to identify the most suitable segmentation scale in a more objective and reproducible way [121].

When analyzing the fluxgate gradiometer data from Bourne Park shown in Fig. 16, we visually assessed the suitability of the segmentation, which occurred on two levels. The first level was adapted to the size of the many small bipolar anomalies, probably resulting from shallowly buried iron debris (e.g., parts of farm implements). Objects of a smaller scale were created to preserve the low contrasts often existing between archaeologically relevant features and the background.

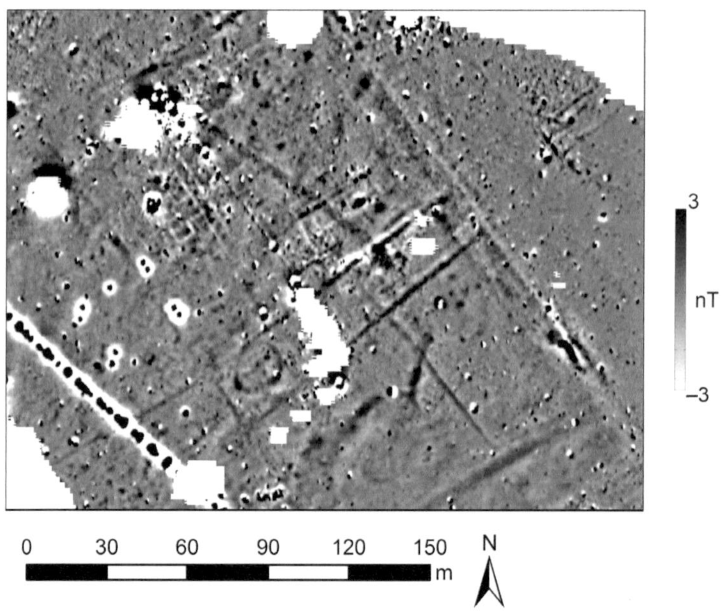

FIG. 16

Extract from the fluxgate gradiometer survey conducted at Bourne Park, showing mainly Roman buildings and enclosures.

Data courtesy of the Canterbury Hinterland Project.

5.3.2 Classification

After segmentation, the created objects can be classified using the combination of attributes that define them, such as geophysical measurement strength, shape, and relationship with other objects. Whether CAOD routines will become useful tools in archaeological interpretation also depends on the question if they are applicable to similar data sets without having to be completely rewritten. For this reason, for the classification of the fluxgate gradiometer data from Bourne Park, we decided to use two recently published classification methods for magnetometer data and to examine if these proved useful after slightly modifying them. For the classification of magnetic anomalies originating from shallow iron debris, Hinterleitner et al. [122] use the ratio between the strength of the anomaly and the separation between its maximum and minimum, with a high ratio indicating probable iron debris. Maxima and minima are described as, respectively, the largest or the smallest value within a certain area, exceeding a certain threshold. A contour is drawn at 30% of the maximum to define the boundaries of the anomaly. When classifying the Bourne Park data, we considered objects with a value above 0.5 nT and below −0.5 nT as maxima and minima, respectively. If these were less than 1.5 m separated from each other, the objects were merged, the difference between the maximum and the minimum was calculated, and the ratio between this difference and the area of the merged anomaly was determined. All anomalies with a ratio >1.3 nT/m^2 were considered a probable anomaly produced by ferrous debris. The value of most of the regions selected in this way was above 5 nT or below −5 nT. To cover the entire anomalies, the selected regions were made to encompass neighboring objects above 2 nT or below −2 nT. This procedure, which roughly applied the existing rule set developed by Hinterleitner et al. [122] and only adjusted the thresholds, allowed the classification of a large majority of the bipolar anomalies probably caused by ferrous debris (Fig. 17B). Areas exceeding 20 nT, not detected by the abovementioned rules, were also classified as originating from modern ferrous objects, such as the pipe in the southwestern part of the image.

For the classification of the remainder of the objects, the rule set developed by Pregesbauer et al. [123] was applied. They use an iterative approach, in which the objects with amplitudes below the 5th percentile and above the 95th percentile are considered as negative and positive anomalies, respectively, and the other objects remain unclassified. For each iteration, the 5th and 95th percentiles are recalculated using the unclassified objects. In the same way, the classification of the Bourne Park data required four iterations, although in the fourth iteration only the objects exceeding the 95th percentile were classified as anomaly. The amplitudes corresponding to the 5th and 95th percentile thresholds are shown in Table 2 for each iteration. Subsequently, all positive anomalies including at least one image segment above the 95th percentile calculated in the first or second iteration were merged. The same occurred for the negative anomalies including at least one segment below the 5th percentile calculated in the first iteration. All other anomalies were discarded. Finally, positive anomalies smaller than ~7.5 m^2 were assigned to a different class, as they can be interpreted as probable pit structures, to be distinguished from larger features

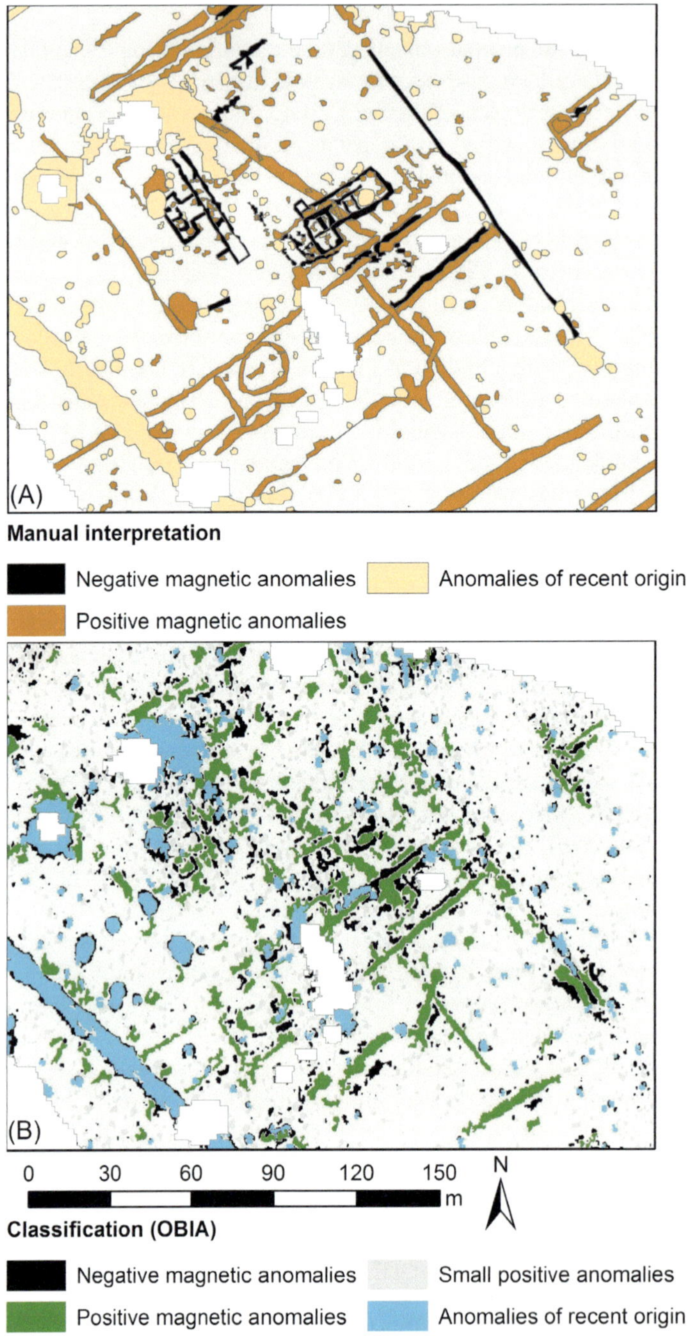

FIG. 17

(A) Manual interpretation of the fluxgate gradiometer data in Fig. 16. (B) OBIA-based classification of the same data.

(A): Courtesy of the Canterbury Hinterland Project.

Table 2 Value of the Thresholds Used in the Different Iterations for the Classification of Negative and Positive Magnetic Anomalies

Iteration	5th Percentile (nT)	95th Percentile (nT)
1	−1.3	2.1
2	−0.9	1.4
3	−0.7	1.1
4	Not used for classification	0.9

The thresholds represent the 5th and 95th percentile of the amplitudes of the unclassified objects.

probably belonging to building structures or ditches (Fig. 17B). When comparing the results of this procedure with the manual interpretation of the magnetometer data (Fig. 17A), it can be seen that most anomalies with archaeological significance were detected, although often in a fragmental way, which reduces the readability of the image. This is clear especially for the weak negative anomalies representing the villa buildings. Moreover, the heterogeneity of the background resulted in the classification of many false positives with the same amplitude as the archaeological features. A quantitative comparison between the manual and the OBIA-based classification resulted in a low COM, COR, and F-measure (45.0%, 35.3%, and 39.6%, respectively). These metrics were calculated in the same way as explained in Section 5.2.

5.3.3 Segmentation and classification based on the integration of different data sets

Further experiments were conducted to find out if the integration of different geophysical data sets can provide images more suited to OBIA. At Bourne Park, the best results were obtained when the object detection was based on two images. The first one contained the first PC computed from three GPR attributes (energy, phase symmetry, and scale parameter of the Weibull distribution), as shown in Fig. 6F, and was used to extract the high-amplitude features from the GPR data, representing mainly building foundations. The classification of these anomalies occurred in two iterations, applying the same rule set as outlined in Section 5.3.2. From each iteration, only the data above the 95th percentile were kept (Fig. 18C).

The second image (Fig. 18A) contained the first PC computed from the fluxgate gradiometer and earth resistance data (Fig. 8A and B) and was used to extract the features characterized by a low resistance and a positive magnetic anomaly (that became negative after the polarity of the magnetometer data had been inverted; see Section 4.2.2). These features represent pits and ditches. Their classification required six iterations; from each, only the anomalies below the 5th percentile were kept. All anomalies including at least one image segment below the 5th percentile calculated in the first iteration were merged. The other anomalies were discarded. As a further step, the classification of the ditches occurred by examining the main direction of the remaining anomalies. Two narrow ranges of directions were defined to extract ditch systems with two different orientations (Fig. 18C). Although a visual analysis seems to indicate that the OBIA-based segmentation and classification have

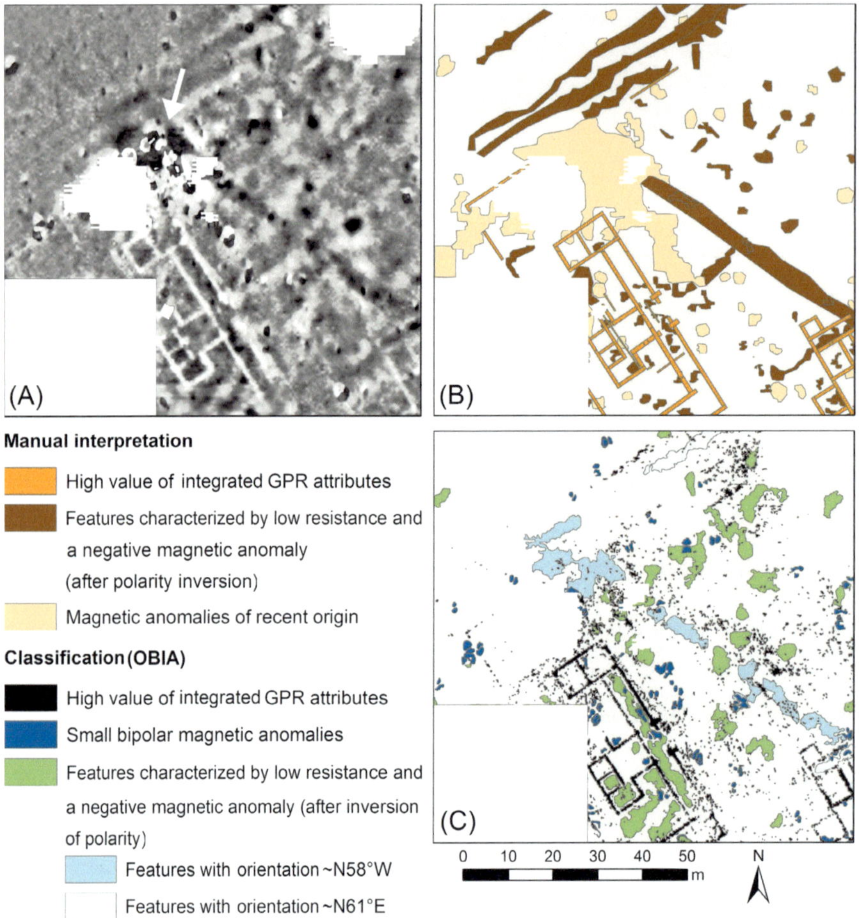

FIG. 18

(A) First PC computed from the fluxgate gradiometer and earth resistance data from Bourne Park (see Fig. 8A and B). (B) Manual interpretation of the data in (A) and in Fig. 6F. (C) OBIA-based classification of the data in (A) and in Fig. 6F.

Fluxgate gradiometer data, earth resistance data, and their manual interpretation courtesy of the Canterbury Hinterland Project.

detected most archaeologically relevant features, also here, the quantitative comparison with the manual interpretation (Fig. 18B) resulted in a low F-measure of 37.0% (COM and COR yielded similar values).

5.4 DISCUSSION

The CAOD methods discussed in the previous sections are not the same as expert archaeological interpretation, but rather tools that can facilitate it. Criticism about the applicability of these methods in complex archaeological contexts is justified, especially when only one type of geophysical data is used, as is demonstrated by the OBIA-based classification of the fluxgate gradiometer data from Bourne Park, where the strength of archaeologically relevant anomalies and background soil heterogeneities is similar and the application of existing classification rule sets proved difficult (Fig. 17B). Technological advances offer the potential, however, of faster and less expensive collection of complementary data. Integration of multisensor data can enhance feature exposure compared with a single data set and ease classification not only by human interpreters but also by computer algorithms. Some trial and error may be required: in the presented data from Bourne Park, extracting the wall foundations from the integrated GPR attribute image and the ditches and pits from a separate image containing the integrated magnetometer and earth resistance data proved more efficient than extracting all anomalies from a single integrated image.

The selection of suitable quantitative evaluation metrics to objectify the debate on the usefulness of CAOD methods and compare different algorithms is difficult. This is linked to the fact that reliable reference data are less easily available in geophysics than in remote sensing ([118]; see also Section 4.5), so that metrics are often based on the comparison with the manual interpretation of the geophysical data only. The complex relationship between computer-aided and manual interpretation is illustrated in Fig. 18. The OBIA-based classification of the magnetometer, earth resistance, and GPR data seems acceptable at least for part of the test area but resulted in a low F-measure. It clearly fails to separate weak, only visually recognizable archaeological features such as the parallel ditches in the northern part of Fig. 18A. Furthermore, the algorithm and the human interpretation result in a different delineation of recently disturbed areas: is the magnetic anomaly near the arrow in Fig. 18A entirely caused by recent ferrous debris, or does it also partially belong to the ditch running in northwestern direction? Here, the discussion moves away from the question which interpretation is the right one to the observation that CAOD and human interpretation operate differently. The human eye and brain assign significance to contrasts only if these have particular shapes or occur in specific spatial patterns that vary throughout the data set. Computer algorithms are less flexible, resulting in a classification with superfluous information. For example, in Fig. 18C, many small areas of high GPR reflection strength are devoid of archaeological meaning, but during classification, they are not easily separable from the building foundations. However, the selection of all features meeting certain criteria has the advantage of a greater objectivity, and as a result, relationships not observed through manual interpretation can become apparent. For instance, the space inside the villa building in Fig. 18C is characterized by a similar combination of low resistance and positive magnetic value as the ditches and pits elsewhere in the image, but this was not perceived as such during manual interpretation because of the adjacent walls with contrasting geophysical properties. This complementary character of human interpretation and computer vision marks the shift that may gradually take place from completely manual delineation toward the use of models based on reduction and classification [28].

A balance needs to be found between the performance of a CAOD workflow and the need for simplicity and wide applicability. As an example, the classification of the ditches in Fig. 18C is a complex operation that can be done in a much more efficient way by a human interpreter. The template matching shown in Figs. 13–15 performs well for the extraction of orthogonal sets of Roman wall foundations, but its workflow is relatively complex, including the determination of several thresholds and the design of classification rules. Even so, highly specific detection strategies can be beneficial if the used parameter values prove to be valid for other sites of the same kind, especially if the data sets are large. The suitability of a particular method therefore depends on the task. As the number of CAOD studies undertaken in archaeological geophysics grows, the most efficient strategies may rise to the surface.

6 CONCLUSION

Uncertainty and subjectivity are inherent to any archaeological interpretation of geophysical data, although these are present on a different scale when interpreting morphologically distinct elements, such as architectural remains from the Roman or medieval period, and when analyzing more amorphous perturbations, for instance, dug features in a prehistoric landscape that could or could not be anthropogenic in nature. There are different ways in which this uncertainty can be approached. First, the potential of combining different types of geophysical data with remote sensing data, and with results from field walking, coring, excavations or legacy data, will continue to grow. Today, in archaeological geophysics, multiple-sensor arrays allow fast prospection of large areas with a high sample density. In the future, speed may even increase, as simultaneous measurements with different geophysical techniques on one platform become more common. Outside the sphere of near-surface geophysics, the same evolution toward a more diverse toolkit resulting in more accurate and precise measurements takes place (airborne laser scanning data with high resolution, the use of RTK GNSS for the registration of finds during field walking, or the use of cone penetrating tests, to give just a few examples). This allows the investigation of landscapes and sites that until recently were hardly accessible, including deeply buried landscapes for palaeolandscape reconstruction. The wider availability and lower cost of the technology, also when it comes to data processing, make it easier to obtain diverse data sets and make the interpretation more robust. Data combination can take the form of side-by-side analysis or data integration, where the application of algorithms developed outside archaeology can open new perspectives. Second, the acknowledgement of the uncertainty and of the fact that not all questions can be answered should encourage us to define the strengths and weaknesses of our prospection methods and to delineate the areas where geophysics can offer added value and answer archaeological questions.

Archaeological interpretation has not kept pace with improved data acquisition and processing, as it still occurs largely manually. In archaeological geophysics, computer-aided object detection is still in its infancy. Nevertheless, it seems beyond doubt that in the near future, certain relatively simple applications such as the detection of small metal objects in magnetometer and EMI data will be used more routinely and with relatively high accuracy. This will save human interpreters' time, which can be invested in addressing more complex problems of interpretation. Because computer-aided detection is still at an early stage, it is difficult to predict how far this evolution will stretch and how quickly these techniques will gain wider acceptance. More studies are needed to assess the applicability of algorithms that have so far been developed for a specific site or landscape. Also, the determination of criteria for quality control is an important objective. It seems likely, however, that more refined algorithms will increasingly assist the human interpreter in the future, without ever taking over completely. Here, the border between data processing and enhancing the exposure of the anomalies, and segmentation (until now often a manual process) seems to fade. The assignment of archaeological meaning to the detected variations by the human interpreter will, however, always remain the last step in the interpretation process.

ACKNOWLEDGMENTS

This work was made possible by a postdoctoral fellowship of the Research Foundation—Flanders (FWO) of Lieven Verdonck and Philippe De Smedt and an IWT (now FWO) PhD grant of Jeroen Verhegge. The GPR prospections were supported financially by the FWO and the Department of Archaeology, Ghent University. The authors would like to thank the senior team of the Canterbury Hinterland Project (CHP): Lacey Wallace (University of Lincoln), Alex Mullen (University of Nottingham), and Paul Johnson (University of Sheffield). The geophysical prospections within the CHP were supported financially by the Faculty of Classics (University of Cambridge), the McDonald Institute for Archaeological Research (University of Cambridge), All Souls College (University of Oxford), and the Roman Research Trust. The GPR prospection at Ammaia occurred within the Radio-Past project (2009–13; EC FP7 Marie Curie action "Industry-Academia Partnerships and Pathways"). The GPR prospection at Interamna Lirenas was undertaken by Ghent University and the University of Cambridge (Faculty of Classics), in collaboration with the British School at Rome and the Soprintendenza Archeologia, Belle Arti e Paesaggio per le Province di Frosinone, and Latina e Rieti, with generous support from the Arts and Humanities Research Council (grant Ref. AH/M006522/1—Beneath the Surface of Roman Republican Cities), the University of Cambridge (Faculty of Classics), and the Comune di Pignataro Interamna.

REFERENCES

[1] C. Smith, Preface, in: C. Smith (Ed.), Encyclopedia of Global Archaeology, Springer, New York, 2014. vii–ix.

[2] R.E. Sheriff, Encyclopedic Dictionary of Applied Geophysics, Society of Exploration Geophysicists, Tulsa, 2002.

[3] K.L. Kvamme, Geophysical surveys as landscape archaeology, Am. Antiq. 63 (2003) 435–457.

[4] S. Campana, Emptyscapes: filling 'empty' Mediterranean landscapes, mapping the archaeological continuum, Archaeol. Polona 53 (2015) 149–152.

[5] C.F. Gaffney, J.A. Gater, P. Linford, V.L. Gaffney, R. White, Large-scale systematic fluxgate gradiometry at the Roman City of Wroxeter, Archaeol. Prospect. 7 (2000) 81–99.
[6] S.J. Keay, M. Millett, S. Poppy, J. Robinson, J. Taylor, N. Terrenato, Falerii Novi: a new survey of the walled area, Papers of the British School at Rome 68 (2000) 1–93.
[7] D. Powlesland, Identifying mapping and managing the unmanageable: the implications of long term multi-sensor research into the archaeology of the Vale of Pickering, Yorkshire, England, in: S. Campana, M. Forte, C. Liuzza (Eds.), Space, Time, Place. Third International Conference on Remote Sensing in Archaeology. Tiruchirappalli, 17–21 August 2009, British Archaeological Reports, International Series 2118, Archaeopress, Oxford, 2010, pp. 9–16.
[8] T. Saey, B. Stichelbaut, J. Bourgeois, V. Van Eetvelde, M. Van Meirvenne, An interdisciplinary non-invasive approach to landscape archaeology of the Great War, Archaeol. Prospect. 20 (2013) 39–44.
[9] G. Verhoeven, C. Sevara, Trying to break new ground in aerial archaeology, Remote Sens. 8 (2016) 918.
[10] N. Linford, P. Linford, A. Payne, First results from a new ground-coupled multi-element GPR array, Archaeol. Polona 53 (2015) 631–635.
[11] C. Meyer, H. Zöllner, D. Pilz, B. Horejs, A. Matthaei, LEA MAX–multi-purpose gradiometer array in the fields of the Kaikos valley (Bergama, Turkey), Archaeol. Polona 53 (2015) 229–232.
[12] P. De Smedt, S. Delefortrie, F. Wyffels, Identifying and removing micro-drift in ground-based electromagnetic induction data, J. Appl. Geophys. 131 (2016) 14–22.
[13] A. Tabbagh, Electrical resistivity and electromagnetism, in: A.S. Gilbert (Ed.), Encyclopedia of Geoarchaeology, Springer, Dordrecht, 2017, pp. 211–219.
[14] L.B. Conyers, J. Leckebusch, Geophysical archaeology research agendas for the future: some ground-penetrating radar examples, Archaeol. Prospect. 17 (2010) 117–123.
[15] A. Schmidt, P. Linford, N. Linford, A. David, C. Gaffney, A. Sarris, J. Fassbinder, EAC guidelines for the use of geophysics in archaeology, Questions to Ask and Points to Consider, Europae Archaeologia Consilium (EAC), Namur, 2015.
[16] J.W.E. Fassbinder, Seeing beneath the farmland, steppe and desert soil: magnetic prospecting and soil magnetism, J. Archaeol. Sci. 56 (2015) 85–95.
[17] N.G. Papadopoulos, P. Tsourlos, C. Papazachos, G.N. Tsokas, A. Sarris, J.H. Kim, An algorithm for fast 3D inversion of surface electrical resistivity tomography data: application on imaging buried antiquities, Geophys. Prospect. 59 (2011) 557–575.
[18] J. Thiesson, A. Tabbagh, F.X. Simon, M. Dabas, 3D linear inversion of magnetic susceptibility data acquired by frequency domain EMI, J. Appl. Geophys. 136 (2017) 165–177.
[19] W. Neubauer, A. Eder-Hinterleitner, 3D-interpretation of postprocessed archaeological magnetic prospection data, Archaeol. Prospect. 4 (1997) 191–205.
[20] A. Schmidt, K. Armstrong, M. van Leusen, Using archaeological models for the inversion of magnetometer data, Archaeol. Polona 53 (2015) 575–578.
[21] D.V. Fitterman, Tools and techniques: active-source electromagnetic methods, in: G. Shubert (Ed.), Treatise on Geophysics, vol. 11, Resources in the Near-Surface Earth, Elsevier, Amsterdam, 2015, pp. 295–333.
[22] D.W. Oldenburg, Y. Li, Inversion for applied geophysics: a tutorial, in: D.K. Butler (Ed.), Near Surface Geophysics, Society of Exploration Geophysicists, Tulsa, 2005, pp. 89–150.

[23] L.B. Conyers, Interpreting Ground-Penetrating Radar for Archaeology, Left Coast Press, Walnut Creek, 2012.
[24] A. Klotzsche, J. van der Kruk, G.A. Meles, J. Doetsch, H. Maurer, N. Linde, Full-waveform inversion of cross-hole ground-penetrating radar data to characterize a gravel aquifer close to the Thur River, Switzerland, Near Surf. Geophys. 8 (2010) 635–649.
[25] A. Aspinall, C. Gaffney, A. Schmidt, Magnetometry for Archaeologists, AltaMira Press, Lanham, 2008.
[26] G.R. Bellini, S. Hay, A. Launaro, N. Leone, M. Millett, Interamna Lirenas e il suo territorio. Indagini archeologiche non invasive 2012, Lazio e Sabina 10 (2014) 195–204.
[27] N. Linford, The application of geophysical methods to archaeological prospection, Rep. Prog. Phys. 69 (2006) 2205–2257.
[28] W. Neubauer, A. Eder-Hinterleitner, Resistivity and magnetics of the Roman town Carnuntum, Austria: an example of combined interpretation of prospection data, Archaeol. Prospect. 4 (1997) 179–189.
[29] P. De Smedt, M. Van Meirvenne, T. Saey, E. Baldwin, C. Gaffney, V. Gaffney, Unveiling the prehistoric landscape at Stonehenge through multi-receiver EMI, J. Archaeol. Sci. 50 (2014) 16–23.
[30] C. Gaffney, J. Gater, Revealing the Buried Past. Geophysics for Archaeologists, Tempus, Stroud, 2003.
[31] P.M. Allison, Pompeian Households: An Analysis of the Material Culture, The Cotsen Institute of Archaeology Press, Los Angeles, 2004.
[32] C. Benech, New approach to the study of city planning and domestic dwellings in the Ancient Near East, Archaeol. Prospect. 14 (2007) 87–103.
[33] E. Fentress, Cosa V: An Intermittent Town. Excavations 1991-1997, Memoirs of the American Academy in Rome Supplementary, vol. 2, University of Michigan Press, Ann Arbor, 2003.
[34] J.W.E. Fassbinder, Magnetometry for archaeology, in: A.S. Gilbert (Ed.), Encyclopedia of Geoarchaeology, Springer, Dordrecht, 2017, pp. 499–514.
[35] A.P. Annan, Electromagnetic principles of ground penetrating radar, in: H.M. Jol (Ed.), Ground Penetrating Radar Theory and Applications, Elsevier, Amsterdam, 2009, pp. 1–40.
[36] A. Tabbagh, What is the best coil orientation in the Slingram electromagnetic prospecting method? Archaeometry 28 (1986) 185–196.
[37] A. Schmidt, Earth Resistance for Archaeologists, AltaMira Press, Lanham, 2013.
[38] M. Dabas, Theory and practice of the new fast electrical imaging system ARP, in: S. Campana, S. Piro (Eds.), Seeing the Unseen, Taylor & Francis Group, London, 2009, pp. 105–126.
[39] L. Verdonck, D. Taelman, F. Vermeulen, R. Docter, The impact of spatial sampling and migration on the interpretation of complex archaeological ground-penetrating radar data, Archaeol. Prospect. 22 (2015) 91–103.
[40] A. Schmidt, A. Marshall, Impact of resolution on the interpretation of archaeological prospection data, in: A. Sinclair, E. Slater, J. Gowlett (Eds.), Archaeological Sciences 1995, Oxbow Books, Oxford, 1997, pp. 343–348.
[41] S. Delefortrie, P. De Smedt, T. Saey, E. Van De Vijver, M. Van Meirvenne, An efficient calibration procedure for correction of drift in EMI survey data, J. Appl. Geophys. 110 (2014) 115–125.

[42] L.B. Conyers, Ground Penetrating Radar for Archaeology, third ed., Altamira Press, Walnut Creek, 2013.
[43] J. Leckebusch, J. Rychener, Verification and topographic correction of GPR data in three dimensions, Near Surf. Geophys. 5 (2007) 395–403.
[44] F. Lehmann, A.G. Green, Topographic migration of georadar data: implications for acquisition and processing, Geophysics 65 (2000) 836–848.
[45] Geoscan Research, Geoplot Version 3.00s for Windows, Geoscan Research, Bradford, 2005.
[46] S. Chopra, K.J. Marfurt, Seismic Attributes for Prospect Identification and Reservoir Characterization, SEG Geophysical Developments Series 11, Society of Exploration Geophysicists, Tulsa, 2007.
[47] W. Zhao, E. Forte, S.T. Levi, M. Pipan, G. Tian, Improved high-resolution GPR imaging and characterization of prehistoric archaeological features by means of attribute analysis, J. Archaeol. Sci. 54 (2015) 77–85.
[48] L. Wallace, A. Mullen, P. Johnson, L. Verdonck, Archaeological investigations of Bourne Park, Bishopsbourne, Kent, 2011–2014, Archaeologia Cantiana 137 (2016) 251–280.
[49] P. Kovesi, Symmetry and asymmetry from local phase, in: Proceedings of the Tenth Australian Joint Conference on Artificial Intelligence, Perth, Australia, 30 November–4 December 1997, 1997, pp. 185–190.
[50] U. Böniger, J. Tronicke, Subsurface utility extraction and characterization: combining GPR symmetry and polarization attributes, IEEE Trans. Geosci. Remote Sens. 50 (2012) 736–746.
[51] M. Doneus, G. Verhoeven, C. Atzberger, M. Wess, M. Ruš, New ways to extract archaeological information from hyperspectral pixels, J. Archaeol. Sci. 52 (2014) 84–96.
[52] P. De Smedt, M. Van Meirvenne, D. Herremans, J. De Reu, T. Saey, E. Meerschman, P. Crombé, W. De Clercq, The 3-D reconstruction of medieval wetland reclamation through electromagnetic induction, Sci. Rep. 3 (2013) 1517.
[53] R. Filzwieser, L.H. Olesen, G. Verhoeven, E.S. Mauritsen, W. Neubauer, I. Trinks, M. Nowak, R. Nowak, P. Schneidhofer, E. Nau, M. Gabler, Integration of complementary archaeological prospection data from a late iron age settlement at Vesterager—Denmark, J. Archaeol. Method Theory 25 (2018) 313–333.
[54] K.L. Kvamme, Integrating multidimensional geophysical data, Archaeol. Prospect. 13 (2006) 57–72.
[55] S. Fitch, V. Gaffney, K. Thomson, Merging technologies: integration and visualisation of spatial data, in: V. Gaffney, K. Thomson, S. Fitch (Eds.), Mapping Doggerland: The Mesolithic Landscapes of the Southern North Sea, Archaeopress, Oxford, 2007, pp. 33–41.
[56] M.S. Watters, Geovisualization: an example from the Catholme Ceremonial complex, Archaeol. Prospect. 13 (2006) 282–290.
[57] M. Grasmueck, T.J. Moser, M.A. Pelissier, J. Pajchel, K. Pomar, Diffraction signatures of fracture intersections, Interpretation 3 (2015) SF55–SF68.
[58] L. Yu, Y. Nie, F. Liu, J. Zhu, Y. Yao, H. Gao, Research on data fusion method for archaeological site identification, in: Proceedings of the 2012 International Conference on Image Analysis and Signal Processing, IASP 2012, Hangzhou, 9–11 November 2012, 2012.
[59] G. Verhoeven, M. Nowak, R. Nowak, Pixel-level image fusion for archaeological interpretative mapping, in: Proceedings of the 8th International Congress on Archaeology,

Computer Graphics, Cultural Heritage and Innovation, Valencia, 5–7 September 2016, 2016, pp. 404–407.
[60] H. Guo, S. Lewis, K.J. Marfurt, Mapping multiple attributes to three- and four-component color models—a tutorial, Geophysics 73 (2008) W7–W19.
[61] A. Schmidt, Visualisation of multi-source archaeological prospection data, in: M. Cucarzi, P. Conti (Eds.), Filtering, Optimization and Modelling of Geophysical Data in Archaeological Prospecting, Milan-Rome, C.M. Lerici Foundation, 2001, pp. 149–160.
[62] S. Piro, P. Mauriello, F. Cammarano, Quantitative integration of geophysical methods for archaeological prospection, Archaeol. Prospect. 7 (2000) 203–213.
[63] H. von der Osten, Different numerical and visual concepts for combining aerial pictures and geophysical mappings, in: J. Bourgeois, M. Meganck (Eds.), Aerial Photography and Archaeology 2003, Archaeological Reports Ghent University 4, Academia Press, Ghent, 2005, pp. 221–226.
[64] J. Verhegge, The forum integrated survey, in: C. Corsi, F. Vermeulen (Eds.), Ammaia I: The Survey: A Romano-Lusitanian Townscape Revealed, Archaeological Reports Ghent University 8, Academia Press, Ghent, 2012, pp. 91–103.
[65] J.J. Gerbrands, On the relationships between SVD, KLT and PCA, Pattern Recogn. 14 (1981) 375–381.
[66] N. Linford, From hypocaust to hyperbola: ground penetrating radar surveys over mainly Roman remains in the UK, Archaeol. Prospect. 11 (2004) 237–246.
[67] M. Misiti, Y. Misiti, G. Oppenheim, J. Poggi, Matlab Wavelet Toolbox 4 User's Guide, The MathWorks Inc., Natick, 2009.
[68] L. Verdonck, F. Vermeulen, R. Docter, C. Meyer, R. Kniess, 2D and 3D ground-penetrating radar surveys with a modular system: data processing strategies and results from archaeological field tests, Near Surf. Geophys. 11 (2013) 239–252.
[69] C. Pohl, J. van Genderen, Remote Sensing Image Fusion: A Practical Guide, Taylor & Francis, Boca Raton, 2017.
[70] P. Pérez, M. Gangnet, A. Blake, Poisson image editing, ACM Trans. Graph. 22 (2003) 313–318.
[71] S. Paul, I.S. Sevcenco, P. Agathoklis, Multi-exposure and multi-focus image fusion in gradient domain, J. Circuits Syst. Comput. 25 (2016). 1650123.
[72] F. Vermeulen, D. Van Limbergen, P. Monsieur, D. Taelman, The Potenza Valley Survey (Marche, Italy). Settlement dynamics and changing material culture in an Adriatic valley between Iron Age and Late Antiquity, Academia Belgica, Rome, 2017.
[73] J. Trachet, S. Delefortrie, M. Van Meirvenne, B. Hillewaert, W. De Clercq, Reassessing surface artefact scatters. The integration of artefact-accurate fieldwalking with geophysical data at medieval harbour sites near Bruges (Belgium), Archaeol. Prospect. 24 (2017) 101–117.
[74] M.G. Canti, F.M. Meddens, Mechanical coring as an aid to archaeological projects, J. Field Archaeol. 25 (1998) 97–105.
[75] J. Schuldenrein, Coring and the identity of cultural-resource environments: a comment on Stein, Am. Antiq. 56 (1991) 131–137.
[76] J.K. Stein, Coring in CRM and archaeology: a reminder, Am. Antiq. 56 (1991) 138–142.
[77] M.R. Bates, A.J. Barham, C.A. Pine, V.D. Williamson, Evaluation strategies for deeply stratified alluvial areas: the use of borehole stratigraphic logs, in: S. Roskams (Ed.), Interpreting Stratigraphy. Site Evaluation Recording Procedures and Stratigraphic Analysis, Archaeopress, Oxford, 2000, pp. 49–69.

[78] R. Dalan, J. Sturdevant, R. Wallace, B. Schneider, S. De Vore, Cutbank geophysics: a new method for expanding magnetic investigations to the subsurface using magnetic susceptibility testing at an Awatixa Hidatsa Village, North Dakota, Remote Sens. 9 (2017) 112.

[79] R. Dalan, B.W. Bevan, D. Goodman, D. Lynch, S. De Vore, S. Adamek, T. Martin, G. Holley, M. Michlovic, The measurement and analysis of depth in archaeological geophysics: tests at the Biesterfeldt Site, USA, Archaeol. Prospect. 18 (2011) 245–265.

[80] T. Missiaen, J. Verhegge, K. Heirman, P. Crombé, Potential of cone penetrating testing for mapping deeply buried palaeolandscapes in the context of archaeological surveys in polder areas, J. Archaeol. Sci. 55 (2015) 174–187.

[81] J. Verhegge, T. Saey, M. Van Meirvenne, T. Missiaen, P. Crombé, Reconstructing early Neolithic paleogeography: EMI-based subsurface modeling and chronological modeling of Holocene peat below the Lower Scheldt floodplain in NW Belgium, Geoarchaeology 32 (2017) 159–176.

[82] J. Verhegge, T. Missiaen, P. Crombé, Exploring integrated geophysics and geotechnics as a paleolandscape reconstruction tool: archaeological prospection of (prehistoric) sites buried deeply below the Scheldt Polders (NW Belgium), Archaeol. Prospect. 23 (2016) 125–145.

[83] T. Lunne, J.J.M. Powell, P.K. Robertson, Cone-Penetration Testing in Geotechnical Practice, Taylor & Francis Group, London, 1997.

[84] M. Heinzelmann, D. Jordan, C. Murer, Amiternum and the upper Aterno valley: a Sabine-Roman town and its territory, J. Roman Archaeol. 23 (2010) 55–83.

[85] M. Millett, Understanding Roman towns in Italy: reflections on the role of geophysical survey, in: P.S. Johnson, M. Millett (Eds.), Archaeological Survey and the City, Oxbow Books, Oxford, 2013, pp. 241–260.

[86] L. Verdonck, F. Vermeulen, C. Corsi, R. Docter, Ground-penetrating radar survey at the Roman town of Mariana (Corsica), complemented with fluxgate gradiometer data and old and recent excavation results, Near Surf. Geophys. 10 (2012) 35–45.

[87] L. Verdonck, D. Taelman, Ground penetrating radar survey at Ammaia, in: C. Corsi, F. Vermeulen (Eds.), Ammaia I: The Survey: A Romano-Lusitanian Townscape Revealed, Archaeological Reports Ghent University 8, Academia Press, Ghent, 2012, pp. 69–81.

[88] A. Wallace-Hadrill, Houses and society in Pompeii and Herculaneum, Princeton University Press, Princeton, 1994.

[89] A. O'Connell, M3 Clonee-North of Kells, Contract 2 Dunshaughlin–Navan, 2009. Report on the archaeological excavation of Lismullin 1, Co. Meath, https://repository.dri.ie/catalog/5425zr47m. Accessed 3 March 2018.

[90] J. Bonsall, C. Gaffney, I. Armit, A decade of ground truthing: reappraising magnetometer prospection surveys on linear corridors in light of excavation evidence 2001–2010, in: H. Kamermans, M. Gojda, A.G. Posluschny (Eds.), A Sense of the Past: Studies in Current Archaeological Applications of Remote Sensing and Non-Invasive Prospection Methods, British Archaeological Reports, International Series 2588, Archaeopress, Oxford, 2014, pp. 3–16.

[91] A. Hesse, Multi-parametric survey for archaeology: how and why, or how and why not? J. Appl. Geophys. 41 (1999) 157–168.

[92] C. Gaffney, Detecting trends in the prediction of the buried past: a review of geophysical techniques in archaeology, Archaeometry 50 (2008) 313–336.

[93] K.L. Kvamme, Geophysical correlation: global versus local perspectives, Archaeol. Prospect. 25 (2018) 111–120.

[94] U. Böniger, J. Tronicke, Integrated data analysis at an archaeological site: a case study using 3D GPR, magnetic, and high-resolution topographic data, Geophysics 75 (2010) B169–B176.

[95] S. Campana, S. Piro, Putting everything together: GIS-based data integration and interpretation, in: S. Campana, S. Piro (Eds.), Seeing the Unseen, Taylor & Francis Group, London, 2009, pp. 325–330.

[96] C. Gaffney, C. Harris, F. Pope-Carter, J. Bonsall, R. Fry, A. Parkyn, Still searching for graves: an analytical strategy for interpreting geophysical data used in the search for "unmarked" graves, Near Surf. Geophys. 13 (2015) 557–569.

[97] C. Pohl, J.L. van Genderen, Review article multisensor image fusion in remote sensing: concepts, methods and applications, Int. J. Remote Sens. 19 (1998) 823–854.

[98] A.E. Kattenberg, G. Aalbersberg, Archaeological prospection of the Dutch perimarine landscape by means of magnetic methods, Archaeol. Prospect. 11 (2004) 227–235.

[99] A. Traviglia, D. Cowley, K. Lambers, Finding common ground: human and computer vision in archaeological prospection, AARGnews 53 (2016) 11–24.

[100] N.P. Sheen, A. Aspinall, A simulation of anomalies to aid the interpretation of magnetic data, in: J. Wilcock, K. Lockyear (Eds.), CAA 93. Computer Applications and Quantitative Methods in Archaeology 1993, British Archaeological Reports, International Series 598, Tempus Reparatum, Oxford, 1995, pp. 57–63.

[101] D.J. Bescoby, G.C. Cawley, P.N. Chroston, Enhanced interpretation of magnetic survey data using artificial neural networks: a case study from Butrint, Southern Albania, Archaeol. Prospect. 11 (2004) 189–199.

[102] H. Becker, Caesium-magnetometry for landscape-archaeology, in: S. Campana, S. Piro (Eds.), Seeing the Unseen. Geophysics and Landscape Archaeology, CRC Press, London, 2009, pp. 129–165.

[103] S. Parcak, Satellite Remote Sensing for Archaeology, Routledge, London, 2009.

[104] R. Bennett, D. Cowley, V. De Laet, The data explosion: tackling the taboo of automatic feature recognition in airborne survey data, Antiquity 88 (2014) 896–905.

[105] C. Sevara, M. Pregesbauer, M. Doneus, G. Verhoeven, I. Trinks, Pixel versus object–a comparison of strategies for the semi-automated mapping of archaeological features using airborne laser scanning data, J. Archaeol. Sci. Rep. 5 (2016) 485–498.

[106] P. Da Pelo, T. D'Orazio, Automatic and semi-automatic approaches to support archaeological trace extraction and digitalization, in: R. Lasaponara, N. Masini, M. Biscione, M. Hernandez (Eds.), Proceedings of the 4th EARSeL Workshop on Cultural and Natural Heritage "Earth Observation: A Window on the Past", Matera, 6–7 June 2013, 2013, pp. 347–362.

[107] J. Leckebusch, A. Weibel, F. Bühler, Semi-automatic feature extraction from GPR data for archaeology, Near Surf. Geophys. 6 (2008) 75–84.

[108] D.C. Cowley, In with the new, out with the old? Auto-extraction for remote sensing archaeology, in: Papers of the First International Conference on Virtual Archaeology, St Petersburg, 2012, 2012, pp. 18–30.

[109] G. Cheng, J. Han, A survey on object detection in optical remote sensing images, ISPRS J. Photogramm. Remote Sens. 117 (2016) 11–28.

[110] K. Kvamme, An examination of automated archaeological feature recognition in remotely sensed imagery, in: A. Bevan, M. Lake (Eds.), Computational Approaches to Archaeological Spaces, Left Coast Press, Walnut Creek, 2013, pp. 53–68.

[111] A. Schneider, M. Takla, A. Nicolay, A. Raab, T. Raab, A template-matching approach combining morphometric variables for automated mapping of charcoal kiln sites, Archaeol. Prospect. 22 (2015) 45–62.

[112] Ø.D. Trier, M. Zortea, C. Tonning, Automatic detection of mound structures in airborne laser scanning data, J. Archaeol. Sci. Rep. 2 (2015) 69–79.

[113] L. Verdonck, Detection of buried roman wall remains in ground-penetrating radar data using template matching, Archaeol. Prospect. 23 (2016) 257–272.

[114] T. Fawcett, An introduction to ROC analysis, Pattern Recogn. Lett. 27 (2006) 861–874.

[115] V. Dey, Y. Zhang, M. Zhong, A review on image segmentation techniques with remote sensing perspective, in: W. Wagner, B. Székely (Eds.), Proceedings of the ISPRS Technical Commission VII Symposium, 100 Years ISPRS. Advancing Remote Sensing Science. Vienna, 5–7 July 2010, XXXVIII-Part 7A, 2010, pp. 31–42.

[116] A. Schmidt, G. Tsetskhladze, Raster was yesterday: using vector engines to process geophysical data, Archaeol. Prospect. 20 (2013) 59–65.

[117] N. Linford, P. Linford, The application of semi-automated vector identification to large scale archaeological data sets considering anomaly morphology, in: B. Jennings, C. Gaffney, T. Sparrow, S. Gaffney (Eds.), 12th International Conference on Archaeological Prospection, Bradford, 12th–16th September 2017, Archaeopress Publishing Ltd, Oxford, 2017, pp. 138–139.

[118] E.G. Ernenwein, Integration of multidimensional archaeogeophysical data using supervised and unsupervised classification, Near Surf. Geophys. 7 (2009) 147–158.

[119] T. Blaschke, G.J. Hay, M. Kelly, S. Lang, P. Hofmann, E. Addink, R. Queiroz Feitosa, F. van der Meer, H. van der Werff, F. van Coillie, D. Tiede, Geographic object-based image analysis–towards a new paradigm, ISPRS J. Photogramm. Remote Sens. 87 (2014) 180–191.

[120] M. Baatz, A. Schäpe, Multiresolution segmentation: an optimization approach for high quality multi-scale image segmentation, in: J. Strobl, T. Blaschke, G. Griesebner (Eds.), Angewandte Geographische Informationsverarbeitung XII, Beiträge zum AGIT Symposium Salzburg 2000, Herbert Wichmann Verlag, Berlin, 2000, pp. 12–23.

[121] L. Drăguţ, D. Tiede, S.R. Levick, ESP: a tool to estimate scale parameter for multiresolution image segmentation of remotely sensed data, Int. J. Geogr. Inf. Sci. 24 (2010) 859–871.

[122] A. Hinterleitner, K. Kastowsky-Priglinger, K. Löcker, W. Neubauer, M. Pregesbauer, V. Sandici, I. Trinks, M. Wallner, Automatic detection, outlining and classification of magnetic anomalies in large-scale archaeological magnetic prospection data, Archaeol. Polona 53 (2015) 296–299.

[123] M. Pregesbauer, I. Trinks, W. Neubauer, An object oriented approach to automatic classification of archaeological features in magnetic prospection data, Near Surf. Geophys. 12 (5) (2014) 651–656.

CHAPTER

Efficiency of the magnetic method in surveying desert sites in Egypt and Sudan: Case studies

7

Tomasz Herbich

Institute of Archaeology and Ethnology, Polish Academy of Sciences, Warsaw, Poland

CHAPTER OUTLINE

1 DESERT SITES IN EGYPT AND SUDAN

Egypt and the northern Sudan are commonly perceived as desert environments hostile to developed settlement, requiring water sources and soils capable of sustaining agriculture and pastoralism. Proper conditions, at least since the climate change in the Neolithic, are met in the Nile Valley and Delta and the Western Desert oases: Bahariya, Siwa, Farafra, Dakhla, and Kharga, as well as Fayum, regarded as an

Innovation in Near-Surface Geophysics. https://doi.org/10.1016/B978-0-12-812429-1.00007-6

oasis despite being adjacent to the Nile Valley and taking water from the river. The cultivation zone in the Nile Valley in antiquity, from a few hundred meters (in Nubia) to more than 20 km (in Middle Egypt) wide, is the result of fertile silt accumulation during seasonal river flooding starting from the early Holocene. In the greater part of the Egyptian Nile Valley, its sides are marked by cliffs that are the edge of limestone beds in Middle and Upper Egypt and sandstone ones (south of Esna/Edfu), commonly referred to as a high desert. An intermediate zone made of sand and gravel, called the low desert, exists between the high desert and the alluvial plain where the Nile flows. It encompasses the upper part of Pleistocene formations filling the valley before the start of the silt accumulation process and the deposits from cliff erosion and the seasonal flow of water down the wadis from the high desert [1] (p. 2). This zone that is devoid of vegetation, thus of desert character, which lies next to the cultivated zone but not threatened with annual inundation by the Nile, was occupied in antiquity, more or less intensively depending on the region.

Desert sites include also oasis settlements situated outside the watered and cultivated areas and coastal sites outside the Nile Delta. Owing to the destruction of irrigation networks in many places in the Western Desert oases, sites occupied in antiquity in relation to agriculture are desert today; the same is true of many sites in Sudan: established within or at the edge of the cultivated zone, they were covered by layers of windblown sand after occupation ceased and are a desert today.

The article presents a selection of archeological sites found in desert areas as defined above: the low and high desert, land that has undergone desertification in the oases and the desertlike seashores.

2 TYPES OF SITES AND THEIR LOCATION

All types of archeological sites can be found in desert areas, the same as on cultivated land in the Nile Valley: cities and villages, religious centers, production sites, cemeteries. The proportions differ however, depending on environmental, economical, and cultural-religious factors. The former set the conditions for permanent settlement in the Nile Valley only outside the flooded zone and in the case of larger cities—nearer to the river, which was the main transport route. Thus, desert cities and villages are almost exclusively in the oases and on the coast and in a very few cases in the low desert within the Nile Valley, usually where the valley was narrow, facilitating access to the river, or close to channels branching off from the Nile. Another category consists of special task settlements, like the villages for workers exploiting natural resources in the desert or building temples and tombs at the desert edge. Next are fortified military settlements. Production centers (glass, pottery, and metal smelting) were located near or within the settlements [2]. The desert was also, owing to its secluded nature, a perfect location for Coptic monastic centers.

The desert fringes of the Nile Valley and the desert above the cliffs were also the main location of burials, for reasons that were at once religious (beliefs) and practical (best preservation of bodies in conditions of low humidity of the surroundings).

With regard to the kings of Egypt, whole complexes related to the tomb and posthumous cult were raised in this zone [3].

Studies of paleolandscapes, present earlier already in European archeology, have also become popular in the archeology of Egypt and Sudan. The physical properties of the Nile silt (see below) create perfect conditions for observing changing river banks and reconstructing networks of canals distributing water from the Nile in desert (or secondarily desertified) areas [4].

3 GEOLOGICAL AND MATERIAL CONDITIONS OF MAGNETIC PROSPECTION

Desert sites in Egypt and Sudan are for the most part made of sand and gravel or else rock waste covering limestone or sandstone massifs, respectively, in Egypt and in both Egypt and Sudan. These are all materials with very low magnetic susceptibility. The only exception are sandstones in some regions of southern Egypt and Nubia, with increased magnetic susceptibility caused by the presence of iron oxides in the binder. Igneous rocks are present only in the Eastern Desert; they contain magnetic iron oxides; hence, it is only in these areas that the bedding has higher magnetic susceptibility.

Dried mud brick made of locally extracted silt was the principal building material throughout the ancient history of Egypt [5]. Silt from the Nile Valley is characterized by strong magnetic susceptibility [6]: $2–5 \times 10^{-3}$ SI in Egypt and up to 8×10^{-3} SI for deposits from the regions of Old Dongola and Kerma in Sudan (according to the measurements taken by the author). Brick production processes, which involved the addition of organic material or sand, changed the susceptibility of the silt bricks, but never was it less than $0.8–1.5 \times 10^{-3}$ SI in Egypt and less 4×10^{-3} SI in Sudan. Mud brick made of local silt other than coming from the Nile alluvia is characterized by a susceptibility not more than $0.2–0.3 \times 10^{-3}$ SI (Fig. 1). Walls of such material can be traced depending on their mass, the depth of deposition below ground surface, and the properties of the matrix surrounding them.

The baked brick, which became common in Egypt from the Roman period (from 30 BC to AD 476) on, had raised magnetic susceptibility due to the effect of thermoremanence [7] (p. 87). Structures made of both kinds of bricks can be traced with the magnetic method, taking advantage of the very strong contrast between the susceptibility of archeological features and that of the surroundings when these features are found in a sand-and-gravel matrix or directly on bedrock (sandstone or limestone). The end outcome is a precise localization of the architecture.

Stone, sandstone, or limestone, a material with just slight magnetic susceptibility, was used for constructing buildings of a sepulchral and sacral nature. In theory, remains of this kind should be invisible in a sand-and-gravel matrix; but in practice, owing to the presence of ash and broken pottery, the magnetic susceptibility of these deposits may be high enough to provide sufficient contrast allowing stone structures to be recorded.

FIG. 1

Medinet Watfa/Philoteris. Wall of sun-dried bricks, bricks made of local silt *(yellow)* and of Nile mud *(dark gray)*. Magnetic susceptibility of sun-dried bricks made of Nile mud (A) and local silts (B) measured with a kappameter. Phot. T. Herbich.

4 PAST AND PRESENT OF DESERT SITE PROSPECTION IN EGYPT AND SUDAN

The first prospection of a site in desert conditions (and the first to be prospected in the Nile Valley) was the research by Albert Hesse at Mirgissa in the part of Nubia located in Sudan, carried out in 1965–67 [8,9]. A proton magnetometer was used for the survey, and the maps for analysis were made by hand. The result was a tracing of regularities in the urban layout of architecture within the fortified enclosure of a Middle Kingdom (2040–1650 BC) fortress. The magnetic properties of Nile silt were first observed during this project.

The next survey was conducted in 1973 by Elizabeth Ralph, whose team from the museum of the University of Pennsylvania introduced the cesium magnetometer to archeology. The object of the investigation was a palace of Amenhotep III (1391–1353 BC) in the desert plain at Malkata in West Thebes. The survey, still using a manual analysis of results, mapped the layout of some walls of mud brick [10].

A team made up of researchers from the Stanford Research Institute in California, in close association with Ain Shams University and the Egyptian Organization of Antiquities, undertook an evaluation of the effectiveness of different geophysical methods in surveying archeological sites, mainly on the necropolis in Giza, Saqqara, and Luxor, located in the high desert [11]. Magnetic prospection was of marginal importance in this project, but even so the potential of the method was noted.

Researchers used a cesium magnetometer to take measurements along single sections cutting identified archeological features and tested the magnetic properties of mud brick under laboratory conditions, accepting in the end the potential of the method (evidently having no idea about Albert Hesse's results).

The next important project to be carried out in the desert, in the funerary complex of Fifth Dynasty (2450–2325 BC) pharaohs in Abusir (between Giza and Saqqara), was by a Czech team headed by Vladimír Hašek [12]. Magnetic research (applying a proton magnetometer) in front of one of the pyramids, initiated in 1979, confirmed the theory about a building of mud brick there. Excavations determined its function as a mortuary temple [13] (pp. 77–78).

In the early 1980s, Egyptian researchers from the geophysical observatory in Helwan tested the magnetic method for the first time. A survey of limited areas in Karanis on the current desert fringes of Fayum Oasis registered anomalies associated with the presence of archeological features (e.g., fireplaces) [14].

In 1987, the author conducted a magnetic prospection to the west of the pyramid of Djoser in Saqqara using proton magnetometers in a differential mode. A series of anomalies was noted, indicating the presence of remains of mud-brick architecture [15]. The highest-amplitude anomaly was verified in excavation as fill of mud bricks, concealing the entrance to the funerary chapel of a noble from the times of the Fifth Dynasty [16] (pp. 16–18). Field research resumed in 1996 using a cesium instrument, located the mud-brick enclosures around the tomb shafts [17].

In 1988–92, Ian J. Matthieson, who was working mainly with the electric resistivity method in Tell Amarna in Middle Egypt, applied the magnetic method to locate ovens, baking areas, and glass manufacturing kilns [18]. That year, he also started a big project in the area of the early Dynastic period (3100–2600 BC) in Gisr el-Mudir temple enclosure in Saqqara, testing different methods to trace archeological features and using the fluxgate gradiometer, already used in Egypt, for the first time after 10 years. Research in the following years on the northern side of the pyramid resulted in the localization of dozens of mastabas with plans typical of the Old Kingdom (2600–2137 BC) [19,20]. But this was already after the magnetic method had conquered Egypt, being applied successfully to survey work both in desert areas (the author was surveying already then in Abydos [21,22]) and in alluvial regions in the Nile Delta (projects in Qantir and Tell el-Dab'a [16,23], pp. 28–30).

In the first years of the 21st century, Czech researchers carried out important geophysical prospection in the Abusir cemetery using the newest generation of cesium magnetometers; they ended up mapping mastabas of the Old Kingdom [24].

Among the projects from the past decade, one should mention research around Asyut [25], which clearly mapped the layout of the settlement and the plan of particular structures and the causeway of the pyramid complex of Amenemhat II in Dahshur [26]. In Sudanese Nubia, a complete plan of the town of Amra from the New Kingdom (1550–1070 BC) was obtained [27].

A full list of magnetic geophysical projects in Egypt and Sudan in the past 20 years is not available due to still patchy publication, and one can hardly estimate the proportions between sites investigated in desert zones (or zones made dry and arid) and those in the floodplain. The number of sites examined by the author and his team in these two countries, 78 in all, does permit, however, certain observations of a statistical nature. Of the number mentioned above, 21 are believed to be desert sites (27%), and another 28 (36%) are in regions desertified at a later date (after having been settled); 29 sites lie on Nile alluvia (37%).

5 MEASUREMENT METHODOLOGY

The case studies presented here were all carried out with Geoscan Research fluxgate gradiometers measuring the vertical gradient of magnetic field intensity with resolution up to 0.1 nT (Fig. 2). The sampling grid for the FM18 instrument was 10 × 20 m and for the FM36 and FM256 20 × 20 m. Measurements were taken at eight samples per 1 m^2, taken every 0.25 m along lines half a meter apart. In most cases, a parallel mode was adopted, zigzag mode being used only on Medinet Watfa and Sudanese sites. The results were processed with Geoplot software, the maps printed with Surfer.

6 CASE STUDIES

The 12 case studies discussed here, 10 in Egypt and 2 in Sudan (Fig. 3), are subdivided by the nature of the surveyed archeological sites: urban sites, including cities and villages, palatial centers, workers' villages, monastic settlements, cemeteries

FIG. 2

Magnetic prospection of the el-Amra site using a fluxgate gradiometer Geoscan Research FM256. Phot. T. Herbich.

FIG. 3

Location of sites presented as case studies. Satellite map Google Earth.

and cult places, and industrial centers. Furthermore, research in two settlements in the same geographically consistent region provided testimony of how the magnetic method can help to reconstruct changes in the landscape.

The category of cities and villages is represented by the results of research at el-Amra (geophysical research carried out within the frame of a project by the University of Pennsylvania), in Berenike (American project from the University of Delaware, coorganized first with a Dutch team from Leiden University and then with a Polish one from the Polish Center of Mediterranean Archeology, University of Warsaw). Palatial centers include Usli (Charles University in Prague) and Soniyat (PCMA), whereas Dahshur (German Archeological Institute in Cairo) is discussed as an example of a workers' settlement for individuals tasked with a specific job in a given time. Monastic settlements are represented by the complexes at Bawit (Louvre Museum and French Institut français d'archéologie orientale) and Kellia/Pherme (Mission suisse d'archéologie copte). Cemeteries and cult places are illustrated by the results of research in Abydos, Dayr al-Barsha, and Dahshur. The sites in the low desert in Abydos, spread over a distance of 4 km, are explored by the German Archeological Institute in Cairo (Umm el-Gab) and the joint University of Pennsylvania-Yale-Institute of Fine Arts, New York University supported by Brown University and Michigan University. Geophysical prospection at the cemetery of Dayr al-Barsha are carried out by the Catholic University in Leuven, Belgium, the valley temple in Dahshur as part of the same project as the workers' village. Industrial complexes are investigated by the independent Expedition to Hierakonpolis with archeological verification supported by the Japan Society for the Promotion of Science. As for landscape archeology, the work at Medinet Watfa and al-Qarah al-Hamra is an example. The former was part of a project coordinated by the German Archeological Institute in Cairo; the latter was done within the framework of a joint research project of the University of California Los Angeles, the University of Groningen from the Netherlands, and the University of Auckland, New Zealand.

All the survey work is carried out in cooperation with the Supreme Council of Antiquities (formerly Egyptian Antiquities Organization) in Egypt, with the National Corporation of Antiquities and Museums in Sudan, and with the support of the Polish Center of Mediterranean Archeology, University of Warsaw.

6.1 VILLAGES AND CITIES

6.1.1 El-Amra

El-Amra lies in the low desert west of the Nile, about 10 km south of Abydos. The cemetery on the site was first explored in 1900–01. Deposits ranging from Nagada I through Dynasty 2 (3900–2890 BC), extremely important for understanding the beginnings of the Egyptian state, were uncovered in close to 600 burials. The most recent excavations were carried out in 2007, including a topographic and surface survey to locate early Dynastic graves not explored since the beginning of the 20th century. Magnetic prospection was conducted on three mounds in the northeastern part of the site, at the edge of the low desert [28] (pp. 111–122). Oval anomalies with lowered magnetic field intensity value, about 4 m more or less in diameter, reflected the pits left by the plunder or excavation of graves; other anomalies of irregular shape corresponded to earth mounds or embankments left from the excavations. However, three clusters of anomalies with a distinctive amplitude of values (in the range −8/+8 nT) on the central mound followed curved lines, apparently outlined, at least in part, by clearly distinct elongated anomalies (Fig. 4). The biggest of the clusters occupies an area of about 3350 m^2; the smaller ones are 500

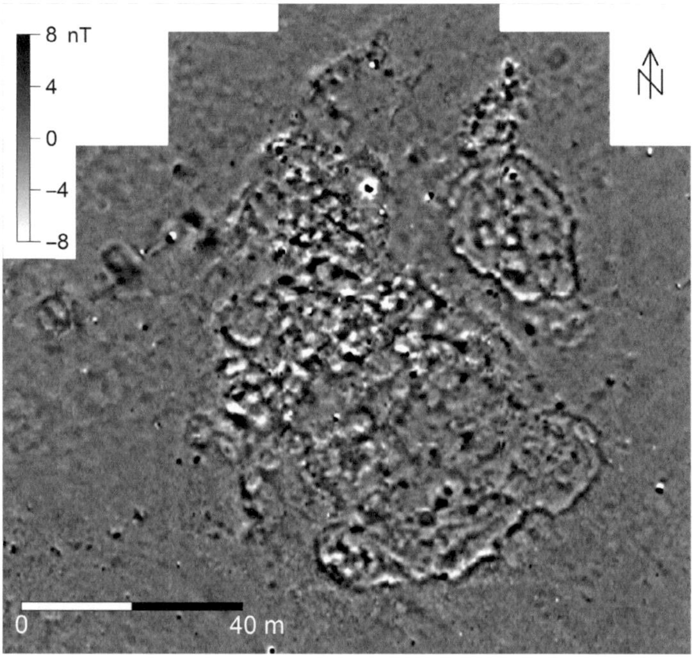

FIG. 4

El-Amra. Extract of the magnetic map with the central and eastern cluster of anomalies corresponding to the settlement remains. Processing T. Herbich.

and 250 m^2. On the ground, the anomalies matched clusters of pebbles, each of the stones 10–15 cm in diameter, surrounded by earth embankments, up to 25 cm high, darker than the surrounding ground. Surface cleaning in place of the anomalies uncovered the tops of walls made of mud mixed with pebbles (Fig. 5).

Thus, an analysis of the magnetic survey results considered in the context of surface remains and features found in shallow excavations led to the identification of the remains of a village of atypical plan but still known from other early sites, namely, an irregular enclosure wall and structures raised in a building technique that called for the emplacement of pebbles in a mud matrix [29] (pp. 12, 137). Enclosures of a parallel kind, dated to Nagada II (3600–3300 BC), are considered as a marker of incipient urbanization [30] (p. 312). The building technique as reconstructed explains the presence of stones on the ground corresponding to walls. Erosional processes caused the mud to be eroded away, leaving stones as the more permanent material visible on the surface. The magnetic map also shows anomalies typical of furnaces or hearths, supporting the idea that it was a settlement. A higher magnitude of response observed for anomalies corresponding to some sections of the wall was due to burnt clay and slag being added to the construction. The settlement was dated to the close of the fourth millennium BC based on pottery found within the village.

FIG. 5

El-Amra. Wall showing a concentration of cobbles where the mud has eroded. Phot. T. Herbich.

Structures of entirely different function and dating were mapped at the north-western edge of the magnetic image as anomalies forming in each case a rectangular outline measuring about 5×7 m (Fig. 4). These were most likely funerary shafts, the anomalies reflecting mud-brick lining. Anomalies recorded in deep depressions presumably corresponded to graves that have been explored. The pottery on the surface is from the Middle Kingdom.

6.1.2 Berenike

The city and port of Berenike in the southern section of the Egyptian Red Sea coast was founded by Ptolemy II near the end of the first quarter of the 3rd century BC for the purpose of working the trade between Egypt and regions lying to the east and south of it. A land route connected it with the Nile Valley. The city as we know it experienced three phases of prosperity: after establishment, during the early Roman period in the 1st to 2nd century AD and starting from the mid-4th century AD. In Ptolemaic times, Berenike was the landing stage for war elephants imported from eastern Africa, which were an important tactical element of the Ptolemaic army. During the Roman period, the harbor was part of a trade network between Alexandria and Rome on the western end and Arabia and India on the eastern one. Its importance in this role was reflected in numerous ancient written sources. It gave up its role in the 6th century when it was ultimately abandoned in favor of other maritime ports in the Mediterranean. The first plan of the ancient city remains dates to the early 19th century, but regular archeological research has not begun until the close of the 20th century [31].

The town occupies about 30 ha lying on a fossil-reef promontory between the mouths of two wadis. The urban mound in the eastern part consists of accumulations rising to 5 m with remains of late occupation visible in the surface. In the western part, the Ptolemaic remains are just under the surface but with no traces on the ground. The geology of the site determined the building materials: most of the architecture consists of irregular cobble-sized elements from the fossil reef and dressed blocks of gypsum anhydrite reinforcing building corners. Linear concentrations of coral heads lie on low sand embankments, indicating the layout of the structures underneath.

The archeological project aimed at methodical excavation of small trenches, hence the interest in noninvasive methods to reconstruct the town plan. Geophysics was included in the survey program but with some reservations, considering the rather low magnetic susceptibility of the building materials known to have been in use in Berenike. The method was tested first in the western part of the town where some slag on the surface suggested industrial activities. Unexpectedly, extensive architecture was mapped, the anomalies corresponding to walls exclusively of magnetic field intensity values lower than that of the surroundings ([16], pp. 50–52; [32]). It thus became clear that the magnetic properties of the surroundings were responsible for the imaging, being apparently rich in minerals containing iron oxides. Excavation subsequently confirmed that the anomalies corresponded to walls of gypsum anhydrite (or their ghost trenches) from the earliest phase of occupation in a given area [33] (pp. 30–44). A magnetic survey of the site became a priority after the project was resumed in 2008, covering by 2015 most of the site surface and being used regularly to plan future excavations, as well as supporting separate topographical analyses of the town and harbor layout [34].

The architectural complex uncovered by the mapping with the magnetic method in the western part was of a defensive nature. It covered a roughly rectangular area, 160 m from north to south and about 80 m from east to west. The outer wall outlines were very clear, along with a complex of rooms on the southern side. Further analyses combined with spot excavations to verify the findings have resulted in a hypothetical phasing of the feature [35]. Archeological evidence has been unearthed to support the theory that wild elephants brought from Africa for the Ptolemaic military establishment were kept in a pen in this area, recorded on the magnetic map, until they were ready to be moved across the Eastern Desert to the training camps in the Nile Valley.

Measurements on the urban mound mapped a small town, which had already been recorded in a ground survey registering visible remains from the mid-4th to the 6th century BC [36] (Figs. 6 and 7). The buildings seem to develop along a decumanus and two cardines [37]. The combination of surface tracing and spot excavations identified the center of the town as a mix of commercial and residential buildings, as well as a Christian ecclesiastical building at the eastern end of the decumanus. The Great Temple, discovered already in the 19th century, stood at the opposite end of the decumanus. The larger size of structures on the outer eastern fringes of this mound was interpreted initially as public buildings, commercial, administrative, or religious in character [38].

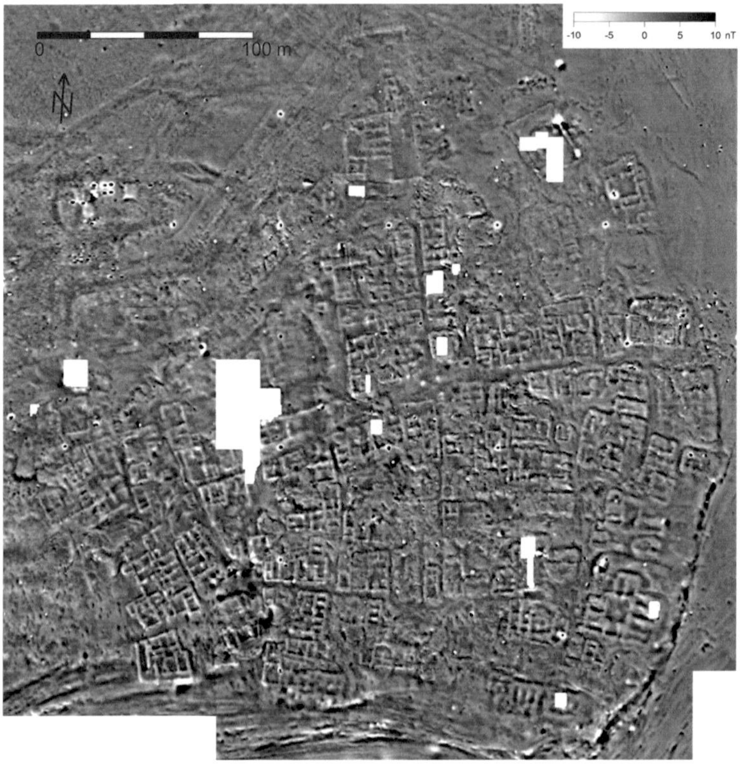

FIG. 6

Berenike. Magnetic map of the town mound in the eastern part of the site. Processing T. Herbich.

Magnetic mapping of this area gave a much clearer picture of individual houses and the insulae that they formed. The Roman layout appears not to have repeated the Hellenistic plan, which is thus for the most part hypothetical, based on what is known of the regular urban planning of other towns of this period. The extent of insulae and the course of streets between them could be traced in minute detail on the geophysical map. The biggest street, decumanus, was of varying width, and the other streets rarely followed a straight line. At the crossing of the decumanus with the eastern one of the cardines, which seems to be a town square close to 20m wide and extending west toward the Great Temple, the magnetic survey has revealed three point anomalies that come together to form a square (the fourth seems to be obscured). An association with a tetrapylon, a structure marking important street crossings in Roman towns, is unavoidable. Assuming that excavations confirm the presence of columns on bases forming such a structure, it would be the first tetrapylon to be discovered in Egypt. The magnetic map also shows evidently different architecture at the eastern edge of the town to the south of the decumanus with buildings of larger size and at

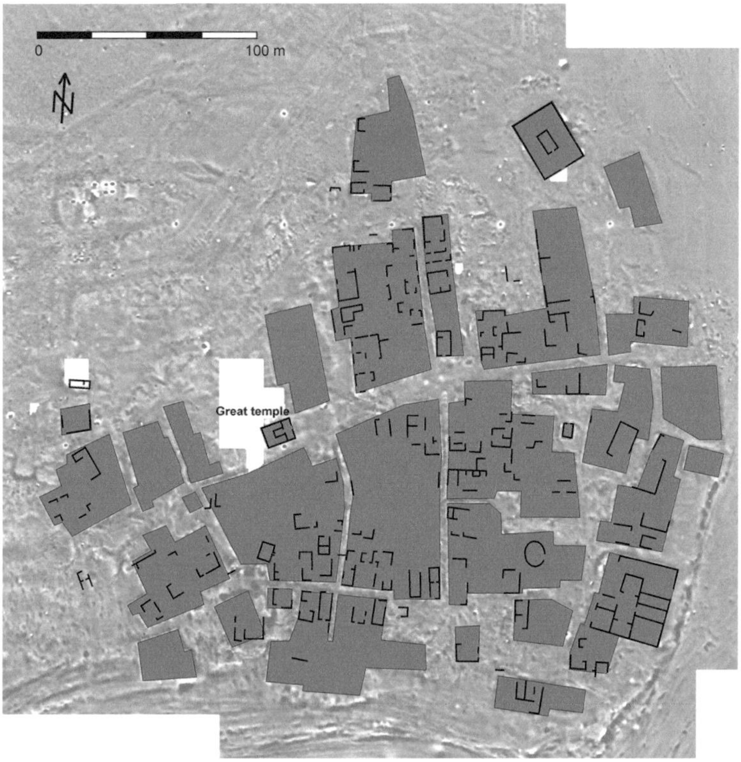

FIG. 7

Berenike. Magnetic map of the town mound with superimposed position of walls recorded during the topographic survey of 1994; shaded boxes show presumed insulae. Processing and drawing T. Herbich.

least partly equipped with large courtyards. The mapping on the southwestern side of the urban mound is detailed to the point of enabling passages between particular units to be identified (Fig. 8). Small-scale excavation in this part confirmed the preciseness of the mapping of individual walls with regard to their actual position (Fig. 9). They also showed that linear anomalies of raised value, separating the built-up area from the sea, could have corresponded to waterfronts artificially formed of, among others, deposits of broken amphorae. Use of amphorae to reinforce waterfronts is known from another Red Sea harbor of Ptolemaic/Roman period in Myos Hormas [39] (pp. 170–171).

Research in the area interpreted as the southwestern bay proved of importance in understanding the functioning of the harbor and the changing conditions of access to the sea. Initially, it was thought that the main harbor of Berenike had been located here with big ships docking at a regular built port [39] (p. 173). However, the magnetic image of the bay in its southern part revealed a series of roughly semicircular

FIG. 8

Berenike. Magnetic map of architectural remains in the southeastern part of the town mound. *White box* marks location of the trench shown in Fig. 9. Processing T. Herbich.

FIG. 9

Berenike. Corner of the house registered by magnetic survey (location of trench; see Fig. 8). Phot. Berenike Project/I. Zych.

anomalies suggesting a receding shoreline. Excavations showed that only small boats could have come into the bay, ferrying in and out the goods loaded on large ships anchored further out to the sea where the anchorage was. The bay quite obviously went out of use as a commercial center in the later 2nd century AD; in the 4th to 6th centuries, all the trading must have taken place along the eastern shoreline of the urban mound in the eastern part of the site, possibly by the reinforced waterfronts mapped by the magnetic survey and taking advantage of large empty spaces like the square area approximately 20 × 50 m in the southeastern section of the town [34].

6.2 PALATIAL CENTRES: SONIYAT AND USLI

The site of Soniyat lies on the right bank of the Nile, between the third and fourth cataract, in an area where the river is flowing latitudinally. Moving upstream, it changes direction to the northeast and because of northerly winds prevented further sailing. The river transport coming from the north had to be replaced by land transport. This influenced the region's significance and development. As an archeological site, Soniyat had been known in the 19th century, when a British traveler noticed ruins of a stone temple there. However, over the next two centuries, the site escaped subsequent travelers' and archeologists' notice, most likely because it was periodically covered by sand dunes. In the years 1997–2001 and in 2016, following the site's rediscovery in 1991, the temple was uncovered and documented ([40,41], pp. 243–249; [42]). The temple was a place of worship for the people of the Kingdom of Kush (from 10th century BC to 4th century AD). After it was uncovered, speculation regarding the ancient name of the settlement revived. It lies in an area currently known as Tegris, which has been connected to the ancient Tergedum, mentioned by Pliny the Elder in a description of Nero's centurion expedition of AD 60 (Naturalis Historia VI, XXXV 184–5) [43]. Another record mentions that in 593 BC, the army of king Psametych II reached a town called *trgb*, which topographically could match Tegris/Tergedum. However, the record also mentions the existence of a royal palace there [42]. In 2013, in order to verify this information, a magnetic survey of the area around the temple was conducted.

As a result of the research, 120 m from the temple, linear anomalies of varying width ranging from 1 to 2.5 m formed a perpendicular layout (Fig. 10). The nature of the anomalies shows that they are a reflection of mud-brick architecture. The complex is surrounded on three sides, on the southwest, northwest, and southeast, by anomalies corresponding to enclosure walls. Measurements have not recorded the northeast wall; it can be concluded from the magnetic image that building remains are covered by superimposed deposits that prevent the registration of structures. The width of the complex is 80 m; the length is at least 65 m. The map shows distinct rows of rooms around courtyards and open spaces between the structure and the enclosure wall. The plan of the structure clearly relates to other royal palaces known in the area [44]. The discovery of the palace strongly reinforces the hypothesis that this is the place mentioned in the description of Psametych's expedition.

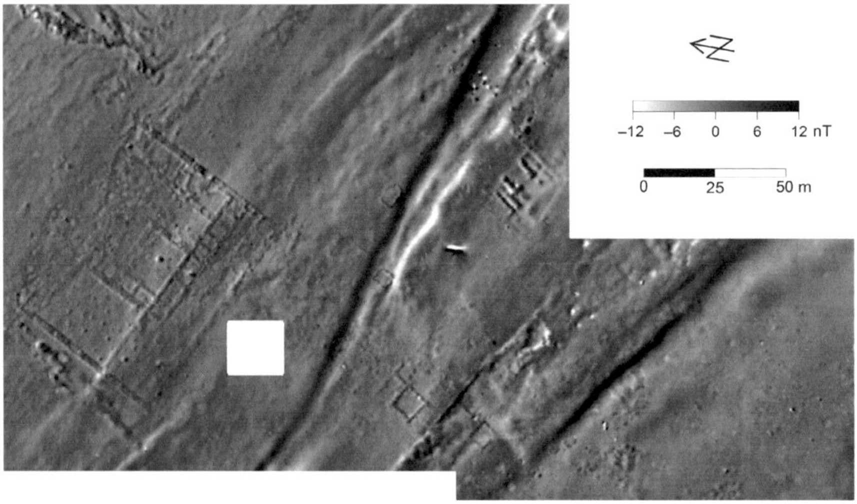

FIG. 10

Soniyat. Magnetic map of the palace building (in the northern part of the site) and the bank of a paleochannel of the Nile (in the southern part). Processing T. Herbich.

Ancient written sources mention that the settlement lay on the Nile. Currently, it is 1.3 km north of the river. The alteration of the riverbed in this place and its shift to the south is confirmed in the descriptions of contemporary travelers. Records of the riverbed's change are reflected in the geophysical image: a longitudinal anomaly, slanting NW-SE in the southern part of the map, can be connected with great certainty to the old river bank. At the time, the palace stood with its front to the river, at a distance of 110 m from the bank, and the temple almost exactly at 90°, at a distance of 60 m from the bank. Another diagonal anomaly running between the temple and the palace could correspond to a channel or shows the earlier stage of the movement of the riverbed. Therefore, the results of the geophysical study allowed not only for locating the structure and partially reading its plan but also for placing it in contemporary environment.

A similar result was obtained through research conducted in Usli. The site, lying 60 km upstream from Soniyat, was discovered during a large surface research project [41]. The presence of fragments of stone blocks and column drums betrayed the existence of temple remains, while pottery collected from the surface allowed it to be dated to the Kush period. There are certain indications that permit the temple to be identified with the one mentioned in the inscriptions on the stela of a Kushite king in 4th century BC [41] (p. 26). Geophysical research documented the presence of mud-brick buildings surrounded by fragments of an enclosure wall (Fig. 11). The structure in the center of the complex, measuring 44 × 30 m, has been interpreted as a palace building [45]. Point anomalies of low magnetic value grouped in regular arrangements of two or four in the side rooms may correspond to the stone bases of columns. Similar anomalies, but of higher amplitude values, in the central hall could

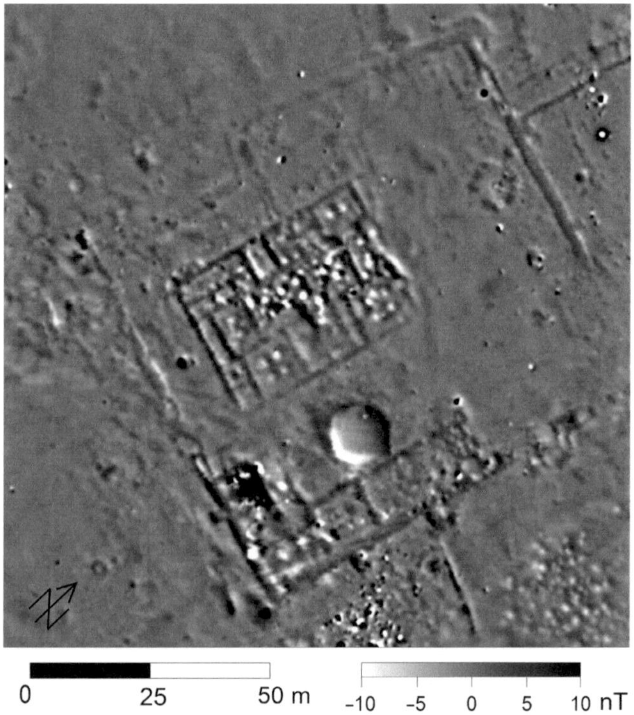

FIG. 11

Usli. Magnetic map showing a palatial building (in the center) and temple (in the south) within an enclosure wall. Processing T. Herbich.

reflect column bases made of material, like baked brick, characterized by higher magnetic susceptibility.

Subsequent excavations have shown that the building on the western side of the palace, within the enclosure wall, is a temple remain. The temple, remains of which were seen on the ground surface when the site was discovered, found no reflection on the magnetic map, presumably because nothing but the stone elements of the architecture have been preserved [46].

6.3 WORKERS' SETTLEMENT: DAHSHUR

The presence of two rectangular installations on the southern side of the Red Pyramid in Dahshur, which dates to the rule of Snefru, founding ruler of the Fourth Dynasty (2600–2450 BC) during the Old Kingdom in Egypt, was observed already in the mid-19th century. The map drawn at the time recorded their dimensions: 165 (east-west) × 60 m (north-south) for the western one and 190 × 60 m respectively for the eastern one. Each of the structures was made up of at least three units ([47], p. I, 35; [48], p. 14). The drawing indicates that they could be traced on

FIG. 12

Dahshur. Area of the workers' village recorded by the magnetic survey, seen from the south. The Red Pyramid of king Snefru in the background. Phot. D. Święch.

the ground without excavation. In the second half of the 20th century, these structures were no longer to be seen (Fig. 12) except in aerial and satellite imagery, and their layout differed substantially from the plan made in the previous century. In functional terms, they were thought to be workmen's housing [49] (p. 16) a hypothesis verified by excavation within a section of the eastern structure. A long room with cross divisions was cleared, yielding large amounts of animal bones and pottery to identify its residential function. Evidence of pottery production and copper and stone tool repairs was also forthcoming. The pottery dated to Snefru's times [50,51] (pp. 291–294). Further investigation, supported by the discovery of a similar complex in the neighborhood of the pyramids in Giza, confirmed the barrack function of these structures, used to house workmen building the nearby pyramids [52] (pp. 306–309).

Geophysical prospection in Dahshur had already a history of its own, having been started by Helmut Becker in 2003 [53]. The mapping revealed a number of mudbrick structures in the cemetery and around the valley temple complex of the Bent Pyramid. The author continued the work in the neighborhood of the valley temple [54] (pp. 207–208), testing at the same time the effectiveness of the method in the area of the workmen's village, where it was expected that the map would locate hearths inside the barracks rather than the run of individual walls that were built of limestone, a material devoid of magnetic properties. In fact, the method recorded all the walls, including those observed in satellite imagery and those not visible on the ground [48]. The results were significantly better in the eastern complex that had not been excavated until now.

Based on the published survey measurements, which seemed to cover at least 80% of the area of the eastern complex, and a comparison of combined satellite and geophysical imagery, the building was suggested to be made up of a block of 17 long parallel (7 × 27 m on average): 12 traced on the magnetic map, 2 in satellite images, and 3 at the extreme eastern border, suggested by the position of an anomaly that was presumed to correspond to the outer enclosure wall [48] (pp. 17–19). The position of hearths in particular units was also identified, always by the short wall or near it; hearths were also located to the south of the rooms, already outside the block, indicating that the building had been entered from the south, from the wadi. Extended investigation of the area to the north and east to cover the whole block of units revised the results, bringing the number of units in one row up to 21 and adding two rooms in a second row in the western part of the complex [55] (p. 17) (Fig. 13). The five units

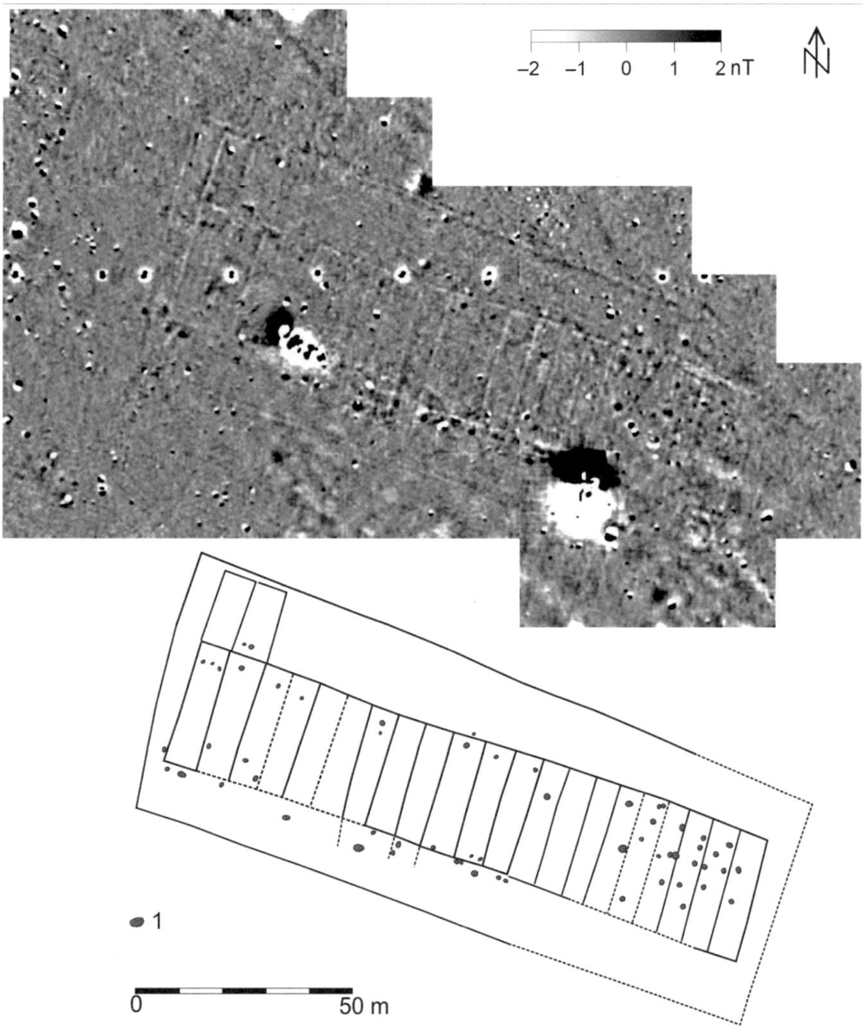

FIG. 13

Dahshur. Above: magnetic map of the eastern section of the workers' village. Below: workers' village reconstructed on the grounds of the results of the magnetic survey and satellite imagery. 1—fire places or kilns. Processing and drawing T. Herbich.

at the extreme eastern edge of the complex contained much more anomalies corresponding to hearths or furnaces than in the remaining rooms. The location of these anomalies was also different, being more likely to be concentrated in the central part and hence suggesting a different function—possibly not residential, but artisanal in character. The full size of the block based on magnetic research is 138 × 27 m, the area enclosed by walls measuring 150 × 60 m; this differs from the measurements taken from 19th century maps.

Geophysical research of the western complex was less clear in its results. Only the wall enclosing the block of rooms on the north and east is clear. The rooms seem to be longer (about 40 m), but the dividing walls are not evident, and there are far less anomalies corresponding to hearths. A series of units was discovered on the southern side of the complex; since they are not present on the 19th century maps, they must have not been visible already by that time. The magnetic image shows the front facades and numerous anomalies corresponding to hearths. Archeological excavations discussed above explored units presumably connected with this complex ([48], p. 17; [55], p. 17).

The image of the building reconstructed on the strength of the collected evidence is very close to the workmen's village discovered by Mark Lehner in Giza, the principal difference being the size of the complex. At Dahshur, there is basically one row of units enclosed by a wall; at Giza, there is a corresponding set of four rows [56].

6.4 MONASTIC SETTLEMENTS

6.4.1 Bawit

The site lies in a low desert, at the edge of the zone of cultivation. Written sources inform a monastic complex being established here in the end of the 4th century, although so far the earliest archeological evidence for occupation is from the 6th century. The end of the monasteries came sometime between the 11th and 12th centuries [57] (p. 229). The mound is elongated, about 40 ha in all, a maximum 930 (NS) × 410 m (EW) in size and 6–7 m high on average (but up to 10 m in places). Archeological excavations between 1901 and 1913 uncovered two churches in the center of the site and a series of monastic units [58]. The architecture is mostly invisible under the mounds of sand (Fig. 14); the tops of walls can be traced in very few places, and there are patches of ground strewn with potsherds. Church architecture employed baked brick faced with limestone, while the other buildings were raised of mud brick made of Nile silt. Earlier work had demonstrated that walls were preserved to a height of 2–3 m. In the late 20th century, an inventory was taken from the excavated structures based on a full range of preserved sources, resulting in an archeological map of the site ([59], pl. I; [60], pl. I–IV; [61], p. 193). It turned out, however, that it was impossible to locate most of the buildings uncovered during the first excavations owing to the absence of a site grid. The only complexes to be recognized in 2002 were the set of structures on the northern edge of the site and the northern church, covering approximately a combined 800 m^2 (i.e., 0.2% of the area of the site) [58].

FIG. 14

Bawit. Northern part of the site, seen from the north. Phot. T. Herbich.

Magnetic prospection covered the entire mound and low areas where traces of the enclosure wall were observed. The objectives of the survey were twofold: to verify the location of structures excavated a hundred years earlier and to create the fullest possible archeological map of the site including excavated and unexcavated structures [61]. Individual features were to be dated based on an analysis of the pottery found on the surface [62].

Magnetic prospection was efficient in mapping architecture on about 60% of the site (Fig. 15); approximately 10% of the surface was too blurred for reading owing to concentrations of potsherds on the surface. In the remaining part (siding with the modern village), it was impossible to obtain any results, presumably due to local farmers digging in search of decomposed mud brick (a material rich in nitrates constituting a perfect agricultural fertilizer) which destroyed this part of the site completely. Comparison of architecture from early excavations and plans recorded by magnetic surveying allowed individual structures to be located again [61] (pp. 175–183) (Fig. 16). The ground surface was carefully observed in search of telltale signs of previous excavations, mainly in the form of shallow depressions filled with windblown sand. Torp's and Golvin's reconstructed map of archeological remains was first superimposed on the magnetic map by matching the position of the sole identifiable structures excavated by Gaston Maspero in the northern part of the site. It was found that the location of other now invisible structures could vary up to several dozens of meters, for example, room XIX of characteristic elongated shape and dimensions equal to 24×7, which matched up very well, overall and in detail

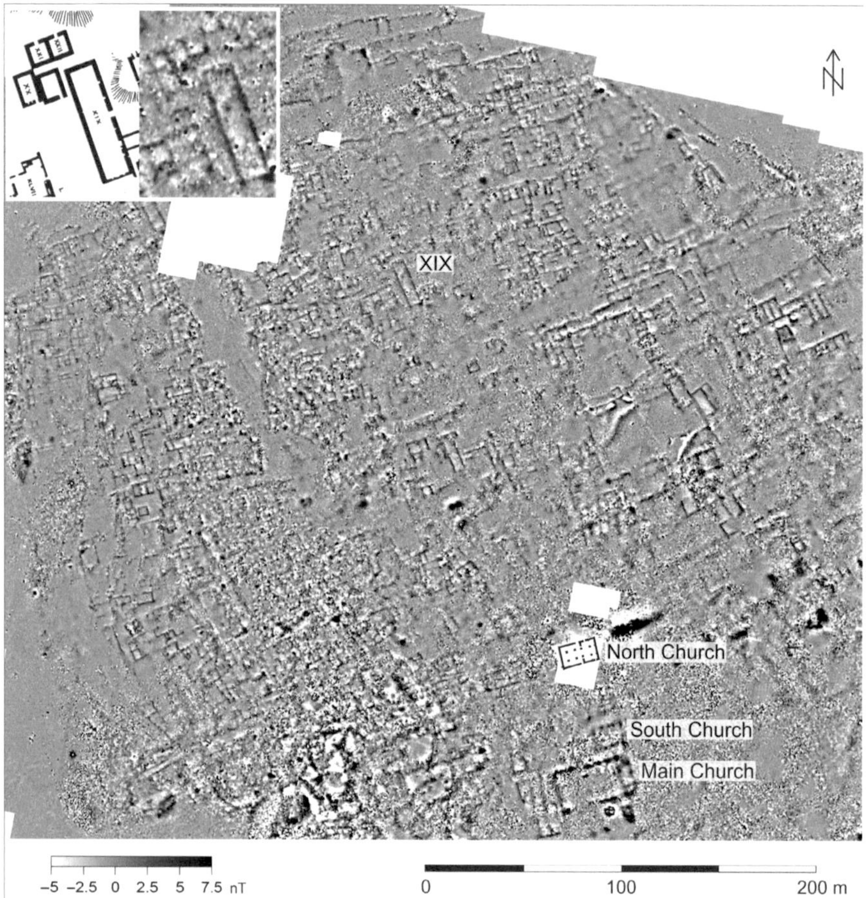

FIG. 15

Bawit. Magnetic map of the northern part of the site. Inset box: building XIX and neighboring structures as mapped by discoverers in the early 20th century (after [60]) compared with a magnetic map of this building and neighboring structures. Processing T. Herbich.

(like the entrance on the long side) with an anomaly registered in a different place. Previously unknown notes of archeologists digging the site in the early 20th century were also helpful in locating archeological features. The maximum error between sketched and real locations of features was close to 100 m.

Exploration on the southern side of the northern church demonstrated how unreliable old archeological plans published from field notes and measurements can be. For example, the distance between the north and south churches, which were excavated almost at the same time, is 16 m (Fig 17, left), while the real distance revealed by magnetic mapping of the remains (and verified by excavation) is 33 m (Fig. 17, middle and right). Moreover, a wall south of the southern church, marked by Chassinat as structure D on his map [65] (pl.VII) (Fig. 17 left), turned out to be part

FIG. 16

Bawit. Map of the northern part of the site reconstructed on the grounds of the magnetic survey results, ground survey, and excavations. 1, extent of the magnetic survey; 2, structures excavated before 2003, standing above ground surface; 3, structures excavated in 2003–16; 4, structures excavated in 1902–13, located on the grounds of the magnetic survey results; 5, structures reconstructed based on the magnetic survey results; 6, kilns; 7, structures registered during the topographical survey (2003); and 8, streets reconstructed based on magnetic survey results. Drawing T. Herbich.

of a wall separating the naos and sanctuary of the biggest (presumably the main) church in Bawit (Fig. 17 right). The structure was almost wholly excavated and found to be more than twice the size of the churches excavated earlier (Figs. 17 right and 18). The high amplitude of values in the case of the anomalies corresponding to the long sides of the church was due to the walls being raised of baked brick and surviving to a height of 1–2 m.

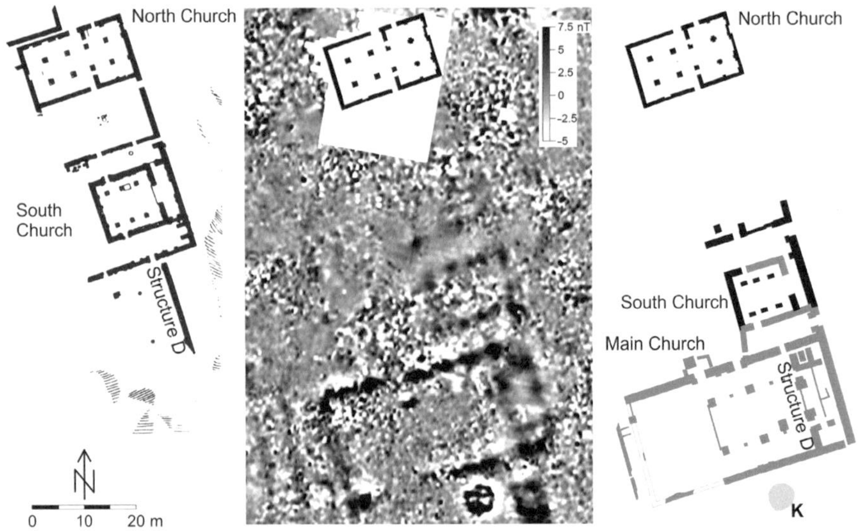

FIG. 17

Bawit. Left: north and south churches and structure D as mapped by the discoverers in the early 20th century (after [60]). Center: magnetic map of the area around the north and south churches. Right: location of the north and south churches and structure D as part of the main church. The south and main churches: structures verified through excavation in light gray, structures marked on the early 20th century maps in black. K—lime kiln. Plan of the north church after ([63], fig. 2), plan of the main church after ([64], fig. 2). Processing and drawing T. Herbich.

Investigation of the church brought interesting data on the depth range of fluxgate gradiometers. The clear anomaly marking the northern edge of the main church was excavated and found to be 4 m below ground surface. Lateral walls of the church, which were reflected in much more distinct anomalies (amplitude up to 20 nT) on the magnetic map, lay at an average depth of 2 m below the surface (Fig. 18).

Magnetic prospection permitted a fragmentary plan of the monastic settlement to be drawn (Fig. 16). The layout of individual buildings could be traced along with the course of streets and location of industrial areas (in the southern part of the site). The urban plan drawn on the basis of an analysis of geophysical results was the first undertaking of its kind in the archeology of Egypt, concerning a site of this size.

6.4.2 *Kellia/Pherme*

In the desert area on the western side of the Delta, in an area of about 100 square km, the remains of a large monastic center of Kellia/Pherme can be found. The center was discovered in 1964, and during 20 years of work, approximately sixteen hundred hermitages have been identified, grouped in five complexes, erected mainly between the 5th and 7th century [66] (pp. 55–82). At the beginning, each hermitage was intended

FIG. 18

Bawit. Main church seen from the east. Dotted line shows the ground surface level over the eastern edge of the church before excavation. Phot. Ifao/Musée du Louvre; Ihâb Mohamed Ibrahim.

for one monk: a few small rooms with a courtyard, surrounded by walls. Over time, the number of rooms increased owing to a change in the rule permitting contacts between monks, previously limited to Sunday services. Land use for agriculture on the borders of the Delta in the past few decades has led to the destruction of most of the complexes.

Most of the hermitages have been preserved as low mounds, with walls just below the surface. The rapid progress of land reclamation forced the archeologists to work in an unusual way: the plans of individual hermitages were made out by observing sand coloration on the surface (in the early morning, the sand above the walls was darker due to moisture), and then, the sand was swept away to show the tops of walls (Fig. 19). Thus, uncovered remains were then marked on the plans. Some hermitages have been unearthed to allow a closer look at the life of monks. They were erected from bricks made from local silt, which contains much less iron oxide than the Nile one. Their magnetic susceptibility varies from 0.05 to 0.5×10^{-3} SI.

After the emergency research ended at the beginning of the 1990s, archeological excavations were renewed in 2006, in the area of two partially preserved complexes: Qusur Hegeila and Qusur Ereima. The purpose of the geophysical research was to verify the accuracy of the plans drawn out a quarter of a century earlier and to determine whether the objects registering on the surface owing to the presence of pottery only were related architecture [67].

In Qusur Hegeila, hermitages whose magnetic susceptibility ranged from 0.3 to 0.55×10^{-3} SI were the most readable on the magnetic maps (Fig. 20 left). In least readable structures, susceptibility of bricks ranged from 0.1 to 0.35×10^{-3} SI. In Qusur Ereima, the susceptibility of bricks ranged from only 0.05 to 0.15×10^{-3} SI that

FIG. 19

Kelia/Pherme. Aerial view of the hermitages in Qusur el-Izeila after excavating the tops of walls. Survey in 1981. Phot. Mission suisse d'archéologie copte.

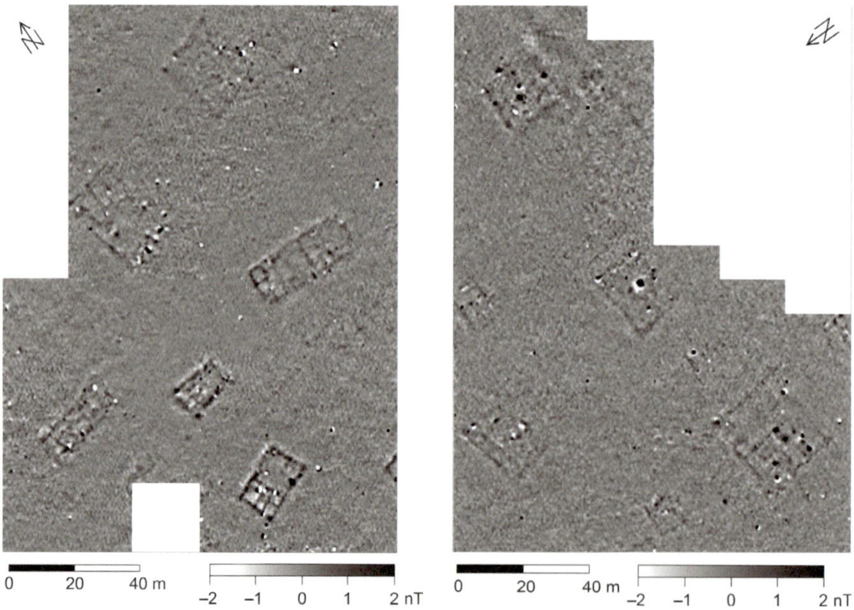

FIG. 20

Kelia/Pherme. Magnetic maps of the hermitages. Left, Pherme north; right, Pherme south. Processing T. Herbich.

however did not prevent the determination of the extent of structures and even elements of the inner layout (Fig. 20 right). In the case of both complexes, measurements allowed to verify the location of individual hermitages in relation to each other, to verify their number in the area researched, and to check some of their dimensions. Additionally, it was possible to identify rooms for household use due to very distinct anomalies corresponding to furnaces and fireplaces.

6.5 CEMETERIES AND CULT PLACES

6.5.1 Abydos

Abydos is one of the most important archeological sites in Egypt because of the number and importance of the archeological finds and because of its importance for Egyptian religion as the main cult place of Osiris. It lies in a vast desert plain, extending from 1 to 2 km in width on the west side of the Nile, between the cliffs of the high desert and the cultivated zone. Monuments include temples connected to the Osiris cult and the pharaohs' funerary cult, the so-called funerary enclosures associated with the funerary rites of the earliest kings of Egypt. Scattered in the plain around it are the funerary enclosures of kings from the earliest dynasties and a series of cemeteries from different epochs, from the Predynastic and early Dynastic through the Late and Greco-Roman periods. A townsite for those servicing the cult places has also been located [68].

Magnetic research was carried out for five different projects between 2001 and 2006, providing the mapping resources for planning explorations, contributing to important discoveries, and revising and supplementing the archeological site maps, which are the result of close to 200 years of excavations [16] (pp. 35–39).

In North Abydos, the features recorded by magnetic survey correspond to structures that, either due to size or typical elements of the plan, can be identified as funerary enclosures from the First and Second Dynasties (3150–2890 BC) [16] (Fig. 21). Excavations, taking place from 2003, have confirmed this identification, including the determination that one of these enclosures was for Aha, the first king of the First Dynasty. Another enclosure, of an unidentified ruler, cannot be fully excavated as it lies for the most part under a modern cemetery, but the survey among the modern graves permitted the full extent of this complex to be established [70]. Measurements in an open area within the modern necropolis uncovered another rectangular structure, its external dimensions being approximately 45 × 27 m (Fig. 21 inset box). The outer walls are fairly distinct, and inside, there are a series of parallel linear anomalies. This turned out to be a large building of the Ptolemaic period, the long anomalies corresponding to the highest part of arched vaulting covering the long rooms, lying at a depth of approximately 1 m below the ground surface [71] (Fig. 22). In other parts of the site, the magnetic image of tombs and funerary chapels raised of mud brick was so precise that it allowed dating of individual features based on characteristic elements of plan [21].

In the so-called Middle Cemetery, dozens of funerary structures were recorded (mainly from the Middle Kingdom to the Greco-Roman Period), only a few of these

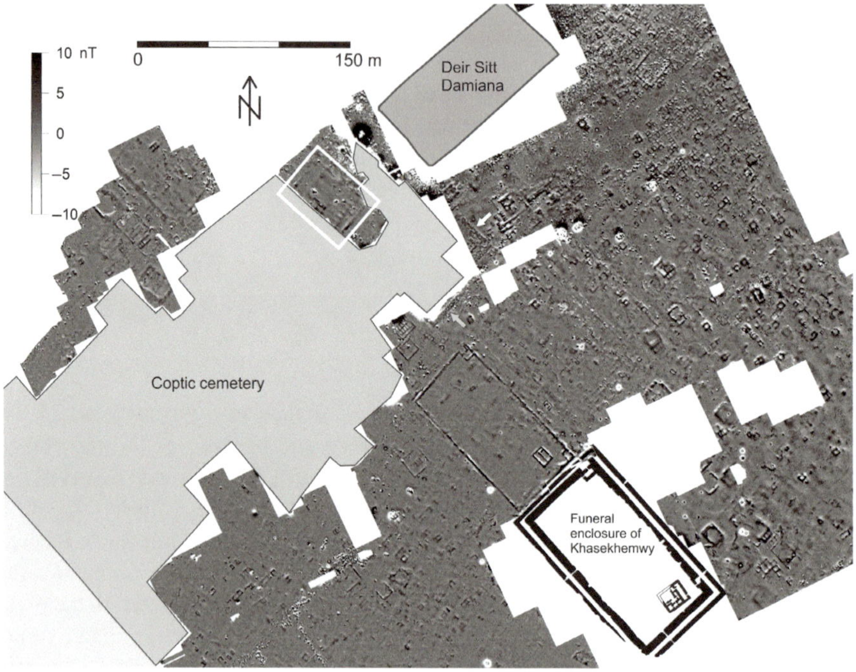

FIG. 21

North Abydos. Magnetic map of the royal funeral enclosures and cemetery. White box marks an anomaly corresponding to the structure shown in Fig. 22. White arrow marks the funeral enclosure of king Aha; *gray arrow* marks the southeastern section of the enclosure of an unidentified king. Drawing and processing T. Herbich. Map of the Khasekhemwy enclosure after ([69], fig. 2).

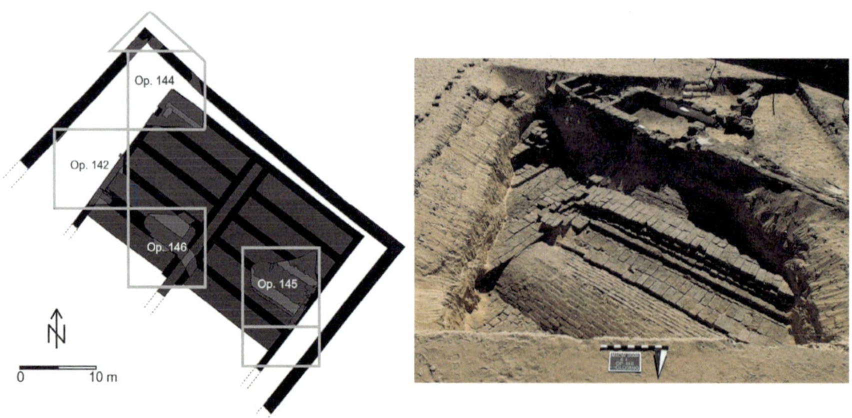

FIG. 22

North Abydos. Large vaulted building registered by the magnetic survey. Left: reconstruction of the building plan based on the magnetic results and excavation (outlines of trenches marked in *gray*); right: part of the building in Op.145. Plan and phot. Brown University Abydos Project/L. Bestock.

having any traces on the surface (Fig. 23). The survey also rediscovered an Old Kingdom mastaba with the outer dimensions 25 × 27 m, excavated by Richard Lepsius in the first half of the 19th century and then covered by sand during excavations in the later part of the century [72].

Investigations in South Abydos gave equally good results. An unknown necropolis was located within the Middle Kingdom mortuary complex of Senwosret III (Twelfth Dynasty), lying in the low desert on the southern side of the temple [73] (Fig. 24). The shape of the anomalies suggested burial shafts with mud-brick lining. These were initially dated to the period of the temple and the associated townsite *Wah-Sut*, but excavations in 2012–14 indicated that all six of the uncovered tombs were from the New Kingdom and connected with a settlement that has yet to be discovered. The tops of these shafts, which were from 1 to 1.5 m high, lay just 10–20 cm under the surface, drawing very distinct anomalies on the magnetic map. The shape of the anomalies represented well the different types of tomb, comprising a shaft leading to the vault that consisted of from one to three chambers [74] (pp. 360–421). Moreover, to the south of the townsite *Wah-Sut*, the survey located remains of regular architecture extending beyond the area of the town reconstructed based on past excavations [22].

The survey in front of the T-shaped enclosures around the tomb of Senwosret III mapped structures excavated at the beginning of the 20th century; renewed excavations in 2013 added new data, leading to the conclusion that the complex of tombs before mastaba S10 belonged to an unknown local dynasty from the Second Intermediate Period and that the number of tombs was different from that noted by the early 20th-century excavators [75] (Fig. 25). The chamber for the bark, which is very well delineated on the magnetic map, was reexplored in detail. The differentiated distinctness of the anomalies outlining the chamber was found to correspond to the different depth at which the tops of walls were located below ground surface. The vertical walls (about 0.55 m thick) of the chamber with remains of arched vaulting were 1.6–2 m high, resting about 3 m below ground surface (Fig. 26). In the front part of the chamber (on the western side), where the anomaly was more distinct, the vaulted ceiling of the chamber was preserved up to a meter below the surface [76].

Further south, an unknown pylon before the mortuary temple of Ahmose, first ruler of the New Kingdom, was discovered. Measurements around the cenotaph pyramid of Tetisheri, Ahmose's grandmother, excavated in the early 20th century, located a structure with corner rooms, 68.5 × 86.5 m in dimensions [77]. The pyramid thus turned out to be part of a much larger and more elaborate complex. Moreover, the presence of an enclosure wall, which the early archeologists had missed, was suggested by the text on a stela inside the funerary chapel of Tetisheri [16] (pp. 37–38). Excavations made to verify this discovery found that the tops of walls, which were preserved 0.7–1 m high, were just 10–20 cm below ground surface and that mummified animals were buried in the corner rooms. In effect, the textbook presentation of this complex (e.g., [3], p. 210), current in the literature for years, had to be changed substantially as a result of the combined geophysical and archeological reexcavation projects.

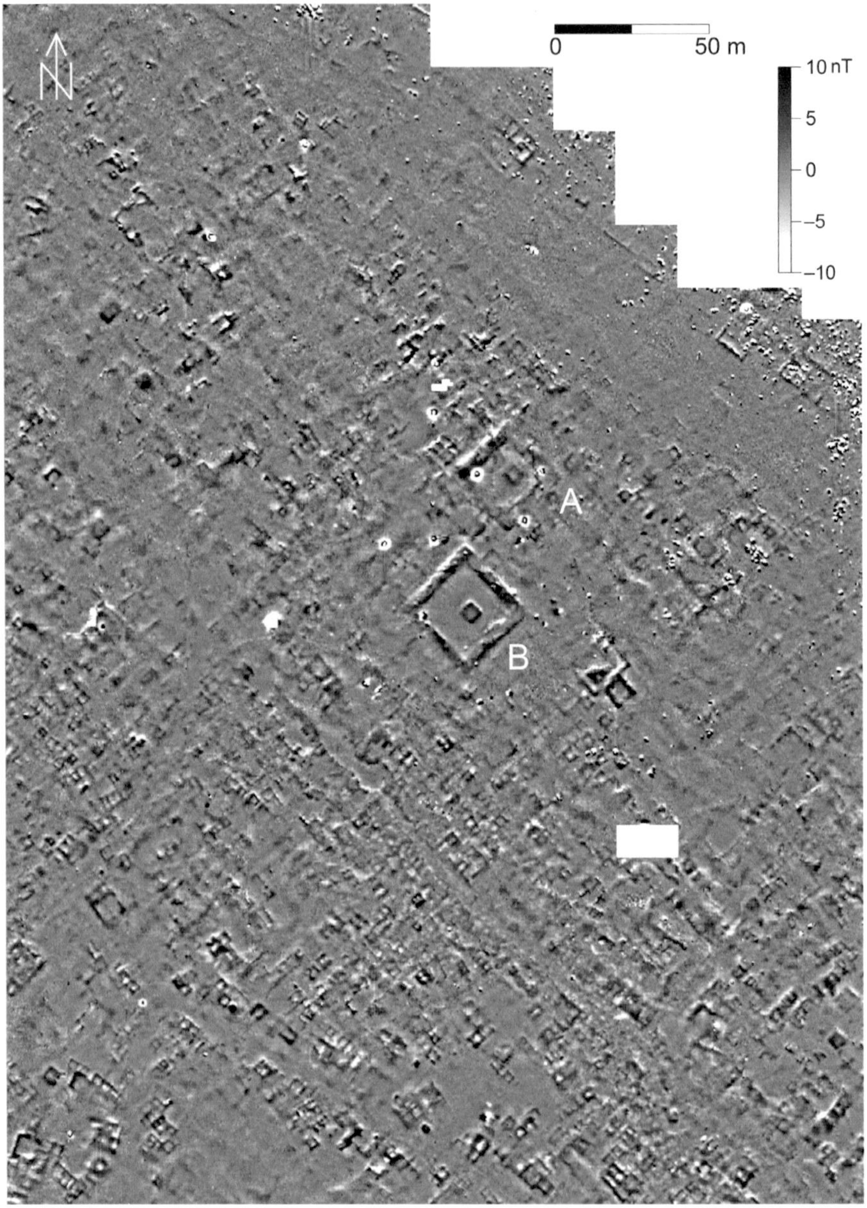

FIG. 23

Abydos Middle Cemetery. Magnetic map of the necropolis. A, mastaba of Iuu; B, mastaba of Weni. Processing T. Herbich.

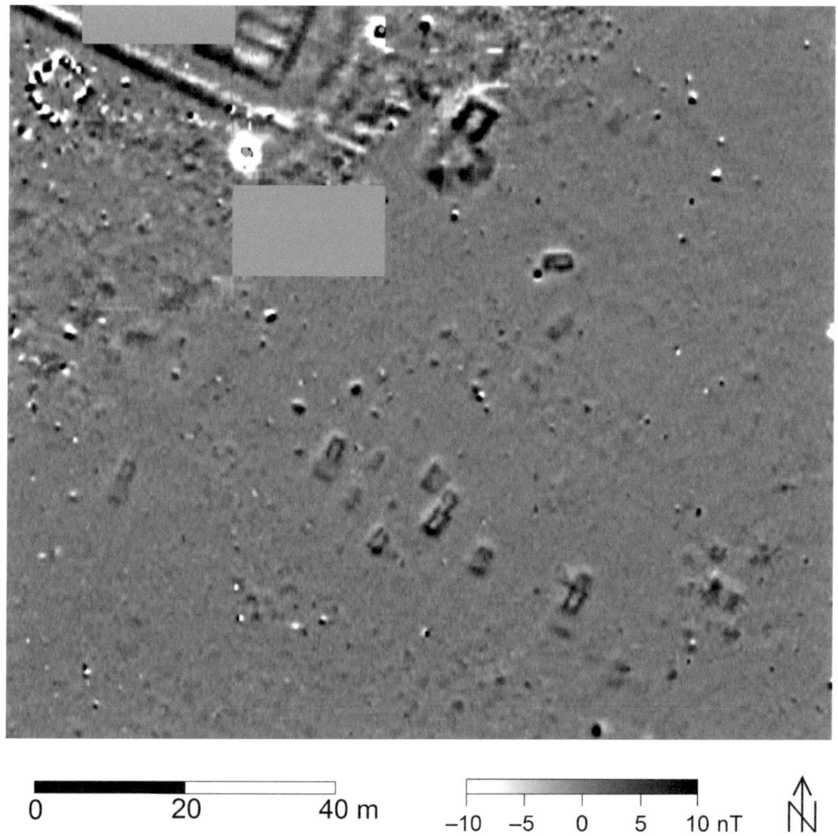

FIG. 24

South Abydos. Magnetic map of the necropolis neighboring on the mortuary temple of Senwosret III (at top left). Processing T. Herbich.

The prospection carried out in cemetery U from the Predynastic period turned out to be a perfect training ground for recording mud-brick structures under layers of sand and gravel. The site had already been excavated, uncovering tombs of local rulers from the second half of the fourth millennium BC [78] (pp. 109–110). The tombs had been documented and reburied following the end of the excavations. The magnetic survey was intended to verify that all the tombs in this cemetery had actually been located, which was done originally by observing a slightly different density of the sand cover above the burial chamber. Research established the borderline conditions for recording mud-brick features in a sand matrix with a fluxgate gradiometer. In the case of tombs with walls up to 1 m high and bricks laid lengthwise in one row (approximately 12 cm wide), the instrument was incapable of picking up anything, if the depth was more than 0.6 m. Walls laid of bricks crosswise (combined thickness of about 40 cm) were detectable by the instrument up to a depth of approximately 1 m [79].

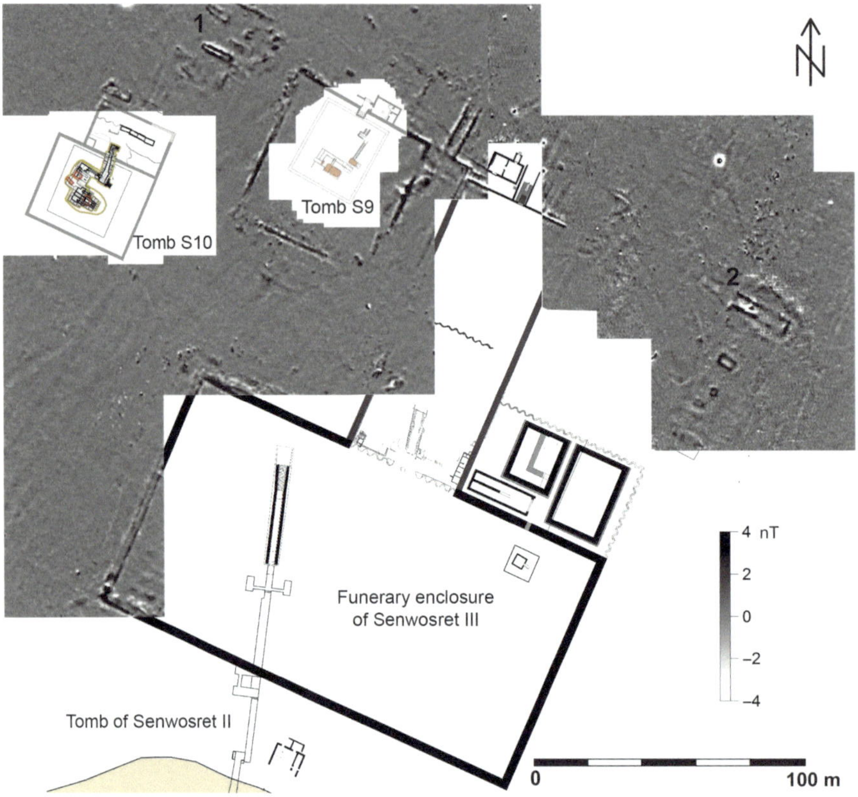

FIG. 25

South Abydos. Magnetic map superimposed on the overall plan of the Senwosret III tomb enclosure and necropolis. S9 and S10, 13th Dynasty royal tombs. 1, Necropolis of the Second Intermediate Period. 2, Royal boat building. Plan after Wegner [76] (fig. 3, p. 5). Processing T. Herbich.

FIG. 26

South Abydos. Royal boat building in the mortuary complex of Senwosret III. Phot. J. Wegner, Penn Museum, University of Pennsylvania.

6.5.2 Dayr al-Barsha

Dayr al-Barsha in Middle Egypt is known foremost for the complex of rock-cut tombs of nomarchs (local governors) from the Middle Kingdom, located near the top of the cliffs forming the western edge of the desert that extends between the Nile Valley and the Red Sea [80]. The tomb of Djehutihotep is perhaps the best known for the relief scene of the transport of a colossal statue decorating its walls. The scene is instructive in terms of reconstructing the technological lore of ancient Egypt. The cemetery that covers a few square kilometers on the low desert is much less well known. The tombs, which date from all periods from the Old Kingdom through late antiquity, were excavated in the early 20th century, but no plan was ever published, either of the individual tombs or the complex as a whole [81] (pp. 237–243). All that can be said based on excavation notes is that the structures were lined with mud bricks and had shafts going down several meters, typical of the Middle Kingdom. The dead were laid to rest with their grave furnishings in chambers at the bottom of the shafts. Today, the landscape of this cemetery consists of pits of all shapes and sizes, surrounded by archeological dumps, the tops of mud-lined shaft walls still visible in places. One of the tombs excavated in 1915 stands open, its triple shaft exposed [82] (p. 16, fig. 5).

The research project initiated in 2002 was aimed at establishing the overall organization of the cemetery and reconstructing social stratification [81] (pp. 243–247). The magnetic method of prospection was an obvious choice owing to the extent of the cemetery, as well as the use of Nile silt as a building material. Two areas at the mouth of Wadi Nakhla that cuts through the cliff edge, separated by a modern canal controlling flash floods from the wadi, proved to be the most informative regarding the plan of the cemetery and the nature of the funerary architecture. In the southern zone 9, measurements recorded rectangular structures corresponding to mud-lined shafts that were occasionally to be seen at the bottom of large pits (attesting to clearing in modern times) or in areas showing no signs of modern disturbance [16,82] (Figs. 27 and 28). Narrow linear anomalies, either parallel or perpendicular to the shaft axis of symmetry, indicated that groups of tombs were found inside walled enclosures, thus separated out from the rest of the cemetery. A number of oval anomalies with raised magnetic field intensity values, mainly 2–3 m in diameter and observable throughout the cemetery grounds, were interpreted after the first season of archeological excavations as traces of robbers' pits, often filled with silty material that was characterized by higher magnetic susceptibility than the sand-and-gravel matrix in which they were deposited.

The excavations in 2002–07 gave a key to interpreting the magnetic map, confirming the interpretation in some cases and in others supplying new information on how to read the magnetic imagery. In the case of the anomalies in the northern part of zone 9, it was quickly confirmed that rectilinear anomalies correspond to sections of walls constituting the funeral enclosures and their intensity was dependent on the state of preservation of individual structures and their depth below ground surface (Figs. 28–30). Walls were just under the surface; hence, the magnetic survey picked up even sections preserved as the lower part of bricks merely a few centimeters high.

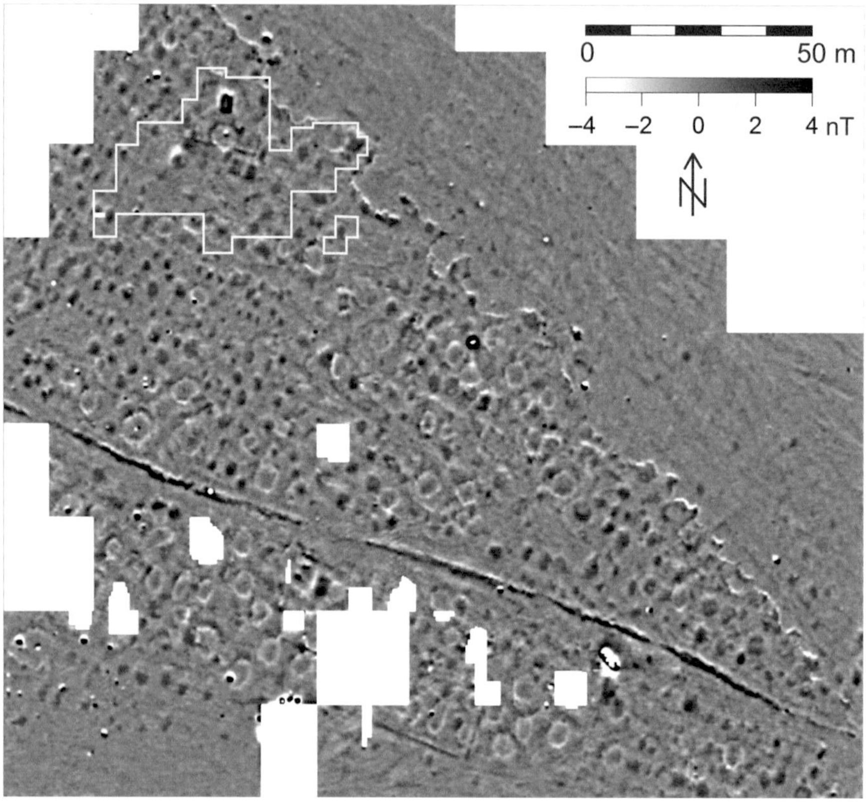

FIG. 27

Dayr al-Barsha. Magnetic map of zone 9. White lines mark the area excavated in 2002–05. Processing T. Herbich.

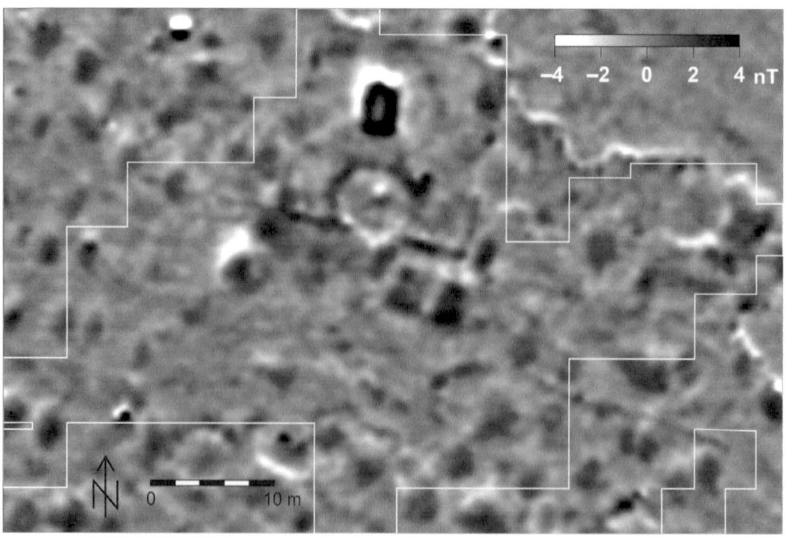

FIG. 28

Dayr al-Barsha. Fragment of the magnetic map shown in Fig. 27. White lines mark the area excavated in 2002–05. Processing T. Herbich.

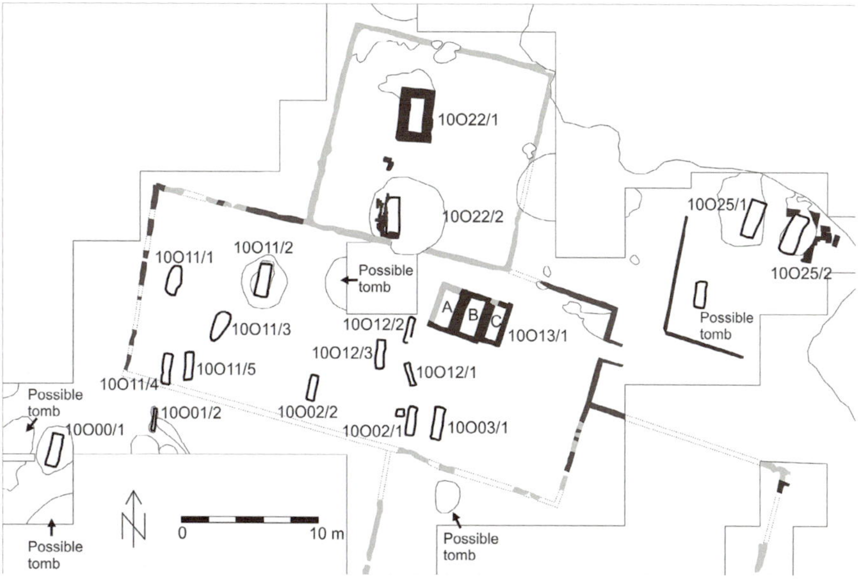

FIG. 29

Dayr al-Barsha. Plan of tombs excavated in 2002–05. Drawing T. Herbich after [83] (p. 401, fig. 9).

FIG. 30

Dayr al-Barsha. Southeastern corner of the enclosure wall registered by the magnetic survey around tombs 10022/1 and 10022/2, seen from the east. Top of mud-brick lining of the shaft of tomb 10022/1 visible at top right. Phot. Dayr al-Barsha Project, Leuven Catholic University.

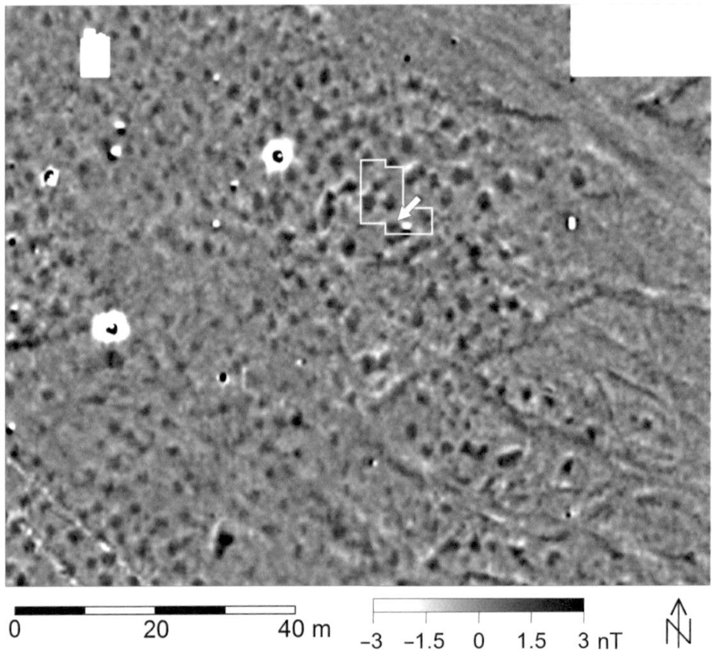

FIG. 31

Dayr al-Barsha. Magnetic map of zone 8. White line marks the area excavated in 2007–08. White arrow marks the anomaly corresponding to the tomb shown in Fig. 32. Processing T. Herbich.

This occurred only in the case of bricks laid crosswise to the wall face (header bond); if laid lengthwise (stretcher bond), they were not registered [16,82].

Excavations also confirmed that the identically oriented, rectangular anomalies of similar size (approximately 2×4 m) corresponded to mud-lined burial shafts. Chambers on the south side of the shafts were not detected: they had no mud-brick lining and nothing in the grave furnishings (which consisted mostly wooden coffins for the mummified remains and a few ceramic vessels) that could cause a measurable change in the intensity of the magnetic field [83] (pp. 401–408).

The oval anomalies initially thought to reflect robbers' pits turned out to reflect silty material deposited in shallow depressions that formed when the fill of the burial shafts underneath subsided; silt accumulated subsequently in these hollows. Excavations in zone 8 illustrated this phenomenon, where each one of the verified anomalies corresponded to the fill of a shaft leading to a chamber (Figs. 31 and 32). The tombs had no elements of mud architecture that a magnetic survey would have been able to detect [84] (pp. 447–452).

The magnetic survey also mapped a substantial structure that was not reflected in ground relief. It was a long linear anomaly, very distinct, up to 2 m wide and running

FIG. 32

Dayr al-Barsha. Shaft of tomb corresponding to the anomaly marked by the white arrow in Fig. 31. Phot. Dayr al-Barsha Project, Leuven Catholic University.

for a length of 210 m across the investigated area (Fig. 27). Excavations revealed that it corresponded to a Middle Kingdom road built of Nile silt; several layers of mud showed that the road had been repaired several times [82] (pp. 22–24). Water erosion had removed all trace of it in the section directly by the mouth of Wadi Nakhla and alongside the edge of the cliff, but it was evident that the road had once connected the quarries in the upper parts of Wadi Nakhla with the cultivated zone in the Nile Valley. Presumably, blocks from the quarries were transported down this road, just as in the image depicted inside the tomb of Djehutihotep: a statue weighing perhaps 80 t, being moved on a sleigh that was pulled over a surface ensuring slipperiness thanks to water poured on the mud in front of the sleigh [85].

A combined analysis of a ground survey and magnetic prospection, verified by point excavations, permitted a reconstruction of the overall organization of the necropolis. It was arranged around the road that cut across it from east to west. The more elaborate tombs in terms of their architecture lay closer to this road; they had mud-lined shafts and chambers and were grouped within enclosures surrounded by mud-brick walls. Farther from the road, the burials were simpler, mostly rock-cut shafts and chambers without mud-brick lining. Different kinds of tombs were found within the enclosures: both the shafted kind either with or without mud-brick lining (traced on the magnetic map) and the shallow pit kind, although the latter, less than a meter below the surface, did not cause any change of magnetic field intensity and were thus not detectable by the magnetic survey.

6.5.3 Dahshur valley temple

So-called valley temples were part of the funeral complexes of Egyptian kings where the dead rulers were worshipped. The temples were located at the edge of the low desert, close to the alluvial plain, connected to the pyramid by a causeway. Unlike the complexes built later, the valley temple of the Bent Pyramid of king Snefru, which was discovered in 1951, stood halfway between the pyramid and the edge of the desert, about 800 m from the pyramid and 700 m from the alluvial plain [86].

When the temple was first excavated, a structure was discovered against the northern side of the sanctuary, oriented at a slightly different angle than the main building. It was constructed of mud brick, its walls 2.5 m thick, and had a courtyard in the northern part, entered through three entrance rooms in the southern part; the excavator considered it a storeroom of some kind from the Middle Kingdom period [86]. Later excavations in the early 20th century interpreted it as a royal vestry [87]. Doubts remained and the next project that was initiated involved magnetic prospection of the temple surroundings [55] (p. 17). These revealed the presence of numerous anomalies typical of structures formed of Nile silt (Fig. 33 left).

The anomalies located to the north of the sanctuary, surrounding the mud-brick building from the north and east, were interpreted as the remains of a 5 m-thick enclosure wall. Discrete anomalies not exceeding 1 m in diameter were thought to reflect pits filled with mud serving as a kind of natural pot for planting trees and bushes. Excavations fully confirmed this idea; combined with the results of the magnetic prospection, they allowed the complex to be reconstructed and gave grounds for a reconsideration of its function [89].

Enclosures of this size, wall thickness, and plan find parallels in funeral enclosures known from Abydos, Hierakonpolis, and Saqqara [90] (pp. 162–181). The layout of the building inside the enclosure (reconstructed solely from the excavations as the geophysical plan was not clear due to vestigial preservation of the walls in the part not covered by later architecture) did not suggest a residential structure, whereas the wall thickness indicated a building of some grandeur and importance. The presence of a garden is unique compared with other enclosures of this kind. Its plan could be reconstructed based on the magnetic survey results and excavations (Fig. 33 right). Plants were grown in pits up to a meter deep, running in rows along the walls of the enclosure and leaving an empty courtyard to the north of the building (Fig. 34). The roots of trees and bushes were found in the mud inside some of the pits; a study of the botanical remains identified palms and sycamore trees. Parts of the ground between the pits, characterized by slightly elevated values of magnetic field intensity, corresponded to mud deposits presumably intentionally introduced to grow

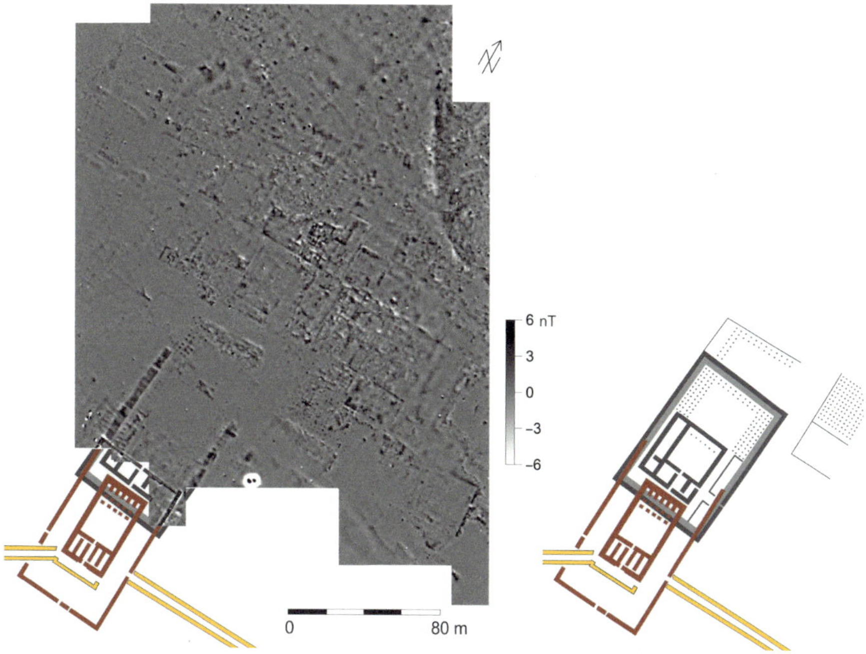

FIG. 33

Dahshur. Left: magnetic map superimposed on the plan of a mud-brick enclosure and valley temple. Right: reconstruction of the mud-brick enclosure with mud-brick building and garden and stone valley temple. Foundation of the earlier phase of the enclosure wall in *gray*; *dark gray* marks the later phase of the enclosure wall and brick building; in *brown*, the valley temple, and in *yellow*, the causeways. Processing and drawing T. Herbich (temple and brick building plan after [88],. p. 7, fig. 2).

FIG. 34

Dahshur. Plant pits filled with Nile mud in the foreground, the foundation of the first phase of the enclosure wall in the back, seen from the northeast; the Bent Pyramid of Snefru in the background. Phot. DAI/J. Pinke.

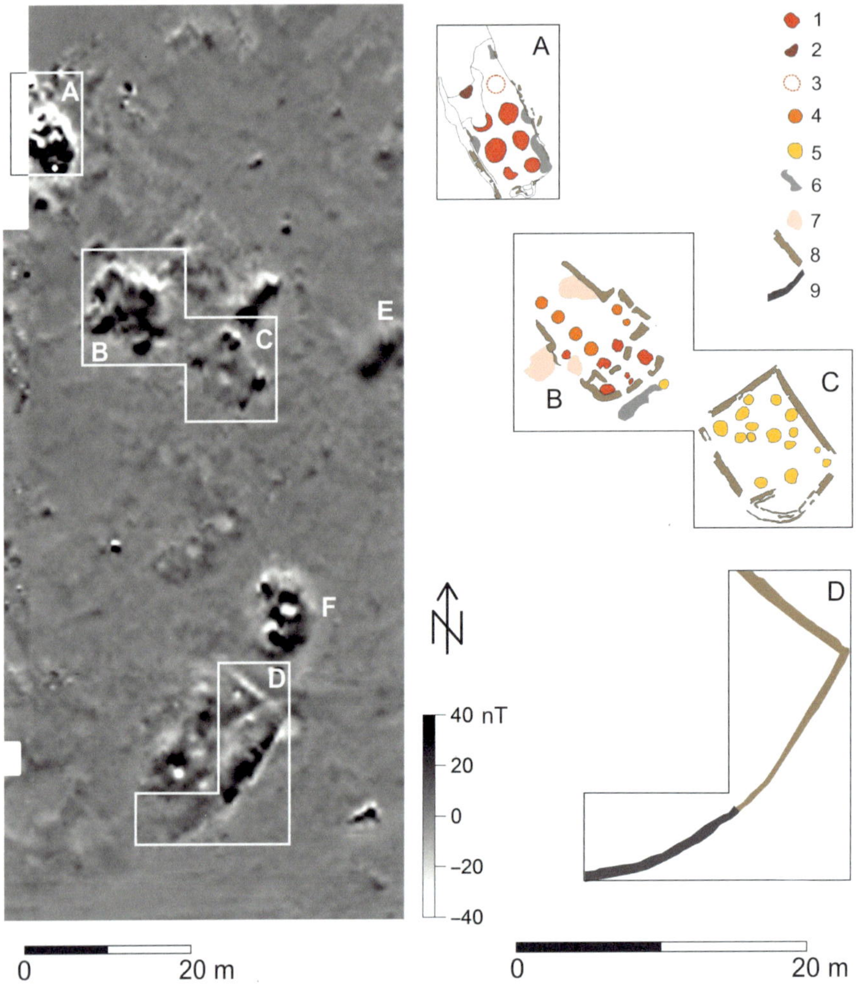

FIG. 35

Hierakonpolis. Locality HK11. Left: magnetic map. White lines mark the outlines of trenches A–D as shown on the right. Right: map of excavated structures. 1, Kilns; 2, kiln excavated in 1979; 3, hypothetical kiln; 4, remains of vats set over fireplaces; 5, fireplaces; 6, walls of burnt mud; 7, burnt soil; 8, walls of mud; and 9, wall of stones. Processing and drawing T. Herbich, after ([91], figs. 4, 8, 12; [92,93]).

smaller plants. The presence of a garden requiring watering with water brought from the Nile and the huge enclosure wall also pointed to the special importance of the complex. Stratigraphic studies combined with other evidence demonstrated that the mud-brick building inside the enclosure preceded the stone valley temple; once the precinct was reduced from the southern side, the enclosure and temple functioned together, after which the enclosure was dismantled [94] (pp. 5–7). While it existed, the brick building was repeatedly renovated and altered, all in the reign of one ruler, that is, Snefru. This suggested to the excavator that "the building in Dahshur may be interpreted as a ceremonial building that the king used on special occasions on which he had to leave his ordinary residence, possibly in the course of rejuvenating rituals" [95] (p. 126). A new and narrower wall replaced the dismantled enclosure, and the garden was extended toward the north. The change was evidenced on the magnetic map by anomalies corresponding to rows of pits in which bushes must have been planted. It is not clear how long the garden functioned in this new shape [88].

The atypical situation of the valley temple of the Bent Pyramid at such a distance from the alluvial zone was explained with the help of Helmut Becker, responsible for an extensive geophysical project carried out in Dahshur in 2003–08. He used first a fluxgate gradiometer and then a dual cesium magnetometer measuring at the same time the full intensity of the magnetic field along two lines [96]. The second section of the causeway that he discovered in the last year of research led from the valley temple toward the Nile, but it did not reach the alluvial plain. Instead, it terminated in a harbor built in the lower end of the wadi. Both the causeway and the wall surrounding the harbor were of mud brick [97]. The southern section of the harbor enclosure wall was easily traced on the magnetic map. Drilling established the depth at which the remains were found and excavations confirmed that the wall, which was about a meter thick, was at a depth of 4 m under windblown sand. The survey in the area of the harbor contributed new data on the depth reached by fluxgate gradiometers. The map traced both the northern section of the harbor wall at a depth of about 3.5 m and the southern section of the wall, which was previously recorded with a cesium gradiometer [79].

6.6 PRODUCTION CENTERS: HIERAKONPOLIS

Intermittent archeological excavation since the late 19th century established Hierakonpolis as a regional power center in the fourth millennium BC, a probable seat of rulers of a local kingdom prior to the unification of upper and lower Egypt into one state organism [98,99]. Magnetic prospection carried out on three separate occasions between 1998 and 2010 in localities of an industrial and settlement character (respectively, sites HK11 and HK24) yielded results that have helped in reconstructing industrial processes in the Predynastic period, providing also the earliest material evidence of these processes in Egypt and in the world in some cases ([16,100], pp. 40–43; [101,102]). Excavations to verify structures recorded on the magnetic map have also given the rare opportunity to establish relations between archeological features and their magnetic image.

The survey on site HK11C covered an area that was previously excavated, where one of the earliest updraft kilns ever discovered was investigated [103]. A number of dense anomalies distinct from the surrounding area (up to the range of −50/+70) were noted (Fig. 35 left). The area marked as A touched upon the uncovered furnace, hence the idea that the other anomalies (B, C, and F on the map) were a record of pottery kilns [16] (pp. 41–42). All these anomalies were excavated between 2003 and 2016 in view of the importance of the site for the history of technology, pottery industry, and craft specialization [91]. These features were not visible on the surface, but production processes were suggested by the presence of ash and burnt soil appearing just below the surface.

A remarkable semisubterranean firing complex was uncovered on the spot of the anomaly marked as A (Fig. 35 right). It consisted of eight heavily burnt circular features, each containing circles of fire bars (rods of clay) in an upright position, of differentiated height, installed to support large ceramic cooking pots (vats). The walls of the rectangular complex were made of kiln debris: fire bars, clay slabs, and potsherds, cemented together with mud. Thick deposits of charcoal and ash were found along the walls. The reddish color of the walls, floors, and fire bars indicated repeated burning at high temperatures and application of heat to the entire complex. Its function as a brewery was established by analogy with a similar complex uncovered in 1989 in another part of the site (HK24A) [104].

On the spot of anomaly B, where there was no evidence of any kind of industrial activity on the surface, a platform of burnt mud, 30 cm high, was discovered. It was furnished with three pits along the edge, each 1 m in diameter. Vessels were fired in these pits, the source of heat being somewhere outside. Debris was not removed; hence, the quickly rising ground level forced the platform to be raised and new furnaces to be built in order to ensure access to the wind at the higher level. At least three levels of kilns were recorded. Further pottery kilns were excavated in the section to the north, followed by a series of well-preserved freestanding vats, up to 40 cm high and 80 cm in diameter (Fig. 36). Unlike the brewery on the spot of anomaly A, the vats here were supported on stones and potsherds [105]. The residues in the vats contained malted emmer wheat, indicative of beer production. The remarkably well-preserved complex is a unique example of an integrated industrial area for grain-based foods (including beer) and potmaking. Radiocarbon testing of the samples gave a date in the turn of Nagada I and II (3600 BC) period, making it the oldest known brewery in the world and the oldest set of pottery kilns ever discovered in Egypt [106].

The anomaly marked as C (Fig. 35) corresponded to a building with walls made of mud brick, preserved to a mere 15 cm above the ground. Accumulated layers of burnt debris with concentrations of ash, potsherds (from Nagada II, 3600–3300 BC), scorched stones, and charcoal pointed to activities related to the use of fire inside the structure. Excavations uncovered 14 hearths from 70 to 90 cm in diameter, whereas examination of the faunal remains demonstrated meat and fish preparation on an industrial scale. Judging by the percentage distribution of animal remains, the prepared food was consumed elsewhere [91].

The proximity of these three complexes to a cemetery from the same period (locality HK6) inclined researchers to assume that beer and food production, combined with the production of vessels for transport purposes, was associated with funerary rites [92].

FIG. 36

Hierakonpolis. Locality HK11. Remains of a brewery excavated in trench B, seen from the east. Phot. Hierakonpolis Expedition.

In the case of the complex located on the spot of anomaly D (Fig. 35), the evident limits of the ruins on the northeast and southeast corresponded to the freestanding walls of a large building, preserved up to 80 cm in places in the northern part and no more than a brick or its part in the eastern part. The southern, curving section of the wall was constructed of stone. Pottery finds placed the structure in the Nagada IIB–IIC period. Anomalies inside the complex reflected concentrations of ashes mixed with charcoal; by the stone wall, they corresponded to a mud-coated floor covered with brown soil that contained sherds, lithics, vessel fragments, textiles, and organics, all typical of household activities. Radiocarbon dating of the complex on the spot of anomaly B made this the earliest evidence of the use of dried brick for building purposes from Egypt. The conclusion is that the invention of bricks did not come to this region from the north (as the finds from the Nile Delta had seemed to suggest), but that they were an innovation independently introduced by the local community [91,107].

Locating industrial activities was also the aim of the survey carried out in the part of the site closer to the Nile Valley, in locality HK24. Two breweries were found there (localities HK24A and HK24B), but surface traces in the form of fragments of fire bars supporting vats indicated that beer production could have taken place also in other sections of the site. At least seven new production complexes were discovered as a result of the prospection, two of these taking on the typical rectangular shape analogous to complexes uncovered in HK11C. Measurements demonstrated also that the remains uncovered in HK24B constituted only part of the set [102]. Altogether, 16 pits for heating vats were recorded (about 0.6 m deep), covered with a thick deposit of "kiln debris." Radiocarbon dating of the complex placed it in the middle Nagada II period. Research revealed that clusters of fire bars in the western part of the investigated area did not match up with the site of a brewery discovered about 10m to the north where there were no surface traces of any kind of activity) [108].

6.7 LANDSCAPE RESEARCH

The potential of the magnetic method of prospection for researching changes in the landscape is best exemplified by a case study of two sites in Fayum Oasis. Fayum is hardly a typical desert oasis irrigated by the underground Nubian aquifer, being instead supplied with water from the Nile brought in by the Bahr Jusuf canal, a side branch of the Nile, which enters the oasis through an inlet 1.5–2.5km wide and 10km long. The oasis is subjected therefore to the same conditions as the floodplains in the Nile Valley [109] (p. 171). The ground falls away to the northwest, the difference in levels between the Nile Valley and the northwestern edge of the oasis being about 70m. The water level in Qarun lake (Lake Moeris in antiquity), the main water reservoir in the oasis, demonstrated significant changes over the ages. In the Paleolithic, the lake filled the entire basin forming the oasis. Intentional limiting of water inflow during the Middle Kingdom period reduced it to just a fourth of its size by the Roman period, naturally occupying the lowest-lying area in the northwestern part of the oasis. These changes determined the extent of the cultivated zone and areas conducive to settlement [110]. Changes in the lake levels as confirmed by geomorphological studies, including seasonal influx, changed settlement conditions. The desertification of fringe zones progressing since late antiquity helped to preserve archeological sites in these regions. These changes can now be studied with archeological means, including the magnetic method, as demonstrated by the research carried out on two sites, Medinet Watfa (ancient Philoteris) and al-Qarah al-Hamra, lying on the opposite shores of the lake.

6.7.1 Medinet Watfa/Philoteris

Philoteris was one of the villages that were established in northwestern Fayum in the times of Ptolemy II (284–247 BC) who initiated an irrigation network intended to create new areas for cultivation to be settled by Greek settlers coming to Egypt in consequence of the division of the empire left by Alexander the Great. The villages, of which Philoteris was one, lay on the southern side of the western end of Moeris lake, on ground that was irrigated by water coming 30km from the center of the oasis, in a canal that was specially built for the purpose. The canal network in the Philoteris area had attracted scholarly attention already in the early 19th century, but the evidence, including remains of water-collecting reservoirs, became more evident once aerial images of the region were produced [111]. Excavations carried out in the end of the 19th century were aimed largely at finding papyri to discover the name of the settlement; thus, the first regular excavations started in 2000 had to carry out a ground survey mapping the layout of the village and its surroundings.

An examination of the surface pottery gave a date range for the settlement, showing that it was abandoned in the 4th century AD [112]. The extent of the inhabited area was determined as 430×250 m; typical features were noted, like a temple, baths, houses, stores, and cemetery. A separate complex, identified as a temple, was recorded 250m to the northwest of the village [113]. Building materials included limestone blocks and dried mud bricks, the latter made of mud rich in Nile silt and local silt of a characteristic strong yellow color and much lower magnetic susceptibility (no more than 0.3×10^{-3} SI). In many instances, the two kinds were used for building purposes simultaneously (Fig. 1).

The network of basins and canals was traced, supplementing the ground survey with satellite imagery; the results seemed to be an excellent reconstruction of the water economy around the village. The canal supplying the village with water was identified and found to separate on the eastern side into two branches that skirted the town, the northern one filling the water reservoirs and the southern one supplying the village and running further west, toward the next village located 5km away [113]. Two seasons of magnetic prospection supplemented knowledge of the site and its hinterland substantially [114]. In many places, the magnetic map gave a much more detailed image of settlement than that provided by the ground survey, but the main achievement of the prospection was a much improved plan of the canal network (Fig. 37). It was assumed that the good results were due to the layers of Nile silt deposited on the bottoms and sides of the rock-cut canals. Sections that had not been visible in the present ground relief became evident, and it was possible to trace places where the canals had been rebuilt and where bays existed, presumably as anchorages for boats. At one point, the magnetic map revealed a structure that narrowed the canal; it was interpreted as the pillars of a bridge crossing the canal. A complete surprise was the image of a rock-cut canal at least 3m deep, running west toward yet another village established by Ptolemy II, but evident at the magnetic map only for about 200m beyond the boundaries of Philoteris. The canal had evidently been prepared with much effort, but the lack of Nile silt in its bed indicated that it was used for a short time, if at all. The canal as it stands can be construed as proof of a failing system of water supply.

Excavation in the village in 2012–14 confirmed the presence of Nile silt deposits in the canals, uncovered the bridge pillars and gave an idea of the installations that regulated the water level in the reservoirs [115]. A complex observed to the northwest of the village, initially interpreted as a temple, was identified as a Roman army camp.

6.7.2 Al-Qarah al-Hamra

The site of al-Qarah al-Hamra lies on the opposite side of the lake to Medinet Watfa and was located by a ground survey carried out on the eastern shores on Lake Qarun/Moeris. The only trace of archeological remains in this locality was a limited number of potsherds scattered on a flat, sandy surface. Magnetic research in the area of the clustered pottery mapped a small regularly laid out village, particularly distinct in the southern and western parts (Fig. 38). The nature of the anomalies points to dried mud

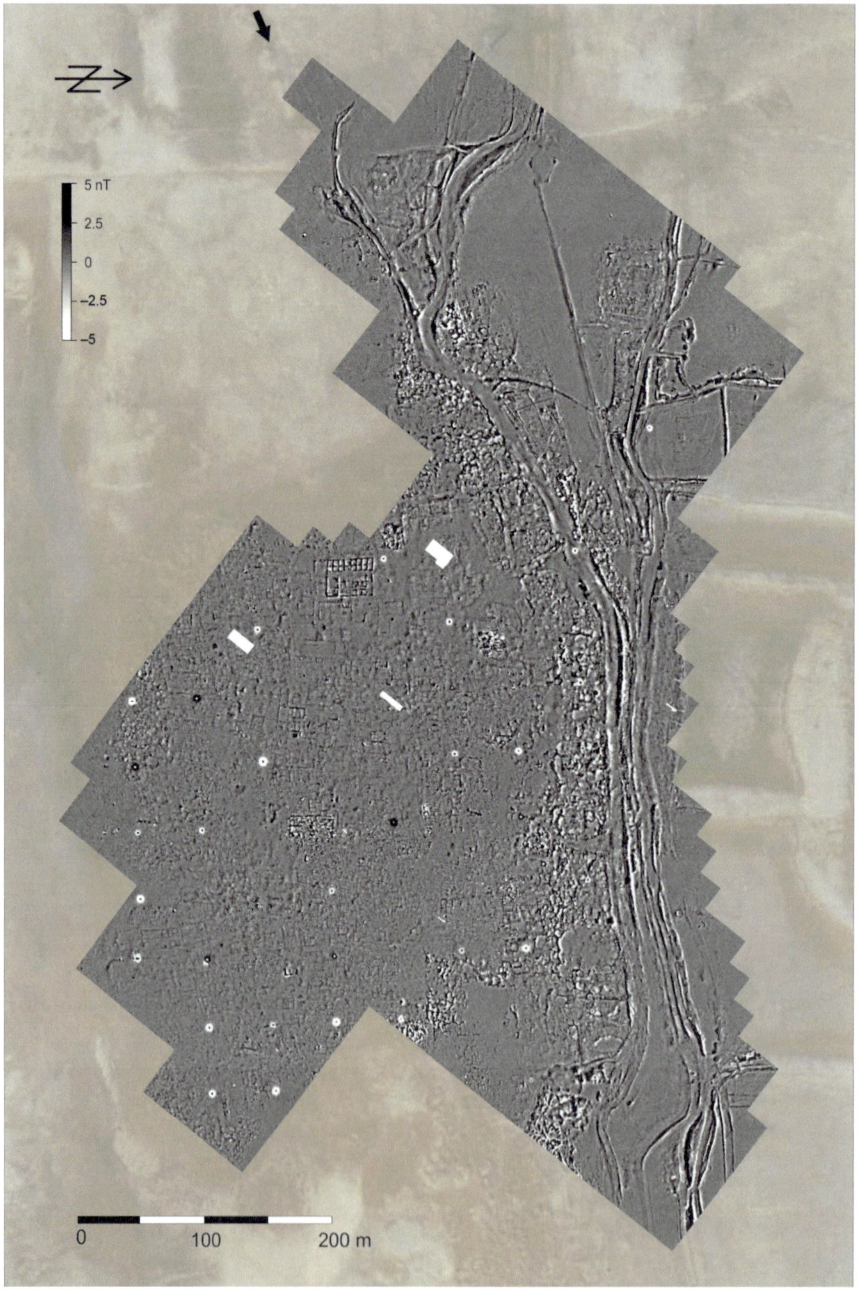

FIG. 37

Medinet Watfa/Philoteris. Magnetic map superimposed on a satellite image. Processing T. Herbich, satellite picture Google Earth.

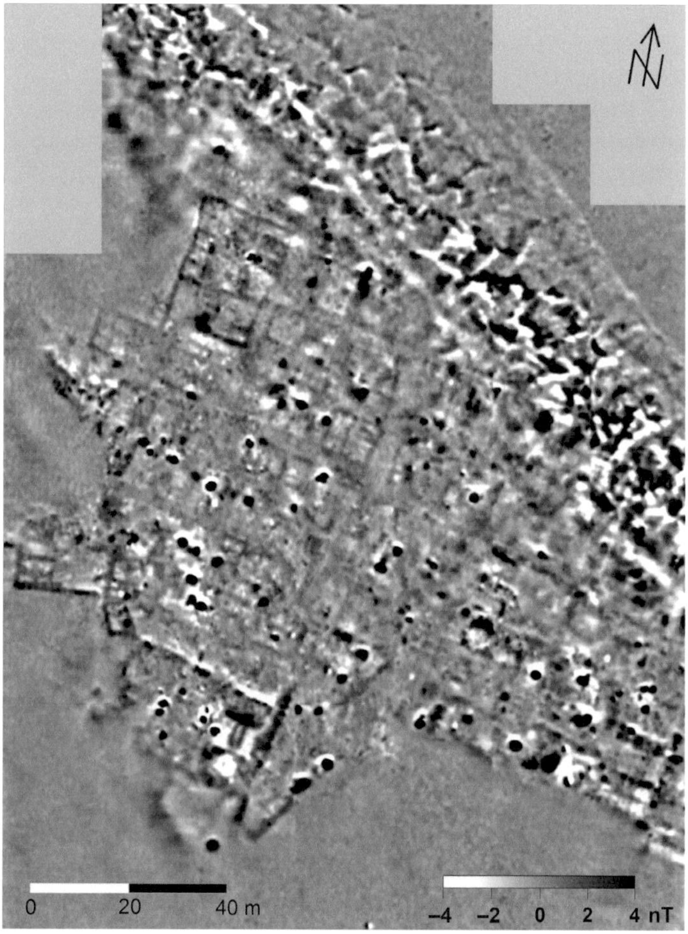

FIG. 38

Al-Qarah al-Hamra. Magnetic map. Processing T. Herbich.

brick made of Nile silt as the building material. The rectilinear street grid is evident, and the clear plan permits identification of the function of particular buildings. For example, a series of parallel rectangular structures located north of an *E*-W street in its western section exemplify typical Roman shops. The plan of the westernmost structure in the village identifies it as a temple with a courtyard and three long parallel rooms opening onto it. Oval anomalies up to 2 m across, showing high value amplitudes, observed in many areas but always within the village, should be interpreted as hearths or furnaces. Slightly larger anomalies presenting a similar value range, 3–4 m in diameter and located at the southern edge of the village, may have been pottery kilns (the prevailing northerly winds in Egypt are the reason why production installations are usually located to the south of settlements) [73].

Excavations positively verified two of the oval anomalies interpreted as hearths or furnaces; Nile silt was also identified as the building material *par excellence* for the village architecture [116].

A set of irregular anomalies of high value amplitudes that were observed in the northeastern part of the surveyed area clearly did not reflect any kind of architecture. These anomalies dropped in intensity toward the northwest where they were also cut off, the magnetic image of the area being completely undisturbed as in the sections to the south and west of the village. It is currently assumed that these reflect a slow process of water erosion. The village was located on the northern lakeshore, close to the water, on a slope descending toward the lake. Rising water in the lake dissolved the mud-brick architecture letting the silt be deposited along the shore. The new shoreline once the water level had been stabilized is indicated by the line corresponding to the northeastern extent of the anomalies. Excavations confirmed that the village was under water for some time [73]. Partly dissolved mud-brick walls and layers of mud attest to this. A film of calcium carbonate, which is formed by algae in shallow alkaline water, was noted on potsherds and stone blocks [116] (p. 52).

The site was abandoned in the 4th century as suggested by the pottery evidence. Thus, we have proof of the same processes taking place on two sides of the lake: on the southwestern side of the lake, water ceased to be supplied to the higher ground located there, whereas on the northeastern edge, rising levels in the lake led to the destruction of the village. These two facts can be related to a cataclysm that affected Fayum in the 4th century AD, causing an observable change in settlement patterns. The most logical explanation of these changes is the destruction of a dam regulating the water supply from the Nile to the oasis. Once this had occurred, it became impossible to transport water to the western part of the oasis, even as the uncontrolled flow to the lower parts of the depression caused a rise of water levels in the lake, destroying villages that were located too close to its edge [109].

These two case studies show how geophysical research was able to contribute to a study of changing landscapes in a region of several hundred square kilometers: desertification that followed depopulation of certain areas due to the lack of water and destruction of settlements as a result of excess water in other areas.

7 SUMMARY AND CONCLUSIONS

Following the case studies presented above, it is time to discuss methodological issues connected with the application of the magnetic method of prospection on desert sites and to reflect on geophysical interpretation of results in the light of 20 years of experience with surveying sites in such a location.

Desert sites in Egypt and Sudan differ from those in the alluvial plain in their susceptibility to eolic erosion due to the low compactness of surface layers. The ground relief and content of surface layers on such sites is a factor of erosional processes, and close observation of both may assist in the interpretation of magnetic maps. It is prerequisite, however, to study and document the ground relief and to follow well-considered strategies of archeological verification of anomalies that demonstrate characteristic and repeatable characteristics corresponding to specific archeological features.

The case of Bawit (6.4.1) is an excellent example of how the study and documentation of ground relief can help to interpret magnetic results. The site had been excavated more than a hundred years ago, and the architecture discovered then was gradually covered by sand, either as a result of backfilling by the excavators or natural processes. The first step was to compare the shape of the anomalies appearing on the magnetic map with existing plans of the uncovered structures. However, the key to understanding whether a structure had been explored previously or was a new, as yet unidentified complex came from examining the composition (content) of the surface layers and the ground relief. On the spot of earlier excavations, the surface scatter of potsherds was minimal, fragments of pottery not being carried easily by winds (unlike the finer sand fraction). Shallow basin-like depressions were also noted (up to 10–20cm deep). Observations of this kind were marked on the geophysical maps, and each location corresponding to an anomaly that suggested the presence of an archeological structure was photographed. For example, more than 400 photo images were taken in the course of the prospection at Bawit, which covered 50ha. A special system adopted the images being taken in a way that allowed them to be easily located with respect to the corners of squares in the survey grid used for the survey (20 × 20m). This photo material turned out to be of key importance for the final interpretation of particular anomalies and determining the actual location of structures known from earlier excavations.

Dayr al-Barsha turned out to be an example of the trouble encountered when verifying the results of geophysical prospection by means of archeological excavation. Dozens of oval anomalies, from 1 to 3m on the long axis, were documented. The smaller ones were examined during the first season of excavations in the course of verifying an anomaly that proved to be a road with a mud surface cutting across the cemetery. These turned out to be shallow pits filled with erosional silty deposits. No relation to the graves was noted; hence, these pits were interpreted as robbers' pits dug in search of tombs. The regularity of these anomalies supported this interpretation, assuming methodical looting [82]. Excavation, 6years later, which explored another cluster of anomalies of this kind (with a long axis up to 3m), demonstrated that the hollows were the result of fill (practically indistinguishable from the surroundings) subsiding inside burial shafts, whereas the silty deposits that caused the anomaly were properly interpreted as result of natural erosional processes [84]. Therefore, there is a close albeit indirect connection between the anomalies and tombs. In this case, understanding the anomaly was dependent on a critical approach to earlier hypotheses and planning excavations in a way that will aid in interpreting the map.

The choice of instrument and measurement methodology is also of importance for the results of geophysical research. Comparing results obtained in the same area with fluxgate gradiometers and cesium magnetometers, the latter measuring total magnetic field, leads one to the conclusion that in the case of structures at a shallow depth, better effects are obtained with the vertical gradient rather than the total magnetic field. An example of this is the magnetic map of the settlement north of the valley temple at Dahshur. The map obtained with a cesium instrument indicated human activity, but it was the fluxgate instrument that discerned the remains of a village with an irregular plan and evidence of gardens in the form of pits filled with mud for growing plants. Similar results were obtained in the area to the east of the Red Pyramid [96] (p. 132).

Most of the surveys carried out in desert sites require manual transport of the measuring instruments. Uneven ground (also due to earlier excavation of a site) and not very compact surfaces exclude the use of multichannel systems on wheels. It is the author's experience that such systems could be used on less than a quarter of sites in a desert environment and even then usually not on all of the site. On some sites, the unevenness of the ground also made the distance between probes important in the case of using a gradiometer. The most effective were instruments with a distance between probes equal to 0.5 m (like in case of FM by Geoscan Research, Fig. 2).

The last observation concerns an issue that is briefly described as theory versus practice. Depositional and postdepositional processes at a site often lead to unforeseeable results, which could affect the effectiveness of geophysical methods. Moreover, sites exist in specific geologic circumstances, which can also influence geophysical results. The survey in Berenike (6.1.2) was a good example. Theory had it that gypsum anhydrite blocks and coral-reef heads were undetectable by the magnetic method in a sand environment. The layers surrounding and covering the ruins, which were not taken into account when evaluating the effectiveness of the method, due to their mineral composition had proved to have sufficient magnetic susceptibility to make the method extremely effective for the recording of architecture and for planning future archeological excavations. Had theory taken priority over practice in this particular case, magnetic prospection would not have been undertaken, involuntarily depriving Egyptian archeology of important discoveries.

ACKNOWLEDGMENTS

The projects discussed in the paper would have been much more difficult to carry out without the logistic support of the Polish Centre of Mediterranean Archaeology, University of Warsaw, specifically its Research Center in Cairo, as well as colleagues participating in the fieldwork, especially Artur Buszek, Krzysztof Kiersnowski, Jakub Ordutowski, Robert Ryndziewicz, and Dawid Święch. One of the instruments used for the surveys was made available by the Program of Studies of Egiptology PREDE-CONICET of Buenos Aires within the framework of an agreement signed with the PCMA UW.

REFERENCES

[1] R. Said, The Geological Evolution of the River Nile, Springer-Verlag, New York, 1981.
[2] B. Kemp, Ancient Egypt, Anatomy of Civilization, Routledge, London, 2006.
[3] A. Dodson, S. Ikram, The Tomb in Ancient Egypt, Thames & Hudson, London, 2008.
[4] T. Herbich, Geophysical methods and landscape archaeology, J. Egypt. Archaeol. 41 (2012) 11–14.
[5] B. Kemp, Soil (including mud-brick architecture), in: P. Nicholson, I. Shaw (Eds.), Ancient Egyptian Materials and Technology, Cambridge University Press, Cambridge, 2000, pp. 78–103.
[6] A. Weller, A. El-Bassiony, Magnetic properties of archaeological building materials from Egypt, Near Surface 2006—Helsinki, 4–6 September 2006, PO45, 2006.
[7] J.W.E. Fassbinder, Seeing beneath the farmland, steppe and desert soil: magnetic prospecting and soil magnetism, J. Archaeol. Sci. 56 (2015) 85–95.
[8] A. Hesse, Mesures et interprétation en prospetion géophysique des sites archéologiques du Nil, Prospezioni Archeol. 2 (1967) 43–48.
[9] A. Hesse, Introduction géophysique et notes techniques, in: J. Vercoutter (Ed.), Mirgissa I, Paul Geunther S.A., Paris, 1970, pp. 51–121.
[10] E.K. Ralph, Magnetometer survey at Malkata, Egypt, MASCA News Lett. 9 (2) (1973) 3–5.
[11] L.T. Dolphin, A.H. Moussa, G. Mokhtar, Application of Modern Sensing Techniques to Egyptology, SRI International Radio Physics Laboratory, Menlo Park, 1977.
[12] V. Hašek, F. Obr, M. Verner, Application of geological and geophysical methods in archaeological investigations of ancient Egyptian remnants in Abusir, Przegląd Archeologiczny 35 (1988) 5–47.
[13] V. Hašek, Methodology of Geophysical Research in Archaeology, Archaeopress, Oxford, 1999.
[14] A.G. Hussein, Magnetic prospecting for archaeology in Kom Oshim and Kiman Faris, Fauoum, Egypt, Zeitschrift für Ägyptische Sprache und Altertumskunde 110 (1983) 36–51.
[15] K. Myśliwiec, T. Herbich, Polish research at Saqqara in 1987, Études et Travaux 17 (1995) 177–203.
[16] T. Herbich, Archaeological geophysics in Egypt: the Polish contribution, Archaeol. Polona 41 (2003) 13–55.
[17] J. Fassbinder, H. Becker, T. Herbich, Magnetometry in the Desret Area West of the Zoser, S Pyramid, Saqqara, Egypt, in: H. Becker, J. Fassbinder (Eds.), Magnetic Prospecting in Archeological Sites, Monuments and Sites VI, ICOMOS, Munich, 2001, pp. 64–65.
[18] I.J. Mathieson, The development of geophysical survey at Tell el-Amarna, Memphis and Saqqara necropolis, in: S. Ikram, A. Dodson (Eds.), Beyond the Horizon: Studies in Egyptian Archaeology and History in Honour of Barry J. Kemp, Supreme Council of Antiquities, Cairo, 2009, pp. 241–243.
[19] I. Mathieson, J. Dittmer, The geophysical survey of North Saqqara, 2001–2007, J. Egypt. Archaeol. 93 (2007) 79–93.
[20] I. Mathieson, A. Tavares, Preliminary report of the National Museums of Scotland Saqqara survey project, 1990–91, J. Egypt. Archaeol. 79 (1993) 17–31.
[21] T. Herbich, D. O'Connor, M. Adams, Magnetic mapping of the northern cemetery at Abydos, Archaeol. Polona 41 (2003) 193–197.

[22] T. Herbich, J.W. Wegner, Magnetic survey at South Abydos: revising archaeological plans, Archaeol. Polona 41 (2003) 200–204.
[23] E. Pusch, H. Becker, J. Fassbinder, Wohnen und leben oder: wietere Schritte zu einem Stadtplan der Ramses-Stadt, Ägypten und Levante 9 (1999) 155–170.
[24] R. Křivánek, M. Bárta, Geophysical prospection in South Abusir, Egypt, 2002, Aracheol. Polona 41 (2003) 220–223.
[25] M.C. Flossmann-Schütze, Les maisons-tours de l'association religieuse de Touna el-Gebel, in: S. Marchi (Ed.), Les maisons-tours en Égypte durant la basse époque, les périodes ptolémaïque et romaine, Université Paris-Sorbonne, Paris, 2014, pp. 9–31.
[26] T. Abdallatif, A.E. El Emam, M. Suh, I.A. El Hemaly, H.H. Ghazala, E.H. Ibrahim, H. H. Odah, H.A. Deebes, Discovery of the causeway and the mortuary temple of the Pyramid of Amenemhat II using near-surface magnetic investigation, Dahshour, Giza, Egypt, Geophys. Prospect. 58 (2010) 307–320.
[27] N. Spencer, S. Hay, Amra West: remote sensing at a pharaonic town in northern Sudan, in: P. Johnson, M. Millet (Eds.), Archaeological Survey and the City, Oxbow Books, Oxford, 2013, pp. 176–201.
[28] J.A. Hill, T. Herbich, in: R.F. Friedman, P.N. Fiske (Eds.), Life in the cemetery: late predynastic settlement at el-Amra, Egypt at its Origins 3. Proceedings of the Third International Conference "Origin of the State. Predynastic and Early Dynastic Egypt," London, 27th July–1st August 2008, Orientalia Lovaniensia Analecta 205, Leuven, 2011, pp. 109–135.
[29] M.A. Hoffman, The Predynastic of Hierakonpolis—An Interim Report, Egyptian Studies Association 1, Cairo University Herbarium, Western Illinois University, Giza, Macomb, 1982.
[30] F. Hassan, R.G. Matson, Seriation of predynastic potsherd from the Nagada region (Upper Egypt), in: L. Krzyżaniak, M. Kobusiewicz (Eds.), Late prehistory of the Nile Basin and the Sahara, Studies in African Archaeology, 2, Poznań Archaeological Museum, Poznan, 1989, pp. 303–315.
[31] S. Sidebotham, Berenike, in: K.A. Bard (Ed.), Encyclopedia of the Archaeology of Ancient Egypt, Taylor & Francis, London and New York, 1999, pp. 170–172.
[32] T. Herbich, Magnetic survey, in: S.E. Sidebotham, W.Z. Wendrich (Eds.), Berenike 1999/2000. Report of the Excavations at Berenike including Excavations in Wadi Kalalat and Siket, and the Survey of the Mons Smaragdus Region, Cotsen Institute of Archaeology, University of California, Los Angeles, 2007, pp. 22–29.
[33] S.E. Sidebotham, Excavations, in: S.E. Sidebotham, W.Z. Wendrich (Eds.), Berenike 1999/2000. Report of the Excavations at Berenike including Excavations in Wadi Kalalat and Siket, and the Survey of the Mons Smaragdus Region, American University in Cairo Press, Los Angeles, 2007, pp. 30–165.
[34] I. Zych, T. Herbich, Magnetic prospection in the service of uncovering the Hellenistic and Roman port of Berenike on the Red Sea in Egypt, Archaeol. Polona 53 (2015) 95–118.
[35] M. Woźniak, J.K. Rądkowska, in: In search of Berenike of the Ptolemies, The Hellenistic fort of Berenike, Its Localization, Form and Development (part one), Polish Archaeology in the Mediterranean 23/1 (Research 2001), 2014, pp. 505–526.
[36] F.G. Aldsworth, S.E. Sidebotham, W.Z. Wendrich, The town site: survey, plan and description, in: S.E. Sidebotham, W.Z. Wendrich (Eds.), Berenike 1994, Preliminary Report of the 1994 Excavations at Berenike (Egyptian Red Sea Coast) and the Survey of the Eastern DesertResearch School of Asian, African and Amerindian Studies, Leiden, 1995, pp. 13–20.

[37] H. Barnard, W.Z. Wendrich, Survey of Berenike, in: S.E. Sidebotham, W.Z. Wendrich (Eds.), Berenike 1999/2000. Report of the Excavations at Berenike including Excavations in Wadi Kalalat and Siket, and the Survey of the Mons Smaragdus Region, Cotsen Institute of Archaeology, University of California, Los Angeles, 2007, pp. 4–21.
[38] S.E. Sidebotham, W.Z. Wendrich, Interpretative summary and conclusion, in: S.E. Sidebotham, W.Z. Wendrich (Eds.), Berenike 1999/2000. Report of the Excavations at Berenike including Excavations in Wadi Kalalat and Siket, and the Survey of the Mons Smaragdus Region, University of California, Los Angeles, 2007, pp. 368–374.
[39] S.E. Sidebotham, M. Hense, H.M. Nouwens, The Red Land, American University in Cairo Press, Cairo, 2008.
[40] B. Żurawski, Pliny's "Tergedum" discovered, Sudan & Nubia 2 (1998) 74–81.
[41] B. Żurawski, in: Survey and excavations between old dongola and Ez-Zuma, Nubia II. Southern Dongola Reach Survey 1, Neriton, Warsaw, 2003.
[42] B. Żurawski, Where was the "Land of Shas"? Some comments on the Psmatik II's expedition to the south caused by the recent archaeological discoveries in the Southern Dongola Reach, 2018, forthcoming
[43] K.-H. Priese, Zur Ortliste der römischen Meroe-Expeditionen unter Nero, Meroitica 1 (1973) 123–126.
[44] M. Baud, La maison de roi: le palais, in: M. Baud (Ed.), Méroé, Un empire sur le Nil, Museée du Louvre, Paris, 2010, pp. 241–243.
[45] M. Bárta, L. Suková, V. Brũna, The latest explorations at Usli, Northern Province, Sudan & Nubia 17 (2013) 66–69.
[46] M. Bárta, J. Turek, Usli. An prahu říše boha Amona, Pražské Egyptologické Studie 14 (2015) 104–114.
[47] C.R. Lepsius, Denkmaeler aus Aegypten und Aethiopien I, Nicolaische Buchhandlung, Leipzig, 1897.
[48] N. Alexanian, T. Herbich, The workmen's barracks south of the Red Pyramid at Dahshur, MDAIK 70/71 (2015) 13–23.
[49] D. Arnold, Überlegungen zum Problem des Pyramidenbaus, MDAIK 37 (1981) 15–28.
[50] H. Ernst, Die Arbeitsiedlung, in: R. Stadelmann, N. Alexanian, H. Ernst, G. Heindl, D. Raue (Eds.), Pyramiden und Nekropole des Snefru in Dahschur, MDAIK 49 (1993), pp. 291–294.
[51] R. Stadelmann, Die Pyramiden des Snofru in Dahschur. Zwiter Bericht über die Ausgrabungen an der nördlichen Steinpyramide, MDAIK 39 (1983) 225–241.
[52] R. Stadelmann, N. Alexanian, Die Friehöfe des Alten und Mittleren Reiches in Dahschur, MDAIK 54 (1998) 293–317.
[53] H. Becker, Archaeological prospecting by caesium magnetometry, in: N. Alexanian, H. Becker, M. Müller, S.J. Seidlmeyer (Eds.), Die Residenznekropole von Dahschur, MDAIK 62 (2006), pp. 25–27.
[54] T. Herbich, Geophysical surveying in Egypt: periodical report for 2013–2015, Archaeol. Polona 53 (2015) 206–212.
[55] N. Alexanian, F. Arnold, T. Herbich, S. Müller, in: Dahschur, Ägypten: Die Arbeiten der Jahre 2014 und 2015, e-Forschungberichte 2016 des Deutschen Archäologischen Instituts, 3 (2016) 12–20.
[56] M. Lehner, The pyramid age settlement of the southern mount at Giza, JARCE 39 (2002) 27–74.
[57] G. Hadji-Minaglou, Recent excavations at Bawit, in: G. Gabra, H. Takla (Eds.), Christianity and Monasticism in Middle Egypt, British Academic Press, Cairo/New York, 2015, pp. 229–240.

[58] D. Benazeth, Recherches archéologiques à Baouit: un nouveau départ, Bulletin de la Société d'Archéologie Copte 43 (2004) 9–24.

[59] H. Torp, Le monastére copte de Baouît. Quelques notes d'introduction, in: Acta Ad Archaeologiam Et Artium Historiam Perentinentia, vol. 9, 1981, pp. 1–8.

[60] Golvin, Baouit. Plan général. Parties A, B et C. In: Clédat J. Le monastère et la nécropole de Baouît (notes mises en œuvre et éditées par D. Bénazeth et M.-H. Rutschowscaya), MIFAO 111 (1999) 439–445.

[61] T. Herbich, D. Bénazeth, Le kôm de Baouît: étapes d'une cartographie, BIFAO 108 (2008) 165–204.

[62] S. Marchand, G. Marouard, M. Mossakowska-Gaubert, F. Calament, C. Meurice, in: V. Ghica (Ed.), Survey du monastere d'Apa Apollo de Baouit: Premieres donnees archeologiques et analyses preliminaires de la ceramique et du verre des ermitages dits 'de la montagne, Actes du colloque international Ermitages d'Égypte, Cairo, 24–26 January 2009, IFAO, Cairo, 2018 forthcoming.

[63] D. Bénazeth, L'église de l'archange Michel à Baouit, J. Copt. Stud. 15 (2013) 3–20.

[64] G. Hadji-Minaglou, Le site monastique de Baouît et son église principale: état des lieux à l'issue de la campagne de fouilles de 2013, in: Cahiers de la Bibliothèque Copte 21 (Études coptes 14), 2016, pp. 47–62.

[65] É. Chassinat, Fouilles à Baouît I/I. Mémoires de l'Institut français d'archéologie orientale 13, (1911).

[66] M. Capuani, Christian Egypt, The American University in Cairo Press, Cairo, 2002.

[67] T. Herbich, H.D. Brooks, S.J. Davis, A geophysical survey of ancient Pherme: magnetic prospection at an early Christian monastic site in the Egyptian Delta, JARCE 43 (2007) 129–137.

[68] J.W. Wegner, Abydos, in: D.B. Redford (Ed.), The Oxford Encyclopedia of Ancient Egypt, The American University in Cairo Press, Cairo, 2001, pp. 7–12.

[69] M.D. Adams, D. O'Connor, S.H. D'Auria, The Shunet el-Zebib at Abydos: architectural conservation at one of Egypt's oldest preserved royal monuments, in: Offerings to the Discerning Eye. An Egyptological Medley in Honor of Jack. A. Josepson, Brill, Leiden, Boston, 2010, pp. 1–8.

[70] L. Bestock, The evolution of royal ideology: new discoveries from the reign of Aha, in: B. Midant-Reynes, Y. Tristant (Eds.), Egypt at Its Origins 2, Orientalia Lovaniensia Analecta, vol. 172, 2008, pp. 1091–1106.

[71] M.D. Adams, L. Bestock, Preliminary report on the 2008–2009 season of Institute of Fine Arts, New York University Expedition to Abydos, (2009) (Unpublished).

[72] T. Herbich, J. Richards, The loss and rediscovery of the vizier Iuu at Abydos. Magnetic survey in the middle cemetery, in: E. Czerny, I. Hein, H. Hunger, D. Melman, A. Schwab (Eds.), Timelines. Studies in Honour of Manfred Bietak, Orientalia Lovaniensia Analecta, vol. 149/1, 2006, pp. 141–149.

[73] T. Herbich, in: S. Piro (Ed.), Geophysical surveying in Egypt: recent results, VI Conference on Archaeological Prospection, Rome. Proceedings, Rome, 2005, pp. 1–5.

[74] K.M. Cahail, In the Shadow of Osiris: Non-Royal Mortuary Landscapes at Abydos South During the Late Middle and New Kingdoms, Ph.D. thesis, University of Pennsylvania, 2014.

[75] J. Wegner, Royal funerary equipment of a king Sobekhotep at south Abydos: evidence for the tombs of Sobekhotep IV and Neferhotep I? JARCE 51 (2015) 123–164.

[76] J. Wegner, A royal boat burial and watercraft tableau of Egypt's 12th Dynasty (c. 1850 BCE) at South Abydos, Int. J. Naut. Archaeol. 47 (2016).

[77] S.P. Harvey, Abydos, in: G.L. Stein (Ed.), The Oriental Institute 2002–2003 Annual Report, University of Chicago, Chicago, 2003, pp. 15–25.

[78] G. Dreyer, Abydos, Umm el-Qa'ab, in: K. Bard (Ed.), Encyclopedia of the Archaeology of Ancient Egypt, Routledge, London, 1999, pp. 109–114.

[79] T. Herbich, How deep can we see? Practical observations on the vertical range of fluxgate gradiometers when surveying brick structures in the Nile Valley, in: M. Jucha, J. Dębowska-Ludwin, P. Kołodziejczyk (Eds.), Aegyptus est imago caeli. Studies presented to Krzysztof M. Ciałowicz on His 60th Birthday, Institute of Archeology of the Jagiellonian University, Kraków, 2014, pp. *255–265*.

[80] D. Kessler, Bersheh, in: D.B. Retford (Ed.), The Oxford Encyclopedia of Ancient Egypt, vol. 1, The American University in Cairo Press, 2001, pp. 177–179.

[81] H. Willems, M. de Meyer, D. Depraetere, C. Peeters, S. Hendrickx, T. Herbich, D. Klemm, R. Klemm, L. Op de Beeck, M. Depauw, Preliminary report of the 2002 campaign of the Belgian mission to Deir al-Barsha, MDAIK 60 (2004) 237–283.

[82] T. Herbich, C. Peeters, Results of the magnetic survey in Deir al-Barsha, Egypt, Archaeol. Prospect. 13 (2006) 11–24.

[83] H. Willems, M. de Meyer, C. Peeters, S. Vereecken, D. Depraetere, T. Dupras, L. Willams, T. Herbich, G. Verstraeten, G. van Loon, A. Delattre, Report of the 2004–2005 campaigns of the Belgian mission to Deir al-Barsha, MDAIK 65 (2009) 377–432.

[84] H. Willems, Report of the mission of the Katholieke Universiteit Leuven to Dayr al-Barsha 4 March–26 April 2007, Annales du Service des Antiquités de l'Égypte 84 (2010) 429–465.

[85] H. Willems, C. Peeters, G. Verstraeten, Where did Djehutihotep erect his colossal statue? Zeitschrift für Ägyptische Sprache und Altertumskunde 132 (2) (2005) 173–189.

[86] A. Fahkry, The Monuments of Snefru at Dahshur, vol. I, U.S. Government Printing Office, 1959, Cairo.

[87] R. Stadelmann, Neue Forschungen and der Knickpyramide. Die Kampagnen des DAI Kairo in den Jahren 2005 und 2006, Sokar 14 (2007) 8–11.

[88] F. Arnold, Der Pyramidentempel als Ritualschauplatz. Eine Spurensuche im Tal der Knickpyramide, Sokkar 33 (2) (2016) 6–15.

[89] F. Arnold, Ein Garten des Königs Snofru in Dahschur, Sokar 29 (2) (2014) 6–13.

[90] D. O'Connor, Abydos. Egypt's First Pharaohs and the Cult of Osiris, American University in Cairo Press, Cairo, 2009.

[91] M. Baba, R.F. Friedman, Recent excavations at HK11C, Hierakonpolis, in: M.D. Adams (Ed.), Egypt at Its Origins 4, Orientalia Lovaniensia Analecta, vol. 252, Peeters, 2016, pp. 179–205.

[92] M. Baba, Another portion of potatoes: excavation at KH11C in 2016, Nekhen New 28 (2016) 10–11.

[93] I.H. Takamiya, Beer after sheep? HK11C squares A6-A7 in 2007, vol. 19, Nekhen News, 2007, p. 28.

[94] N. Alexanian, F. Arnold, The complex of the Bent Pyramid as a landscape design project, 2017, in: M. Ullmann (Ed.), 10. Ägyptologische Tempeltagung. Ägyptische Tempel zwischen Normierung und Individualität, Harrassowitz Verlag, Wiesbaden, 2016, pp. 1–15.

[95] F. Arnold, A ceremonial Building of king Snofru at Dahshur, 2017, in: M. Bietak, S. Prell (Eds.), Ancient Egyptian and Ancient Near Eastern Palaces I, Contributions

to the Archaeology of Egypt, Nubia and the Levant, 5 Austrian Academy of Science Press, Vienna, 2018, pp. 113–124.

[96] H. Becker, Caesium-magnetometry for landscape archaeology, in: S. Campana, S. Piro (Eds.), Seeing the Unseen. Geophysics and Landscape Archaeology, CRC Press, London, 2009, pp. 129–165.

[97] N. Alexanian, Die Gestaltung der Pyramidenanlagen des Snofru in Dahšūr/Ägypten, in: I. Gerlach, D. Raue (Eds.), Sanktuar und Ritual. Heilige Plätze im archäologischen Befund, Verlag Marie Leidorf GmbH, Rahden, 2013, pp. 159–169.

[98] R. Friedman, Hierakonpolis, in: D.B. Retford (Ed.), The Oxford Encyclopedia of Ancient Egypt, vol. 2, The American University in Cairo Press, 2001, pp. 98–100.

[99] M.A. Hoffman, Egypt Before Pharaos: the Prehistoric Foundations of Egyptian Civilization, Knopf, New York, 1979.

[100] T. Herbich, Magnetic survey at Hierakonpolis, Nekhen News 10 (1998) 17–18.

[101] T. Herbich, The geophysical survey, Nekhen News 11 (1999) 17.

[102] T. Herbich, Back to magnetometry: 2010 survey, Nekhen News 22 (2010) 18–19.

[103] J.F. Harlan, Excavations at locality 11C, in: M.A. Hoffman (Ed.), The Predynastic of Hierakonpolis—An Interim Report, Egyptian Studies Association, vol. 1, Cairo University Herbarium, Western Illinois University, Giza, Macomb, 1982, pp. 14–25.

[104] I.H. Takamiya, Firing installations and specialization: a view from recent excavations at Hierakonpolis locality 11C, in: B. Midant-Reynes, Y. Tristant (Eds.), Egypt at Its Origins 2, Orientalia Lovaniensia Analecta, vol. 172, Peeters, 2008, pp. 187–202.

[105] M. Baba, Pottery production at Hierkonpolis in the Naqada II period: towards a reconstruction of the firing technique, in: R.F. Friedman, P.N. Fiske (Eds.), Egypt at its Origins 3, Orientalia Lovaniensia Analecta, vol. 205, Peeters, 2011, pp. 649–670.

[106] M. Baba, Up against the walls at HK11C, Nekhen News 23 (2011) 22–24.

[107] M. Baba, Industrial scale food preparations at HK11C, Nekhen News 27 (2015) 12–13.

[108] I.H. Takamiya, The inside story: the installation and granary at HK24B, Nekhen News, 2011, pp. 20–21.

[109] C. Römer, The Nile in the Fayum. Strategies of dominating and using the water resources of the river in the oasis in the Middle Kingdom and the Graeco-Roman period, in: H. Willems, J.-M. Dahms (Eds.), The Nile: Natural and Cultural Landscape in Egypt, Transcript Verlag, Bielfeld, 2017, pp. 171–191.

[110] F. Hassan, G. Tasie, Modelling environmental and settlement change in the Fayum, J. Egypt. Archaeol. 29 (2006) 37–39.

[111] P. Davoli, L'archeologia urbana del Fayyum di età ellenistica e romana, Generoso Procaccini, Napoli, 1998.

[112] D.M. Bailey, in: C. Römer (Ed.), Surface pottery form Philoteris (AD 2000-3), Philoteris in the Themistou Meris. Report on the archaeological survey carried out as part of the Fayum survey project. Zeitschrift für Papyrologie und Epigraphik, vol. 147, 2004, pp. 301–305.

[113] C. Römer, Philoteris in the Themistou Meris. Report on the archaeological survey carried out as part of the Fayum Survey Project, Z. Papyrol. Epigr. 147 (2004) 281–305.

[114] T. Herbich, C. Römer, in: W. Neubauer, I. Trinks, R.B. Salisbury, C. Einwögerer (Eds.), Magnetic research of an ancient village and its water resources: the case of Philoteris/Medinet Watfa in Oasis Fayum, Egypt, Archaeological Prospection. Proceedings of the 10th International Conference on Archaeological Prospection, Vienna May 29–June 2 2013, Vienna, 2013, pp. 261–264.

[115] P. Kopp, Canals, Wells and Basins: Excavation in Philoteris in 2012 and 2014, (2014) (Unpublished Report, DAI Archive).

[116] H. Barnard, W.Z. Wendrich, B.T. Nigra, B.L. Simpson, R. Cappers, The fourth-century AD expansion of the Graeco-roman settlement of Karanis (Kom Aushim) in the northern Fayum, J. Egypt. Archaeol. 101 (2015) 51–67.

FURTHER READING

[117] A. Aspinall, C. Gaffney, A. Schmidt, Magnetometry for Archaeologists, Altamira Press, Lanham, 2008.
[118] S. Campana, S. Piro (Eds.), Seeing the Unseen. Geophysics and Landscape Archaeology, CRC Press/Balkema, Leiden, 2009.
[119] C. Gaffney, J. Gater, Revealing the Buried Past, Geophysics for Archaeologist, Tempus, Stroud, 2003.

CHAPTER

Advanced SFCW GPR systems

8

Egil Eide*, Neil Linford†, Raffaele Persico‡,§, Jacopo Sala¶

Department of Electronic Systems, Norwegian University of Science and Technology, Trondheim, Norway Geophysics Team, Historic England, Portsmouth, United Kingdom† Institute for Archaeological and Monumental Heritage IBAM-CNR, Lecce, Italy‡ International Telematic University Uninettuno (UTIU), Rome, Italy§ 3D-Radar AS, Trondheim, Norway¶*

CHAPTER OUTLINE

1 INTRODUCTION

A step-frequency continuous waveform ground-penetrating radar (SFCW GPR) is a frequency-domain radar system based on a waveform consisting of a succession of sine waves at increasing frequency [1]. In particular, the term “continuous wave” specifies that the transmission and reception of the signal at each frequency occurs simultaneously, unlike other step-frequency systems, such as gated GPRs [2]. The main concept behind this particular technology is to obtain a large bandwidth and a good signal-to-noise ratio (so to have a good penetration of the signal) without

Innovation in Near-Surface Geophysics. https://doi.org/10.1016/B978-0-12-812429-1.00008-8

radiating a large amount of power. The recorded raw data therefore represent the frequency response of the subsurface, and an inverse Fourier transform can be applied to obtain a synthetic trace in time domain.

This chapter gives a brief overview of the SFCW technique applied to GPR, providing some insight into the characteristics of the hardware, the properties of the waveform, the relevant parameters for data collection, and the aspects related to the processing and postprocessing of the data. There are indeed not only similarities between SFCW GPR and impulse GPR but also substantial differences. Focusing on this diversity, the authors' ambition is to reveal and eliminate some common misconceptions about the SFCW technique and provide an adequate background to get the most out of the data collected with this kind of GPR.

2 OVERVIEW OF THE STEP-FREQUENCY TECHNIQUE

A step-frequency is a radar waveform consisting of a train of sine waves whose frequency increases linearly in discrete steps. The transmitted signal can be represented by a staircase function in a frequency versus time plot, as shown in Fig. 1, where each step of the staircase corresponds to a sine wave with a specific frequency. The frequency f_n of the nth sine wave is given by

$$f_n = f_0 + (n-1)\Delta f, \quad n = 1, 2, \ldots, N \tag{1}$$

where f_0 is the minimum frequency, Δf is the size of the frequency step, and N is the total number of sine wave steps. The total effective bandwidth is given by

$$B_{eff} = (N-1)\Delta f \tag{2}$$

The amount of time the system spends on a single frequency is often called dwell time, referred to as t_{dw} in Fig. 1. The lower boundary value for the dwell time is given

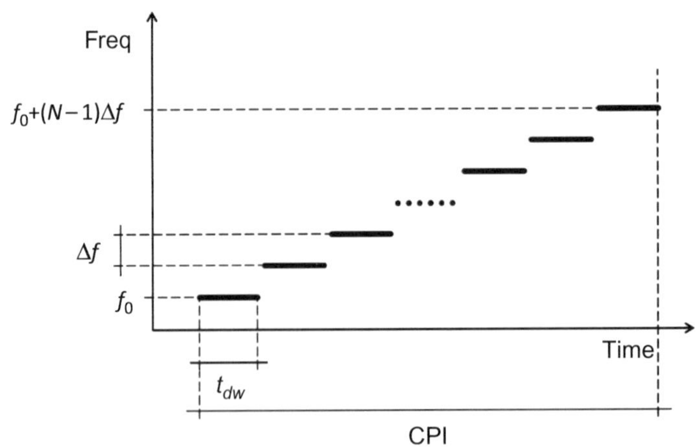

FIG. 1

Staircase function representing the step-frequency radar waveform.

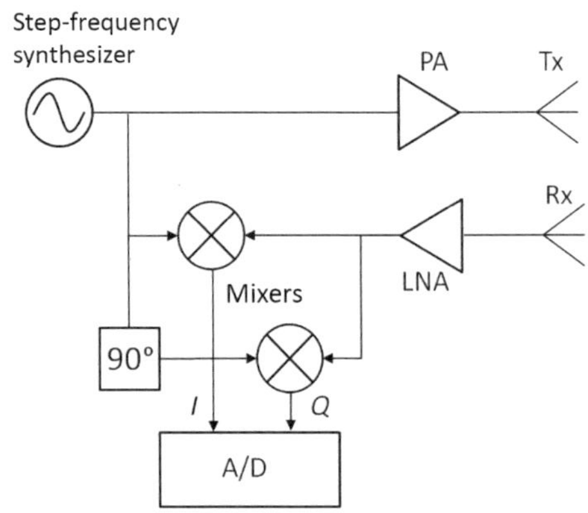

FIG. 2

Simplified step-frequency radar super-heterodyne block diagram.

by the period of the sine wave at the minimum frequency. The total integration time, also burst time or coherent processing interval (CPI), is the total time extent in which the signal relative to a single radar trace is collected for data processing. It is therefore equal to the amount of time required to collect a complete waveform:

$$CPI = Nt_{dw} \tag{3}$$

Fig. 2 shows a simplified, basic block diagram for a bistatic step-frequency radar. The step-frequency synthesizer is responsible for the change in frequency in the transmitted waveform. At the end of the receiver stage, the signal is finally divided into the in-phase (I) and quadrature (Q) channels (or components) that are then passed through an A/D converter and, step by step, build up into the spectral content of the frequency-domain raw data.

The received signal S_n at the nth frequency can be written as

$$S_n = I_n + jQ_n \tag{4}$$

Or the equivalent

$$S_n = A_n \exp(j\varphi_n) \tag{5}$$

where A_n is the magnitude of the nth frequency sample[1] and φ_n is its phase. A time-domain radar trace can be obtained applying an inverse discrete Fourier transform (IDFT) to the recorded frequency-domain data. More details and considerations regarding the frequency to time conversion will be given later on in the chapter.

[1]Please note that in this context, sine wave, frequency samples, and harmonics are interchangeable terms.

3 SETTING OF THE SYSTEM FOR CORRECT MEASUREMENTS

The fundamental waveform parameters for a step-frequency radar are the ones shaping the staircase function in Fig. 1:

- t_{dw}: dwell time
- Δf: frequency step
- N: number of frequency samples

Depending on the particular SFCW GPR being used, the operator can have the possibility to control these parameters when deciding the settings for data acquisition. The following is a brief overview of the step-frequency waveform parameters that should help understanding their influence on the data being collected.

3.1 DWELL TIME

As already mentioned, the dwell time is the time the system spends transmitting and receiving a single sine wave. This parameter influences the quality of the frequency-domain data: a longer dwell time corresponds to a higher signal-to-noise ratio, hence *cleaner* frequency samples. Common values for the dwell time in commercially available systems are at least one order of magnitude higher than the longest sine wave period of the step-frequency waveform. In order to determine a tangible improvement in terms of depth of penetration, the dwell time should be augmented following an exponential curve, which in turn would increase dramatically the time associated with a complete waveform and therefore slow down the advancement rate for data collection. Moreover, the soil itself is what finally determines the penetration of the GPR signal. Although increasing the dwell time does bring some benefits in the final radar image in terms of clarity of the signal at depth, the main advantage of increasing this parameter is to make the signal (at any fixed frequency) more resistant to interference from external radio sources, as will be shown in Section 5.2.

3.2 FREQUENCY STEP

The frequency step is the increment in frequency between successive sine waves in the waveform and corresponds to the sampling in frequency of the raw frequency-domain data. As such, given the properties of the Fourier transform, the frequency step defines an important quantity for the synthetic time-domain data: the amount of time that can be uniquely resolved, for example, through IDFT or IFFT, before the signal repeats itself. This unambiguous time interval is given by

$$T = \frac{1}{\Delta f} \tag{6}$$

And the equivalent range of nonambiguity is

$$R = T\frac{c}{2\sqrt{\varepsilon_r}} = \frac{c}{2\Delta f\sqrt{\varepsilon_r}} \tag{7}$$

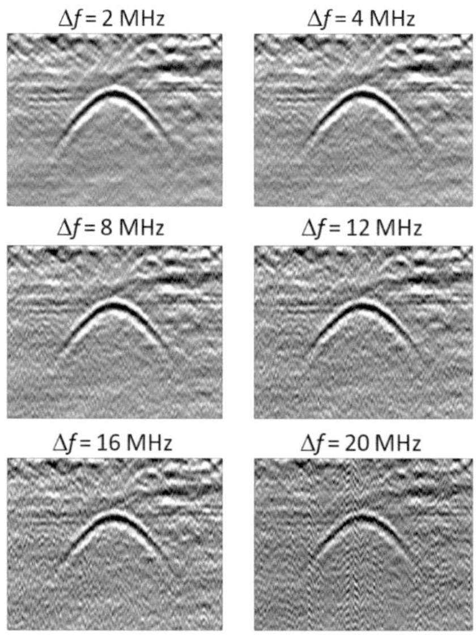

FIG. 3

Radar profile calculated from frequency-domain data that have progressively decimated to simulate the effect of different values for the frequency step. The data were collected originally with a 2 MHz frequency step.

where c is the speed of light in free space, approximately equal to 3×10^8 m/s and ε_r is the dielectric constant characterizing the subsoil. It is therefore important to take into consideration the maximum depth of investigation when deciding the settings for the frequency step. Although, strictly speaking, the frequency step is not a parameter affecting the quality of the waveform, it does affect the time-domain reconstruction. This is easy to understand if one considers the nature of the frequency to time conversion and contextualizes the data in GPR applications. The Fourier transform is an integral transform: each and every point in time domain is resolved as an integration over the spectrum. Since the attenuation in the soil is frequency-dependent (generally, the soil acts as a sort of low-pass filter), having a denser spectrum, that is, a lower frequency step, yields better results in the time-domain image, especially at depth, than would be achieved with a waveform covering the same bandwidth but characterized by a higher frequency step and therefore less populated with data. This is illustrated in Fig. 3, where the same radar profile in time domain has been calculated simulating the effects of different values for the frequency step by introducing a decimation factor in the original raw frequency data.

3.3 NUMBER OF STEPS

The number of steps defines together with the frequency step the effective bandwidth for the step-frequency waveform. The synthetic time resolution and the relative range resolution are given by

$$\triangle T = \frac{1}{B_{eff}} = \frac{1}{(N-1)\triangle f} \tag{8}$$

$$\triangle R = \triangle T \frac{c}{2\sqrt{\varepsilon_r}} = \frac{c}{2(N-1)\triangle f \sqrt{\varepsilon_r}} \tag{9}$$

Instead of setting explicitly the number of steps, it is more common to alter this parameter indirectly by giving the values for the minimum and maximum frequency for the waveform. Restricting the bandwidth can be interesting when one is concerned about a particular subset of frequencies. For example, the maximum frequency could be reduced when the shallower subsurface is not significant for the purposes of the survey. However, it should be noted that the only practical advantage of this approach would be to achieve a higher maximum speed for data collection (i.e., fewer sine waves would be transmitted and received within one waveform); the same "effect" on the data can be achieved during postprocessing by simply disregarding the unwanted frequency samples during frequency to time conversion or by applying a band-pass filter on the frequency-domain data before the inverse Fourier transform. Furthermore, it is often difficult to predict exactly the absorption characteristics of the area under investigation, and mapping the frequency of the subsurface over a wide range of frequencies leads to greater flexibility for the reconstruction of the radar image.

4 FREQUENCY TO TIME CONVERSION

Let's assume for simplicity a situation with a single isotropic scatterer buried in a homogeneous medium. The nth generated signal, the sine wave at frequency f_n, can be written as

$$u_n(t) = A \exp(j2\pi f_n t) \tag{10}$$

Let's also assume that the transmitted power, A, is constant over the spectrum. The received signal at the same frequency is

$$s_n(t) = A_n \exp\left(j2\pi f_n \left(t - \tau_{tgt}\right)\right) \tag{11}$$

where τ_{tgt} is the total time the signal uses to travel from the transmitter, to the target, and finally to the receiver, and A_n depends on the size of the target, the dielectric contrast it has with the surrounding medium, its distance from the antenna, its position with respect to them, and the attenuation characteristics of the subsoil. The phase of the relative nth frequency sample is therefore (see Eq. 5)

$$\varphi_n = -2\pi f_n \tau_{tgt} \tag{12}$$

The time-domain trace is obtained summing over all the frequency components by the means of an inverse Fourier transform. Fig. 4 shows plots for frequency and time domain for an example with a single reflector. As can be seen, the spectrum of the received signal is not constant, essentially because of the finite size of the reflector.

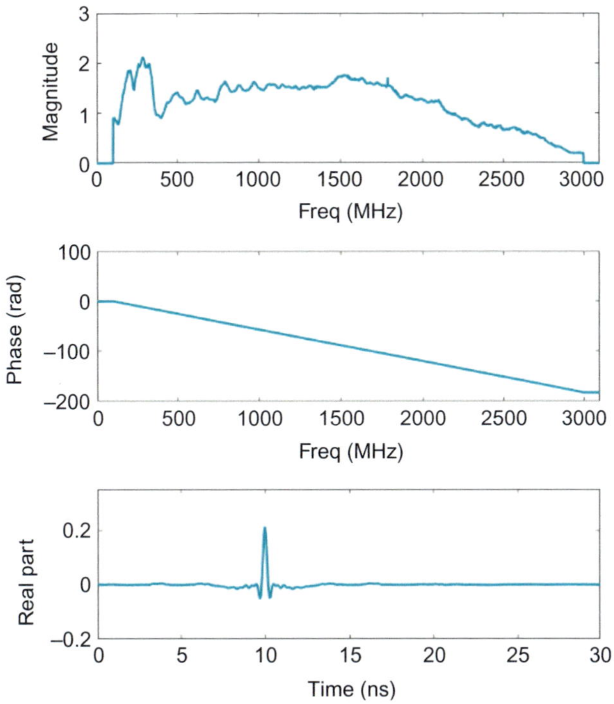

FIG. 4

Simulated data for an arrival from a single reflector at 10 ns two-way travel time from the radar.

In a real situation, the sine waves at different frequencies will be the composition of various contributions from layers and objects populating the subsurface and, if present, external radio signals interfering with the measurements. Consequently, Eq. (12) becomes

$$s_n(t) = \sum_i A_{n,i} \exp\left(j2\pi f_n(t-\tau_i)\right) + A_{n,env} \exp\left(j\varphi_{n,env}\right) \tag{13}$$

where the index i represents the various contributions from the different reflectors and the last terms, with subscript *env*, summarize environmental noise and interfering signal at frequency f_n.

One interesting thing to notice is that all the information relative to the time of the different arrivals is contained in the phase spectrum. The phase itself can provide enough information for the reconstruction of a time-domain profile. Fig. 5 shows a comparison between a radar profile and the same obtained after having replaced all the measurements of the amplitude in the frequency-domain data to a value of one (a sort of full *spectral whitening*). The noise level has of course increased, but the information regarding the various arrivals has remained intact.

Through inverse Fourier transform, the frequency response of the subsurface, the data recorded by the SFCW GPR, is converted into an impulse response, a synthetic

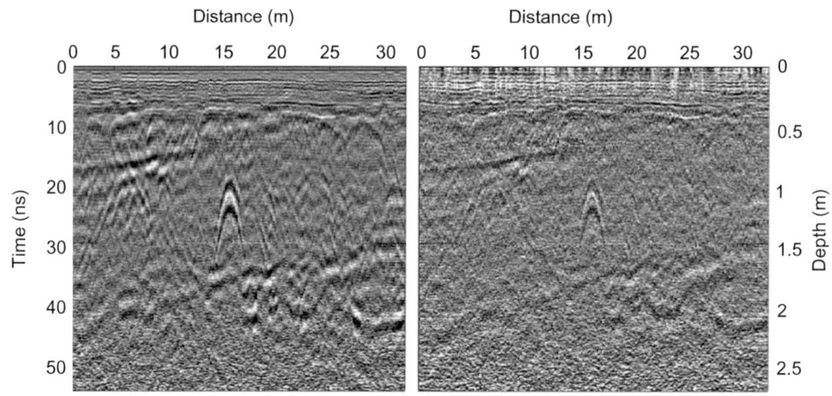

FIG. 5

Original time-domain section (left) and the same profile calculated after replacing the amplitude of the frequency data with ones, preserving only the phase (right).

time series representing the sequence of arrivals from the subsurface. Conceptually, one can think of the frequency to time conversion as a software operation that emulates an impulse GPR: data representing the subsurface are in fact being *probed* by the numerical representation of an impulse. Even when no impulse representation is explicitly applied, the frequency-domain data are being implicitly multiplied by a rectangular window (all ones), which is equivalent to a correlation with a *sinc* function in time domain. It is fairly common during postprocessing of step-frequency data sets to shape the amplitude spectrum through a tapering window. This allows the control of the trade-off between sharpness of the central lobe (resolution) and amplitude of the side lobes (ringing) of the pulses in time domain. Window functions that are commonly used include Hamming, Hanning, Kaiser, Tukey, and, of course, boxcar or rectangular. Depending on the particular application and the type and depth of the targets to be detected, one can decide to use a wide window (which gives a very sharp main lobe but has the drawback of strong side lobes) or a narrow one (low secondary lobes but a wider central lobe). Another thing to consider is that changing the shape of the window might also have an impact on the visibility of deeper features in the time-domain image; narrower windows reduce the contribution from the lower portion of the bandwidth, that is, those frequency components that can possibly travel to greater depths. The possibility to shape the power spectral density, interacting directly with the raw data in frequency domain, demonstrates the great flexibility provided by a SFCW GPR during the postprocessing of the data. It is possible to obtain radically different time-domain images starting from the same raw data. The characteristic trade-off between resolution and depth of penetration is, in the case of impulse-based GPRs, something that the GPR operator needs to resolve before the data collection. The questions to be addressed are related to the general scope of the survey campaign, the size of the targets and their depth, and the type of soil to be investigated. These will give indications regarding the optimal central frequency of the equipment to be used for the survey. For step-frequency systems, while it is true that obtainable depth and resolution are finally given by the system survey parameters, as described earlier in this chapter, one could argue that the choices influencing the balance between resolution and investigation depth have been pushed more into the postprocessing phase. Fig. 6 displays two radar profiles obtained from the same data set. One has been produced using the full bandwidth, which for this survey was set to be from 100 to 2500 MHz, while the second is the result of considering only the components between 100 and 800 MHz. The latter is an obvious improvement when considering the visibility of deeper reflectors, but at the same time, there's a clear loss of resolution at shallow depths.

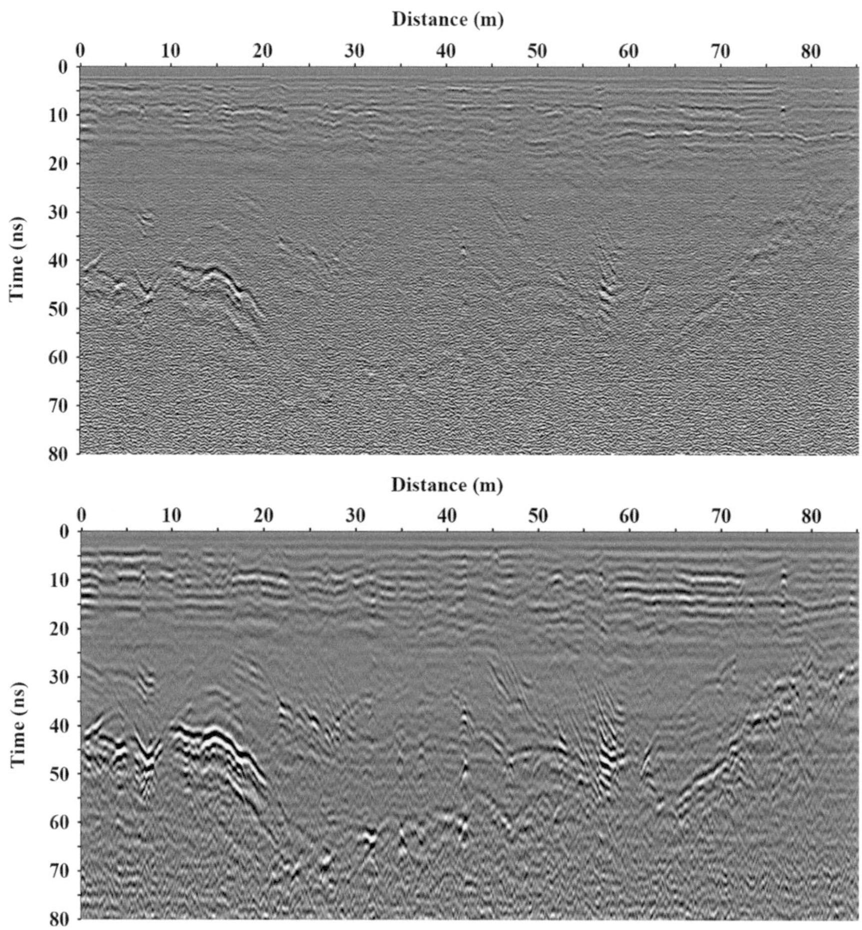

FIG. 6

Time-domain sections obtained using the full available bandwidth, 100–2500 MHz, (top) and the subset 100–800 MHz (bottom).

With the intention of optimizing resolution at different depths, one possibility is to divide the total time span of interest into a number of intervals and obtain for each interval a time-domain realization by selecting the optimal range of frequencies for that depth. The various sections can then be stitched together in a single image for further processing and interpretation. Another approach is based on the introduction of a geometric factor in the IDFT, shaping dynamically the power spectral density and in particular diminishing the contribution of the high-frequency components at greater depths. We recall here the definition of the IDFT:

$$s_k = \frac{1}{N}\sum_{n=0}^{N-1} S_n \exp(j2\pi nk/N) \tag{14}$$

The modified IDFT can be written as

$$s_k = \frac{1}{N}\sum_{n=0}^{N-1} S_n \exp((-\alpha + j2\pi)nk/N) \tag{15}$$

The introduction of α determines a gradual low-pass filter. An example of this second approach applied to the same data of Fig. 6 is shown in Fig. 7.

5 HARDWARE OF STEP-FREQUENCY SYSTEMS

5.1 3D-RADAR SYSTEM

The GeoScope step-frequency GPR manufactured by 3D radar AS is one of the few step-frequency GPRs that are commercially available on the market today. The system was first introduced in 2004 and has been further developed over the years.

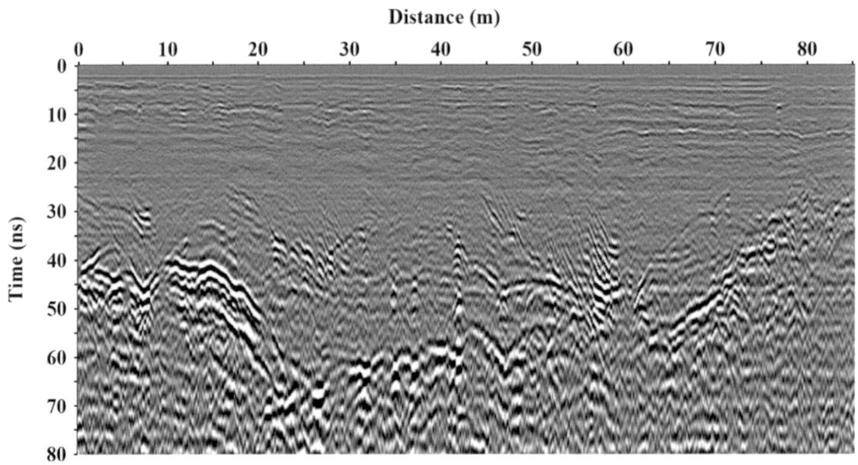

FIG. 7

Time-domain section calculated through a modified IDFT, with the introduction of a geometric factor that diminishes the contribution of high-frequency components as depth increases.

The current version, the GeoScope Mk4, is the fourth-generation step-frequency GPR from the company, and it is still considered as the fastest step-frequency system on the market. By using a digital frequency source instead of traditional phase-locked loop technology, the GeoScope can generate waveforms from 100 MHz to 3 GHz with a user-selectable frequency step providing from 150 to 1500 frequencies. The step-frequency radar has a dual-channel coherent receiver that means that the whole waveform length (typically a few milliseconds) is used as 100% efficient integration time. By comparison, impulse GPRs use stroboscopic sampling with significant loss of energy. Fig. 8 shows an overview of the GeoScope GPR system.

The GeoScope radar unit contains the frequency synthesizers, transmitter and receiver circuits, and the embedded computer for real-time control and data processing. It is controlled by a laptop PC through an Ethernet connection.

The GeoScope is designed to operate with an electronically scanned antenna array containing up to 41 antenna elements. The antenna elements are arranged in a linear array as shown in Figs. 9 and 10 where the transmitting and receiving antennas are offset with respect to each other. During the survey, the radar combines the transmit/receive antenna sequentially to obtain a number of profiles (or channels) over a swath width ranging from 1 to 3 m as shown in Fig. 8. The dual-channel receiver architecture allows the system to collect two survey lines simultaneously. In the standard (minimum offset) antenna scanning sequence, the GeoScope sequentially transmits/receives the full step-frequency waveform on each antenna pair.

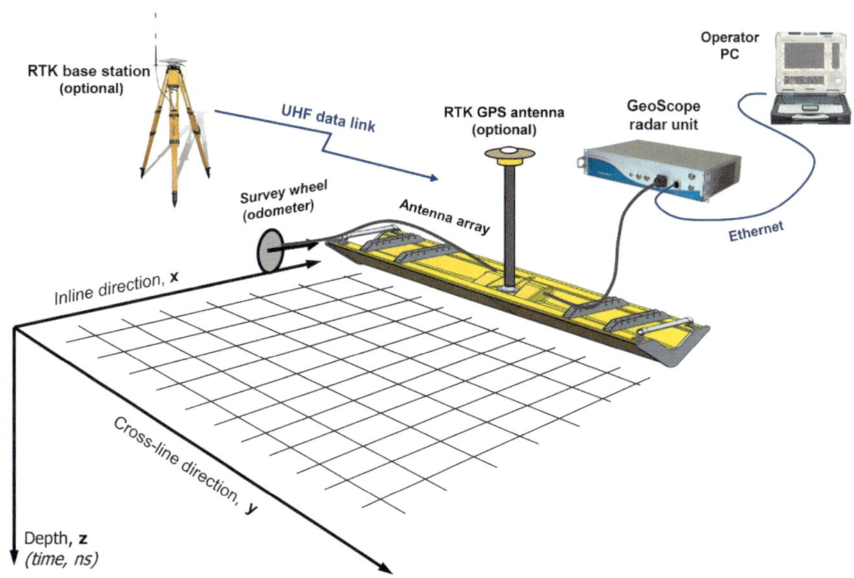

FIG. 8

3D radar GeoScope GPR system overview.

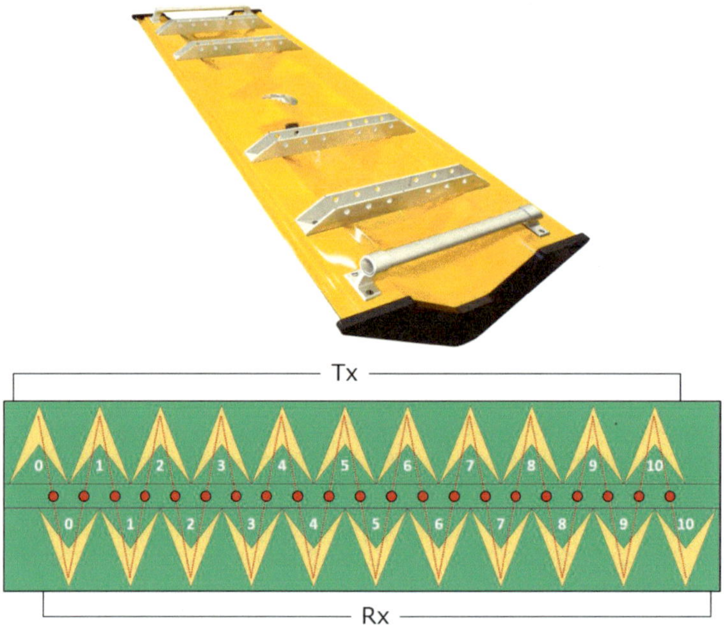

FIG. 9

3D radar DX array (top) and layout of the antenna elements for the model DX1821 (bottom).

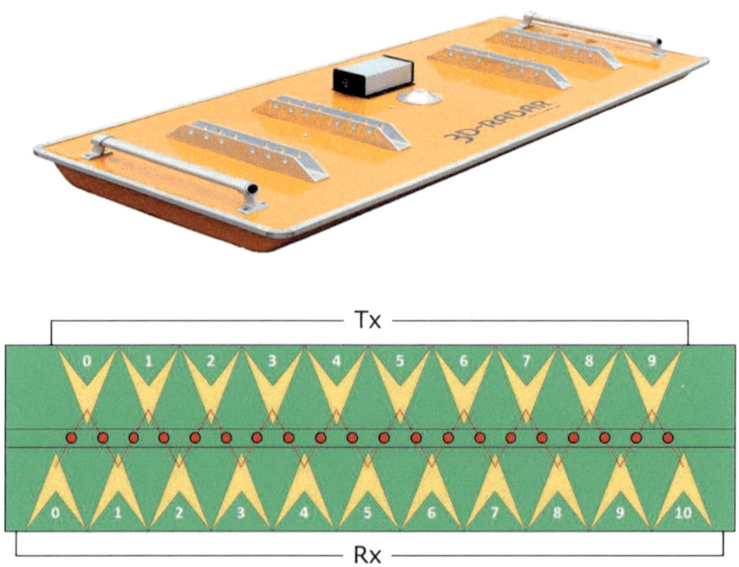

FIG. 10

3D radar DXG array (top) and layout of the antenna elements for the model DXG1820 (bottom).

The antenna array contains the antenna elements, RF-multiplexing circuits for electronic scanning, and an embedded calibration system that continuously monitors and deconvolves the radar system response. Due to the self-calibration of the system, there is no time drift in the system eliminating the warm-up time that is often found in impulse GPR systems.

The wideband nature of the GeoScope step-frequency radar covering a continuous frequency range from 100 MHz to 3 GHz makes it important to design the antenna array such that it covers as much of the frequency range as possible. The DX-series air-coupled antenna typically cover the frequency range from 200 MHz to 3 GHz with a low-frequency roll-off below 400 MHz. The DX-series antenna system consists of air-coupled bow-tie monopole pairs as shown in Fig. 9. The air-coupled antenna array can be operated at elevations from 30 cm off the ground allowing high-speed road inspection surveys.

The 3D radar DXG-series of ground-coupled arrays use the same type of boomerang-shaped monopole elements as the DX series of air-coupled antenna array, but it is optimized for operation directly onto the ground surface. For shallow imaging of concrete rebars and small objects, it is important to keep the offset-distance between the transmitter and receiver antenna as small as possible. Therefore, the monopoles are arranged such that the feed points are located as close as possible to simulate a zero-offset antenna structure. Fig. 10 shows the layout of the DXG array seen from the bottom side. Just like the air-coupled arrays, the antenna are scanned sequentially using RF multiplexers.

In both DX and DXG arrays, for each transmitter position, two receiver antenna are used simultaneously for reception. The separation between two neighbor transmitter antenna is 15 cm, but since the receiver array is offset compared with the transmitter array, the effective line (or channel) separation between the lines of the array is 7.5 cm.

The real-time 3D view (RT3D) of the GeoScope GPR, shown in Fig. 11, gives the user an instant view of data from every channel from a 3D radar antenna array while collecting data in the field. By displaying simultaneous views of the in-line profile (B-scan or radargram), crossline profile, and a horizontal time slice at a selected depth, RT3D makes field work much easier and enables the user to view and recognize targets while surveying as the objects are showed in real time.

5.2 RECONFIGURABLE SYSTEMS

At this point, we introduce an innovative development implemented in the recent years, namely, the reconfiguration of the system antenna [3]. Indeed, this reconfigurability is not a precise protocol and is not necessarily applicable only to step-frequency systems. However, the reconfiguration in frequency domain appears to be more directly feasible and, indeed, both an Italian patent [4] and above all a prototype of step-frequency reconfigurable GPR system has been already implemented some years ago and is in the equipment of the Laboratory of Applied Geophysics to Archaeological and Monumental Heritage of the Institute for Archaeological and

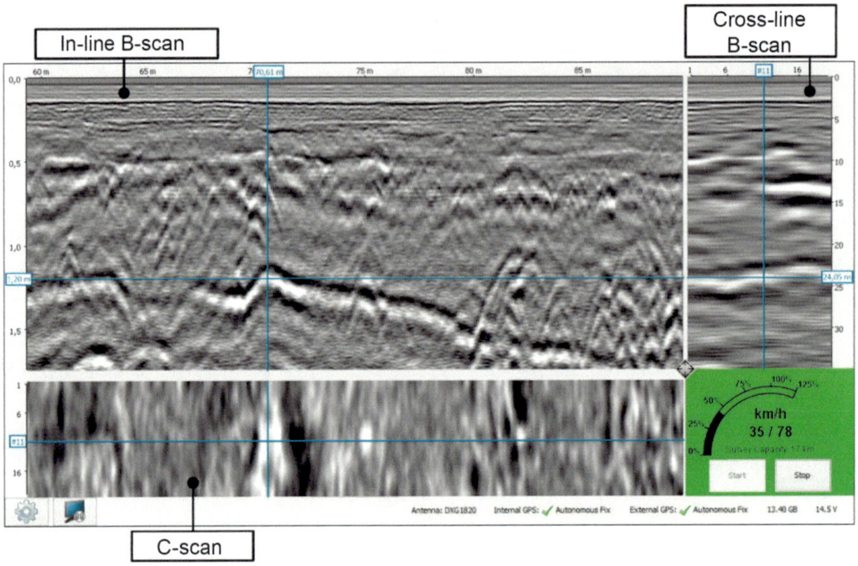

FIG. 11

3D radar GeoScope data acquisition software with real-time 3D view.

Monumental Heritage IBAM-CNR (http://www.ibam.cnr.it/en/risorse-e-servizi/laboratori). The system was implemented in 2009–10 as a collaboration between IBAM-CNR, IDS Corporation and University of Florence within the project AITECH, financed by the Puglia region with European funds (http://www.aitechnet.com/ibam.html). Subsequently, the system was tested in several sites in Italy, Norway, and Malta [5–8], in particular within a collaboration between IBAM-CNR and the University of Bari Aldo Moro, related to a Ph.D. research program. During this time, the system underwent development within the framework of the European Cost Action TU1208 with regard to the possibility to implement in the field an algorithm for the rejection of undesired interferences from other electromagnetic sources [9,10], described in the following text in more detail. Research activities are still on-going, in particular with regard to the possibility to visualize the data in real time, which is not possible using the current prototype at the moment. This research is being performed through student degree thesis activities that are being carried out in collaboration with the International Telematic University Uninettuno (UTIU).

In particular, the implemented system allows the acquisition of three B-scans at the same time centered on three different central frequencies, approximately equal to 120, 250, and 500 MHz. The three antennas are not physical but "equivalent," achieved from a unique couple of antenna with two series of synchronized switches implemented with PIN diodes along the arms, according to the scheme represented in Fig. 12.

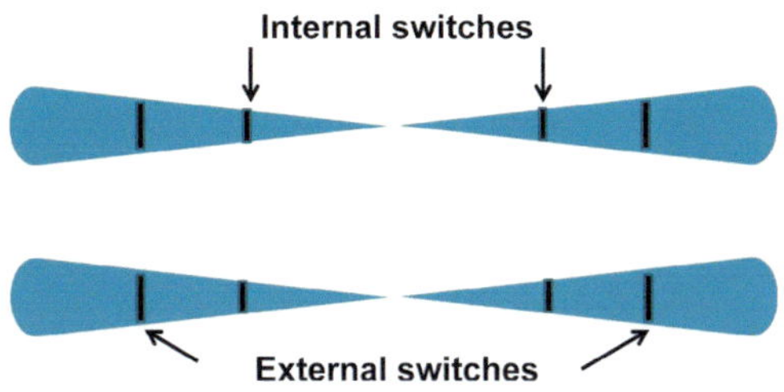

FIG. 12

Scheme of the reconfigurable antenna.

Courtesy of Radio Science, AGU.

In particular, with reference to Fig. 12, when both the internal and external series of switches are set to off, so that they prevent the passage of current through them, we have executed an electronic cut of the antenna that become physically shorter and are better adapted to radiate higher frequencies (band centered on approximately 500 MHz in this case). When the internal switches are set to on and the external are set to off, the antenna assume a medium size, suitable for the transmission of intermediate frequencies (i.e., a band about centered on 220 MHz). If both the internal and external switches are set to on, then the antenna assume their maximum available length suitable to transmit lower frequencies (a band about centered on 120 MHz).

Most commercial GPR systems have at most two couples of antenna with different bands. Moreover, they are two couples of physically different antenna, which implies that more space is needed to house the footprint of each antenna element. In this case, the current prototype can certainly be improved from an electromagnetic point view, for example, in the first instance through modifying the shape of the arms of the antenna. In particular, the antenna of the reconfigurable system, apart from the presence of the switches, are essentially two bow-tie antennas. It is well known that the input impedance of bow-tie antenna depends on the flare angle of the arms of the antenna [11], so that in order to optimize the matching of the antenna with the feed line (or line to the receiver), antenna operating at a higher frequency should have a larger flare angle. In the prototype, the flare angle is constant whatever the configuration of the switches. On the other hand, the detached parts of the antenna, that is, those beyond the switches when they are set off, are not irrelevant, because some parasitic currents develop in them, which might influence both the radiation pattern and the input impedance. The influence of these phenomena is not excessive, because the detached parts are in turn divided into two separate components at high frequency (when both the internal and external switches are set to off), which makes them relatively small in terms of wavelength. At any rate, a systematic study of all these phenomena has not been carried out yet and probably could drive an improvement of the performance of the antenna of the system [12]. Even so, the current system allows possibilities discussed above, for example, Fig. 13 shows three images collected simultaneously with the three antennas in a real case with the B-scans

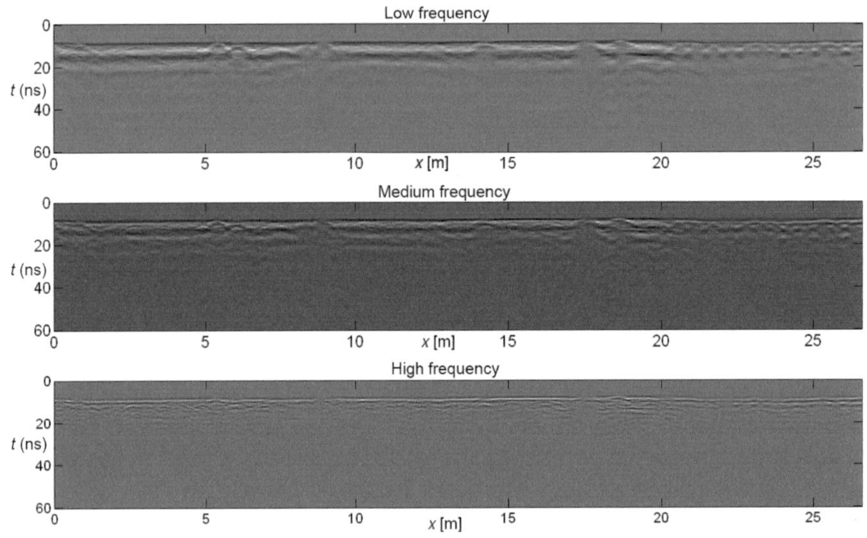

FIG. 13

Three processed B-scans simultaneously gathered by the reconfigurable GPR system.

gathered over a road in the city of Lecce. Fig. 13 shows the anomalies of interest were quite shallow, as was expected in this case (even if minimal processing has been performed on the data in the figure). However, it is often the case that the depth to the anomalies of interest is unknown or that there are several anomalies of interest at different depth levels (this can happen, particularly in archaeological sites). In such cases, it can be useful to have at disposal several data sets simultaneously, because as shown in the previous section, the lower frequencies can penetrate better in the soil, at the expense of a coarser resolution.

Please note that, in Fig. 13, for each of the three antennas, the entire sweep from 50 MHz to 1 GHz has been done, with frequency step 2 MHz. The band of the reconfigurable antenna performs a natural filter within this comprehensive range.

Another feature of the reconfigurable system is that it can prolong the dwell time of the harmonic components in a selective and programmable way. This has revealed to be potentially useful for the rejection of electromagnetic interference, especially narrow band interference. In particular, it is well known that prolonging the dwell time of the harmonic signals of step-frequency signals amounts to an improvement of the signal-to-noise ratio [13]. However, an extension of all the measured harmonic signals also increases the time needed to take the datum, and this increases the time required for the measurement campaigns, increasing the costs, and, potentially, leads to the deterioration of the data if the survey is rushed and velocity of the human operator is not kept low enough to maintain an adequate spatial sample density.

Selecting the frequencies where possible interference has occurred and can be mitigated through altering the dwell time requires analysis to be performed in the field, because in general we do not know a priori the disturbed bands. In Refs. [10,11], a method to perform this task in the field is identified through the variance of the frequency samples. In order to better specify this parameter, let us remember that, customarily, the value of the amplitude and phase of the current received tone in the current position is retrieved by a heterodyne demodulation receiving chain, which essentially corresponds to the flow diagram of Fig. 2, apart for some complications due to the use of an intermediate frequency. So, apart from details available, for example, in Ref. [13], let's say that at each frequency a sinusoidal signal is radiated and that without losing generality we represent this as $\cos(2\pi ft)$, and a sinusoidal signal is received at the same frequency but with different amplitude and phase that we represent as $A\cos(2\pi ft+\varphi)=I\cos(2\pi ft)+Q\sin(2\pi ft)$, where

$$\begin{cases} I = A\cos(\varphi) \\ Q = -A\sin(\varphi) \end{cases} \tag{16}$$

I and Q are said to be the in-phase and quadrature component of the signal and are worked out by the demodulation of the received signal at each frequency f_n. An important point to be outlined is that, customarily, a step-frequency is not based on a single demodulation operation but performs it N times (N is chosen by the manufacturer), so that the retained values of the in-phase and quadrature components indeed are the algebraic averages of the N gathered samples[2]:

$$\begin{cases} I = \dfrac{I_1 + I_2 + \cdots I_N}{N} \\ Q = \dfrac{Q_1 + Q_2 + \cdots Q_N}{N} \end{cases} \tag{17}$$

A measure of "how much disturbance" has occurred for the considered frequency can be given by the variance of the N repeated samples of the in-phase and quadrature samples, namely,

$$\begin{cases} VAR(I) = \dfrac{I_1^2 + I_2^2 + \cdots I_N^2}{N} - \left(\dfrac{I_1 + I_2 + \cdots I_N}{N}\right)^2 \\ VAR(Q) = \dfrac{Q_1^2 + Q_2^2 + \cdots Q_N^2}{N} - \left(\dfrac{Q_1 + Q_2 + \cdots Q_N}{N}\right)^2 \end{cases} \tag{18}$$

As is well known, the variance is given by the average value of the squares of the samples minus the square of the average value of the samples. Indeed, the definition is based on the statistical average rather than the algebraic average. However, lacking further information, here, the two quantities are assimilated to each other. The variance quantifies how much a random quantity is spread about its average value. In the case at hand, it is a suitable parameter for setting the selective prolongation of the dwell time, because ideally the samples should have the same value. Let us also

[2]Please note the difference between Eqs. (4), (17). In Eq. (4), the index n stands for the nth I and Q components, that is, the in-phase and quadrature components at the nth frequency within the range spanned by the system. In Eq. (17), the attention is focused on any single frequency, and the index is meant to list repeat samples of in-phase and quadrature components gathered at that frequency.

outline that making use of the variance of the samples is conceptually more correct than switching on the receiver while switching off the transmitter. In fact, the problem is not the pure amount of the disturbing signal but rather its effect on the GPR signal while in use, which is of course more linked to the signal-to-noise ratio at the given frequency, where with the term "noise," we mean any spurious signal beyond the reflection from buried targets. Usually, the variance of the samples increases meaningfully in the presence of interference. To quantify this, we have labeled a "vector of the interferences" quantity:

$$v(f) = \max_{X} (VAR(I) + VAR(Q)) \tag{19}$$

where X is the abscissa along the B-scan and f is the frequency. In fact, the variance of course varies from frequency to frequency because the source of interferences is concentrated on some frequency more than one another and varies also with respect to the position, because the source of interference is closer or farther from the GPR while it moves, and the exposure of the GPR to the interference might be more or less direct depending on the position of the instrument. Therefore, the sum of the variances of the in-phase and quadrature samples is a function of two variables, automatically discretized into a matrix depending on the frequency and on the position. Taking the maximum along the position provides a function of the frequency, automatically discretized in a practical case as a column vector depending on a frequency index.

Of course, in order to measure the vector of interferences, an initial calibration B-scan has to be taken. This can be done making use of the first B-scan already collected, or alternatively, one can move the instrument along a zig-zag path within the area to be prospected, so that more possible positions are accounted for in the maximization of Eq. (19).

After estimating the vector of interference, we have programmed a prolongation of the default dwell time essentially proportional to it for each frequency. More precisely, after visualizing the vector of the interferences, the human operator chooses an integer number M. Then, the dwell times versus the frequency will be prolonged time a factor, given by

$$w(f) = \min\left(10, ceil\left(M\frac{v(f)}{\max(v(f))}\right)\right) \tag{20}$$

where the function "*ceil*" indicates the first integer number higher or equal to the argument.

The formula indicates that essentially, the dwell time at the most disturbed frequency is prolonged by a factor equal to M and the other ones proportionally less, dependent on the vector of interferences. The function ceil, however, prevents the possibility shortening of any dwell time lower than its default value. Finally, the maximum limit of the extension of any dwell time is equal to 10 times its default value in order to avoid excessive extension of the time needed for the measurements (indeed there are also hardware issues related to the reconfigurable system that prohibits the possibility to prolong the dwell times beyond this limit).

To make clear what we mean with Eq. (20), if T is the default dwell time (customarily, it is the same for all the tones), then $w(f)T$ is the dwell time of the harmonic tone at frequency f after the reconfiguration of the dwell times. There is some degree of arbitrariness in the choice of the factor M, of course, but we do not have at the moment a criterion for the optimal choice of this parameter, and so, we leave it to the experience of the human operator, at least until future possible studies might drive us to a more optimal (if any) choice automatically driven by the data. Finally, please note that the factor is meant to be referred to each antenna, so that indeed, three factors have to be chosen if all the three equivalent configurations of the antenna are exploited simultaneously. In particular, if some of the antenna is deemed not to need any reconfiguration of the dwell times, it is sufficient to choose $M=1$ for that antenna as can be easily verified from Eq. (20).

Let us now show an example. The reconfiguration of the dwell times has been applied in order to reject an interference caused on purpose by means of a PRC transceiver working (as usually commercial transceivers do) in a narrow band centered at 446 MHz. The experiment has been performed in Florence by the Florence Engineering. This example has been already published in Ref. [9] and is represented with permission from the international journal *Radio Science*. In Fig. 14, the vector of interference for each of the three antenna is represented. In particular, LB stands for "low-frequency band," MB stands for "medium-frequency band," and HB stands for "high-frequency band." As can be seen, the MB and HB antenna are affected by the interference of the transceiver, whereas the LB antenna are also sensitive to a disturbance at about 100 MHz, probably related to an FM broadcast radio transmitting antenna present in one of the neighbors of the laboratory of the Florence Engineering s.r.l. In order to mitigate the effect of the interference in the case at hand, we have set $M=10$ for all the antennas, which has generated the factor of enlargement versus the frequency represented for each of the three antennas in Fig. 15. As can be seen, the most disturbed tone has been prolonged by a factor 10, whereas the tones on the radio FM bands have been enlarged by a factor between two and four only with reference to the LB antenna. The other tones have preserved their original default duration.

In Fig. 16, the B-scan gathered with the MB antenna on which the vector of the interference has been evaluated is shown. In the rectangle, the effects of the interferences are shown. We have taken a B-scan along the same path but making use of the reconfigured dwell times. The result is shown in Fig. 17, where the effects of the interferences have disappeared.

Indeed, the default dwell time of the prototype is already quite long by itself (about 100 μs), whereas commercial step-frequency systems customarily use a lower dwell time (see the previous section). If the default dwell time is shorter, of course, the effect is expected to be stronger and the reconfiguration more meaningful. Please also note that the reconfiguration of the dwell time allows to mitigate the effect of the interference without filtering the data, which means without losing anything of their informative content.

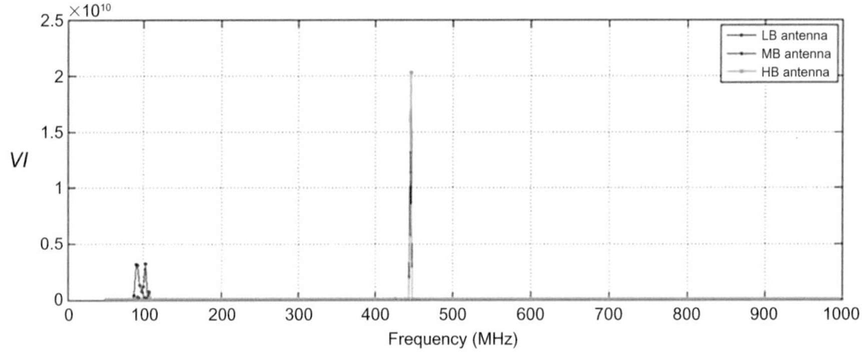

FIG. 14

Vector of interference versus the frequency for each of the three equivalent antennas.

Courtesy of Radio Science, AGU.

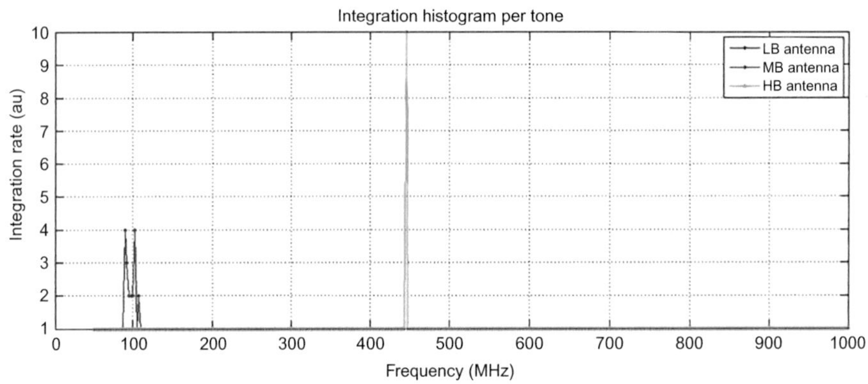

FIG. 15

Extension factor $w(f)$.

Courtesy of Radio Science, AGU.

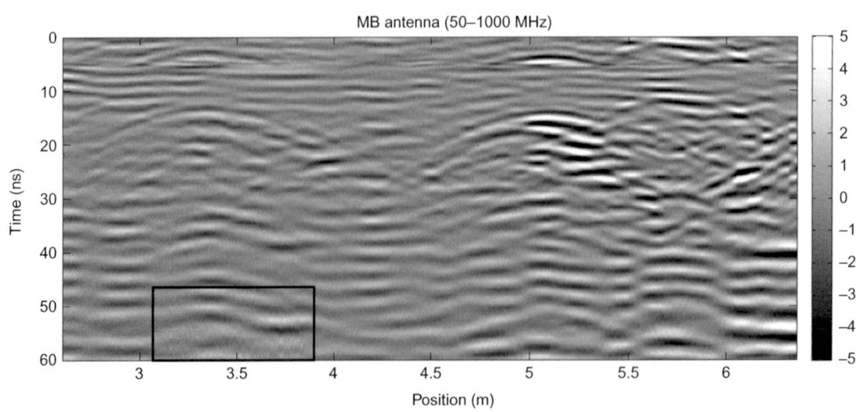

FIG. 16

Calibration radargram for the reconfiguration of the dwell time with the MB antenna.

Courtesy of Radio Science, AGU.

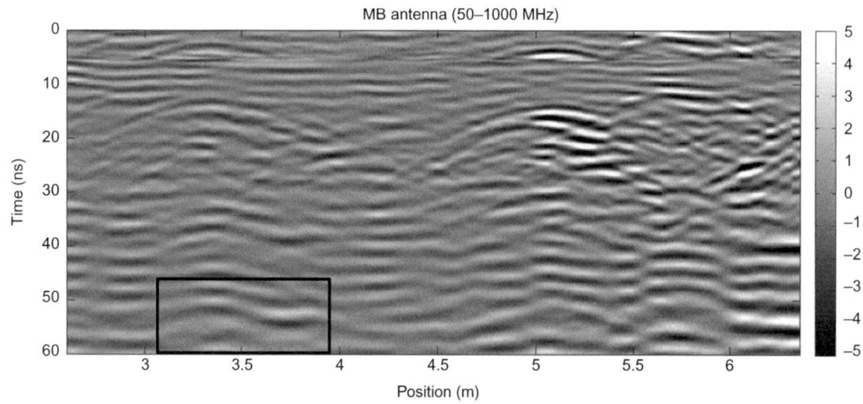

FIG. 17

B-scan gathered with the MB antenna after the reconfiguration of the dwell times.

Courtesy of Radio Science, AGU.

6 USE OF A CONTINUOUS WAVE STEP-FREQUENCY GPR SYSTEM AND MULTIELEMENT ANTENNA ARRAY FOR ARCHAEOLOGICAL SURVEYS

Let us now show some applications in the field of SFCW GPR systems. Multichannel GPR instrumentation has proved to be very effective for near-surface archaeological surveys, where the speed of acquisition and close sample intervals using vehicle-towed systems can produce highly detailed images of the buried remains over unprecedented areas of coverage. Previous surveys using single- or dual-channel GPR systems have demonstrated the need for a crossline sample spacing approaching the Nyquist frequency limit of less than 0.1 m for the mapping of archaeological remains, and the availability of multichannel systems allows data coverage at an increasingly large, landscape scale [14–16]. Many of the systems in use are based on impulse technology and have delivered excellent results when applied to archaeological sites with an appropriate center frequency ground-coupled antenna, selected to match the soil conditions, depth, and spatial resolution required to image the remains. The case studies shown here have been collected with commercially available continuous wave step-frequency (SFCW) GeoScope GPR system, manufactured by 3D radar, utilizing a range of initial prototypes and production versions of a ground-coupled antenna array (see Fig. 18). Some comparison is also made to results obtained with the SFCW system and an air-coupled antenna array, together with comparative data collected with a Sensors & Software pulseEKKO 1000 impulse GPR.

SFCW radar systems potentially offer a broad energy distribution across a wide frequency bandwidth, although when used with an air-coupled antenna over ground

FIG. 18

The GeoScope vehicle-towed GPR system in use with the ground-coupled DXG1820 antenna 20 channels array (spaced 0.075 m apart) towed behind a light-weight ATV, with a GNSS receiver for positional control.

with a conductive surface layer, typical of many archaeological sites in the United Kingdom, the signal can often contain a high-amplitude initial reflection from the microtopography of the site surface. While appropriate data processing can help to address this issue [17], encouraging results were also obtained from a five-channel G0605 prototype ground-coupled antenna array, first deployed to allow survey to be conducted between the standing stones of Stonehenge [18]. Fig. 19 shows a single comparative profile where data from the wider area survey collected with a V1821 air-coupled array overlap with a shorter line of data collected with the ground-coupled antenna.

The data demonstrate a good correlation between the two antenna designs throughout the 75 ns two-way travel time window where dipping reflectors from the underlying geology have been identified.

Following this successful trial, a full-width 22-channel G1922 version of the ground-coupled antenna was introduced, for use with the MkIV GeoScope console. This system offered faster acquisition across a wider frequency bandwidth (from 60 MHz to 3 GHz) with a crossline 0.075 m spacing between the individual elements in the array. Field tests over the Roman remains at Silchester corroborated the results from the earlier prototype (Fig. 20), demonstrating an increased depth of penetration at the site compared with the previous air-coupled array and a similar signal-to-noise ratio to an original data set collected with a single-channel pulseEKKO 1000 impulse GPR using a 450 MHz center frequency ground-coupled antenna.

Site conditions, particularly soil moisture levels, will have varied between the three surveys conducted at the Silchester test site, and in addition, the single-channel

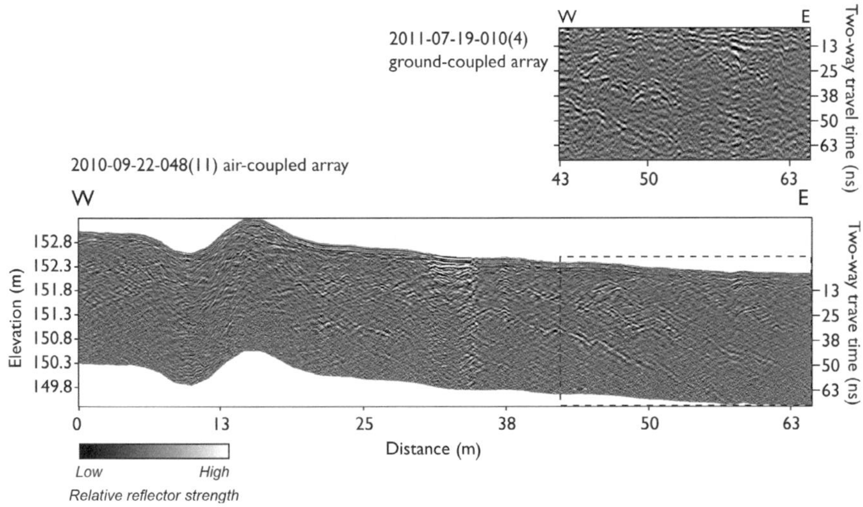

FIG. 19

Topographically corrected profile of a single channel collected across the ditch and bank and Stonehenge (shown at 13 m) using a V1821 air-coupled antenna array. The inset data show a comparative section of the profile collected with a prototype G0605 ground-coupled antenna (no topographic correction applied).

survey was collected using a channel spacing of 0.5 m. However, despite these variations, the data show a good correlation between the recorded anomalies with the physically larger antenna arrays coping well with the uneven surface conditions, reducing the influence of decoupling along the survey lines shown in the initial single-channel data set.

Fig. 20B shows Wigner-Ville energy time-frequency analysis of data collected with the different systems over a prominent buried wall of the Romano-Celtic temple readily identifiable in all three data sets. While the frequency distribution of energy is similar between the systems when compared against two-way travel time, both the ground-coupled SFCW and impulse antenna demonstrate a broader return of energy across the frequency range beyond 30 ns.

Further field tests were conducted with the G1922 antenna array over a range of sites, including a Roman villa, formal postmedieval garden remains, and a medieval farmstead to assess the response of the ground-coupled antenna to more challenging site conditions, particularly where water saturated, conductive soils persist. Fig. 21 shows an amplitude time slice from a complex of earthworks defining a shrunken medieval/postmedieval settlement at Catridge Farm, Lacock, Wiltshire, where standing water was encountered on the surface of the site, presumably perched over the clayey soils that have developed here [19].

Although good signal penetration was only obtained to approximately 25 ns, the near-surface data produced high-amplitude anomalies related to former building

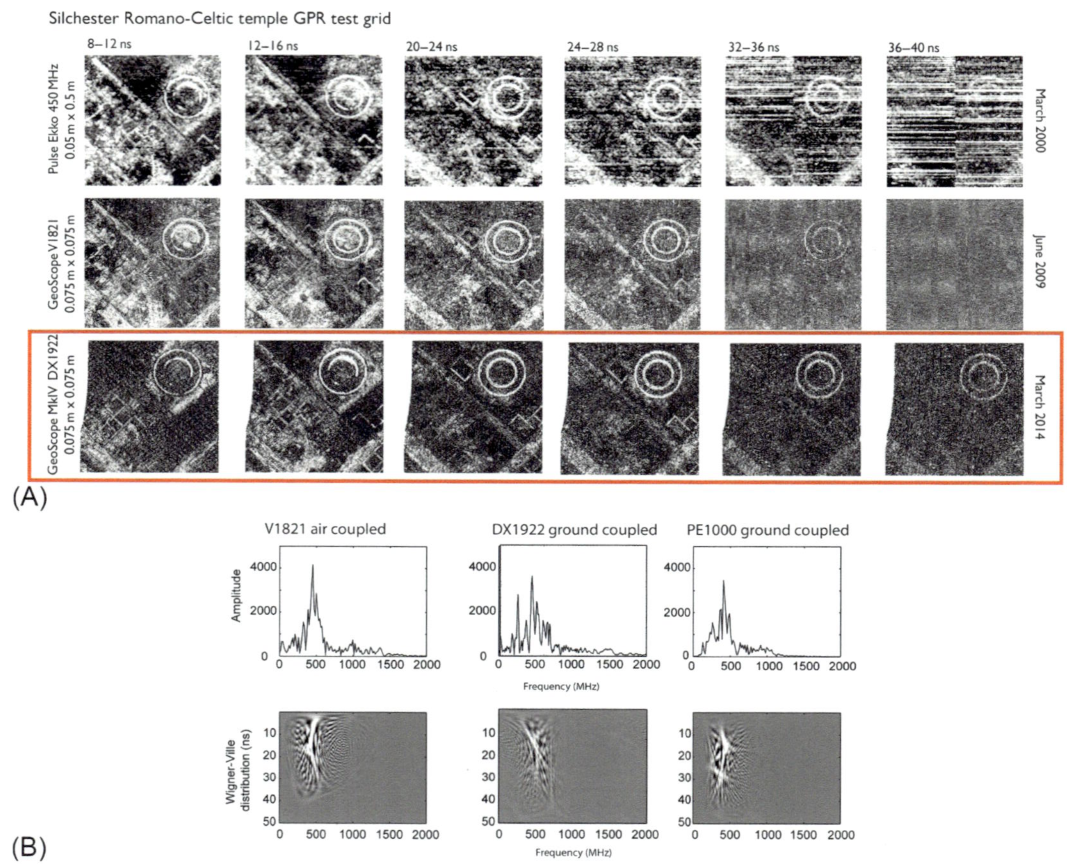

FIG. 20

(A) Summary amplitude time slices collected over the Roman-Celtic temple at Silchester comparing the response between a ground-coupled impulse GPR, an air-coupled SFCW GeoScope system, and the GeoScope with a ground-coupled array. (B) Frequency spectrums and Wigner-Ville distributions derived from these data for the same systems.

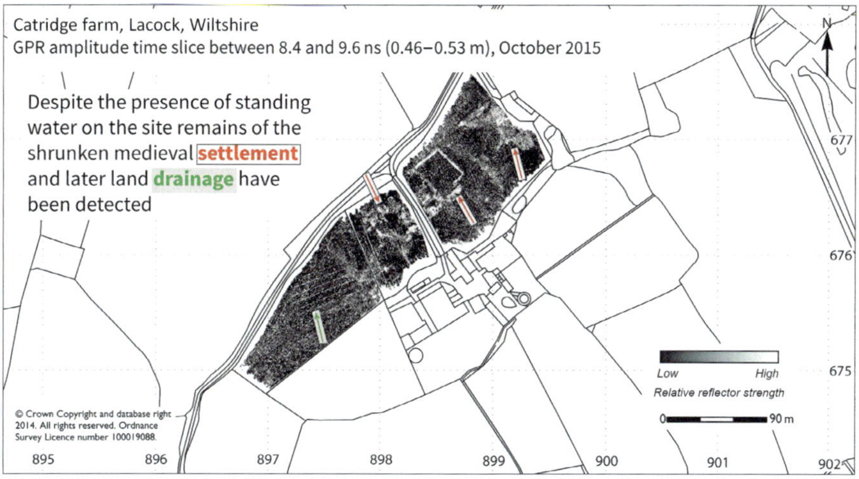

FIG. 21

Despite the presence of standing water on the site, remains of the shrunken medieval settlement and later land drainage have been detected with the ground-coupled G1922 antenna array.

platforms and wall footings associated with the main enclosure surrounding the settlement, enhancing the evidence available from the analytic survey of the surviving earthworks. A pattern of land drains has also been detected to the west of the site, presumably in an attempt to promote better drainage across the site.

A full production DXG1820 version of the ground-coupled antenna became available for field work in 2015 offering optimization of the individual element design to aid the recovery of weak return signals. While the cross channel spacing of 0.075 m was maintained in the new antenna, the number of channels was reduced to 20 providing a 1.5 m swath width during acquisition. A trial survey was conducted over the Warblington Roman Villa, Havant, Hampshire, to provide a field test for the new DXG1820 antenna array [20]. Fig. 22 shows the results from the 2.4 ha survey that successfully identified Roman building remains that correspond well with the previous earth resistance coverage of the site undertaken by volunteers from the Chichester and District Archaeology Society. In addition, the GPR survey has revealed some further structural detail in areas of the site where building rubble appears to obscure the earth resistance results and has uniquely identified lines of Roman post holes, presumably stock or enclosure fences. Despite the waterlogged, clay-rich soil conditions at the site, the GPR has successfully recorded significant anomalies to an approximate depth of 2 m and suggests that the main villa building has developed through a more complex series of phases that may, in part, be identified through the realignment of the walls. Some evidence for differing construction methods is also found through the increased depth of the wall footings to the south of

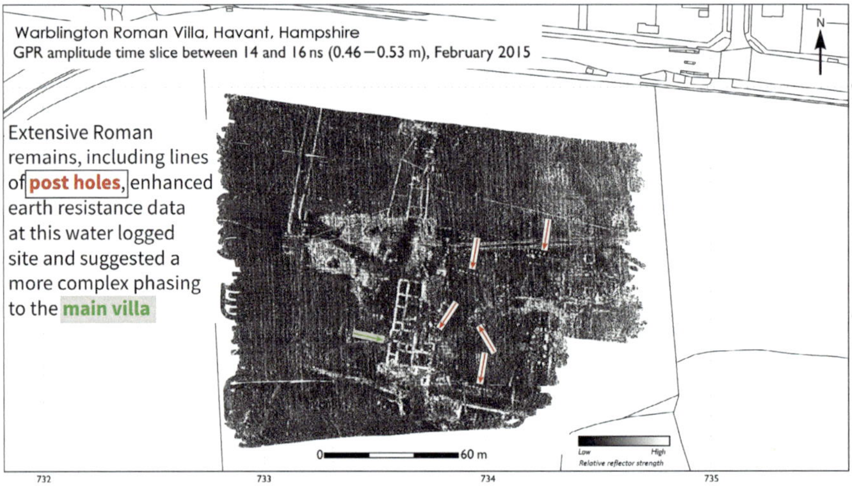

FIG. 22

Extensive Roman remains, including lines of post holes, enhanced earth resistance data at this waterlogged site and suggested a more complex phasing to the main villa DXG1820. An animated version of these data can be viewed here: youtu.be/_9FbEmcFToE.

the building complex, perhaps indicating that this phase of the building was designed with more substantial wall footings to support an upper story.

One advantage of vehicle-towed array-based GPR system is the ability to cover quite large areas rapidly, once the equipment has set up and prepared for acquisition. For example, initial coverage targeted over the remains of the forum at Wroxeter Roman City, Shropshire, was extended to cover a field in the adjacent farm yard that had not been covered by the extensive previous magnetic survey of the entire city [21]. This proved fortuitous as it confirmed the continuation and survival of the forum buildings into an area beyond archaeological excavations of the site. Some additional activity, including a possible small rectangular shrine, was also revealed immediately beyond the forum (Fig. 23).

Ambient RF noise sources can impact upon data quality when in close proximity to local transmitters operating within the bandwidth of the GPR system. Sources of interference may include close proximity to mobile phone transmitters, site communication systems, and RF frequency modem from a GNSS base station used to broadcast correctional data for the positional control of the survey in progress.

Fig. 24A shows the effect of a GNSS modem, broadcasting at nominal frequency of 450 MHz, on an extract of the surrounding GPR data. In this case, the GNSS base station and modem was situated on the top of an extant Bronze Age barrow to maximize the broadcast range between two towed instrument systems operating in the wider vicinity. A regular pattern of RF interference partially obscures the identification of two further, circular barrow ditches but can be much improved by suppressing

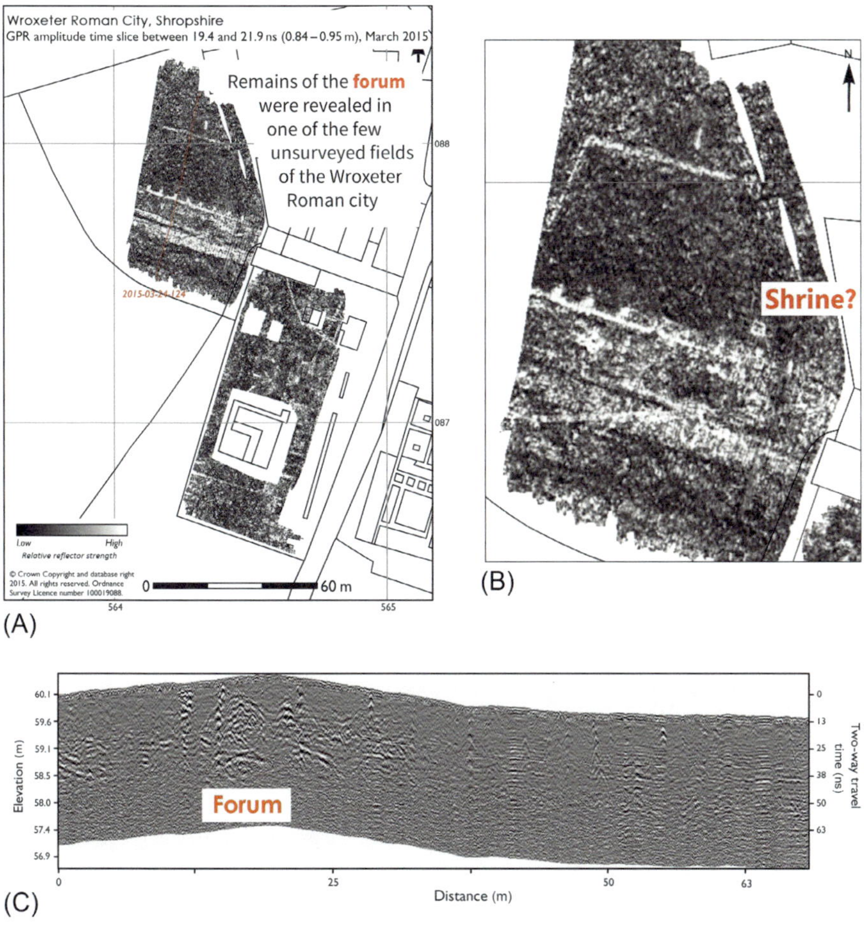

FIG. 23

(A) Remains of the forum were revealed in one of the few unsurveyed fields of the Wroxeter Roman city, including (B) a small shrine type building appearing at a deeper time slice between 24.3 and 26.7 ns (1.05–1.16 m). A single topographically corrected profile (C) is also shown across the walls of the forum remains.

the specific, limited range of frequencies associated with the GNSS modem much improving the resultant processed data (Fig. 24B).

Application of vehicle-towed systems to very large area surveys has provided unprecedented levels of coverage, often revealing anomalies that would not previously have been recognized within such data. For example, a large-scale survey over 25 ha at West Amesbury, approximately 1 km south east of Stonehenge, revealed an extensive range of anomalies related to the underlying geomorphology, modern services, and significant archaeological activity (see Fig. 25 and Ref. [22]).

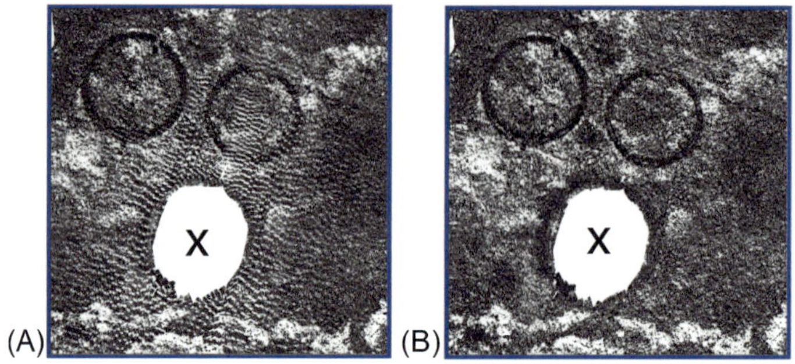

FIG. 24

(A) RF interference from a GNSS base station radio modem situated on raised barrow at "x" has obscured the data quality within the immediate radius of this 60 × 60 m extract from the full data set. (B) Removing a limited number of frequencies close to the 450 MHz operating frequency of the modem from the SFCW data set allows the noise to be suppressed.

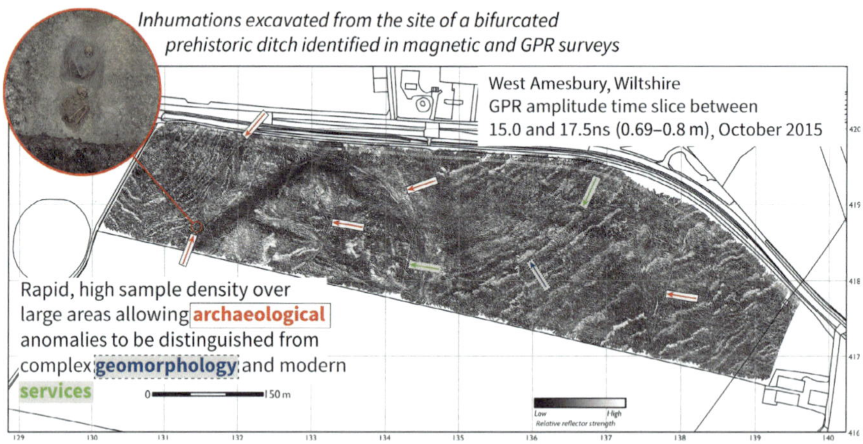

FIG. 25

Rapid, high sample density coverage over large areas, matching the extent of magnetic survey, was conducted at West Amesbury close to Stonehenge. Subsequent archaeological excavation targeted a bifurcating ditch and revealed a double inhumation dating from the mid-Bronze Age.

Limited excavation following both the GPR and a cesium magnetometer survey targeted a junction with bifurcated ditch, also known from aerial photographic records of crop marks. This excavation revealed a quite unexpected double inhumation with two, intercut mid-Bronze burials found within the course of the ditch.

Of course, considerable care must be taken when considering such retrospective interpretation of the original GPR data; however, this does illustrate the detail that may be derived when a high sample density is applied. This can present a particular challenge when attempting to consistently analyze and interpret very large array data sets and attempts to utilize semiautomated algorithms may well prove increasingly necessary. Edge detection algorithms can rapidly abstract closed polygons from a user-defined amplitude threshold and have been demonstrated to be very effective when applied to GPR raster data sets [23–25]. The potential archaeological significance of the vector objects is then determined through a consideration of the morphology and continuity between multiple data sets. A particular application is made to the location of pit-type anomalies, which are often highly numerous across a landscape, yet can be time consuming to interpret through manual extraction of each individual response from multiple layers within a data set [26–28].

Defining specific constraints, in terms of the accepted size of a potential pit-type feature, likely depth and whether it presents a circular outline can assist in the classification of the anomalies. In the case of the GPR data considered here, the initial georeferenced raster amplitude time slices were first converted to a binary image, based on a user-defined amplitude threshold [29], which is then processed to extract connected objects based on size. This can be used to remove either very small objects, more likely to represent noise, or those too large to be representative of a pit-type feature. Further binary image processing is applied to close gaps within the objects based on an appropriate structuring element, for example, a disc shape for detecting pit-type responses and rectangular-shaped element for linear walls. Exterior boundaries of the binary objects are then traced to abstract a set of closed vector polygons.

Properties of the vector polygons, including the area enclosed and size of the perimeter, can be used as a further basis for the classification of circular, pit-type anomalies. A simple approach to determining whether an anomaly is approximately circular or not can be achieved through estimating the apparent radius of the circle based first upon the area enclosed by each vector polygon and then by the perimeter. If the ratio of the two estimates for the radius approaches unity, then the anomaly is more likely to be circular, although in practice some relaxation of the morphological classification may be necessary to obtain the best results from real data [30]. Combining the morphological classification of vector polygons with other appropriate constraints, in terms of the size and depth of potential pit-type anomalies, can be used to assist the selection (Fig. 26A). A similar approach can be applied to target other anomaly types, for example, by considering the rectilinear morphology and orientation of the vector polygons to identify Roman structural remains (Fig. 26B) and with GPR data to consider the continuity of selections made between amplitude time slices.

7 CONCLUSIONS

In this chapter, the current state of the art of step-frequency GPR systems has been described, and while we do not claim to have provided an exhaustive treatment of the theory or an extensive literature review, we have aimed to provide evidence for the unique features of the step-frequency GPR systems in comparison with pulsed

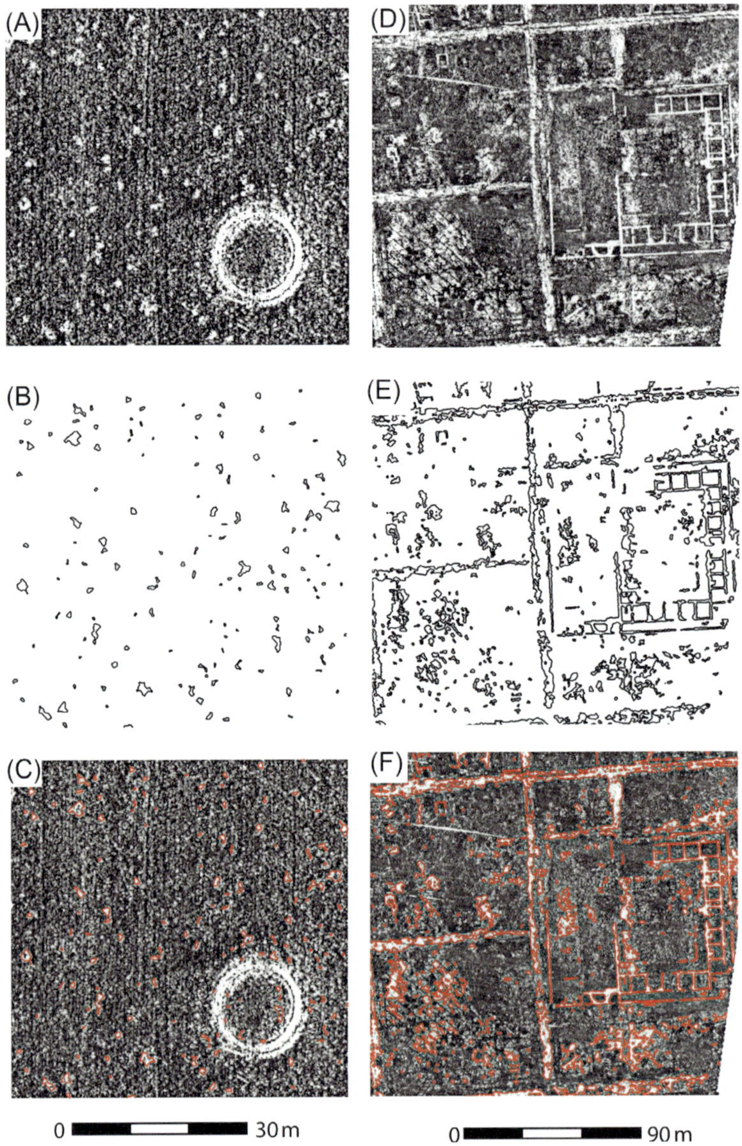

FIG. 26

Identification of approximately circular, pit-type anomalies shown as (A) a grayscale image of the GPR time slice, (B) vector outlines of the abstracted pit-type anomalies prior to size classification, and (C) an overlay of the vector outlines superimposed on the GPR data. A similar process for the identification of Roman structural remains without any morphological constraints is also shown as (D) a grayscale image of the GPR time slice, (E) vector outlines of the abstracted structural remains, and (F) an overlay of the vector outlines superimposed on the GPR data. The color overlays used for (C) and (F) are available in the electronic version.

systems. In particular, step-frequency GPRs can allow a larger frequency band to be achieved together with an improved signal-to-noise ratio, dependent on the dwell time of the harmonic tones. This produces a high effective bandwidth but with a narrow instantaneous bandwidth, and hence, a high analog to digital sampling rate is not required to adequately record the signal. Furthermore, the peak power is smaller compared with impulse GPRs. In addition, step-frequency systems allow the application of continuously variable filters with respect to depth, thus exploiting only the (better penetrating) lower frequencies in order to image the deeper layers, where the resolution is degraded, due to physical reasons, with respect to the information recovered from higher frequencies in the shallower layers.

Step-frequency systems may also allow innovative solutions to be more easily implemented than pulsed systems, as demonstrated here by the use of reconfigurable antenna versus the frequency, and the suppression and rejection of interference through either limited band stop filtering or the use of an adaptive dwell time. Several case studies have also been presented, in relation to cultural heritage sites, where at least part of the theoretical discussion has been shown to hold in the field in comparison with pulsed systems, under often challenging soil conditions.

REFERENCES

[1] R. Persico, Introduction to Ground Penetrating Radar: Inverse Scattering and Data Processing, John Wiley & Sons, Inc., Hoboken, 2014.

[2] M.H. Jol (Ed.), Ground Penetrating Radar: Theory and Applications, Elsevier, Amsterdam, 2009.

[3] R. Persico, G. Prisco, A reconfigurative approach for SF-GPR prospecting, IEEE Trans. Antennas Propag. 56 (2008) 2673–2680.

[4] N. Masini, R. Persico, F. Soldovieri, Sistema GPR stepped frequency reconfigurable, Italian Patent no. 0001395231 of September 5th, 2012.

[5] R. Persico, M. Ciminale, L. Matera, A new reconfigurable stepped frequency GPR system, possibilities and issues: applications to two different cultural heritage resources, Near Surf. Geophys. 12 (6) (2014) 793–801.

[6] L. Matera, M. Noviello, M. Ciminale, R. Persico, Integration of multisensor data: an experiment in the archaeological park of Egnazia (Apulia, Southern Italy), Near Surf. Geophys. 13 (6) (2015) 613–621.

[7] L. Matera, M. Ciminale, R. Persico, E. Eide, J. Sala, A. Anderson Stamnes, in: Application of conventional and non-conventional GPR systems in the Viking remains of Torget, Proc. of 8th International Workshop on Advanced Ground Penetrating Radar IWAGPR, Florence, Italy, 2015.

[8] R. Persico, S. D'Amico, L. Pajewski, V.P. Gracia, in: Ground-penetrating radar prospection at the Jesuits' church in Valletta, Malta, Proc. of NSG Conference, Barcelona, Spain, September 4–8, 2016.

[9] R. Persico, D. Dei, F. Parrini, L. Matera, Mitigation of narrow band interferences by means of a reconfigurable stepped frequency GPR system, Radio Sci. 51 (2016) https://doi.org/10.1002/2016RS005986.

[10] R. Persico, G. Leucci, Interference mitigation achieved with a reconfigurable stepped frequency GPR system, Remote Sens. 8 (2016) 926, https://doi.org/10.3390/rs8110926 (open access).
[11] C.A. Balanis, Advanced Engineering Electromagnetics, John Wiley & Sons, Inc., Hoboken, 1989.
[12] R. Persico, N. Romano, F. Soldovieri, Design of a balun for a bow tie antenna in reconfigurable ground penetrating radar systems, Prog. Electromagn. Res. C 18 (2011) 123–135.
[13] D.A. Noon, Stepped-frequency radar design and signal processing enhances ground penetrating radar performance (Ph. D. Thesis), Department of Electrical & Computer Engineering, University of Queensland, Australia, 1996.
[14] A. Booth, N. Linford, R.A. Clark, T. Murray, Three-dimensional, multi offset ground-penetrating radar imaging of archaeological targets, Archaeol. Prospect. 15 (2) (2008) 93–112.
[15] N. Linford, P. Linford, L. Martin, A. Payne, Stepped-frequency GPR survey with a multi-element array antenna: results from field application on archaeological sites, Archaeol. Prospect. 17 (3) (2010) 187–198.
[16] C. Gaffney, V. Gaffney, W. Neubauer, E. Baldwin, H. Chapman, P. Garwood, H. Moulden, T. Sparrow, R. Bates, K. Löcker, A. Hinterleitner, I. Trinks, E. Nau, T. Zitz, S. Floery, G. Verhoeven, M. Doneus, The stonehenge hidden landscapes project, Archaeol. Prospect. 19 (2012) 147–155.
[17] J. Sala, N. Linford, Processing stepped frequency continuous wave GPR systems to obtain maximum value from archaeological data sets, Near Surv. Geophys. 10 (1) (2012) 3–10.
[18] N. Linford, P. Linford, A. Payne, Stonehenge Monument Field and Barrows, Wiltshire, Report on Geophysical Surveys, September 2010, April and July 2011, 2012 (English Heritage Research Reports Series 34/2012).
[19] N. Linford, P. Linford, A. Payne, E. Caswell, Catridge Farm, Gastard, Lacock, Wiltshire: Report on Geophysical Survey October 2014, 2015 (Historic England Research Reports Series 4/2016).
[20] N. Linford, Warblington Roman Villa, Havant, Hampshire: Report on Geophysical Survey, February 2015, 2015 (English Heritage Research Reports Series 82/2015).
[21] N. Linford, P. Linford, Wroxeter Roman City, Shropshire, Report on Geophysical Survey, March 2015, 2015 (Historic England Research Reports Series 81/2015).
[22] N. Linford, P. Linford, A. Payne, Southern WHS Survey, West Amesbury Farm, Wiltshire, Report on Geophysical Surveys, August 2015, 2015 (Historic England Research Reports Series 95/2015).
[23] A. Schmidt, G. Tsetskhladze, Raster was yesterday: using vector engines to process geophysical data, Archaeol. Prospect. 20 (1) (2013) 59–65.
[24] L. Verdonck, Detection of buried Roman Wall remains in ground-penetrating radar data using template matching, Archaeol. Prospect. 23 (4) (2016) 257–272.
[25] J. Leckebusch, A. Weibel, F. Bühler, Semi-automatic feature extraction from GPR data for archaeology, Near Surv. Geophys. 6 (2) (2008) 75–84.
[26] P. Linford, An automated approach to the analysis of the arrangement of post-pits at Stanton Drew, Archaeol. Prospect. 12 (3) (2005) 137–150.
[27] Ø. Trier, L. Pilø, Automatic detection of pit structures in airborne laser scanning data, Archaeol. Prospect. 19 (2) (2012) 103–121.

[28] A. Schneider, M. Takla, A. Nicolay, A. Raab, T. Raab, A template-matching approach combining morphometric variables for automated mapping of charcoal kiln sites, Archaeol. Prospect. 22 (1) (2015) 45–62.

[29] N. Linford, P. Linford, L. Martin, A. Payne, Geophysical Evidence for Assessing Plough Damage (Študijné zvesti Archeologického ústavu SAV, 41), (2007) pp. 212–213.

[30] Ø. Trier, S. Larsen, R. Solberg, Automatic detection of circular structures in high-resolution satellite images of agricultural land, Archaeol. Prospect. 16 (1) (2009) 1–15.

CHAPTER

Putting it all together: Geophysical data integration

9

Kenneth L. Kvamme*, Eileen G. Ernenwein†, Jeremy G. Menzer‡

Department of Anthropology, University of Arkansas, Fayetteville, AR, United States Department of Geosciences, East Tennessee State University, Johnson City, TN, United States† Environmental Dynamics Program, University of Arkansas, Fayetteville, AR, United States‡*

CHAPTER OUTLINE

Innovation in Near-Surface Geophysics. https://doi.org/10.1016/B978-0-12-812429-1.00009-X

"Data integration" is a term that has come to be widely employed in the literature of archaeological geophysics, but it is not unique to that enterprise. It is common to many fields of study, so goals, procedures, and results vary widely. The term has a surprisingly long history and is linked with a variety of synonyms depending on the field of inquiry, including "data fusion"; "image merging"; "image integration"; or more simply as "combined," "complementary," or "composited" information, particularly in geographic information system (GIS) contexts [1]. There tends to be broad agreement about what it means, however, and the term has even become common thanks to "big data" and the millions of internet searches that require integrated content [2]. In these contexts, Lenzerini [3] provides suitable definition as "the problem of combining data residing at different sources, and providing the user with a unified view of these data."

In geophysics and satellite remote sensing, the perspective narrows with a focus on spatial contexts, maps, and imagery. In remote sensing, multiband manipulations are nearly as old as the discipline itself where band ratios, vegetation indexes, and principal component transformations all fundamentally yield integrations of various data types [4]. Pohl and van Genderen [1] (p. 823) indicate that the goal is "to obtain more information than can be derived from each of the single sensor data [sets] alone ('1+1=3')," with goals of increasing information content and raising the reliability of interpretations. Objectives include greater data accuracy and utility, improved classification performance, increased confidence, and reduced ambiguity of interpretations [5].

With its long tradition of data fusion, satellite remote sensing recognizes several levels of integration. The lowest occurs at the *pixel level* where numerical combinations are applied to merge individual sensor data [6]. Fusions between higher spatial-resolution panchromatic imagery and lower-resolution multispectral data for "pan sharpening" of the latter currently receive large focus [7], but band differences, ratios, and supervised classification functions of multiple bands are equally common. A more complex approach to fusion occurs at the *feature level*, where objects (e.g., roads, buildings, and trees) are extracted from sensor data based on numerical properties, shape, or neighborhood characteristics using segmentation methods or other techniques. Extracted objects are then combined to generate an integrated outcome, often through the use of neural network decision trees or statistical approaches [8].

1 DATA INTEGRATION IN ARCHAEOLOGICAL GEOPHYSICS

Data integration has become an important topic in archeology generally (e.g., Ref. [9]), and it dominates the literature of archaeological remote sensing from aerial and satellite platforms [10,11]. It has also seen tremendous interest in archaeological geophysics, early demonstrated in the first issue of *Archaeological Prospection* (1994) that includes three papers that integrate findings between multiple geophysical surveys or with aerial results. It is no wonder that Piro et al. [12] (p. 212) are enthusiastic, stating that "Although the cost of an integrated approach is likely to exceed the cost of any single method survey, the benefit is so high that the cost-to-benefit ratio of the whole operation drastically lowers and makes the financing effort well rewarded." Since that time, considerable methodological advancements have been achieved, as the following pages testify. In them, a myriad of data integration methods that include one or more ground-based geophysical surveys are examined and reviewed, and a general classification of approaches is offered.

A recent special issue of *Near Surface Geophysics* dedicated to "integrated geophysical archaeological investigations" defines it as pertaining "to developments and applications that make use of more than a single method, resulting in complementary data sets for improved imaging and archaeological interpretation of the data describing buried archaeology" [13] (p. 519). In general, we follow this definition, although we show that successful fusions have also occurred between different characteristics of *single* geophysical data sets. In the following pages, we examine fusions between aspects of the same data set; between different types of geophysical data; and between geophysical and other information, including aerial photography (AP), LiDAR, archaeological surface collections, geochemistry, and satellite imagery. We also explore and categorize the wide variety of approaches that have been applied for successful integrations. Finally, we follow with a series of case studies that illustrate a sampling of methods undertaken and new areas potentially to be explored.

2 ARCHAEO-GEOPHYSICAL DATA

Until recently, most geophysical surveys in archeology have been conducted in grid blocks composed of parallel transects along which instrumentation is moved manually to acquire measurements. Data are obtained at highly variable sampling intervals (e.g., between 0.04 and 1 m) depending on the geophysical method and instrument employed. Transects are typically separated by a short distance (usually 0.5–1 m) to ensure relatively uniform spatial coverage of broad two-dimensional (2D) spaces. Many instruments yield only a single measurement at a locus, making the resultant data 2D. However, some techniques acquire multiple measurements at a locus from various depths, permitting generation of three-dimensional (3D) data volumes. Multiple depth data at a single locus (e.g., downhole measurements) yield one-dimensional (1D) information [14]. In the past decade, 2D and 3D methods of data acquisition have begun to be acquired by cart systems that hold large arrays of instruments propelled by motorized vehicles with location controlled by global navigation satellite systems (GNSS) [15]. In addition to larger areas of coverage, narrower transect spacing is achieved yielding unsurpassed spatial resolutions. Schneidhofer et al. [16] report 0.04×0.08 m in one such survey.

Archaeogeophysicists work with a wide variety of methods, but only a subset is commonly employed in data integration studies. They include earth resistivity methods that occur in two principal forms. Apparent electrical resistivity (ER) surveys use four-probe systems, usually for lateral 2D investigations at a constant target depth, which may be varied. On the other hand, electric resistivity tomography (ERT) surveys generally utilize many more electrodes that can generate, after inversion, insightful 2D pseudosections in the vertical plane or 3D information through a volume of the subsurface [17]. Magnetic surveys, also 2D, are perhaps most common in archaeological applications. They are usuallyperformed with magnetic gradiometers (MG) that compute differences in measurements between top and bottom sensors in order to eliminate constant diurnal changes of the earth's magnetic field. Less frequent are total magnetic field (TMF) surveys [18]. Electromagnetic induction (EMI) instruments quantify apparent electrical conductivity (EC) through different volumes depending on coil separation, orientation, frequency, and soil conditions [19]. They also yield near-surface magnetic susceptibility (MS) data. Finally, ground-penetrating radar (GPR) transmits microwave energy into the earth to record the amplitudes and two-way travel times of resultant reflections from subsurface discontinuities in dielectric properties. The result is a 2D reflection profile in the *vertical* plane often referred to as a *radargram*, but when multiple transects are parallel and close together, the data form a 3D volume [20]. Software can then be employed to extract "slices" of data from the same window of time or depth in each radargram. These slices are then concatenated to form a 2D horizontal plan of GPR reflections, similar in form to native 2D data sets [21,22]. Other methods (self-potential, seismic techniques, and geochemistry) are used less frequently. All reveal various characteristics of the subsurface. Standard references offer additional theoretical or technical details of geophysical principles and instrumentation [23,24].

2.1 DATA PREPROCESSING

Most field-collected geophysical data, regardless of type, require extensive processing to remove or reduce the effects of instrument noise, technology deficiencies, or operator errors. It is well known that EMI instruments are subject to drift, for example [25]. MG data acquired with fluxgate systems usually illustrate heading errors. ER data that yield data "spikes" resulting from poor probe contacts and grid imbalances are common due, in part, to moisture changes between survey dates. Extensive compilations of corrective procedures have been cataloged to eliminate or reduce such unwanted artifacts and deficiencies (e.g., Ref. [26]). GPR data, acquired in *vertical* profiles, must be assembled, gained, filtered, and subjected to various other operations (migration, Hilbert transform, deconvolution, and terrain correction) before presentation, and time or depth slices must be generated to realize 2D horizontal plan views [20]. Complex inversion procedures are similarly required for ERT data [27]. Finally, various "enhancements" may be employed to improve 2D planimetric data, including high-pass filters to remove geologic or other background trends, low-pass filters to reduce noise and consolidate anomalies, fast Fourier transforms to reduce such periodicities as plow marks, and interpolation to generate different (usually higher) data densities that reduce pixilation and that might be required for merging with other data sets [28].

In the following pages, it is assumed that preprocessing of geophysical data sets has been accomplished such that unwanted components have been removed and desired enhancements have been made. Moreover, unless otherwise specified, it is assumed that GPR and ERT data have been reduced to 2D horizontal depth slices, suitable for integration with other 2D data sets. Distinct sections specifically address integrations of 1D and 3D data.

3 GOALS OF DATA INTEGRATION

Why exactly are geophysical data integrations undertaken in archeology? What do practitioners hope to gain? What specific advantages arise through the practice? These are all significant questions, but the answers vary depending on the goals of a project.

3.1 MORE COMPLETE VISUALIZATIONS

Perhaps the commonest goal of data integration is to achieve better and more complete visualizations of the subsurface. Many geophysical methods record independent dimensions of the subsurface [29]. For example, magnetic properties may be unrelated to conductivity characteristics so that an MG survey may reveal hearths and other burned features but not adobe walls, while a lateral resistivity survey may not respond to burned features but readily indicate those walls owing to moisture variations. Any single map generated by these methods will contain only part of the archaeological evidence, but when the data are combined, a fuller picture of the subsurface may be achieved [30]. Two or more geophysical methods applied to the same area offer the potential of visualizing more features hidden below the surface. The same can be said of integrations between geophysical and other prospection data. For example, geophysical plan maps fused with aerial imagery showing crop marks or LiDAR illustrating archaeological terrain variations contribute to a more holistic view [31].

Geophysical methods frequently guide destructive excavations, and clear visualizations of the subsurface facilitate this task [13]. Moreover, the Valletta Convention [32] specifies that noninvasive methods should be employed whenever possible to preserve sites of cultural heritage.

3.2 DATA REDUCTION AND SIMPLIFICATION

Geophysical surveys might be connected with other forms of air or space prospection and information from traditional archaeological surface collections. A host of images depicting findings can result, making it tedious and inefficient to review each, compare similarities and differences, and collate elements that might be archaeologically significant. Integrating such findings into a single more parsimonious view offers an efficient means to quickly review what is known. Moreover, researchers unfamiliar with prospection methods may not easily interpret resultant imagery. Certain forms of object-based displays, such as vector polygons representing interpreted archaeological feature types (see below), can make complex data sets of high dimensionality easily understood.

3.3 CONTEXT

Geophysical surveys frequently benefit by being placed within a larger landscape context in order that a sense of space and place may be realized around a revealed settlement or monument. This has been achieved by superimposing geophysical results atop a larger-area aerial, satellite, or LiDAR image of the nearby region (e.g., Ref. [33]). In so doing, the viewer might better realize how revealed defensive systems integrate with surrounding landscape features such as the steep slopes and the possible relevance of waterways to a settlement's locus.

3.4 IMPROVE ACCURACY OF SUBSURFACE FEATURE IDENTIFICATIONS

A related goal focuses on the identification of what subsurface features actually represent. In other words, through a multidimensional fusion of geophysical (and other) information, the likelihood increases specific identifications of subsurface targets. For example, four small MG anomalies in a square pattern surrounding a larger central one centered in an 8–10 m^2 area of high ER with a meter-wide linear extension to one side that points to a Coalescent tradition lodge in the Great Plains, the United States, with its four support posts surrounding a central hearth and compact earthen floor with linear entryway [34].

Part of the identification process considers characteristics other than physical properties. Anomaly sizes, their shapes, and relationships with other anomalies and their sizes and shapes are nearly as important to interpretation. This circumstance arises because regular geometric shapes—squares, rectangles circles, and even straight lines—are generally indicative of human constructions and rarely result from natural processes, which tend to exhibit irregular forms [35]. For instance, a small circle seen in geophysical imagery likely points to a house or other built structure; rows and columns of small, systematically spaced rectangular anomalies likely signify a graveyard in certain Western societies.

3.5 A "TEST" OF OTHER DETECTION METHODS

Some geophysical surveys are undertaken to validate potential archaeological features revealed, for example, by crop marks seen in aerial imagery [36]. A similar circumstance arises when one geophysical survey confirms another. Hesse [37] indicates the importance of anomalies in extant geophysical data that can serve as a basis for testing the accuracy of new geophysical devices. Novo et al. [38] utilize this idea when data from a "STREAM X" multichannel GPR system are compared with several prior geophysical data sets.

3.6 ESSENTIAL DATA NEEDS AND IMPROVED GEOPHYSICAL UNDERSTANDING

Certain forms of geophysical data processing or analysis actually *require* the integration of geophysical data with ancillary information. The most common is topographic correction of GPR radargrams that require surface elevations along the length of each profile. Linear profiles are "bent" to match terrain curvature, permitting improved interpretations of reflection geometries and increased understanding of anomalous zones or the lack of them [39]. Because high soil conductivity limits GPR penetration, researchers have also conducted EC or ER surveys in the near surface to ascertain whether conductivity is sufficiently low to warrant subsequent GPR investigations or to better understand its responses [40].

Geophysical interpretations are sometimes *improved* through integration with elevation data. Magnetic, resistivity, and other anomalies may be caused in part by topographic variations, such as soil mounding and small holes (e.g., from looters), or by slope breaks [24] (p. 349); their merging with elevation data can greatly facilitate their interpretation.

Other forms of ancillary data are useful and perhaps even essential to geophysical interpretations. The consideration of modern infrastructure can aid anomaly interpretation through the revelation of buried pipelines, power lines, or other utilities. Historic aerial photographs might show former roads, buildings, fence lines, garden spaces, trees, and the like. This is useful when examining recently changed landscapes or historic settlements that no longer exist, because they help explain observed anomalies [41].

4 THE ROLE OF GIS

GIS are software for managing spatially distributed information [42]. Consequently, they are ideal for diverse spatial data sets from an archaeological site or region, including geophysical data. Multiple data sets may be placed within GIS, permitting their ready visualization and manipulation. Different geophysical modalities may be easily compared in separate windows or overlaid with one toggled on and off or made semitransparent, facilitating anomaly comparisons. Moreover, geophysical data may be examined against other information, such as aerial or satellite imagery, LiDAR, or surface artifact mappings. Indeed, Campana [43] (p. 22) observes that "entry of the data into an archaeological GIS is the basis of any attempt at integration of the information." Yet, although there may be a perception that placing diverse information within GIS automatically makes them "integrated," it is emphasized that GIS are merely data management tools. The fusion of information occurs only when specific GIS methods are applied to combine two or more data sets.

GIS offer countless benefits to the analysis and manipulation of spatial information and numerous tools for various forms of integration. Most geophysical data inherently occur in a *raster* format, meaning that measurements occur in a matrix of rows and columns (where columns might correspond to field transects and rows to individual samples). GIS permit routine data management operations such as registration and rectification—the spatial "matching" or aligning of geophysical data with other prospection information into a common projection and coordinate space, an absolute prerequisite for any form of fusion. Measurements can be interpolated or resampled to different (usually higher) spatial resolutions (e.g., to match other data sets or reduce pixilation). Moreover, most raster GIS include a host of image processing routines for image enhancements, including modifiers for contrast and brightness, low- and high-pass filters, and edge detectors [4]. For data integration, GIS include software for digitizing, or *vectorization*, of anomalies interpreted as significant. In GIS parlance, vectors may be classified as points, lines, or polygons; the latter employed to enclose areas of any shape. Display capabilities include the simultaneous portrayal of multiple vectors, perhaps color coded by geophysical method [44]. They also permit combined displays of raster data, in distinct windows, in the same window with transparent layers on top, as a color composite (one data set as red, another blue, and third green), or perhaps with one data set as a pseudo three-dimensional view (usually a ground surface) with a second image overlaid as a color ramp (e.g., Ref. [45], Plate 3). Additionally, with geophysical (and other) information in raster format, map or image calculators, common GIS components, permit mathematical manipulations for their numerical integration. Each of these operations is illustrated in the sections below. GIS do pose constraints, since they are principally designed for the handling of 2D data sets, limiting their usefulness when working with volumetric data (e.g., GPR, seismic, and ERT; see Ref. [46], p. 329).

5 INTEGRATIONS BY DATA TYPE

Data integration can occur at many different levels, with a variety of geophysical data types, and in one, two, or three spatial dimensions. Some examples integrate various aspects of a single data set, while more often, others combine results from two or more geophysical methods. In recent years, it has become increasingly common to link geophysical results with air or space remote sensing or with other ground-based forms (e.g., surface artifact densities), largely facilitated by the growth of GIS for managing all spatial information within a site or region.

5.1 ONE GEOPHYSICAL DATA SET

The integration of one geophysical data set with itself may seem a contradiction, yet examples exist. Conyers [47] makes convincing arguments why the interpretation of GPR time slices *must* be integrated with the profiles from which they were generated.

Vertical profiles aid understanding and classification of anomalies seen in horizontal plan views (e.g., by their "thickness" and reflection complexity), and should they indicate dipping stratigraphy, they can insure generation of slice maps that parallel rather than crosscut stratigraphic horizons. Although most researchers generally recognize these points, they may too often be overlooked, especially as new vehicle-driven multiantenna GPR arrays now generate hectares of coverage, making thoughtful interpretations of individual anomalies against numerous profiles less likely (e.g., Ref. [48]).

The 3D complexity of GPR means that integrations become possible along the vertical dimension. The simplest are manually drawn vectors representing interpreted features in each 2D horizontal slice showing significant elements at each depth, as Trinks et al. [49] offer from Viking Age Birka, Sweden. Quantitative methods also permit integration of GPR anomalies from multiple time slices, in what Goodman et al. [50] describe as "GPR overlay analysis." This method merges the largest amplitude anomalies in individual slice maps with those from others to form a composite plan of the most pronounced anomalies. Linford [45] presents a different tactic that accomplishes much the same thing. He subjects a series of GPR time slices depicting a Roman villa in the United Kingdom to a principal component analysis (PCA). The first component yields anomalies related to the villa as the major axis of variance, while subsequent components largely describe background noise.

Closely related are integrations between the same kind of geophysical data but made at different times. Several examples exist in ER surveys where data collected from the same area but at different seasons are compared. Kvamme [51] shows the results of resistivity surveys in a northern Great Plains village, the United States, made in two different summers, one wet and one dry, which reveal different elements of a fortification system. Schmidt et al. [52] show repeat resistivity surveys of the same area performed once a month over a 16-month period indicating gradations in the visibilities of anomalies but pronounced differences between seasons of the year. A PCA transformation illustrates fusion of common features in the first component.

5.2 INTEGRATION OF MULTIPLE 2D GEOPHYSICAL AND NONGEOPHYSICAL DATA SETS

Commonly used 2D geophysical methods include MG, MS, EMI, and ER. Other types of 2D data are also used, including depth or time slices from ERT and GPR surveys, aerial and satellite imagery, topography, artifact density, and soil maps. The following review of integrations follows increasing dimensionality between one or more 2D geophysical and nongeophysical data sets and myriad combinations of such data.

In an early example, Doneus and Neubauer [33] illustrate the advantages of combining aerial photographs and topography with MG data for the study of Neolithic enclosures in Austria. The first two allow identification of archaeological sites at the landscape scale, while MG provides greater detail in targeted locations. Two recent technological advances augment this basic approach. First, geophysical sensors are increasingly deployed on multisensor carts for rapid survey at greatly expanded scales that typically cover hectares. Second, the spatial resolution of satellite sensors has increased to match the level of aerial photography, thus increasing the supply of panchromatic imagery and offering multispectral and even hyperspectral bands. The scales have therefore moved in tandem—ground-based geophysics can now efficiently cover enormous areas, while there is practically no limit to satellite coverage.

Multiple case studies demonstrate the integration of a single, landscape-scale 2D geophysical data set with terrain, aerial, and satellite imagery. Linford et al. [15] show that rapid, vehicle-towed MG survey is both complementary to aerial photography and airborne LiDAR and in some cases a substitute, as illustrated by several landscape-scale projects in the United Kingdom. Saey et al. [53] integrate EC and MS data from a vehicle-towed multireceiver EMI sensor with historic APs and a terrain model to visualize a WWI battlefield and surrounding landscape in western Belgium. In the Tavoliere of southern Italy, Ciminale et al. [54] use APs, MG, and multispectral satellite imagery to map and reconstruct Neolithic sites and the associated paleoenvironment in an 18 km^2 area. In all these examples, the geophysical data reveal important and complementary rather than duplicate archaeological features when compared with aerial and satellite imagery. Johnson and Haley [55] suggest, however, that cultural features discovered by MG, EMI, and other geophysical methods might also be teased out of multispectral bands, using features visible in geophysical data to train a statistical model and apply it more broadly (this study receives more focus below).

Maps of artifact density and soil properties from regularly spaced soil samples have also been integrated with 2D geophysical data. At the Hill Farm site in Texas, Perttula et al. [56] plot animal bone and ceramic sherd densities to assess the intensity of occupation for individual houses identified in MG data and better understand other magnetic anomalies. Similar advantage is gained by integration of multielement chemical analysis of soil samples taken at regular intervals. Dirix et al. [57] use the spatial clustering of elements to distinguish natural (e.g., bedrock-related) from anthropogenic sources of magnetic anomalies at ancient Sagalassos, Turkey.

Integrating two or more 2D geophysical data sets is relatively widespread. An early example by Neubauer and Eder-Hinterleitner [58] combines MG and 2D ER data through arithmetic and color combinations to reveal the layout of Roman Carnuntum in present-day Austria (discussed more fully below). Indeed, there are a great many examples of MG data being integrated with other 2D geophysical data, including with EMI [59–62], with EC [30], and with both EMI and 2D ER [63]. At the UNESCO World Heritage Site of Songo Mnara, Tanzania, Welham et al. [62] discover a "hidden majority" of the population as revealed by MG and EMI and confirmed by excavation. The areas between extant stone ruins were previously assumed to be open space but in fact are dense residential areas built of wattle and daub. In North America, Clay [64] demonstrates the effectiveness of MG and EC used in concert at sites in Ohio, Kentucky, and Mississippi to detect aspects of historic masonry structures, Mississippian mounds and houses, and prehistoric earthworks.

Integration of multiple 2D geophysical data sets with one or more 2D nongeophysical data sets has also proved worthwhile [44,65–67]. The combination of MG, MS, and topography at KoBulawayo, Zimbabwe, creates a detailed plan of the site including King Lobengula's late-19th-century Royal Enclosure [65]. At a Middle Neolithic tool production site in northeastern Sweden, Viberg et al. [67] use MS measurements from two different instruments in conjunction with soil phosphate analysis to locate a midden deposit containing fire-cracked rock and waste from a seasonal occupation.

5.3 INTEGRATING DATA TO INCLUDE THE VERTICAL OR DEPTH DIMENSION

Integration becomes more complex when a vertical or depth component is added, creating a third spatial dimension. This includes soil probes and cores, downhole measurements, GPR and seismic reflection profiles, ERT pseudosections, and point clouds produced by LiDAR and photogrammetry. The century-old tradition of communicating results with static 2D, often gray scale figures is a serious and sometimes prohibitive constraint. Visualizing 3D data is greatly enhanced by animation, a substantial departure from how we have become accustomed to communicating ideas in academia. Despite the fact that animations can be embedded within PDF documents—the standard digital format for scholarly publications—this is rarely done [68]. Instead, authors are occasionally given the option to include animations and videos as supplemental, online-only content. An attempt to publish animations of GPR data as rotating isosurface renderings and videos of increasing slice depth was made with the first and only article in the online, peer-reviewed journal *e-tiquity* [69]; the journal was subsequently canceled due to pushback in the community, chiefly because the HTML article could only be consumed while online. The following review catalogs ways that practitioners of archaeological geophysics have sought to integrate and communicate 3D data, including the use of 2D GPR, ERT, and seismic profiles; horizontal time or depth slices; curves and graphics depicting downhole geophysical measurements and soil properties; and use of perspective graphics including fence and block diagrams, isosurfaces, 3D volumes, and point clouds.

Given that GPR and ERT data are collected as 2D profiles, it is natural that they would be integrated directly in that format. Tsokas et al. [70] assess the structure of the subterranean Eupalinian aqueduct in Greece by integrating GPR and ERT profiles collected along the walls and ceiling from the tunnel interior. Wall thickness is measured by GPR, while ERT assesses the width of excavation behind them in select locations. Wilken et al. [71], working at Katarínka monastery in Slovakia, also use ERT and GPR but in this case collected along the same transects. The use of ERT inversions as color transparencies overlaid on radargrams forms an effective fusion and leads to the interpretation of a previously unexplained feature as a water well or food storage pit. Another method that can be depicted in 2D vertical sections is downhole MS (DMS). Kassabaum et al. [72] study the deep internal structure of two large platform mounds at the Feltus Mound complex in Mississippi using ERT and DMS transects placed to crosscut targeted portions of the mound.

Several new midden deposits are discovered by ERT, while DMS reveals a burned surface, cooking pits, and probable construction berms.

In most cases where 3D methods are used, data are collected in closely spaced transects, and horizontal time or depth slices are produced [36,73–75]. In a great many cases, these 2D versions are used at the exclusion of the original profile data, typically to facilitate comparison and integration with other 2D data. At Fort Ancient, a hilltop enclosure in the Ohio Valley, Burks [76] illustrates the approach through the discovery and analysis of Moorehead Circle, a finding he describes as "one of the most complex architectural features ever found at a Hopewell ceremonial centre." While 19th-century maps and a LiDAR-derived terrain model provide regional and environmental context, MG, ER, and a GPR slice describe multiple aspects of the circle. MG shows an oval enclosure with a southern gateway and strongly magnetic central feature; ER delineates prepared surfaces; and GPR reveals an outer circle of posts, prehistoric trenches, and a limestone pavement.

Another approach to integrating information in the vertical dimension is the extraction and analysis of soil cores [77–80] and use of downhole sensors [72,81–83] in conjunction with geophysical and other prospection approaches. Henry [84] describes a multistage research design that begins with broadscale geophysical survey, followed by the use of downhole sensors to better understand selected features, and finally the use of in situ soil measurements to map excavation profiles. This is demonstrated at the LeBus Circle site in Kentucky, where the entire earthwork is surveyed with MG and a portion with MS. A host of features are then further investigated with DMS, followed by in situ MS mapping of excavation profiles. Integration of these data greatly improves interpretation of the 2D MG and MS surveys [84]. This multistage approach, in which features identified by at-surface geophysical surveys are further explored using slightly more invasive core extraction and downhole logging, leads to integration of multiple 2D and 3D data sets for greatly improved interpretations, even in the absence of excavation. At the Early Bronze Age settlement of Fidvár in southwestern Slovakia, for example, Nowaczinski et al. [77] integrate ERT pseudosections with a host of soil properties (sedimentology, grain size, Munsell soil color, MS, total organic carbon, chemical composition, and elemental composition) from multiple cores to precisely define lateral and vertical ditch geometries without excavation.

A relatively recent development is the use of in situ and downhole reflectance spectroscopy. Similar to multispectral scanners deployed on satellites and aircraft, downhole multispectral sensors use light reflectance and absorption to characterize soil chemistry. Dalan et al. [81] make the case that downhole geophysics and spectroscopy provide much more detail in the vertical dimension than is possible with at-surface geophysical surveys, chiefly because they can identify thin layers and often penetrate to greater depths. They demonstrate this through examination of a fortification ditch and multiple earth lodges at the Biesterfeldt site in North Dakota, where they integrate at-surface GPR, ERT, MG, and ER data with downhole MS, magnetic viscosity, TMF, capacitance, ER, and direct-push color (red, green, blue, and near-infrared light reflection). Matney et al. [83] further explore this approach using in situ subsurface soil spectroscopy in a truck-mounted soil probe system that also measures soil conductivity and probe insertion force. Their work shows that often ambiguous features in at-surface geophysical data can be defined and better understood with subsurface probing. They integrate these measures with at-surface MG and ER surveys, chemical analyses of core samples, and traditional excavations at two protohistoric villages in Kansas.

As previously mentioned, 3D visualization is severely limited by conventional static 2D illustrations. Volumes of data are displayed in perspective views as opaque blocks ([85], figs. 6–8; [86], fig. 9) and isosurfaces ([87], fig. 8; [78], fig. 6; [88], fig. 6; [89], fig. 11), although they are almost always limited to single modalities. Fence diagrams, graphic displays of crisscrossing vertical sections displayed in a perspective view, are also used. In most cases, it is simply to fuse two or more spatial dimensions of a single modality, such as GPR ([90], fig. 9a and b; [91], fig. 13; [92], fig. 13) or ERT ([93], fig. 4), but in some cases, fence diagrams include 3D vector drawings of known architecture or interpreted features ([94], fig. 4b; [95], fig. 5d). Multiple modalities are also shown in this manner. Forte and Pipan [96] use a fence diagram of GPR reflection profiles intersected by a horizontal seismic tomogram to effectively visualize the interior of a Late Bronze Age burial mound in northern Italy. Closely related to fence diagrams are block diagrams, in which a surface map is combined with one or more cross sections. Nowaczinski et al. [97] use a block diagram to display MG data draped over a terrain surface, with an ERT profile showing the vertical dimension. The figure fuses the three types of information and very clearly illustrates two Early Bronze Age houses and a ditch at Fidvár (Slovakia).

The fusion of multiple 3D data sets continues to be a challenge on multiple fronts. GIS-like software seems essential, but few exist that are capable of true 3D data handling and measurement capabilities. If multiple data sets are successfully fused in 3D, the challenge then becomes one of communication, once again running into the constraints of static 2D figures required for publication. At the UNESCO World Heritage site of Tiwanaku, Bolivia, Cothren et al. [98] integrate millions of data points from total station surveys, long-range and short-range terrestrial laser scanning, photogrammetrically derived terrain model and orthomosaic, GPR, EMI, MG, and a 1911 topographic map. They successfully coreference all data to a common 3D coordinate system within a single software, resulting in a "multiscale, multitemporal, and multisource" 3D model consisting of all input data as ASCII point clouds [98] (p. 96). Our Case study 2 illustrates similar fusions of multiple combined 3D data sources.

To date, the most successful visualizations of integrated 3D data, we argue, have made use of 3D modeling [80,99]. The refinement of complex 3D data into 3D objects facilitates clear communication of outcomes, particularly in conventional 2D figures. Panisova et al. [95] combine total station, terrestrial laser scans, microgravity, and GPR data in the form of 3D polygonal models at the Church of St. George in Slovakia. Four medieval crypts and a previously unknown wall foundation are modeled from the geophysical data, while the aboveground details of the church are modeled with laser scanning and total station mapping. Viberg et al. [99] utilize a similar approach but link GPR with photogrammetry into a single 3D model of the Church at St. Lawrence in Sigtuna, Sweden. Perhaps the most impressive example of 3D fusion and visualization is the work of Margaret Watters at the Catholme Ceremonial Complex in the United Kingdom, published more than a decade ago [100]. She integrates ER and GPR as individual 3D objects in a single visualization software, facilitating integration and making it possible to produce 2D illustrations that effectively communicate the findings ([100], figs. 3–7).

6 METHODS OF 2D INTEGRATION

Most geophysical surveys continue to prioritize 2D or "area" coverage of broad regions. Methods of data integration in satellite remote sensing occur at the pixel level or feature level, and the same is true in archaeological geophysics, although the goals, methods, and focus each domain receives vary widely. Feature-level integrations tend to dominate, yet unlike satellite remote sensing where objective feature-defining algorithms dominate, features in archaeogeophysics tend to be *subjectively* interpreted and manually generated in the form of vector drawings. Often, researchers merely illustrate multiple prospection data sets, displaying them separately, side by side, or perhaps through overlays. Dozens of such papers include "integration" in their titles, yet any combination of data occurs only in associated text, probably challenging notions held in other disciplines of what constitutes data fusion.

Whatever form of 2D data integration is pursued, it is assumed that all preprocessing has been accomplished and that absolute accuracy in the coregistration of individual data components has been achieved, so they align perfectly in space (Fig. 1). This requirement may not seem a large problem for multiple geophysical surveys acquired from the same field grid. Yet, it can be an enormous issue when rectifying and registering aerial or space-based information to that grid.

6.1 BASIC INTEGRATIONS OF MULTIDIMENSIONAL DATA

Presentations of multiple geophysical and related data sets occur in many ways, from two-dimensional gray scale or color mappings in separate figures or side by side, through simple to complex overlays of various types. This group forms the most common approach to the presentation of multimethod surveys, where integration of findings occurs through argumentation and without graphic, mathematical, or other manipulations of data or imagery. This pathway is illustrated in column 1 of Fig. 1.

6.1.1 Separate or side-by-side displays

Most multimethod projects go no further in their integration efforts than side-by-side displays coupled with in-text discussions of similarities and differences in findings. A sampling of such studies illustrates the general approach. Wide-area MG and GPR

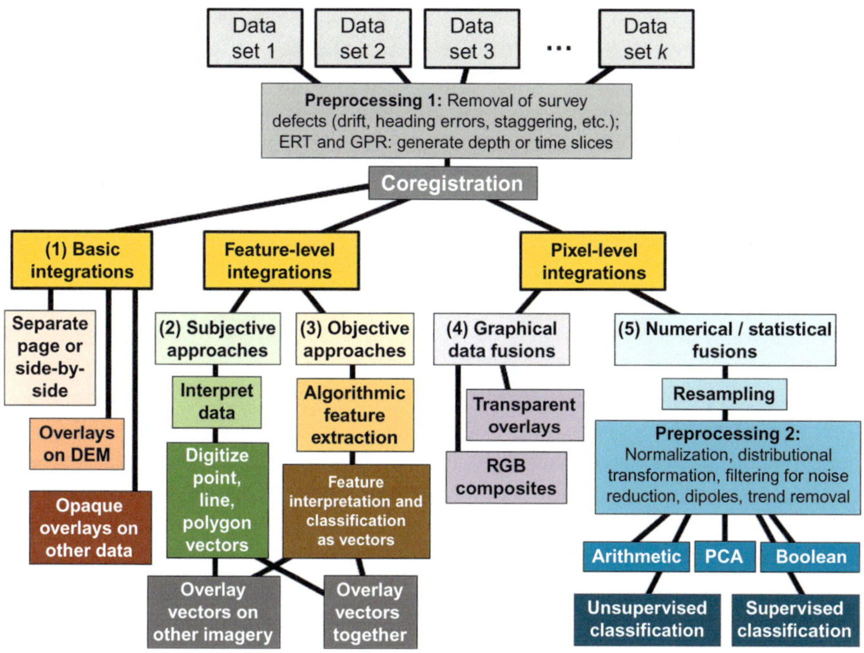

FIG. 1

Flowchart showing modes and pathways to various forms of 2D data integration.

maps of Roman Flavia Solva, Austria, are illustrated together and permit detailed maps of the city's buildings and road network [101]. Conyers [102] is more introspective at Roman Verulamium, United Kingdom, with a comparison of four GPR depth slices against MG results that indicate greater architectural detail and depth information in the former, with evidence of iron objects and additional walls due to a burning event or magnetic building materials in the latter. Side-by-side portrayals of EC, ER, MG, and MS results recorded over fired Caddo houses from the 1400 CE Tom Jones site, in Arkansas, are compared by Lockhart [63]. Electrically conductive and magnetically susceptible clays immediately above electrically resistant and nonmagnetic shallow chalks enabled large patterned contrasts caused by their ritual burning. Side-by-side displays are given in case studies below (Figs. 3B–M, 6A, B, and 8A–F).

Similar applications occur in marine geophysics. For instance, Quinn et al. [103] describe work associated with the frigate *La Surveillante*, which was sunk in 1796 in the ill-fated attempt by the French Armada to land in Bantry Bay, on the southeast coast of Ireland. A digital elevation model (DEM) combines surface elevations and bathymetric depths that illustrate the Bay and wreck location. A local TMF mapping and cross section through the wreck shows a 25 nT range in the bipolar magnetic field resulting from the frigate's iron cannon, ammunition, chains, fastenings, and anchor. In addition, sonographs from 100 to 500 kHz side-scan sonars successfully image the wreck, and a subbottom chirp profile yields a vertical resolution of 20 cm. Together with information from divers and coring, the combined data place the ship in a shallow depression on a solid hard layer surrounded by 2 m of fine-grained anoxic sediments. Moreover, wooden material is preserved, wreckage is concentrated due to a calm seafloor environment, and a detailed map of the remains is offered along with a model of processes associated with its sinking and subsequent deterioration.

In recent years and especially with the growth of GIS, air or space imagery and other data are included in integrated knowledge bases. For example, Campana et al. [104] illustrate crop marks associated with a medieval mound in Tuscany and then compare its mapping against MG, a GPR slice map, ER surveys at three depths, and surface distributions of pottery. While many results are parallel, MG shows an inner ditch system to be highly magnetic, perhaps due to increased organic matter and domestic occupation debris. On the other hand, GPR indicates a circular wall around the innermost enclosure, and ER shows higher resistivity in this area where test excavations later confirm a structure. An unusual comparison between multiple ground-based geophysical (MG, lateral and downhole MS, EC, ER, GPR profiles, and depth slices) and other data (vegetation and shadow markings in color aerial imagery, thermal infrared (TIR), and high-resolution DEM) is offered by Kvamme [51]. The ability of each method to reveal 11 archaeological feature types from the late prehistoric fortified village of Double Ditch, in North Dakota, is rated from poor to excellent.

6.1.2 Image integrations through overlays

Opaque overlays. More advanced integrated portrayals are achieved when a geophysical mapping is draped opaquely atop another image, whether geophysical or from another form of prospection. The overlay might be slightly smaller in area, permitting a view of survey results in the context of its local landscape when placed over a DEM or satellite view. Alternatively, the purpose may be the "matching" of anomalies along the edges of the smaller overlay corresponding with anomalies in a larger image beneath. Novo et al. [38] provide an excellent example with an image of half a Gallo-Roman temple in Normandy obtained through a GPR depth slice overlaid on a larger ER image that portrays the other half and indicates perfect correspondences between anomalies representing walls and other structural elements. Henry et al. [105] similarly drape EC, MS, and GPR slice maps over a larger background MG image from an Adena burial mound, in Kentucky. MG gives excellent definition of an exterior ditch and suggestions of interior mortuary features, with numerous anomalies that match those in the overlay. EC better shows lobes of an interior embankment, however, while MS displays regions of topsoil eroded from the mound, and GPR indicates anomalies supporting indications of mortuary features and construction phases. Lockyear and Shlasko [36] do nearly the same thing with a site-wide MG map of Roman Verulamium, United Kingdom, above which lie smaller images of ER, MS, or GPR slice maps.

Overlays of multiple data sets using different display formats. Other display formats besides flat 2D images are possible, most notably pseudo-3D views that are particularly useful when DEM generated through LiDAR or other means serves as a base over which geophysical mappings are draped. Such portrayals readily facilitate visualization of a survey's context within a landscape. An excellent illustration is given by Saey et al. [53], who develop a "fused electromagnetic metal prediction" index from four EC data sets (discussed below under pixel-level integrations). They overlay it on a DEM of 3.2 ha of the Ypres salient, a major conflict zone of the Great War, showing spatial relationships between the landscape and war features, including iron-lined trenches and other metallic distributions. The 3D representation reveals the trench system sited just below the uppermost part of a ridge, which, for strategic reasons, was preferred over lower lying areas. Technological developments aid in such approaches. Gaffney [46] employs a three-sensor fluxgate array linked with real-time GNSS in the Cyrenaica Archaeological Project. Movement of the cart simultaneously records elevations along with MG data, permitting construction of a DEM upon which MG imagery is ultimately draped.

Geophysical overlays on DEMs can show correspondences between anomalies and subtle archaeological terrain variations. A high-resolution DEM of previously described Double Ditch [51] clearly shows the defensive ditches from which the site is named, while superimposed MG data expose two previously unknown ditches (Fig. 2A). Their relationships with adjacent observation platforms and the inner ditches revealed by the DEM effectively convey the nature of the village's outer defenses. A similar mapping illustrates a *subsurface* DEM, cleverly developed by Saey et al. [80] at the site of two concentric circular ditches of likely medieval age, discovered through AP cropmarks in Belgium. GPR travel time velocities are integrated with multifrequency EC data to model the thickness of ditch infilling above a sandy subsoil in a two-layer soil system, which is then subtracted from surface altitudes. The result is a subsurface DEM of the former ditch surface, over which GPR time slices are draped.

Very different overlay approaches are also possible. Forte and Pipan [96], for example, investigate a 4.5 m-high Bronze Age burial mound in northern Italy and acquire GPR profiles from 100 to 250 MHz antennae over the tumulus. Penetration is insufficient to visualize the burial chamber due to the attenuation of radar energy, however. Seismic tomography is therefore undertaken to generate a velocity map that reveals a burial chamber beneath the mound at the original ground level. An oblique view of this map with superimposed terrain-corrected GPR transects effectively illustrates the upper-layered mound stratigraphy imaged by GPR and the lower burial chamber well below a zone of meaningful reflections.

Another type of overlay displays variations in a prospection data set through isoline contours draped over a background image. Fig. 2B shows 5 cm elevation contours describing surface depressions over prehistoric houses seen in MG data that better illustrate their full sizes, rounded rectangular forms, central hearths, and surrounding corn storage pits at mid-15th-century Huff village, North Dakota. Erosion has caused the surface depressions to be significantly smaller than true house sizes

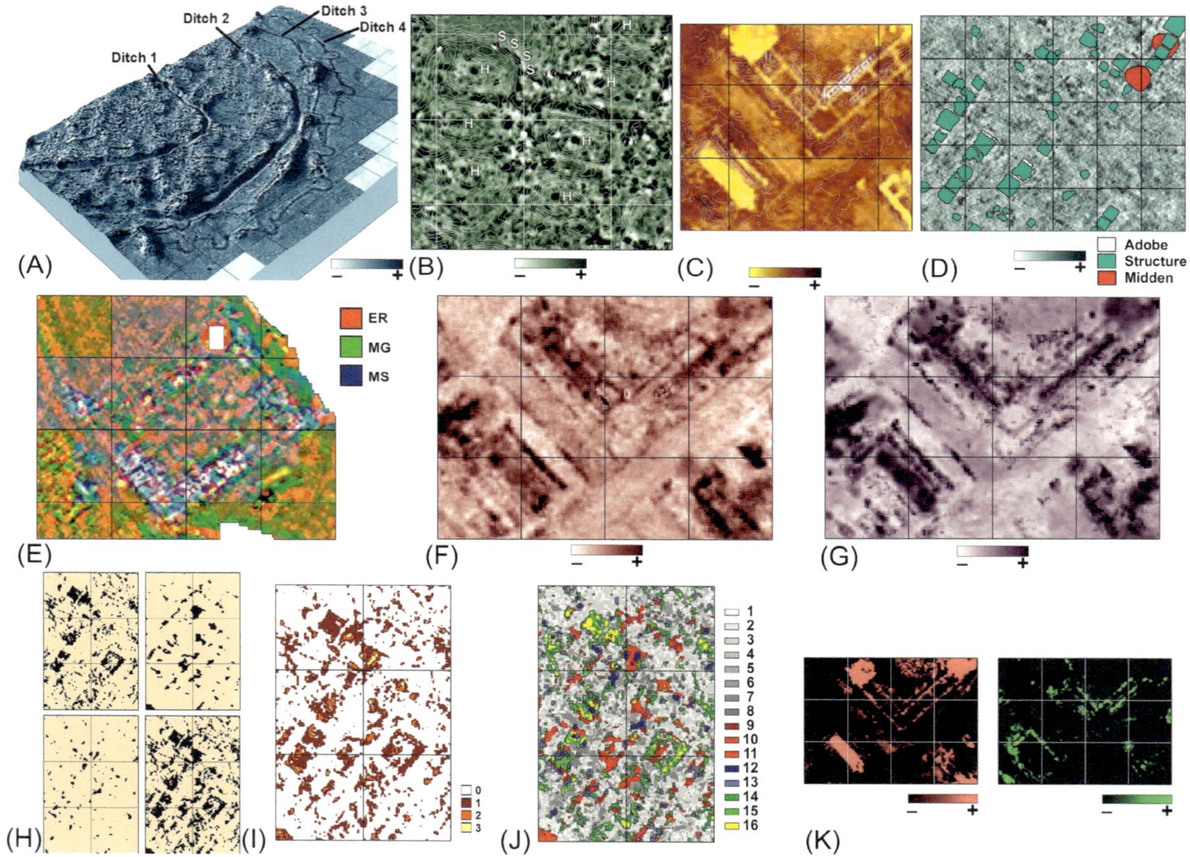

FIG. 2

Forms of data fusion: (A) MG data revealing two additional fortification ditches (with bastions) draped over an elevation surface showing two ditches, observation platforms, and house depressions at Double Ditch; (B) elevation contours depicting surface depressions over rounded rectangular houses seen in MG data, each with a central hearth (H) and surrounded by corn storage pits (S) at Huff village; (C) EC contoured over ER at Army City; (D) interpreted and vectorized polygons at Pueblo Escondido; (E) RGB composite of Fort Clark Trading Post; (F) summation and (G) maximum of normalized ER, GPR, MG, and MS at Army City; (H) principal anomalies in binary form for GPR, MS, MG, and their Boolean union (upper left to lower right) at Pueblo Escondido, (I) sum of binary layers in (H); (J) *k*-means cluster analysis at Pueblo Escondido; and (K) logistic regression probability surfaces for concrete/brick (left) and pipelines (right) at Army City. *Grids lines* in all figures are 20 m apart.

(these data are more fully explored in Case study 3). Similarly, EC anomalies associated primarily with buried metal pipelines are contoured over ER data that reveal brick and concrete floors, walls, and footings at early 20th-century Army City, a Great War military support town in Kansas (Fig. 2C; these data are further examined in Case study 4; see Ref. [75] for a site-wide view). A different integrative overlay employs photogrammetry to more completely illustrate the 11th-century ruins of the Church of St. Lawrence in Sigtuna, Sweden [99]. Recognizing that the standing ruins (a tower) would leave a hole in their GPR slice map, photogrammetry is applied to high-resolution photos to generate a 3D model of it. This model is superimposed over an architectural plan of buried walls and foundation stones revealed by GPR to complete the layout in a seamless 3D perspective view of all elements of the church, below- and aboveground. Case study 2 offers a similar result (Fig. 5).

6.2 FEATURE-LEVEL INTEGRATIONS

The definition of likely archaeological or other features in geophysical and associated data leads to feature-level integrations. In current practice, such features are generally defined subjectively through visual interpretations of the data and their manual digitization into point, line, or area vectors, through GIS. Interpreted features thus become *objects* corresponding to specific archaeological targets (e.g., houses, hearths, and walls) or other cultural or natural features not of primary interest (e.g., paleochannels, pipelines, metallic debris, and plow marks). Vectors derived from prospection data sets may be draped over other data for purposes of insight or comparison, leading to a form of fusion. More commonly, vectors derived from multiple prospection data sets may be simultaneously displayed and overlaid to form a single image of all interpretations, forming a strong graphic for portraying the integrated, interpreted knowledge of a site. This pathway is illustrated in column two of Fig. 1.

Object-based research in satellite remote sensing routinely employs advanced algorithms and machine learning to automatically define features of interest in imagery, such as buildings, roads, and trees [8,106]. These undertakings have been slow to develop in archaeogeophysics, in part because of typical noisiness of the data, including irregularities and incomplete revelations of archaeological targets (see below).

6.2.1 Interpretive vectors overlaid on prospection imagery

Vectors generated from one data set and overlaid on another can offer powerful interpretive potential. Linford et al. [15], for example, take vectorized interpretations of archaeological features seen in aerial imagery at the Kitridding Hill earthwork and a rectilinear ditch system at Little Chalfield, both in the United Kingdom, and overlay them on large-area MG surveys "to illustrate the benefits of a combined aerial and ground-based approach to mapping the historic environment." Similar results are described by Scardozzi et al. [107] at Hierapolis of Phrygia, Turkey, but their vectorized interpretations are overlaid on the MG and GPR slice maps from which they were derived only to guide the viewer to significant features. On the other hand, Masters and Stichelbaut [66] vectorize frontline and communication trenches, machine gun nests, bunkers, mine craters, and a field of barbed wire seen in period APs from a WWI battlefield at Ploegsteert, Belgium. These vectors are then draped over an image of a 16 ha MG survey that confirms many of these interpretations but which also reveals additional battlefield features not visible in wooded areas of the time, where camouflage was employed or where features were obliterated from shelling. Shell craters, now mostly leveled, are also seen magnetically as dipoles. These features, also vectorized, offer a more complete combined view of the battlefield.

6.2.2 Interpretive vectors overlaid with each other

Frequently, interpretive vectors derived from multiple prospection surveys are combined in a single map that summarizes all known or inferred archaeological knowledge about a site or landscape. At Roman Altinum, in the Venice Lagoon, Mozzi et al. [108] utilize aerial and satellite data along with targeted MG, ER, GPR, and EMI surveys that yield vector interpretations from each data set, which are combined to produce a detailed plan of the city spanning 14 ha. Vector polygons and lines are employed by Ernenwein [74] to portray interpretations of pit house and other structural features, adobe walls, and middens from MS, MG, and multiple GPR slices from the 13th-century Pueblo Escondido, New Mexico (Fig. 2D; shown over a regression-fused image of the data). In Europe, Asăndulesei [44] vectorizes interpretations from AP, ER, LiDAR, MG, and TMF surveys at an Neolithic Cucuteni settlement in Romania, combining the results in a single image with a LiDAR background. Schneidhofer et al. [16] similarly integrate archaeological features and paleoenvironmental interpretations, including beach deposits, colluvium, near-surface bedrock, and palaeochannels, through vectorization of MG surveys and GPR slices at Viking Age Gokstad, in Norway.

6.2.3 Automated object-based feature methods

Few studies in archeology have yet to employ advanced algorithmic or object-based methods [8] to define potential archaeological or other features. The few that do employ this technique use satellite, aerial, or LiDAR data [109–111] or only a single geophysical data source. Leckebusch et al. [112], for example, describe regression algorithms for automatic extraction of 3D models from GPR volumetric data, while Fitch et al. [113] define paleochannels and other subsea features through 3D solid modeling and segmentation methods applied to a vast seismic database from the North Sea Palaeolandscapes Project. Although this pathway to data fusion remains to be developed, one scenario is that it begins with feature extraction that is then followed by interpretation and classification into archaeologically meaningful categories, illustrated in column three of Fig. 1.

Only a few papers describe object-based feature extractions that utilize multiple geophysical methods. An early study by Watters [100] produces solid-model objects derived through segmentation and isosurface rendering based on 3D ERT and GPR data using Amira medical imaging software. Results yield 3D objects representing small and large pits, furrows, and post holes hidden beneath the surface, together with volumetric data that imparts significant information about the Catholme

Ceremonial Complex, United Kingdom. More recent approaches by Nowaczinski et al. [78] produce reconstructions of prehistoric pits at Bronze Age Fidvár, Slovakia, as 3D models based on ERT methods. In addition to excellent renderings of form, classifications by shape, length, width, depth, and volume are obtained for nearly 30 pits. Case study 3 further explores this domain.

6.3 PIXEL-LEVEL INTEGRATIONS

Pohl and van Genderen [1] indicate two primary pathways for pixel-level fusions in satellite remote sensing. One they refer to as "color-related" that includes a variety of image display techniques, and the other consists of statistical-numerical combinations. In the former, each data set is treated as an image that forms a picture of spatial variations. A variety of advanced computer graphic methods then combine or composite the multiple images into a single portrayal, forming effective fusions of the data. Statistical-numerical combinations utilize mathematical functions that alter the data numerically, generating outcomes that can be very different from individual inputs but which share their information content.

6.3.1 Computer graphic integrations

Graphic fusions of prospection data at the pixel level include transparent overlays and color composites (column four of Fig. 1) because display values, whether in color or gray scale, truly change within each pixel. The same cannot be said of previously discussed graphic overlays where one image opaquely covers another. The goal remains the same; imagery that simultaneously shows elements of several dimensions in such a way that more information is portrayed than in any individual input. Contemporary software commonly permits graphic integrations of data at multiple spatial resolutions (pixels of different size).

Transparent overlays. The draping of one or more transparent images atop a base image is a form of fusion that has achieved considerable use because it permits simultaneous viewing of multiple data sets. In Mongolia, Lin et al. [114] overlay transparent geophysical data on drone-acquired aerial imagery. Strong anomalies in transparent MS data over one aerial image probably point to buried metal objects associated with the loci of visible surface stones of a Bronze Age tomb. At another site, they overlay transparent MG data on an aerial image of a line of stones associated with a Turkish tomb (*babal*). The former indicates a series of large square features caused by soil changes not visible on the surface.

An innovative application of a transparent overlay is given by Linford [45] who illustrates a mosaic floor from the Bignor Roman Villa, United Kingdom, over a GPR depth slice approximately 0.28–0.32 m below it with reflections that suggest a hypocaust below. High-amplitude reflections probably result from the *pilae* supporting the floor, while low-amplitude regions point to void spaces. Even more unusual is an application by Watters [100] who laments the 2D limitations of GIS in handling 3D GPR and ERT data at the Catholme Ceremonial Complex, United Kingdom. Instead, 3D Amira medical imaging software is employed to visualize the data as *voxels* with various degrees of transparency and to produce solid-model objects. Subsurface features are thus imaged through combinations of transparency and solid modeling. Although these examples illustrate the effectiveness of this form of fusion, Kvamme [75] warns that superimposing more than four images tends to produce a "muddy" appearance but nevertheless goes on to composite six geophysical data sets, each depicting different aspects of the historic Army City settlement.

Color composites are conceptually similar to transparent overlays, but with several important differences. A maximum of only three images may be employed in combinations that can yield a large suite of colors (most current computer monitors permit at least 2^{24}, about 17 million). Resultant color mixings may be explained by red-green-blue (RGB) color theory [4], which offers distinct interpretational benefits. Locations bright with primary colors indicate places with high measurements in each of the corresponding images, while duller values point to low values. Moreover, cyan indicates large measurements in the green and blue images, yellow shows high values in green and red, magenta indicates the same for red and blue, white means high measurements in all three, while black indicates the converse, with low measurements in all. Obviously, gradations of color indicate less extreme measurements in the various inputs.

With $n = 3$ possible inputs, there are $n! = 6$ possible combinations of RGB, and merely changing them can yield markedly different views of the data (see Ref. [115], for an illustration). However, with more than three data sets, the number of possible permutations rises sharply, with ${}_nP_3 = n!/(n-3)!$ possible, making the exploration of all choices onerous. Keay et al. [116] fuse MG, ER, and GPR in nearly a hectare of an urbanized section of Roman Portus, Italy, observing that "the RGB image provided a mechanism for interpreting positive and negative features and allowed visualization of subtler features that might otherwise go undetected." An RGB composite of ER, MG, and MS surveys from the Fort Clark Trading Post (1832–61), a later center of the North American fur trade in North Dakota, is given in Fig. 2E [117]. While ER (assigned to the red band) shows a number of interior walls, the exterior palisade, and courtyard features, MG (assigned to the green band) and MS (assigned to the blue band) reveal corner bastions (left and right corners) and the principal buildings within the fort due to sandstone foundation blocks of high MS and the large numbers of iron artifacts within the rooms. A somewhat different approach is offered by Böniger and Tronicke [118] where an RGB composite of a GPR "energy" slice map, absolute MG, and terrain slope gradient is placed transparently over a fourth dimension portraying GPR "similarity" (variability of reflection strength) in gray scale, which contributes subtle details of the 30×40 m study area. The result effectively shows buried architectural elements beneath the Palace Garden of Paretz, Germany.

6.3.2 Challenges to numerical/statistical integrations

Ground-based geophysical applications face numerous challenges to their quantitative integration at the pixel level. One problem is that of divergent spatial formats and resolutions. Some geophysical data might yield high spatial sampling densities (e.g., $200\,m^{-2}$ traces for GPR) and others very low (e.g., $1\,m^{-2}$ for ER), while aerial imagery may vary enormously, thanks most recently to low-flying drones that can yield centimeter-level information [119]. Worse, data densities may not be uniformly distributed. MG data, for example, might offer $10\,m^{-1}$ samples along transects, with transects separated by 0.5 m, yielding rectangular pixels. Although divergent spatial resolutions can more easily be employed in graphic pixel-level fusions, arithmetic combinations of numerical data (e.g., by GIS calculators) typically require resampling of the data to a common spatial resolution. This could mean the downsampling of some data (equivalent to the throwing away of information) and the upsampling of others (through interpolation), neither of which is desirable, to an "average" resolution for the fusion. Greater challenges to numerical fusions at the pixel level arise from certain peculiar characteristics of geophysical data that may require additional preprocessing (pathways for numerical fusions are given in column five of Fig. 1).

Focus in distribution tails

Unlike most forms of quantitative information collected in applied research where interest typically lies in the central tendencies of bodies of data, in geophysics, interest generally centers on less frequent anomalous measurements in the *tails* of distributions. From a statistical standpoint, this means that analyses might be based on the rare and unusual using methods designed for the abundant and typical, leading to new research challenges that perhaps require alternative strategies.

Noise

Anomalous measurements are caused not only by features of anthropogenic origin, the typical targets of investigation, but also often by natural characteristics of the environment, including topographic, soil, or moisture changes, paleochannels, rodent burrows, isolated rocks, and tree roots, to cite a few. In addition, anomalies of cultural origin may be recent (e.g., pipelines, plow scarring, and metallic debris) or derive from periods not of interest (e.g., a recent historic settlement above a prehistoric one). If we liberally regard noise as anything that reduces the clarity of a desired signal [18] (p. 77), then anomalies stemming from natural sources, unwanted anthropogenic sources, electronic noise, and operator errors all contribute, and the task becomes one of the separating anomalies of interest from others. This undertaking can be extremely difficult when anomalies arising from targets of interest are in the minority and if they lack regular geometric expressions (as squares, circles, rectangles, and straight lines) or systematic spatial repetition characteristic of human constructions [35]. It is primarily because of noise that machine recognition algorithms prove so difficult and why subjective procedures for feature definition dominate in archaeological geophysics.

Other distributional issues

Magnetic data are bipolar, and their distributions are often extremely leptokurtic (long tails). GPR data tend to exhibit right skewness, while outliers are common in lateral ER surveys stemming from "data spikes" (extreme positive or negative outliers caused by poor probe contacts). MG and EMI data exhibit similar bipolar outliers stemming from the presence of metals. A single target represented by extreme positive *and* negative measurements (whether a hearth, ceramic pipe, or metallic object) may show a visual (and subjective) correlation between two data sets, but from a statistical standpoint, a low correlation coefficient will likely be realized, yielding a spurious result of little use to subsequent numerical integrations. Moreover, skewed data distributions and outliers in general negatively impact statistical indexes (e.g., means, variances, and correlations), introducing difficulties to the use of advanced multivariate methods, cornerstones of data integration methodologies.

Lack of correlation

Many physical properties of the subsurface represent independent or nearly independent dimensions. The magnetic susceptibility of soils may be unrelated to their conductivity, while the latter may be independent of changing dielectric properties, for example [29]. Although regions of high conductivity may disperse GPR energy and inhibit reflections, in low conductivity zones, GPR responses can be highly variable, meaning that anomalies seen in EC and GPR may not well correlate. Furthermore, the nature of data acquired by field instrumentation further contributes to reduced correlations. MS surveys reveal only the induced component of soil magnetism from very shallow sources, giving a result often different from MG. Because EC is the theoretical inverse of ER, strong correlations between them have been reported [120], but depending on coil separation and frequency in EC devices and interelectrode distances in ER surveys, unequal soil volumes may be evaluated resulting in poor relationships. Low correlations between prospection data sets bring to question the applicability of a suite of fusion methods common to satellite remote sensing, where highly related satellite bands are common and where fusion methods based on correlational structures, such as PCA, are standard practice [4]. In a large-area survey, Ernenwein [121] actually shows near-zero Pearsonian correlations between MG, MS, EC, and GPR data sets (max $r=0.156$). A similar outcome is demonstrated between GPR, MG, and ER by Ogden et al. [122] with max $r=0.21$. Kvamme [75] found much the same between five ground-based geophysical data sets; however, the low correlations were improved after transformations to reduce skewness and outliers such that 4 of 10 pairwise correlations reached approximately $r=0.3$ with two others showing $r>0.2$, still unimpressive but sufficient for PCA (see Case study 4 for further evidence).

6.3.3 Approaches to numerical/statistical integrations

Additional preprocessing

In order to address the previous shortcomings, further preprocessing may be required for effective numerical integrations. These procedures lie in several distinct domains:

- *Normalization of measurement scales.* Measurements yielded by various geophysical devices sometimes differ by orders of magnitude (e.g., ±10 nT vs 120–300 Ω m). Unless normalized to a common range (0–1 is often preferred), larger scales will overwhelm lesser ones in many forms of integration. Several options have been explored including division by maximum value [58] or standardization to z-scores [75].
- *Elimination of bipolar anomalies.* MG and EMI data sets commonly yield bipolar anomalies in the presence of ferrous or other metals, which confuse integrations because the same target offers extremely divergent positive and negative measurements. In such cases, absolute values have been considered [121]. Case study 3 offers a less drastic algorithmic solution for the identification and elimination of small MG dipoles resulting from metallic litter.
- *Transformation of data distributions.* As previously noted, skewed data distributions and outliers negatively impact statistical indexes and advanced multivariate methods. Logarithmic, power, or other functions may be employed to reduce skewness and outlier magnitudes and achieve distributional forms closer to Gaussian. A power function of the form $x' = x^b$ works well, where x' is the new value, x the old, and b a constant. With $b < 1$, the right tail is reduced and the left increased, while $b > 1$ generates the reverse; the effect becomes more pronounced as b departs from unity.
- *Filtering to reduce noise.* Low-magnitude noise in geophysical data comes from soil and moisture variations, rodent work, vegetation patterns, and the like. A simple low-pass filter that replaces each measurement with the average of its neighborhood (perhaps Gaussian weighted) is one way to subdue it. Periodic noise from plow marks, which may obfuscate visualization of archaeological targets, may be reduced through the fast Fourier transform [34].
- *Filtering to reduce broad trends.* Some geophysical data sets illustrate broad regional trends stemming from geologic or other environmental circumstances. For example, soil conductivity may gradually decrease as one moves away from moist soils adjacent to a river. High-pass filters are generally employed to reduce their effects [17] (p. 147).

Arithmetic integrations

One of the earliest papers to combine multiple geophysical surveys numerically is by Neubauer and Eder-Hinterleitner [58], who explore various mathematical blendings of ER and MG to visually enhance the pooled information from a 0.64 ha portion of a large public building (possibly the *curia*) in Roman Carnuntum, Austria. Each data set is normalized to a 0–1 range such that linear anomalies (representing walls) in both data sets have similar low values. However, before ratioing, the data are normalized to a 1–2 range to avoid very small denominators or divisions by zero. The results of summation, difference, multiplication, and division operations yield very different views of the combined data and a better understanding of cross correlations between the methods.

Piro et al. [12] undertake slightly different procedures in their investigation of several ancient sites in the vicinity of Rome. MG, GPR time slice, and ERT depth slice are obtained in each area and subjected to a function that quantifies the absolute difference of values from an average undisturbed background and then normalizes each data set through division by the maximum. Two functions are then applied: a simple average and a product. They found (p. 212) "noticeably enhanced" discrimination where the integrations better define "position, extension, depth, thickness and physical characteristics of any anomalous body." A basic summation of ER, MG, MS, and GPR slice from a portion of Army City [75]) is illustrated in Fig. 2F with the data normalized by standardization to z-scores. Also illustrated is the *maximum* per pixel of the data sets (Fig. 2G). Although these fusions each show the village structure and its components more completely, the maximum function appears to more greatly emphasize low-level noise (the input data for these fusions are illustrated in Case study 4, Fig. 8B, C, E, and F). Ogden et al. [122] apply the *minimum* function to the three geophysical data sets at Roman Portus, which they found better defines voids between structures and other negative anomalies.

Matera et al. [115] are more analytic in their algebraic combinations of MG, pulsed GPR (PGPR), and stepped frequency GPR (SGPR) slice maps in the vicinity of a large Roman structure in southern Italy. Each data set is first normalized to a 0–1 range (subtraction of the minimum followed by division by the maximum), and then, the four possible integrated mappings are produced: MG + PGPR + SGPR, MG − PGPR − SGPR, MG + PGPR − SGPR, and MG − PGPR + SGPR (those generating the negatives of these results are ignored). Next, five test areas of several square meters, each centered on a prominent anomaly of archaeological origin, are established, and values from these targets and adjacent background values absent of anomalous indications are extracted. The archaeological and background samples are then subjected to two indexes developed in satellite remote sensing for enhanced discrimination of crop marks. A "sensitivity index" is the normalized difference between the mean archaeological and mean background values, while a "separation index" quantifies the level of separation between two distributions [123]. A simple scatterplot of these indexes clearly indicates the composite MG − PGPR − SGPR gives best performance in anomaly discrimination.

A completely different data integration that does not seek to better visualize multisensor variations is described by Saey et al. [53]. Rather, their goal is to consolidate sensor outputs into a single index that best identifies metals in a WWI battlefield, the previously described fused electromagnetic metal prediction index. It is a linear composite of four EC measurements from a dual-frequency device operating in horizontal and vertical dipole modes.

Binary and Boolean methods

Boolean methods begin with binary maps (composed of 0–1 integer data) typically generated through reclassification of continuous geophysical measurements into an "anomaly" class versus a "background" class, based on thresholds that isolate extreme measurements. At Pueblo Escondido, Ernenwein [121] presents a Boolean union of binary maps derived from GPR slices, MG, and MS that depicts the site's principal anomalies (Fig. 2H). Kvamme [75] found that simply summing the binary maps is more insightful, especially when color coded. With k inputs, the sum varies from $0-k$, with zero indicating undisturbed background (no anomalies), one the Boolean union, and k the intersection. This technique is applied to Ernenwein's [121] binary maps with the result given in Fig. 2I. Ogden et al. [122] interpret the binary sum of anomalies as a "confidence map" at Roman Portus because it portrays the number of geophysical methods that indicate any particular anomaly.

Statistical integrations: Principal components and related methods

The use of PCA to fuse correlated GPR slice maps and repeat ER surveys was discussed earlier, as were problems associated with its application to largely independent prospection data. At Army City, after distributional transformations to reduce skewness and outliers in six geophysical data sets, PCA yields two significant components with eigenvalues greater than unity that together explain 49% of the variance [75]. These components are then subjected to a *varimax* rotation that redistributes the variance to clarify the pattern of loadings (correlations) of the individual variables. The first factor (28.5% of variance) forms a composite principally of MS, MG, ER, and a GPR slice that maps robust building floors, walls, areas of intense burning from the conflagration that consumed the town in 1920, and street gutters into which materials of high MS eventually washed. The second component (20.5% of variance) shows highest loadings in EC and TIR, primarily mapping buried water and sewer pipes and cooler pipe trenches, cellars, and warmer zones corresponding to shallowly buried brick and concrete floors. Case study 4 illustrates standard and local PCA results from this site (Fig. 8J–L).

Statistical integrations: Unsupervised classification

Unsupervised classification procedures offer the promise of objective anomaly assignment into potentially meaningful subsurface classes based on similarities of geophysical responses. A k-means cluster analysis [4] of six geophysical dimensions at Army City yields a number of insights. A $k=2$ class solution divides the region into classes representing "archaeological" anomalies versus background (as determined by subsequent test excavations and detailed analyses). At $k=3$, the previous anomalous class is divided in two, showing "positive" (floors and walls) and "negative" elements (pipelines). The $k=4$ result next divides the background into two classes, apparently based on TIR relationships, one of which seems to correspond with built-up cultural deposits or former garden spaces. Finally, $k=6$ best represents important anomaly classes ranging from brick and concrete floors, to walls, burned features, street gutters, and pipelines [75].

A more detailed, if less insightful, analysis based on k-means clustering is presented by Ernenwein [121] at Pueblo Escondido. The $k=2$ solution also divides the region into anomaly and background classes, but anomalous areas seemed too large and overlapping with regions thought to represent the background. It is not until $k=5$ classes are generated that a class corresponding to clear archaeological features is indicated. At $k=16$, numerous apparently archaeological classes occur (Fig. 2J illustrates results in a portion of the site), and analyses against the Boolean union of all anomalies (described earlier) offer important insights (see Fig. 2H, lower right). Eight clusters clearly partition the background with only 0%–13% overlap with defined archaeological anomalies. Five other clusters match defined anomalies with 100% accuracy, while three agree 31%–60% of the time. While it was hoped that these eight clusters were candidates for meaningful subsurface classes, comparisons against the raw data show they merely represent weak, moderate, and strong anomalies in the three data sets. Similar results are obtained by Ogden et al. [122] using *Isodata* clustering [4] at Roman Portus. Similar to the foregoing, their $k=2$ solution maps all anomalies against undisturbed background, with GPR apparently dominating. At $k=4$, three classes indicate positive anomalies that correspond only with descending orders of anomaly magnitudes, but one class did clearly isolate negative anomalies.

Statistical integrations: Supervised classification

Greater interest has been generated by the possibilities of supervised classification methods [4] for prospection data fusions. They are generated by "training" loci of known class membership (such as anomaly presence or absence or specific classes of archaeological features revealed through excavation). Summary statistics are generated for each class that characterizes multivariate tendencies in geophysical and other responses. With training sites representing samples of targets of interest, the classification functions then map, often probabilistically, other loci with similar characteristics. The resultant mapping not only represents a fusion of geophysical inputs but also provides a prospecting tool for locating potential archaeological features with geophysical and other properties similar to those exhibited by the training sites. Moreover, classification accuracy may be assessed by observing rates of correct predictions at locations of known class membership.

At Pueblo Escondido, Ernenwein's [121] training sites include archaeological anomalies revealed by 74 test excavations. Although the desirability of defining such meaningful archaeological classes as adobe walls, pit houses, middens, patios, and hearths common to such sites is realized, careful consideration suggested those types are best defined by their forms, sizes, and contexts within a settlement rather than specific geophysical responses. Consequently, four *geophysical* training classes apparent in excavations and associated with clear anomalies are defined: "burned in situ" (hearths and other burned features such as structures leaving a remanent magnetic field), "enhanced" (features with high accumulations of MS materials, such as middens), "nonmagnetic" (unburned deposits or features with low MS, including floors and adobe walls), and "background" (locations without anomalous indications). Although other supervised classifiers were evaluated, the Mahalanobis D^2 statistic [124], a multivariate standardized distance, most robustly discriminates the classes among the GPR, MS, and MG dimensions. The resultant mapping clearly reveals known and likely nonmagnetic pit houses and walls, enhanced MS zones (probably middens but including burned structures), and burned hearths and wall segments in correct contexts within pit houses. Moreover, classification accuracy computed on the training sites is excellent with burned features 86% correct, enhanced features 93% correct, nonmagnetic features 85% correct, and background locations 100% correct, pointing to the prospecting potential of this approach.

A supervised classification that focuses on more interpretable archaeological classes is given by Kvamme [125] using six ground and aerial geophysical data sets from historic Army City. Training sites are established in 104 small excavations that define five archaeological classes: concrete/brick, pipes, gutters, "other" class (a catchall for all other cultural deposits), and a background class absent of archaeological features. A multinomial logistic regression classifier [126] is chosen for its generally robust performance. Four linked functions are generated, one for each archaeological class, that offer interpretability through the absolute sizes of their coefficients (the data are standardized). For example, the coefficients show quantitatively that ER is 1.2 times more influential than MS and 3.6 times more important than GPR in defining concrete and brick features, with other data sets contributing negligibly. A logistic transformation permits probability surfaces (based on six-dimensional fusions) for each archaeological class; those for concrete/brick and pipelines in a portion of the town are illustrated in Fig. 2K (Fig. 2C shows the actual EC and ER data). High probability areas offer clear portrayals of locations possessing geophysical properties similar to the training sites and therefore form optimal prospecting maps for similar features. Classification accuracy, based on the training data, is shown to be good with 91% correct for the background, 81% for concrete/brick, 78% for pipes, and 61% for gutters but a low 41% for the "other" class, not surprising due to its mixed content of physical and chemical properties.

Predicting geophysical properties with air and space data

Several researchers have presented an interesting variant of the foregoing. Johnson and Haley [127] (p. 41) observe that

> *Since geophysical survey is generally time-consuming and therefore poorly suited for broad area survey and site discovery, one possibility would be to link geophysical survey data to the generally more broadly available airborne and satellite data. If, for example, you could find broad signatures for buried features in the airborne data, you would be able to target specific locations for intensive geophysical survey.*

This approach was pioneered at the Hollywood mound site, a late prehistoric ceremonial center in western Mississippi [55]. MG data, which identify many well-fired burned houses at this settlement, are reclassified to positive and negative elements (both greater than one standard deviation from the mean) with the remainder neutral background. These classes serve as training sites in several 20 m^2 test blocks centered on prominent archaeological features. It is unclear why negative elements of dipolar features were expected to correlate with vegetation and soil patterns in air and space imagery, however. Aerial and satellite data include 27 bands from the ATLAS and ADAR sensors, historic panchromatic images, a color infrared image, and the Ikonos satellite, ranging from short wavelength visible through TIR. Many of these data sets, flown in different seasons and years, variously show prehistoric mounds and houses through crop markings, even though they are no longer visible on the surface due to intensive leveling and plowing. The data are coregistered to the geophysical grid and resampled to a uniform 0.5 m spatial resolution. A stepwise canonical discriminant function [124] reduces the 27 bands to 18 significant ones. With three classes, two discriminant functions are generated, with the first accounting for 81% of the variance and principal contributions from the blue and red Ikonos bands and the "green" false-color infrared band. Classification of the training sites indicates a good overall correct rate of 66%, and the three-class predictive mapping generally corresponds with actual MG classes. A later study [127] revisits these data, but this time, a two-class MG solution is pursued based only on robust positive anomalies versus background areas. In this case, classification accuracy rises to 74%.

A recent paper expands on this idea. Agapiou et al. [128] employ ground measurements from a handheld spectroradiometer (with response from 350 to 1050 nm) together with a GPR survey of 0.26 ha of the Vészto-Mágor Tell, Hungary. The reflectance values of the spectroradiometer are then recalculated to simulate the GeoEye-1 satellite based on the relative response filters of its sensor, and a normalized difference vegetation index is calculated. NDVI and 10 GPR depth slices (each 20 cm thick to a depth of 2 m) are each classified into 10 classes by *k*-means clustering to reduce computation time. The probability of each NDVI class "matching" GPR classes at each depth is estimated, apparently by the proportion of overlap between it and each of the GPR classes. The outcome forms predictive maps for the presence of GPR anomalies (potential archaeological targets) based on NDVI values. The probability matrix is then applied to a near-contemporaneous high-resolution GeoEye-1 satellite image where NDVI is calculated and also subjected to a 10 class *k*-means classification. The results yield 10 "pseudo-GPR" depth slices. One such slice, representing 0.9–1 m below the surface, is compared with the corresponding actual GPR slice that shows a number of similar anomalies, although differences are also apparent and no indexes of correlation or correct prediction are offered. The authors conclude with a demonstration that pseudo-GPR imagery can be expanded to entire regions based on wide-area GeoEye-1 coverage.

7 CASE STUDIES

The following sections present a series of case studies that illustrate some of the foregoing methods of geophysical data integration. Several also present new directions and possibilities for fusions that may offer new interpretive and procedural advantages when prospection data sets are considered together.

7.1 CASE STUDY 1: FEATURE- AND PIXEL-LEVEL INTEGRATIONS

Historic Cherokee towns in southeast Tennessee and western North Carolina are well documented along tributaries of the Tennessee, Chattahoochee, and Savannah rivers [129]. Very little is known about this time period some 100 km away in northeastern Tennessee. Investigations at several sites in this region began in 2012. The Runion site makes a good example of what can be done with data fusion in the early stages of a project, while questions are still forming, and excavations are being planned. We hypothesize it is an ancestral Cherokee town, based on geophysical surveys and a small collection of artifacts found on the ground surface and eroding out of the river bank.

Six hectares of the site were surveyed with MG in early 2017 using a SENSYS MXPDA, a pushcart system with five fluxgate gradiometers. Results from these data show a small village of square houses measuring 6–7 m on a side, with one four times the size of the others, suggesting that it is a Cherokee town house, a large communal structure (Fig. 3; we refer to this large structure as a town house hereafter but realize this may be incorrect). Large curvilinear features visible in MG may indicate a fortification system that surrounds the village on three sides, while the river would have served as natural protection along the eastern margin. As a means of testing other geophysical methods at the site, one 30×30 m grid was set up over the 11×11 m structure and surveyed with a GSSI SIR4000 GPR unit with a 400 MHz antenna and an EM38-MK2 EMI device. All data were collected in parallel transects spaced 0.5 m apart.

Several questions can be addressed from the geophysical surveys over the large structure. Which methods are most valuable at this site? What type of information does each one provide? What can we learn about the town house from these data? Which methods of data fusion will likely offer the maximum amount of insight moving forward?

The data (Fig. 3) clearly show that MS, GPR, and MG detect the town house. MG (Fig. 3M) shows very strong magnetic features around the perimeter and what is likely a central hearth. Lineations trending N-S are plow marks, which appear to have cut into the structure. MS provides a very different view. The shallower MS component (Fig. 3B) shows negative readings over the main part of the house, while the deeper MS data (Fig. 3C) show high readings in the same general area. This is likely due to a polarity shift, a common occurrence in EM instruments [130,131]. The EC results do not show the town house clearly (Fig. 3D–E), but low conductivity, more pronounced in the shallower component, coincides with a portion of the structure. GPR slices also clearly show the house, with various views of the walls and features in the interior and a second structure to its north.

Preparations for data fusions include inverting the scale of the EC data so that positive values are associated with known features. In addition, the shallow MS data set is inverted to match the deeper component with positive values for the town house. Dipolar anomalies in MG are "fixed" by taking the absolute value of all figures falling below -2 sd.

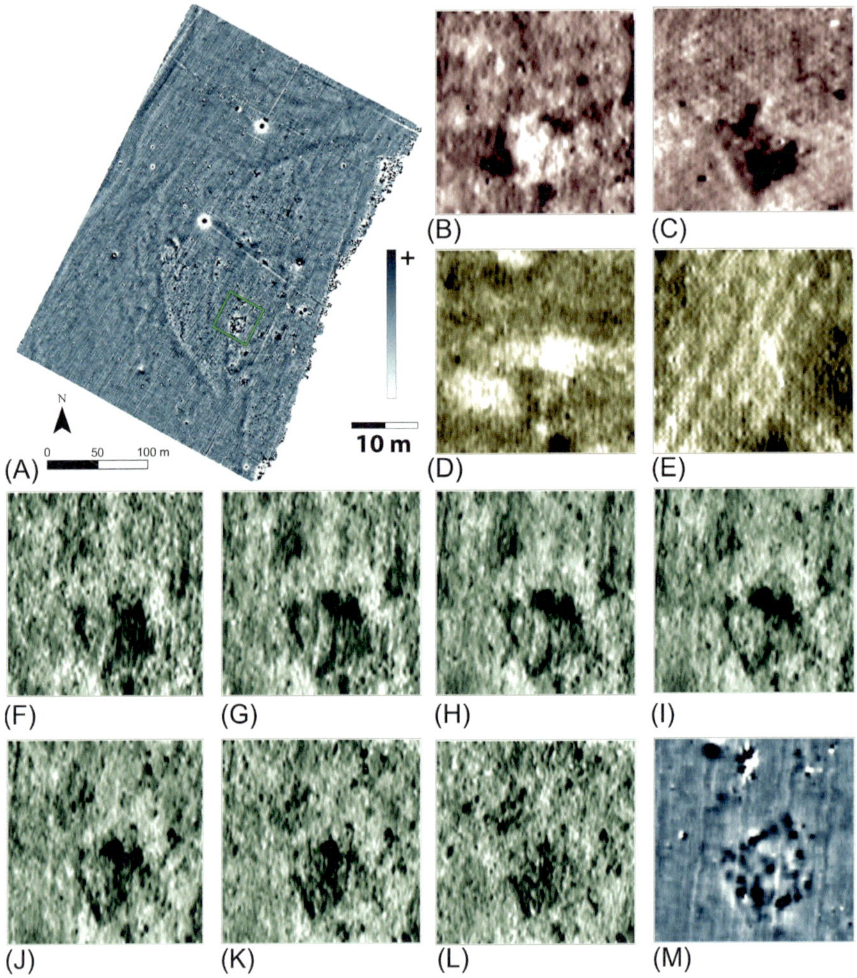

FIG. 3

All data collected to date at Runion: (A) MG with 30 × 30 m test survey area indicated; (B–E) MS at 0.5 and 1 m coil separations followed by EC in the same order; (F–L) GPR slices at the following depths, in cm: 32–35, 35–39, 39–42, 42–46, 46–49, 49–53, and 53–56; and (M) MG extracted from (A).

In order to gain a more complete picture of the town house and surrounding area, all data are reclassified into binary sets using a 1 sd threshold. These sets are then added together to produce a binary sum (Fig. 4A), which clearly indicates both structures. This result may be too heavily influenced by the GPR slices, however, because they well outnumber all other inputs. A weighted binary sum is therefore completed,

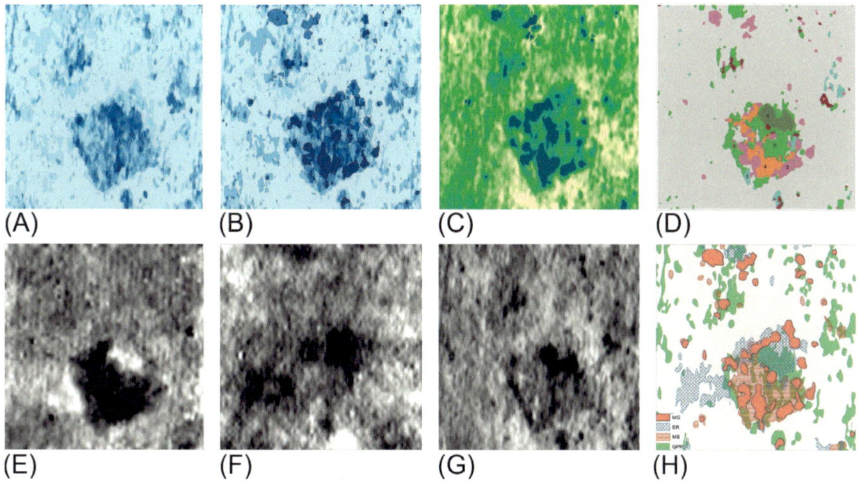

FIG. 4

Feature- and pixel-level fusions of Runion geophysics: (A) sum of all binary inputs; (B) weighted sum of all binary inputs; (C) weighted sum of standardized, continuous inputs; (D) unsupervised classification of all input layers using cluster analysis; (E) mean of both MS components; (F) mean of both EC components, after inversion; (G) mean of all GPR slices; and (H) vector overlay of binary versions of (E), (F), (G), and MG.

giving 25% weight each to GPR, MS, EC, and MG. This gives a slightly more complete view of the entire data set (Fig. 4B). The central hearth revealed by MG is now more visible, and a similar feature in the smaller structure may also point to a hearth. Similarly, a weighted sum of the original, continuous data is also computed (Fig. 4C).

The data are then used as inputs for unsupervised classification where a variety of classifiers are examined with mixed results. The best outcome (Fig. 4D) is given by a histogram peak clustering algorithm that identifies seven classes [132]. At this early stage, it is difficult to interpret these classes, but it is clear that cluster two is related to the central hearth in the town house, and so, other portions of this class may also be burned features. Overall, these classes show that features and fill within the town house tend to be more magnetic, perhaps burned, in the south and west, and the northeast corner is filled with material that is highly resistive and strongly reflective in GPR.

Fusions E–G in Fig. 4 are averages for MS (the shallow MS scale was inverted first), inverted EC (theoretically equivalent to ER), and seven GPR slices. These not only are fusions in their own right but also are used along with the MG data to create a vector overlay (Fig. 4H) through an *objective* approach. All four inputs are reclassified into feature presence/absence categories, and the results are converted to vector polygons. After the background polygon is removed, all polygons with an area $<0.25\,m^2$ are also removed, resulting in "cleaner" versions of the data. Polygon edges are improved by a curve-fitting algorithm that smooths sharp angles. Finally, they are overlaid and given different fill patterns and colors.

Overall, the process of manipulating all input data and exploring different ways to combine them add much to the overall understanding of the site and the information collected thus far. It is clear that EMI and GPR provide valuable information and should continue to be used at the site. These results should be integrated with the existing MG data to create mappings that combine all findings to help distinguish different types of deposits and features. Clearly, the town house has a central hearth and features in the southwest corner that exhibit strongly magnetic properties and in the northeast that are highly resistive. Excavations will likely target these features or others like them, and this information will be used to interpret the rest of the data. They may also be employed as training sites for subsequent supervised classifications.

7.2 CASE STUDY 2: POINT CLOUD FUSION

7.2.1 Introduction and background

Two medieval tunnels, known as Souterrain 1 and Souterrain 2, lie just below the surface at Maison de la Châtaigne (Chestnut House), in the village of Mourjou, France. Their branched subterranean morphology includes niches and silos characteristic of the medieval age, likely from the 11th to 14th centuries [133] (pp. 62–64). The tunnels, carved into highly weathered granite, are entered through collapsed areas on the surface, which permitted traditional mapping in 1998, but today, Souterrain 2 is closed for safety concerns. Souterrain 1 remains open and is occasionally visited by tourists. Aerial photogrammetry, terrestrial laser scanning, and GPR are combined to map the tunnels and surrounding landscape in 3D, determine their spatial relationships, and search for additional passageways and original entrances.

7.2.2 Methods

In January 2016, a terrestrial laser scanning system documented the accessible interior of Souterrain 1. The following summer, over 500 overlapping digital aerial photographs covering the entire property (about 1 ha) were acquired using a DJI Phantom 3 Advanced quadcopter. Additional photographs of areas under tree canopy were taken from the ground. Twelve ground control points were placed throughout the survey area with a RTK GNSS system. Agisoft *PhotoScan Professional* software was used to align 368 photographs and produce a point cloud of the ground surface and aboveground features, a bare-earth DEM, and a digital orthophoto.

GPR surveys conducted in 2015–16 employed a Geophysical Survey Systems, Inc. SIR4000 with 200 and 270 MHz antennae. Transects were spaced 1 m apart and covered 0.28 ha. GPR data processing included DC drift compensation, range gains, band-pass filters, and background removal. A ground velocity of $0.09\,\mathrm{m\,ns^{-1}}$ was estimated through hyperbola fitting, and the radargrams were terrain-corrected by the DEM. Fully processed data volumes of each GPR block were then exported as

ASCII point clouds as were terrestrial laser scans acquired from inside the tunnel and the aboveground photogrammetric point cloud. In other words, all point cloud data were combined into a single 3D environment using *CloudCompare* (http://www.danielgm.net/cc/), a free and open-source software.

7.2.3 Results

Although the GPR surveys were certainly conducted over the tunnels, Souterrain 1 was not initially visible in the data. The integration of terrain, GPR, and subterranean laser scans of Souterrain 1 within the same 3D environment allows measurements to be made between features of interest in all three data sets, helping us to focus on where to look. Although the GPR time window is sufficient to record reflections to approximately 5 m below surface, attenuation limits depth to about 1.5 m. The subterranean laser scan reveals that the shallowest portion of Souterrain 1 lies 2.5 m below the surface, beyond effective GPR penetration, so it cannot be detected. Souterrain 1 likely continues under the adjacent house captured in the surface photogrammetric scan (Fig. 5A). Souterrain 2, however, is shallower and is confirmed by GPR (Fig. 5B–D). It likely extends beyond the survey area. No other tunnels or medieval entrances are found.

The fusion of 3D GPR data with point clouds of above- and belowground features proves very effective for visualization and interpretation of the tunnels at Maison de la Châtaigne. The ability to see the spatial relationship between each data set and directly measure distances between features is especially critical to the understanding of why Souterrain 1 is not visible and facilitates visualizing Souterrain 2. Communicating these results as static, 2D figures remains a challenge.

7.3 CASE STUDY 3: AUTOMATIC FEATURE RECOGNITION

As geophysical surveys become progressively larger [16,134], it is increasingly necessary to consider and implement methods for automatic feature recognition, simply because it becomes too difficult and time-consuming for human interpreters to examine and analyze the tens of thousands of anomalies commonly revealed in large surveys. Such undertakings have been implemented in archaeological applications of aerial and satellite remote sensing [109–111], but few can be found in ground-based geophysics (however, see Ref. [112]).

The following exercise examines the problem of automatic identification of archaeological features of a well-understood Great Plains village in North Dakota, based on the integration of information from a DEM and MG survey and using relatively simple GIS operations. The site is Huff village, ancestral to the Mandan tribe, which dates to the mid-15th century. It contains earthen-walled houses of rounded rectangular shape surrounding a central plaza with an adjacent large ceremonial house, all known from subtle depressions in the surface and archaeological excavations described by Wood [135]. The 5 ha village is surrounded on three sides by a fortification ditch containing uniformly spaced bastions, with the remaining side

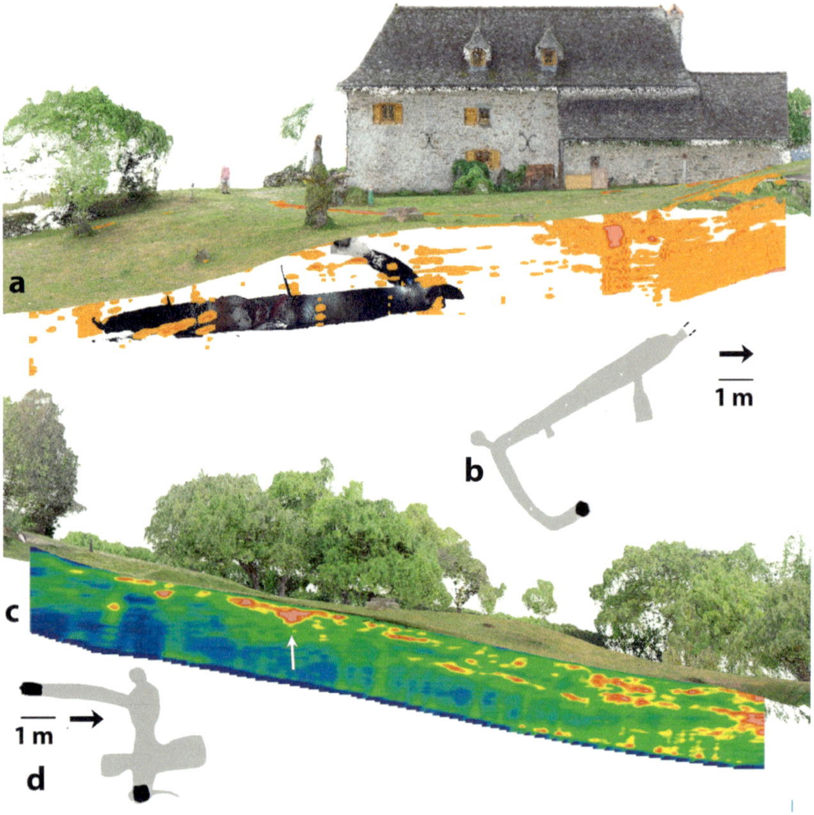

FIG. 5

Integrated data at Maison de la Châtaigne: (A) Souterrain 1 laser scan integrated with surface point cloud and GPR displayed at >75% reflection amplitude; (B) plan view Souterrain 1; (C) GPR and surface point clouds showing Souterrain 2 in cross section, indicated by arrow; and (D) Souterrain 2 plan view map.

adjacent to the Missouri River. The DEM was obtained by robotic total station with subcentimeter accuracy [136] and an average point spacing of 0.35 m; interpolation through inverse distance weighting generated a raster with 0.25 m spatial resolution. MG was acquired with a Bartington 601 dual fluxgate gradiometer with 0.125 m sample spacing in transects separated by 0.5 m. These data are resampled to match the DEM's spatial resolution.

Most of the archaeological features detectable through prospection in Northern Plains villages may be identified in these data sets. Imbedded within the DEM's surface is information about house locations, their shapes and sizes, and the defensive ditch, bastions, borrow pits, and central plaza. MG holds information pointing to the locations of hearths and ubiquitous 1.5–2 m deep pits employed for corn storage, which tend to be highly recognizable by MG because they are filled with magnetically enriched settlement soils after abandonment. A portion of the village showing houses, hearths, and storage pits is illustrated in Fig. 2B.

7.3.1 Preprocessing

Huff exhibits a uniform slope that drops about 4.5 m across the village's width, which is eliminated through a 7.5 m radius high-pass filter that "flattens" the landscape. Surface depressions (houses, ditches, and bastions) thereby become negative in value with positive results representing raised berms that once formed walls surrounding houses or excavated earth from the defensive ditch, all centered about zero-valued "neutral" ground (Fig. 6A). Small idiosyncrasies in this surface from animal burrows, a few looters' holes, and unfilled early-20th-century archaeological excavations are reduced through repeated application of an adaptive box filter (replacing deviant values with local means) and application of a narrow-radius (0.5 m) low-pass filter [4]. In the MG data (Fig. 6B), iron objects that litter the site are deposited by hundreds of visiting tourists each year. The robust dipolar anomalies they cause represent a form of noise that is reduced as follows: (1) Large negative poles (<4 nT) are isolated, (2) large positive poles (>4 nT) are next defined, (3) a distance buffer of small radius (<1.5 m) is generated around the negative poles, (4) positive poles within that buffer are joined with negative ones through a Boolean union, and (5) the resultant dipolar areas are set to zero, the approximate MG mean.

7.3.2 Recognizing houses

Significant depressions in the flattened DEM below a threshold of −0.06 m define 171 *negative* polygons, which include houses, ditches, borrow pits, and remnants (after smoothing) of animal dens and looters' holes. A *compactness ratio* is calculated for each: $C=(A_p/A_c)^{1/2}$, where A_p is the area of a polygon and A_c is the area of a circle having the same perimeter as the polygon [124] (p. 356; Fig. 6C). Compactness ratios between 0.65 and 0.9 are taken to represent the range of rounded rectangular house shapes with those beyond that range eliminated. In addition, with house depressions typically measuring 6×10 m (about 60 m^2), polygons less than half and those more than twice that area are eliminated. Although these steps leave 119 house candidates, many false positives are evident owing to other depressions, likely from soil borrowing (Fig. 6D). Raised berms that surround houses (but not borrows) next receive focus. They are composed of sediments excavated from floors and former wall materials. Areas in the flattened DEM above 0.04 m are defined as *positive* polygons (Fig. 6E, dark polygons). Spatial proximity is then employed to ultimately define houses as those polygons with significant berm areas (>20 m^2) within a radius of 4 m, which yields 90 potential houses (Fig. 6F). Three lack the central hearths required of all houses (revealed by MG, see below), but two are known to be houses with hearths removed by excavations [135]. The single false positive therefore represents a borrow pit (arrow, Fig. 6G).

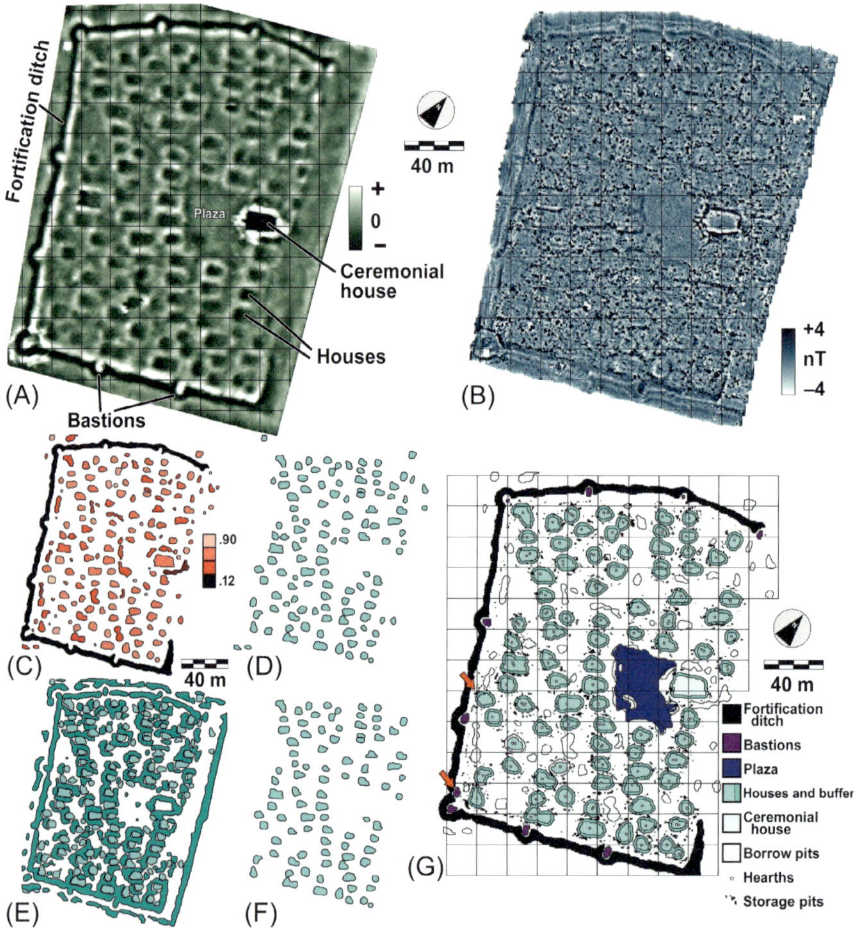

FIG. 6

Data and processing results for feature extraction at Huff village: (A) DEM after trend removal, (B) MG data, (C) negative polygons from DEM coded by compactness ratio *C*, (d) $C > 0.65$ and $30\,m^2 < area < 120\,m^2$, (E) positive polygons from DEM superimposed on (D), (F) negative polygons with $>20\,m^2$ of positive polygon in (E) within 4 m, and (G) final features recognized for all classes.

7.3.3 Recognizing the ceremonial house

Each Mandan village contains a larger structure for certain ceremonies, termed a "ceremonial house" [135]. This feature is identified by returning to the negative polygons (Fig. 6C) and simply selecting the largest one with a compactness ratio >0.65 (Fig. 6G). It measures approximately $220\,m^2$.

7.3.4 Recognizing the fortification ditch

This feature is defined as the negative polygon (Fig. 6C) with the lowest compactness ratio ($C=0.118$, due to its great linearity) or the largest area (2967 m^2). The ditch surrounds the village, and well outlines each bastion, important for their definition below (Fig. 6G).

7.3.5 Recognizing borrow pits

Earth borrowing was a major activity in Plains villages because the sides of the houses were formed of built-up earth; in later periods, the roofs of *earth lodges* were also earth covered [135]. The negative polygons (Fig. 6C) *minus* the fortification ditch, houses, and polygons $<6\,m^2$ (to eliminate small looters' holes or animal dens) represent likeliest candidates (Fig. 6G).

7.3.6 Recognizing bastions

Bastions are elevated above the surrounding ground, are oval-shaped, measure about 6×7 m, and are surrounded on three sides by the fortification ditch (Fig. 6A). The complexity of this feature requires template matching [110]. A donut template is applied with a 7 m diameter hole coded as +1 and surrounding ring of 4 m width as -1 (Fig. 7A). When matched over a bastion in the DEM and cross multiplied, the central region yields a high positive product (owing to positive elevations), and the negative ring further increases the sum (from the negative elevations in the ditch). This template is centered in turn over each cell of the DEM causing high sums over bastions but unfortunately in other places as well due to high points associated with berms near the ditch and houses (Fig. 7B). After it is reclassified to indicate highest values (dark polygons, Fig. 7C), those with high proportions of overlap within a 1.5 m buffer of the fortification ditch (light polygon, Fig. 7C) are retained. The intersection of compactness ratios >0.65, areas between 10 and 40 m^2, and positive elevations >0.15 m identify all nine bastions successfully but yield one false positive owing to an unusual high point resulting from a back dirt pile associated with an early excavation (lower arrow, Fig. 6G).

7.3.7 Recognizing the plaza

The chief characteristic of a Mandan plaza is its flatness and the lack of cultural constructions (easily recognized in Fig. 6A and B). A 4.5 m radius *texture* filter is applied to the DEM that computes a local standard deviation of elevations. Surface depressions and berms generate high variation, while level and featureless spaces indicate low variation. Areas below a threshold of 0.03 sd are identified; the largest homogeneous region must signify the plaza in a densely occupied village. This finding is easily corroborated because plazas are located centrally and adjacent to the ceremonial house, characteristics that this polygon meets (Fig. 6G).

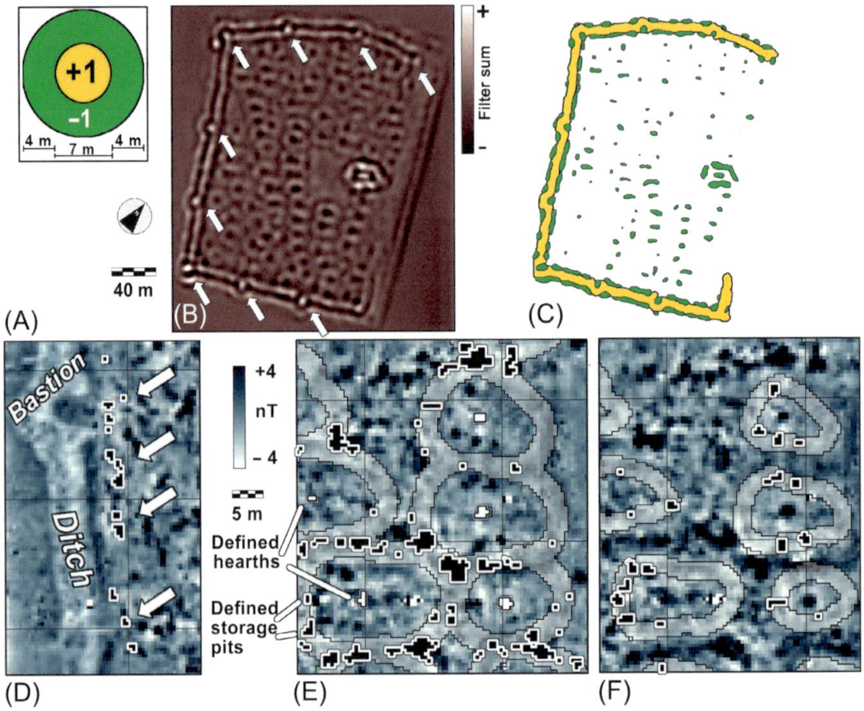

FIG. 7

Methods and results in defining bastions and storage pits: (A) donut filter applied to DEM, (B) result from donut filter with high values showing filter correspondences, (C) highest filter values as small dark polygons and fortification ditch, (D) storage pits near fortification ditch, (E) storage pits surrounding house exterior perimeters, and (F) storage pits near house interior perimeters.

7.3.8 Recognizing hearths

The identification of hearths in villages can be difficult because their associated MG anomalies tend to appear similar in size (1–2 m in diameter), circular form, and magnitude to those associated with storage pits. Hearths always occur centrally within houses, however, and statistical analyses of magnetic field strength consistently show them to be more magnetic than storage pits [137], although their distributions greatly overlap. MG anomalies larger than 3 nT include nearly all hearths. A compactness ratio of $C < 0.65$ permits elimination of those far from circular, and anomalies larger than 6 m^2 are also eliminated as poor candidates (e.g., burned walls and berms with high MS). Spatial context is next considered by computing center points of potential houses (defined via the DEM) and isolating the remaining MG anomalies within 1.5 m of their centers as hearths (Fig. 6G). Results indicate at least one hearth in each defined house (except for the three discussed earlier), although some have several, which agrees with archaeological exposures of auxiliary hearths [135].

7.3.9 Recognizing storage pits

Following foregoing procedures, MG anomalies from 2 to 12 nT (characteristic of their range) are isolated and then selected according to compactness ratios of $C > 0.4$ and areas $< 9\,\mathrm{m}^2$, forming a reduced subset. Spatial context is then employed to zero in on pit locations based on known archaeological patterns. A *fortification ditch pattern* includes 160 pits within 7 m of the interior edge of the ditch (Fig. 7D). A *house exterior pattern* lies adjacent to houses on their outside perimeters. However, surface erosion has caused houses defined by the DEM to be somewhat small compared with their actual floor plans, as revealed by excavations [135]. House polygons are therefore expanded, through buffering, by 2 m. A 3 m buffer is then extended beyond the revised house polygons and intersected with the magnetic subset to define 921 storage pits within these regions (Fig. 7D). Finally, a *house interior pattern* is similar but occurs within houses along perimeter spaces. Utilizing the expanded 2 m perimeter, buffers of 2.5 m are generated on the inside edges of houses and intersected with the magnetic subset to define 373 interior storage pit candidates (Fig. 7D). In total, 1454 subterranean storage pits are recognized within Huff village (Fig. 6G).

7.4 CASE STUDY 4: EXPLORING LOCAL STATISTICS

One recurring issue, much discussed earlier, has been the notion of weak correlation between geophysical data sets. Yet, is this really the case? Here, we revisit early 20th-century Army City, a data set offering six geophysical views of the town: EC, ER, GPR, MG, MS, and TIR. The data were originally acquired at various spatial resolutions and subjected to various normalizing and other transformations (e.g., absolute values of MG were taken to eliminate dipolar anomalies [75]); they are here resampled to a uniform 0.5 m, and only 60×80 m subsets are illustrated for greater detail (Fig. 8A–F). A Pearsonian correlation matrix yields the largest absolute correlation just below $r = 0.3$ (Table 1a), pointing to expected weak relationships, and TIR appears completely independent of the others. However, visual comparisons between the images seem to defy this conclusion with a plethora of obvious and clear correspondences between them. This contradiction is a consequence of applying conventional statistics in spatial contexts. Here, r represents a global or overall correlation between all the 19,200 cells (pixels) in each full image—more or less an average association. Yet, locally, it is obvious that image correlations exist and vary widely. This is the idea behind *local statistics* that permit computation of statistical indexes, regressions, and other models within neighborhoods of various sizes, allowing more nuanced understandings of spatial relationships [138].

In order to more fully explore the nature of correlation between the geophysical data sets, neighborhoods of several sizes are defined, with radii of 0.75 m (including 9 cells), 1.2 m (21 cells), and 1.6 m (37 cells). Pearson's r is calculated within

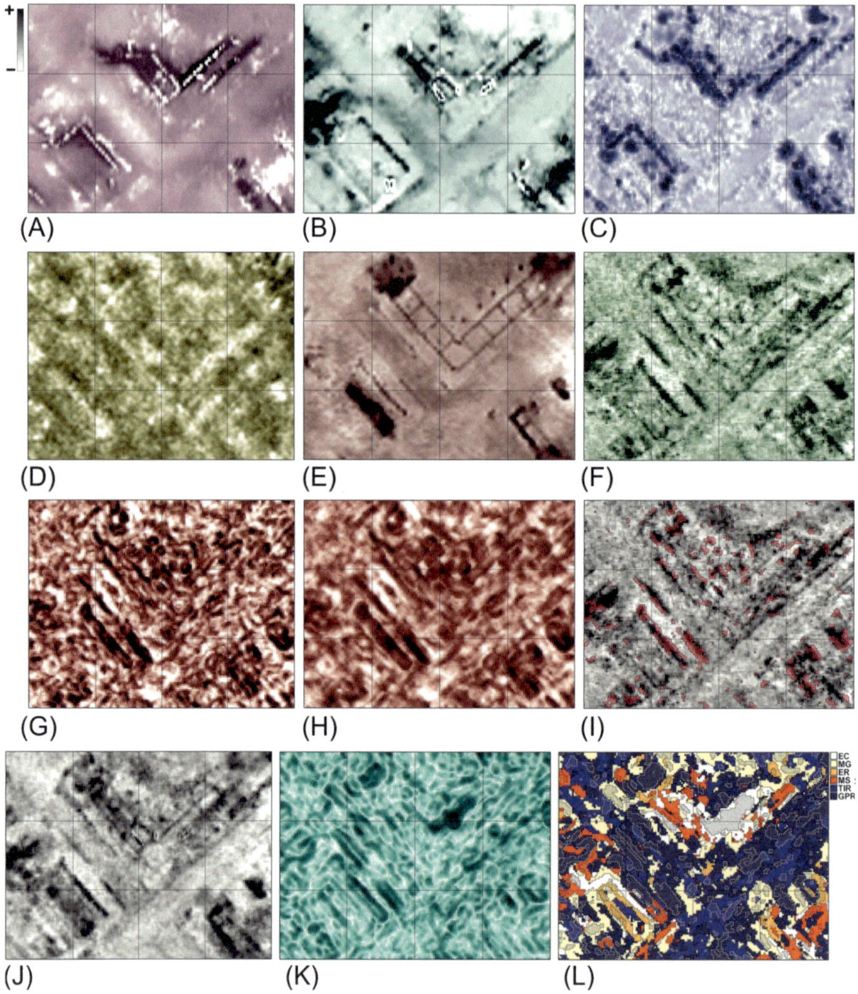

FIG. 8

Geophysical data and local statistical results from Army City: (A) EC, (B) MS, (C) MG, (D) TIR, (E) ER, (F) GPR slice, (G) local *r* between ER-GPR in 1.2 m radius, (H) local *r* between ER-GPR in 1.6 m radius, (I) high ER correlations ($r > 0.7$) in *black* diagonal shading atop GPR, (J) global principal component 1, (K) proportion of total variance of eigenvalue 1 for local PCA, and (L) geophysical method showing highest loading in first local PC with eigenvalues > 0.7 in *white* or *black* diagonal shading. *Black* is high in all *gray* or monochrome scales, grid squares measure 20 m, and north is to the right.

Table 1 Pearsonian Correlation Data for Army City Geophysical Types: (a) Global Correlation Matrix and (b) Local Correlation Data Between ER and GPR for Various Size Neighborhoods

	EC	ER	GPR	MG	MS	TIR
(a)						
EC	1					
ER	0.025	1				
GPR	−0.138	0.281	1			
MG	0.239	0.270	0.207	1		
MS	−0.144	−0.065	0.295	0.199	1	
TIR	−0.049	−0.0117	−0.108	−0.024	−0.031	1

	Minimum	Maximum	Average
(b)			
0.9m radius (9 cells)	−0.983	0.990	0.141
1.2m radius (21 cells)	−0.930	0.969	0.169
1.6m radius (37 cells)	−0.917	0.946	0.187

neighborhoods centered on each cell of the database. To illustrate, the ER and GPR data (global $r = 0.281$; Fig. 8E and F) visually reveal many corresponding anomalies where local correlations should be high and equally important loci of *negative* correlation where one data set illustrates anomalies absent in the other. Local r demonstrates that strong positive and negative correlations do indeed exist between ER and GPR and variation is enormous from extremely negative to highly positive (Table 1b). Mappings for the two larger radii (Fig. 8G and H) illustrate strong pattern and reveal where the greatest relationships of both types exist. Areas of high correlation ($r > 0.7$) are superimposed over the GPR depth slice in Fig. 8I to show its positive correspondences with ER (the lack of space prevents other such presentations).

The multivariate nature of high-dimensional relationships is best explored through PCA, but can a local PCA [139] illustrate more nuanced connections? A standardized global PCA for these data yields a first component with an eigenvalue (variance) of 1.645, accounting for 27.4% of the variance in the database, relatively small numbers because the global correlations are so low (Table 1a). Component 1 (C1) is a linear composite primarily of ER, GPR, and MG, indicated by loadings (0.597, 0.736, and 0.655, respectively) that account for 81% of its variance (Fig. 8J; interestingly, C1 is nearly identical to the simple arithmetic summation of these inputs, plus MS, illustrated in Fig. 2F). A local PCA using a 1.6m radius computes a six-dimensional PCA in the neighborhood of each cell. Here, too, enormously variable results are obtained, only a fraction of which may be reported here. The eigenvalues of C1, for example, are found to account for between 31.1% and 95.6% of the total variance (average, 60.2%), depending on the location (Fig. 8K). In fact, about 21.4% of the region holds C1 eigenvalues accounting for 70% or more of the variance, suggesting that significant areas are inherently one-dimensional. In the context of prospection, this means domination by a small cadre of closely related methods or perhaps only a single one. To investigate this further, the geophysical method that loads most highly on C1 in each cell (and thus its principal driver) is determined and mapped in Fig. 8L. GPR clearly dominates (40.4% of region), followed by TIR (19.0%) and MG (18.2%). ER leads the least (4.0%), with EC and MS also low (both 9.2%). These percentages seem to correspond with the regions of robust anomalies in the primary data mappings (Fig. 8A–F). It is also apparent by comparing Fig. 8L with the basic data that distinct and characteristic areas are dominated by each of the various methods—EC where its anomalies are most robust, MG where it stands out, and so on. Moreover, zones where geophysical responses are truly most "one-dimensional" may be indicated by overlaying areas with eigenvalues >70% atop the prevalent geophysical technique (Fig. 8L).

8 CONCLUSIONS

In this review, the full breadth of data integration pursuits applied to geophysical data has been examined. They range from integrations within one to several geophysical data sets and between geophysical results and other forms of prospection. While the majority are clearly 2D integrations, examples of 1D and 3D pursuits are also noted. Numerous methods of integration are also examined, from feature level to pixel level. While the former, at present, is mostly subjectively defined through manually drawn vectors interpreted from geophysical imagery, computer-driven object-based approaches may soon be more common as a number of examples and a case study illustrate. Pixel-level integrations are numerous, from approaches based on computer graphic composites to a host of mathematical, Boolean, and statistical fusions. In all of this work, one thing has become clear. Interpretations seem to improve as more data sources are consulted and combined in one place.

Many fusion studies illustrate only a goal of showing improved mapping results as guides for future excavations. Other papers are more introspective and attempt to combine knowledge from multimethod surveys to deduce more about specific characteristics of the subsurface. For example, an EC survey might locate low conductivity zones where GPR penetration might be more fruitful. Others combine anomalous indications to make a more complete picture of the subsurface, for example, where certain houses are shown by GPR, while others are revealed by MG. In some instances, data fusions are necessary and integral to data handling and interpretation, such as the critical role elevation data fill for the terrain correction of GPR surveys. Increasingly, as surveys become larger, thoughtful interpretations and comparisons between data sets become impossible owing to large data volumes and thousands of anomalies, making the purpose of multimethod surveys one of producing large-scale subsurface mappings as a basis for site management; particular features and anomalies might be examined and interpreted only later as the need arises.

Commenting on the variety of bewildering pursuits to data integration, Gaffney [46] (p. 329) observes that "some caution must be noted, because although the theoretical benefits are obvious, considerable thought must go into the research design to ensure that each data strand is appropriate for the study, both in interpretative capacity and data quality/density—the worry is that techniques may be 'collected' like stamps, rather than justified from an archaeological perspective." Hesse [37] offers a similar view based on long experience, warning particularly that one should carefully choose appropriate methods for an area of study rather than "a blind systematic use of all the available instruments." We not only echo this view but also realize the wide variety of goals that drive archaeological prospection and data fusion. One should always choose methods with care, but often, there are so many variables impacting outcomes that it is wise to apply methods liberally. Interpretation and fusion of these data give the opportunity to be selective of the results (as in Case study 1). In addition, data integration has many disparate goals. On the one hand, there are those who only want to generate an image that best portrays all possible knowledge of the subsurface of a particular site. They might employ all means of mathematical manipulations, image processing tricks, and computer graphic compositing to achieve those ends. On the other hand, there is the practicing geophysicist who wishes an accurate, unmanipulated view of the subsurface, as close to the raw data as possible, and who views any modifications that alter the data as corruptions of the basic underlying science. Happily, the world of archaeogeophysics now seems large enough to accommodate both points of view.

ACKNOWLEDGMENTS

Original data collection at Huff village was supported by the State Historical Society of North Dakota. Data acquisition at Army City was made possible by a Grant from the Strategic Environmental Research and Development Program, US Department of Defense. SENSYS GmbH kindly provided the MXPDA 5-Channel Magnetometer System for the MG survey at Runion. Funds to support the work at Maison de la Châtaigne were provided by Université Blaise Pascal—Clermont-Ferrand, France. Local statistics in Case study 4 were calculated using the *GWModel* and sp. packages in R.

REFERENCES

[1] C. Pohl, J.L. Van Genderen, Multisensor image fusion in remote sensing: concepts, methods and applications, Int. J. Remote Sens. 19 (1998) 823–854.

[2] A. Halevy, A. Rajaraman, J. Ordille, Data integration: the teenage years, Proceedings of the 32nd International Conference on Very Large Databases, 2006, pp. 9–16.

[3] M. Lenzerini, Data integration: a theoretical perspective, Proceedings of the 21st ACM SIGACT SIGMOD SIGART Symposium on Principles of Database Systems, ACM Press, 2002, pp. 233–246.
[4] T.M. Lillesand, R.W. Keifer, J.W. Chipman, Remote Sensing and Image Interpretation, seventh ed., John Wiley & Sons, Hoboken, NJ, 2015.
[5] R. Rogers, L. Wood, The history and status of merging multiple sensor data—an overview, 1990 ACSM-ASPRS Annual Convention, Denver, CO, Technical Papers, Vol. 4—Image Processing/Remote Sensing, 1990, pp. 352–360.
[6] J. Zhang, Multi-source remote sensing data fusion: status and trends, Int. J. Image Data Fusion 1 (2010) 5–24.
[7] R. Lasaponara, N. Masini, Pan-sharpening techniques to enhance archaeological marks: an overview, in: Satellite Remote Sensing: A New Tool for Archaeology, Springer, London, 2012, pp. 87–109.
[8] T. Blaschke, S. Lang, G. Hay (Eds.), Object-Based Image Analysis: Spatial Concepts for Knowledge-Driven Remote Sensing Applications, Springer, New York, 2008.
[9] M. Van Leusen, Archaeological data integration, in: D. Brothwell, A. Pollard (Eds.), Handbook of Archaeological Sciences, Wiley, New York, 2001, pp. 575–583.
[10] D.C. Comer, M.J. Harrower (Eds.), Mapping Archaeological Landscapes From Space, Springer, New York, 2013.
[11] R. Lasaponara, N. Masini (Eds.), Satellite Remote Sensing: A New Tool for Archaeology, Springer, London and New York, 2012.
[12] S. Piro, P. Mauriello, F. Cammarano, Quantitative integration of geophysical methods for archaeological prospection, Archaeol. Prospect. 7 (2000) 203–213.
[13] M.G. Drahor, G.N. Tsokas, S. Piro, I. Trinks, Foreword, Surf. Geophys. 13 (2015) 519–521.
[14] R.A. Dalan, A geophysical approach to buried site detection using down-hole susceptibility and soil magnetic techniques, Archaeol. Prospect. 13 (2006) 182–206.
[15] N. Linford, P. Linford, A. Payne, Chasing aeroplanes: developing a vehicle-towed caesium magnetometer array to complement aerial photography over three recently surveyed sites in the UK, Surf. Geophys. 13 (2015) 623–631.
[16] P. Schneidhofer, E. Nau, A. Hinterleitner, A. Lugmayr, J. Bill, T. Gansum, K. Paasche, S. Seren, W. Neubauer, E. Draganits, I. Trinks, Palaeoenvironmental analysis of large-scale, high-resolution GPR and magnetometry data sets: the Viking Age site of Gokstad in Norway, Archaeol. Anthropol. Sci. (2016) 1187–1213.
[17] A. Schmidt, Earth Resistance for Archaeologists, Geophysical Methods for Archaeology, AltaMira Press, Lanham, 2013.
[18] A. Aspinall, C. Gaffney, A. Schmidt, Magnetometry for Archaeologists, Geophysical Methods for Archaeology, AltaMira Press, Lanham, 2008.
[19] A.J. Witten, Handbook of Geophysics and Archaeology, Routledge, London and New York, 2006.
[20] L.B. Conyers, Ground-Penetrating Radar for Archaeology, third ed., AltaMira Press, Lanham, 2013.
[21] D. Goodman, Y. Nishimura, J.D. Rogers, GPR time slices in archaeological prospection, Archaeol. Prospect. 2 (1995) 84–89.
[22] D. Goodman, S. Piro, GPR Remote Sensing in Archaeology, Springer, Heidelberg, 2013.

[23] N. Linford, The application of geophysical methods to archaeological prospection, Rep. Prog. Phys. 69 (2006) 2205–2257.

[24] I. Scollar, A. Tabbagh, A. Hesse, I. Herzog, Archaeological Prospecting and Remote-Sensing, Topics in Remote Sensing, Cambridge University Press, Cambridge, New York, Port Chester, Melbourne, Sydney, 1990.

[25] S. Delefortrie, P. De Smedt, T. Saey, E. Van De Vijver, M. Van Meirvenne, An efficient calibration procedure for correction of drift in EMI survey data, J. Appl. Geophys. 110 (2014) 115–125.

[26] M. Ciminale, M. Loddo, Aspects of magnetic data processing, Archaeol. Prospect. 8 (2001) 239–246.

[27] N.G. Papadopoulos, P. Tsourlos, C.B. Papazachos, G.N. Tsokas, A. Sarris, J.H. Kim, An algorithm for fast 3D inversion of surface electrical resistivity tomography data: application on imaging buried antiquities, Geophys. Prospect. 59 (2011) 557–575.

[28] J. Gator, C. Gaffney, Revealing the Buried Past, Tempus, Stroud, (2003).

[29] J.W. Weymouth, Geophysical methods of archaeological site surveying, in: M. B. Schiffer (Ed.), Advances in Archaeological Method and Theory, Academic Press, New York, 1986, pp. 311–395.

[30] R.B. Clay, Complementary geophysical survey techniques: why two ways are always better than one, Southeast. Archaeol. 20 (2001) 31–43.

[31] S. Campana, S. Piro, Putting everything together: GIS-based data integration and interpretation, in: S. Campana, S. Piro (Eds.), Seeing the Unseen: Geophysics and Landscape Archaeology, CRC Press, Boca Raton, 2009, pp. 325–330.

[32] V.M. van der Haas, P. Schut (Eds.), The Valletta Convention, Archaeolingua, Budapest, 2014.

[33] M. Doneus, W. Neubauer, 2D combination of prospection data, Archaeol. Prospect. 5 (1998) 29–56.

[34] K.L. Kvamme, Multidimensional prospecting in North American Great Plains village sites, Archaeol. Prospect. 10 (2003) 131–142.

[35] K.L. Kvamme, Remote sensing: archaeological reasoning through physical principles and pattern recognition, in: A.P. Sullivan (Ed.), Archaeological Concepts for the Study of the Cultural Past, University of Utah Press, Salt Lake City, 2008, pp. 65–84.

[36] K. Lockyear, E. Shlasko, Under the park. Recent geophysical surveys at Verulamium (St Albans, Hertfordshire, UK), Archaeol. Prospect. 24 (2017) 17–36.

[37] A. Hesse, Multi-parametric survey for archaeology: how and why, or how and why not? J. Appl. Geophys. 41 (1999) 157–168.

[38] A. Novo, M. Dabas, G. Morelli, The STREAM X multichannel GPR system: first test at Vieil-Evreux (France) and comparison with other geophysical data, Archaeol. Prospect. 19 (2012) 179–189.

[39] D. Goodman, Y. Nishimura, H. Hongo, N. Higashi, Correcting for topography and the tilt of ground-penetrating radar antennae, Archaeol. Prospect. 13 (2006) 157–161.

[40] M.V. Bongiovanni, M. de la Vega, N. Bonomo, Contribution of the resistivity method to characterize mud walls in a very dry region and comparison with GPR, J. Archaeol. Sci. 38 (2011) 2243–2250.

[41] M.L. Hargrave, Ground truthing the results of geophysical surveys, in: J.K. Johnson (Ed.), Remote Sensing in Archaeology: An Explicity North American Perspective, University of Alabama Press, Tuscaloosa, 2006, pp. 269–304.

[42] J. Conolly, M. Lake, Geographical Information Systems in Archaeology, Cambridge University Press, Cambridge, 2006.
[43] S. Campana, Archaeological site detection and mapping: some thoughts on differing scales of detail and archaeological "non-visibility", in: S. Campana, S. Piro (Eds.), Seeing the Unseen: Geophysics and Landscape Archaeology, CRC Press, Boca Raton, 2009, pp. 5–26.
[44] A. Asăndulesei, Inside a Cucuteni settlement: remote sensing techniques for documenting an unexplored eneolithic site from northeastern Romania, Remote Sens. 9 (2017) 41.
[45] N. Linford, From hypocaust to hyperbola: ground-penetrating radar surveys over mainly roman remains in the UK, Archaeol. Prospect. 11 (2004) 237–246.
[46] C. Gaffney, Detecting trends in the prediction of the buried past: a review of geophysical techniques in archaeology, Archaeometry 50 (2008) 313–336.
[47] L.B. Conyers, Analysis and interpretation of GPR datasets for integrated archaeological mapping, Surf. Geophys. 13 (2015) 645–651.
[48] L.B. Conyers, J. Leckebusch, Geophysical archaeology research agendas for the future: some ground-penetrating radar examples, Archaeol. Prospect. 17 (2010) 117–123.
[49] I. Trinks, W. Neubauer, A. Hinterleitner, First high-resolution GPR and magnetic archaeological prospection at the Viking age settlement of Birka in Sweden, Archaeol. Prospect. 21 (2014) 185–199.
[50] D. Goodman, H. Hiromichi, N. Higashi, Y. Nishimura, M. Tokuda, H.d. Oh, The application of GPR Overlay analysis in archaeological prospection: discovery at the Japanese Imperial Family tombs in Miyazaki Prefecture, Arch. Sci. 1 (2009) 295–297.
[51] K.L. Kvamme, Archaeological prospecting at the Double Ditch State Historic Site, North Dakota, USA, Archaeol. Prospect. 15 (2008) 62–79.
[52] A. Schmidt, P. Linford, N. Linford, A. David, C. Gaffney, A. Sarris, J. Fassbinder, EAC guidelines for the use of geophysics in archaeology: questions to ask and points to consider, EAC Guidel. 2 (2015) 136.
[53] T. Saey, B. Stichelbaut, J. Bourgeois, V. Van Eetvelde, M. Van Meirvenne, An interdisciplinary non-invasive approach to landscape archaeology of the Great War, Archaeol. Prospect. 20 (2013) 39–44.
[54] M. Ciminale, D. Gallo, R. Lasaponara, N. Masini, A multiscale approach for reconstructing archaeological landscapes: applications in Northern Apulia (Italy), Archaeol. Prospect. 16 (2009) 143–153.
[55] J.K. Johnson, B.S. Haley, Using geophysical results in a multivariate exploration of airborne imagery at the Hollywood Mounds, Mississippi, in: J.K. Johnson (Ed.), Remote Sensing in Archaeology: An Explicitly North American Perspective, University of Alabama Press, Tuscaloosa, 2006, pp. 259–267.
[56] T.K. Perttula, C.P. Walker, T.C. Schultz, A revolution in Caddo archaeology: the remote sensing and archaeological view from the Hill Farm site (41BW169) in Bowie County, Texas, Southeast. Archaeol. 27 (2008) 93–107.
[57] K. Dirix, P. Muchez, P. Degryse, B. Mušič, J. Poblome, Integrating multi-element geochemical and magnetic survey at ancient Sagalassos (Southwest Turkey): anthropogenic versus natural anomalies, Archaeol. Prospect. 20 (2013) 233–247.
[58] W. Neubauer, A. Eder-Hinterleitner, Resistivity and magnetics of the Roman town Carnuntum, Austria: an example of combined interpretation of prospection data, Archaeol. Prospect. 4 (1997) 179–189.

[59] P. De Smedt, M. Van Meirvenne, D. Simpson, Multi-signal EMI and geoarchaeology evaluating integrated magnetic susceptibility measurements for archaeological prospection, 9th International Conference on Archaeological Prospection, 2011, pp. 54–57.
[60] D.B. Landry, I.J. Ferguson, S.B. Milne, R.W. Park, Combined geophysical approach in a complex Arctic archaeological environment: a case study from the LdFa-1 site, Southern Baffin Island, Nunavut, Archaeol. Prospect. 22 (2015) 157–170.
[61] D. Simpson, A. Lehouck, L. Verdonck, H. Vermeersch, M. Van Meirvenne, J. Bourgeois, E. Thoen, R. Docter, Comparison between electromagnetic induction and fluxgate gradiometer measurements on the buried remains of a 17th century castle, J. Appl. Geophys. 68 (2009) 294–300.
[62] K. Welham, J. Fleisher, P. Cheetham, H. Manley, C. Steele, S. Wynne-Jones, Geophysical survey in sub-Saharan Africa: magnetic and electromagnetic investigation of the UNESCO World Heritage Site of Songo Mnara, Tanzania, Archaeol. Prospect. 21 (2014) 255–262.
[63] J.J. Lockhart, Tom Jones (3HE40): geophysical survey and spatial organization at a Caddo Mound site in Southwest Arkansas, Southeast. Archaeol. 29 (2010) 236–249.
[64] R.B. Clay, Working with the EM38 earth conductivity meter: geophysical survey at the Hopeton earthwork, Chillicothe, Ohio, May, (2001) Available at http://www.rri.res.in/~prabu/MWA_related/FacilitiesMeet_Seatle_1_2_Nov2010/em38.pdf (last accessed June, 2018).
[65] C. Gaffney, G. Hughes, J. Gater, Geophysical surveys at King Lobengula's Palace KoBulawayo, Zimbabwe, Archaeol. Prospect. 12 (2005) 31–49.
[66] P. Masters, B. Stichelbaut, From the air to beneath the soil—revealing and mapping Great War trenches at Ploegsteert (Comines-Warneton), Belgium, Archaeol. Prospect. 16 (2009) 279–285.
[67] A. Viberg, A. Berntsson, K. Lidén, Archaeological prospection of a high altitude Neolithic site in the Arctic mountain tundra region of northern Sweden, J. Archaeol. Sci. 40 (2013) 2579–2588.
[68] T. Grossman, F. Chevalier, R.H. Kazi, Bringing research articles to life with animated figures, Interactions 23 (2016) 52–57.
[69] L.B. Conyers, E.G. Ernenwein, L.-A. Bedal, Ground-penetrating radar (GPR) mapping as a method for planning excavation strategies, Petra, Jordan, E-tiquity 1 (2002) http://www.saa.org/publications/E-tiquity/e-tiquity.html (last accessed July 2006).
[70] G.N. Tsokas, J.H. Kim, P.I. Tsourlos, G. Angistalis, G. Vargemezis, A. Stampolidis, N. Diamanti, Investigating behind the lining of the Tunnel of Eupalinus in Samos (Greece) using ERT and GPR, Surf. Geophys. 13 (2015) 571–583.
[71] D. Wilken, T. Wunderlich, H. Stümpel, W. Rabbel, R. Pašteka, E. Erkul, J. Papčo, R. Putiška, M. Krajňák, D. Kušnirák, Case history: integrated geophysical survey at Katarínka Monastery (Slovakia), Surf. Geophys. 13 (2015) 585–599.
[72] M.C. Kassabaum, E.R. Henry, V.P. Steponaitis, J.W. O'hear, Between surface and summit: the process of mound construction at Feltus, Archaeol. Prospect. 21 (2014) 27–37.
[73] E. Brizzolari, F. Ermolli, L. Orlando, S. Piro, L. Versino, Integrated geophysical methods in archaeological surveys, J. Appl. Geophys. 29 (1992) 47–55.
[74] E.G. Ernenwein, A geophysical view of Pueblo Escondido: implications for the pithouse to pueblo transition in the Jornada Mogollon, Bull. Tex. Archeol. Soc. 79 (2008) 125–145.
[75] K.L. Kvamme, Integrating multidimensional geophysical data, Archaeol. Prospect. 13 (2006) 57–72.

[76] J. Burks, Geophysical survey at Ohio earthworks: updating nineteenth century maps and filling the 'empty' spaces, Archaeol. Prospect. 21 (2014) 5–13.
[77] E. Nowaczinski, G. Schukraft, S. Hecht, K. Rassmann, O. Bubenzer, B. Eitel, A multimethodological approach for the investigation of archaeological ditches-exemplified by the Early Bronze Age settlement of Fidvár near Vráble (Slovakia), Archaeol. Prospect. 19 (2012) 281–295.
[78] E. Nowaczinski, G. Schukraft, K. Rassmann, S. Reiter, N. Müller-Scheeßel, S. Hecht, B. Eitel, O. Bubenzer, J. Bátora, A multidimensional research strategy for the evaluation of settlement pits: 3D electrical resistivity tomography, magnetic prospection and soil chemistry, Archaeol. Prospect. 22 (2015) 233–253.
[79] V. Pérez-Gracia, J.O. Caselles, J. Clapes, R. Osorio, G. Martínez, J.A. Canas, Integrated near-surface geophysical survey of the Cathedral of Mallorca, J. Archaeol. Sci. 36 (2009) 1289–1299.
[80] T. Saey, S. Delefortrie, L. Verdonck, P. De Smedt, M. Van Meirvenne, Integrating EMI and GPR data to enhance the three-dimensional reconstruction of a circular ditch system, J. Appl. Geophys. 101 (2014) 42–50.
[81] R.A. Dalan, B.W. Bevan, D. Goodman, D. Lynch, S. De Vore, S. Adamek, T. Martin, G. Holley, M. Michlovic, The measurement and analysis of depth in archaeological geophysics: tests at the Biesterfeldt Site, USA, Archaeol. Prospect. 18 (2011) 245–265.
[82] P. De Smedt, T. Saey, E. Meerschman, J. De Reu, W. De Clercq, M. Van Meirvenne, Comparing apparent magnetic susceptibility measurements of a multi-receiver EMI sensor with topsoil and profile magnetic susceptibility data over weak magnetic anomalies, Archaeol. Prospect. 21 (2014) 103–112.
[83] T. Matney, L.R. Barrett, M.B. Dawadi, D. Maki, C. Maxton, D.S. Perry, D.C. Roper, L. Somers, L.G. Whitman, In situ shallow subsurface reflectance spectroscopy of archaeological soils and features: a case-study of two native American settlement sites in Kansas, J. Archaeol. Sci. 43 (2014) 315–324.
[84] E.R. Henry, A multistage geophysical approach to detecting and interpreting archaeological features at the LeBus Circle, Bourbon County, Kentucky, Archaeol. Prospect. 18 (2011) 231–244.
[85] R. Di Maio, M. Fedi, M. La Manna, M. Grimaldi, U. Pappalardo, R. Di Maio, M. Fedi, M. La Manna, M. Grimaldi, U. Pappalardo, The contribution of geophysical prospecting in the reconstruction of the buried ancient environments of the house of Marcus Fabius Rufus (Pompeii, Italy), Archaeol. Prospect. 17 (2010) 259–269.
[86] M.G. Drahor, Integrated geophysical studies in the upper part of Sardis archaeological site, Turkey, J. Appl. Geophys. 59 (2006) 205–223.
[87] N.G. Diamanti, G.N. Tsokas, P.I. Tsourlos, A. Vafidis, Integrated interpretation of geophysical data in the archaeological site of Europos (Northern Greece), Archaeol. Prospect. 12 (2005) 79–91.
[88] E. Rizzo, D. Chianese, V. Lapenna, Magnetic, GPR and geoelectrical measurements for studying the archaeological site of "Masseria Nigro" (Viggiano, southern Italy), Surf. Geophys. 3 (2005) 13–19.
[89] V. Testone, V. Longo, M.C. Bottacchi, P. Mameli, Use of integrated geophysical methods to investigate a coastal archaeological site: the Sant'Imbenia Roman Villa (Northern Sardinia, Italy), Archaeol. Prospect. 22 (2015) 63–74.
[90] M.A. Atya, H. Kamei, A.M. Abbas, F.A. Shaaban, A.G. Hassaneen, M.A. Abd Alla, M. N. Soliman, Y. Marukawa, T. Ako, Y. Kobayashi, Complementary integrated geophysical investigation around Al-Zayyan temple, Kharga Oasis, Al-Wadi Al-Jadeed (New Valley), Egypt, Archaeol. Prospect. 12 (2005) 177–189.

[91] R. Lasaponara, G. Leucci, N. Masini, R. Persico, Investigating archaeological looting using satellite images and GEORADAR: the experience in Lambayeque in North Peru, J. Archaeol. Sci. 42 (2014) 216–230.
[92] T.M. Urban, C. Vella, E. Bocancea, C.A. Tuttle, S.E. Alcock, A geophysical investigation of a newly discovered Early Bronze Age site near Petra, Jordan, J. Archaeol. Sci. 42 (2014) 260–272.
[93] E. Cardarelli, F. Fischanger, S. Piro, Integrated geophysical survey to detect buried structures for archaeological prospecting. A case-history at Sabine Necropolis (Rome, Italy), Surf. Geophys. 6 (2008) 15–20.
[94] E. Cardarelli, A. Godio, G. Morelli, L. Sambuelli, G. Santarato, L.V. Socco, Integrated geophysical surveys to investigate the Scarsella vault of St. John's baptistery in Florence, Lead. Edge 21 (2002) 467–470.
[95] J. Panisova, I. Murín, R. Pašteka, J. Haličková, P. Brunčák, V. Pohánka, J. Papčo, P. Milo, Geophysical fingerprints of shallow cultural structures from microgravity and GPR measurements in the Church of St. George, Svätý Jur, Slovakia, J. Appl. Geophys. 127 (2016) 102–111.
[96] E. Forte, M. Pipan, Integrated seismic tomography and ground-penetrating radar (GPR) for the high-resolution study of burial mounds (tumuli), J. Archaeol. Sci. 35 (2008) 2614–2623.
[97] E. Nowaczinski, G. Schukraft, K. Rassmann, S. Hecht, F. Texier, B. Eitel, O. Bubenzer, Geophysical-geochemical reconstruction of ancient population size—the Early Bronze Age settlement of Fidvár (Slovakia), Archaeol. Prospect. 20 (2013) 267–283.
[98] J.C. Cothren, C.G. Goodmaster, A.B. Barnes, E.E. Ernenwein, A.V. Vranich, W.F. L. Limp, A.P. Payne, Fusion of three-dimensional data at Tiwanaku: an approach to spatial data integration, in: E. Jerem, F. Redő, V. Szeverényi (Eds.), On the Road to Reconstructing the Past, Computer Applications and Quantitative Methods in Archaeology (CAA), Proceedings of the 36th International Conference, Budapest, April 2–6, 2009, Archaeolingua Foundation, Budapest, 2011, pp. 92–98.
[99] A. Viberg, J. Schultzén, A. Wikström, Meshing around: integrating ground-penetrating radar surveys and photogrammetric documentation for the reconstruction of the spatial layout of the Church of St. Lawrence, Sigtuna, J. Archaeol. Sci. Rep. 8 (2016) 295–302.
[100] M.S. Watters, Geovisualization: an example from the Catholme Ceremonial Complex, Archaeol. Prospect. 13 (2006) 282–290.
[101] S. Seren, A. Eder-Hinterleitner, W. Neubauer, S. Groh, Combined high-resolution magnetics and GPR surveys of the Roman town of Flavia Solva, Surf. Geophys. 2 (2004) 63–68.
[102] L.B. Conyers, Ground-penetrating radar mapping using multiple processing and interpretation methods, Remote Sens. 8 (2016) 562.
[103] R. Quinn, C. Breen, W. Forsythe, K. Barton, S. Krooney, D. O'Hara, Integrated geophysical surveys of the French Frigate La Surveillante (1797), Bantry Bay, Co. Cork, Ireland, J. Archaeol. Sci. 29 (2002) 413–422.
[104] S. Campana, M. Dabas, L. Marasco, S. Piro, D. Zamuner, Integration of remote sensing, geophysical surveys and archaeological excavation for the study of a medieval mound (Tuscany, Italy), Archaeol. Prospect. 16 (2009) 167–176.
[105] E.R. Henry, N.R. Laracuente, J.S. Case, J.K. Johnson, Incorporating multistaged geophysical data into regional-scale models: a case study from an Adena burial mound in Central Kentucky, Archaeol. Prospect. 21 (2014) 15–26.
[106] C. Pohl, J.L. Van Genderen, Remote Sensing Image Fusion: A Practical Guide, CRC Press, Boca Raton, FL, 2016.

[107] G. Scardozzi, S. Giese, C. Hübner, Integrated geophysical investigations in Hierapolis of Phrygia (Turkey), Surf. Geophys. 11 (2013) 101–113.

[108] P. Mozzi, A. Fontana, F. Ferrarese, A. Ninfo, S. Campana, R. Francese, The Roman City of Altinum, Venice Lagoon, from remote sensing and geophysical prospection, Archaeol. Prospect. 23 (2016) 27–44.

[109] K.L. Kvamme, An examination of automated archaeological feature recognition in remotely sensed imagery, in: A. Bevan, M. Lake (Eds.), Computational Approaches to Archaeological Spaces, Left Coast Press, Walnut Creek, CA, 2013, pp. 53–68.

[110] Ø.D. Trier, S.Ø. Larsen, R. Solberg, Automatic detection of circular structures in high-resolution satellite images of agricultural land, Archaeol. Prospect. 16 (2009) 1–15.

[111] Ø.D. Trier, L.H. Pilø, Automatic detection of pit structures in airborne laser scanning data, Archaeol. Prospect. 19 (2012) 103–121.

[112] J. Leckebusch, A. Weibel, F. Buhler, Semi-automatic feature extraction from GPR data, Surf. Geophys. 6 (2008) 75–84.

[113] S. Fitch, V. Gaffney, K. Thomson, Merging technologies: integration and visualization of spatial data, in: V. Gaffney, K. Thomson, S. Fitch (Eds.), Mapping Doggerland: The Mesolithic Landscape of Europe, Archaeopress, Oxford, 2007, pp. 33–41.

[114] A.Y.M. Lin, A. Novo, S. Har-Noy, N.D. Ricklin, K. Stamatiou, Combining GeoEye-1 satellite remote sensing, UAV aerial imaging, and geophysical surveys in anomaly detection applied to archaeology, IEEE J. Sel. Top. Appl. Earth Obs. Remote Sens. 4 (2011) 870–876.

[115] L. Matera, M. Noviello, M. Ciminale, R. Persico, Integration of multisensor data: an experiment in the archaeological park of Egnazia (Apulia, Southern Italy), Surf. Geophys. 13 (2015) 613–621.

[116] S. Keay, G. Earl, S. Hay, S. Kay, J. Ogden, K.D. Strutt, The role of integrated geophysical survey methods in the assessment of archaeological landscapes: the case of Portus, Archaeol. Prospect. 16 (2009) 154–166.

[117] K.L. Kvamme, Geophysical surveys as landscape archaeology, Am. Antiq. 68 (2003) 435–457.

[118] U. Böniger, J. Tronicke, Integrated data analysis at an archaeological site: a case study using 3D GPR, magnetic, and high-resolution topographic data, Geophysics 75 (2010) B169–B176.

[119] J. De Reu, J. Trachet, P. Laloo, W. De Clercq, From low cost UAV survey to high resolution topographic data: developing our understanding of a medieval outport of bruges, Archaeol. Prospect. 23 (2016) 335–465.

[120] M.A. Cole, N.T. Linford, A.P. Payne, P.K. Linford, Soil magnetic susceptibility measurements and their application to archaeological site investigation, in: J. Beavis, K. Barker (Eds.), Science and Site: Evaluation and Conservation: The Archaeological Sciences Conference 1993, Proceedings of the Conference, 8–10 September 1993, Bournemouth University, Bournemouth, 1995, pp. 114–162.

[121] E.G. Ernenwein, Integration of multidimensional archaeogeophysical data using supervised and unsupervised classification, Surf. Geophys. 7 (2009) 147–158.

[122] J. Ogden, S. Keay, G. Earl, K. Strutt, S. Kay, Geophysical prospection at Portus: an evaluation of an integrated approach to the interpretation of subsurface archaeological features, Making History Interactive, Computer Applications and Quantitative Methods in Archaeology, CAA2009, Williamsburg, Virginia, Archaeopress, Oxford, 2010, pp. 273–284.

[123] M. Noviello, M. Ciminale, V. De Pasquale, Combined application of pansharpening and enhancement methods to improve archaeological cropmark visibility and identification

in QuickBird imagery: two case studies from Apulia, Southern Italy, J. Archaeol. Sci. 40 (2013) 3604–3613.

[124] J.C. Davis, Statistics and Data Analysis in Geology, third ed., John Wiley & Sons, New York, 2002.

[125] K.L. Kvamme, Archaeological modeling with GIS at scales large and small, Reading Historical Spatial Information from around the World: Studies of Culture and Civilization Based on Geographic Information Systems Data, International Center for Japanese Studies, Kyoto, 2005, pp. 75–91.

[126] D.W. Hosmer, S. Lemeshow, Applied Logistic Regression, second ed., Wiley, New York, 2000.

[127] J.K. Johnson, B.S. Haley, Sensor and data fusion technologies in archaeology, in: V. Lenz-Wiedemann, G. Bareth (Eds.), Proceedings of the ISPRS WG VII/5 Workshop, 18–19 November, Geographisches Institut der Universität zu Köln - Kölner Geographische Arbeiten, Cologne, 2011, pp. 39–47.

[128] A. Agapiou, A. Sarris, N. Papadopoulos, D.D. Alexakis, D.G. Hadjimitsis, 3D pseudo GPR sections based on NDVI values: fusion of optical and active remote sensing techniques at the Vészto-Mágor tell, Hungary, in: C. Papadopoulos, A. Chrysanthi, E. Kotoula, A. Sarris (Eds.), Archaeological Research in the Digital Age, Proceedings of the 1st Conference on Computer Applications and Quantitative Methods in Archaeology Greek Chapter (CAA-GR), Institute for Mediterranean Studies–Foundation of Research and Technology (IMS-Forth), Rethymno, Crete, Greece, 2015.

[129] C.B. Rodning, Temporal variation in Qualla pottery at Coweeta Creek, NC Archaeol. 57 (2008) 1–49.

[130] R.A. Dalan, A review of the role of magnetic susceptibility in archaeogeophysical studies in the USA: recent developments and prospects, Archaeol. Prospect. 15 (2008) 1–31.

[131] A. Tabbagh, What is the best coil orientation in the Slingram electromagnetic prospecting method? Archaeometry 28 (1986) 185–196.

[132] J.A. Richards, Remote Sensing Digital Image Analysis: An Introduction, Springer, Berlin, 1986.

[133] A. Usse, J.-P. Usse, Habitats troglodytiques, souterrains médiévaux et galeries artificielles du Cantal, Bull. Archéol. Région D'Aurillac 7 (1998) 5–89.

[134] C. Gaffney, V. Gaffney, W. Neubauer, E. Baldwin, H. Chapman, P. Garwood, H. Moulden, T. Sparrow, R. Bates, K. Löcker, A. Hinterleitner, I. Trinks, E. Nau, T. Zitz, S. Floery, G. Verhoeven, M. Doneus, The Stonehenge Hidden Landscapes Project, Archaeol. Prospect. 19 (2012) 147–155.

[135] W.R. Wood, An interpretation of Mandan culture history, River Basin Surveys Papers 39, Bureau of American Ethnology Bulletin 198, Washington, DC, 1967.

[136] K.L. Kvamme, E.G. Ernenwein, C.J. Markussen, Robotic total station for microtopographic mapping: an example from the Northern Great Plains, Archaeol. Prospect. 13 (2006) 91–102.

[137] J.R. Bales, K.L. Kvamme, Geophysical signatures of earthlodges in the Dakotas, in: E. P. Pauls, D.C. Roper (Eds.), Plains Earthlodges: Ethnographic and Archaeological Perspectives, University of Alabama Press, Tuscaloosa, 2005, pp. 140–159.

[138] C. Lloyd, Local Models for Spatial Analysis, second ed., CRC Press, Boca Raton, FL, 2010.

[139] P. Harris, C. Brundson, M. Charlton, Geographically weighted principal components analysis, Int. J. Geogr. Inf. Sci. 25 (2011) 1717–1736.

CHAPTER

Ground-penetrating radar for the evaluation and monitoring of transport infrastructures

10

Lara Pajewski*, Simona Fontul†, Mercedes Solla‡

*Department of Information Engineering, Electronics and Telecommunications, Sapienza University of Rome, Rome, Italy**

National Laboratory for Civil Engineering—LNEC, Lisboa, Portugal†

Defense University Center, Spanish Naval Academy, Marín, Pontevedra, Spain‡

CHAPTER OUTLINE

Innovation in Near-Surface Geophysics. https://doi.org/10.1016/B978-0-12-812429-1.00010-6

1 INTRODUCTION

Ground-penetrating radar (GPR) is a nondestructive testing (NDT) technique that uses low-power electromagnetic waves to produce high-resolution images of the subsurface and structures [1,2]. In particular, the GPR instrument transmits a wide-band electromagnetic signal and detects the echoes coming from the subsurface or structure under test. Indeed, discontinuities of electric and magnetic properties in the investigated structure cause reflection and scattering of the electromagnetic waves emitted by the radar. By exploiting signal-processing and imaging techniques, the echoes recorded by the GPR can be transformed into three-dimensional images, which enable seeing into a structure that is opaque to the human eye.

GPR started being used in the 1970s for ice surveys in Antarctica and has since gained broad acceptance internationally, being nowadays successfully and increasingly employed for a great variety of tasks. In civil engineering, GPR can be applied for the inspection of transport infrastructures (roads, highways, airport runways, railways, bridges, and tunnels); for detecting and locating voids, cavities, and buried services (pipes and cables); for monitoring of retaining walls, embankments, and dams; for investigating of buildings and foundations; for mapping soil layers, measuring bedrock depth, and identifying faults and fracture zones in rock (for geotechnical and geologic studies or foundation design); and more. A comprehensive overview on the civil-engineering applications of GPR can be found in [3] (Part II "GPR surveying of pavements, bridges, tunnels and buildings; underground utility and void sensing"); see also the recent review paper [4].

The present chapter is focused on the use of GPR on transport infrastructures. Traditional methods to detect structural deficiencies in transport infrastructures include visual inspection, hammer sounding, chain dragging, and destructive coring—those approaches have been used since the early part of the last century, but they can be inaccurate and in some cases unsafe for the operators (e.g., on busy highways). Various NDT techniques have been developed throughout the last 30 years and are nowadays regularly employed for complementing or replacing traditional inspection methods; GPR is widely recognized as the most powerful and versatile NDT approach. The major strengths of GPR, on which its success in the field of transport-infrastructure evaluation and monitoring is based, are mainly related to the possibility to collect data at high speed, in a noncontact mode and by using mobile acquisition units, and to the reliability and representativeness of measurements. The time needed for a survey is typically one order of magnitude shorter than with other technologies; this allows keeping the obstruction to traffic flow to a minimum, thus confining costs and discomfort to citizens to an acceptable level. Moreover, GPR provides dense and accurate data, with a much higher resolution compared with competing geophysical approaches. The main performance limitations occur in the presence of high-conductivity materials and in heterogeneous conditions causing complicated electromagnetic scattering phenomena. Another drawback is that considerable expertise is necessary to effectively design and conduct a survey; besides, the interpretation of radargrams is generally nonintuitive; thus, specific competences are needed to properly understand measurements and translate them into engineering decision-making data.

Common tasks that can be effectively fulfilled by using GPR on flexible, semi-rigid, and rigid pavement include layer thickness assessment, pavement condition evaluation and identification of damages, and estimation of moisture content, asphalt density gauge and compaction. When GPR is employed on reinforced concrete pavement, it is also possible to map in detail the reinforcing elements and estimate their size; detect zones of honeycombing, voiding, and chloride attack; and identify deteriorated and delaminated regions.

Assessment of ballasted railway tracks by means of GPR is a rather new application, already used in the field with excellent results, but still under study for further development; the main objectives reside in the estimation of ballast thickness, fouling, and moisture content.

GPR diagnoses of concrete bridges comprehend bridge-deck thickness assessment and condition evaluation, detailed mapping of rebar and tendon ducts, and inspection of pillars, along with tasks analogous to those listed for reinforced concrete pavement; note that GPR can also be used for quality control of precast concrete structures, such as bridge-deck beams, before they are employed in constructions. Applications of GPR on masonry bridges encompass detection of hidden arches and ancient profiles, study of building materials and evidences of restoration or reconstruction activities, estimation of moisture content, mapping of cavities and cracks, assessment of masonry thickness (on the bridge pavement and ring stones), and evaluation of bridge foundations. When bridges include timber structures, such as wooden beams, GPR can provide information about moisture content, detection of voids, and identification of zones where termite attack or fungal decay is present.

The main GPR applications in tunnel inspection include thickness estimation of lining and backfill grouting, rebar localization and corrosion assessment, localization of reinforcement and cavities in lining, and estimation of moisture content.

Nowadays, GPR systems typically operate in the 10–10.000 MHz range. The antennas used to emit and receive the electromagnetic signal must have dimensions comparable with the wavelengths of the signals; hence, the frequency range of operation is what basically defines the size of a GPR instrument. Generally, there is a direct relationship between the frequency of the electromagnetic waves emitted by the radar and the resolution that can be obtained. Conversely, there is an inverse relationship between frequency and penetration depth. Hence, high frequencies are used to detect small, shallow targets, and low frequencies are used to detect larger, deeper targets. For the inspection of transport infrastructures, the central frequency of the spectrum emitted by the GPR is commonly between 250 and 2000 MHz, depending on the survey objectives and conditions. With some systems, it is possible to use more antennas simultaneously, operating over different frequency ranges; this is an excellent solution that allows achieving both high resolution and deep penetration in one recording pass, thus reducing the overall survey time and cost.

There are two main types of GPR, both suitable for the investigation of transport infrastructures: impulse and step-frequency systems. Impulse systems are the most widely used; they operate in the time domain and emit a series of short electromagnetic pulses (normally, of 1–10 ns duration). Step-frequency systems are less common; they operate in the frequency domain and emit a series of harmonic waves that frequency is progressively incremented across a broad spectrum in a stepwise fashion. By exploiting the inverse Fourier transform (from frequency to time domain), a step-frequency GPR provides results equivalent to those measured by an impulse GPR. The step-frequency approach is possible because the investigated scenario is regarded as a time-invariant system and the received signal is a linear function of the transmitted one. Impulse systems are typically more sensitive to radio-frequency interference and may require an averaging of measurements to improve the signal-to-noise ratio; they are usually more expensive than step-frequency systems. Step-frequency systems can process a higher energy without increasing the maximum level of the signal; moreover, the thermal noise at their receiver is lower; hence, the signal-to-noise ratio is higher. For measuring the harmonic answer, the technological specifics on the receiver are less severe (the received signal can be sampled at a lower rate). Another advantage of step-frequency systems is the ability to skip frequencies that could interfere with broadcast stations. A drawback of step-frequency systems is that attention must be paid to the aliasing problem, due to the sampling of the harmonic answer: this is resolved by using a frequency step small enough (typically of the order of a few megahertz), which slows down the collection of data.

In order to couple the electromagnetic energy into the investigated structure, GPR antennas can operate in close proximity (ground-coupled antennas), else at a limited distance above the structure (air-coupled antennas). For transport-infrastructure investigations, ground-coupled antennas are normally employed for detailed studies over limited areas, as they allow obtaining higher resolution. Air-coupled antennas, instead, are used for the inspection of wider areas and infrastructures open to traffic, that is, in situations that require acquisition at higher speed.

A GPR can be monostatic, bistatic, or multistatic—as all radar systems. A monostatic system uses the same antenna for both transmitting and receiving electromagnetic signals (nowadays, this type of GPR can be considered as rather obsolete and is seldom used in transport-infrastructure surveys). A bistatic system uses two separate antennas for transmitting and receiving, which are often housed in a single module. Antennas in independent modules represent an interesting technological solution when it may be convenient to place the transmitter and receiver on the two opposite sides of the investigated structure, thus halving propagation and attenuation losses (quality control of precast structures and imaging of pillars and bridge walls). Multistatic systems are composed of multiple spatially diverse monostatic/bistatic radar components, usually with a shared area of coverage. At the present time, commercial multistatic GPR systems are implemented as multiple bistatic systems: they are often called multichannel or multiantenna systems, or GPR systems with antenna arrays. Their main advantage is that they enable faster data collection by increasing the extension of the investigated area per time unit (this is important in surveys where the collection of a large number of profiles is required, such as on a bridge); furthermore, they make it easier for the operator to produce three-dimensional (3-D) images. Commonly, in multistatic GPR systems, there are 4–16 couples of transmitting and receiving antennas, positioned along an axis orthogonal to the direction of acquisition.

Fig. 1 shows examples of GPR systems suitable for the inspection of transport infrastructures. An in-depth overview on GPR systems and antennas for civil engineering can be found in [3] (Part I "GPR instrumentation").

To enable analysis and interpretation of results, GPR data are normally plotted as two-dimensional (2-D) maps, showing the amplitude of the field measured by the receiver as a function of time and position ("B-scans" or "profiles"). Time can be replaced by depth if the propagation velocity in the subsurface or structure is known. Looking at single traces (i.e., "A-scans," data measured in a specific position) can

FIG. 1

Different GPR systems: (A) air-coupled (horn) antennas, (B) ground-coupled antennas, (C) array multichannel system, and (D) ground-coupled antennas mounted in a cart with an odometer wheel.

also be useful in transport-infrastructure applications, for example, to measure distances between pavement layers or estimate the permittivity of upper layers. When several parallel profiles are acquired, as is customarily done on bridges and other reinforced concrete structures, a 3-D matrix of field amplitudes is available, and 3-D images of the investigated scenario can be produced. From a 3-D matrix of field amplitudes, it is also possible and straightforward to obtain horizontal slices of data, that is, plan views of data at designated depths, which are very useful in most civil-engineering applications of GPR.

In some cases, raw GPR data are adequate for the interpretation to be conducted. More often, digital signal processing needs to be applied to the data, as required by the specific application and situation. The steps and algorithms may vary according to each survey and normally involve filtering to remove unwanted noise, gain adjustment to balance signal strengths, migration to remove diffraction effects, corrections for variations in surface topographic elevation, and improvement of graphic display by using suitable color palettes. For complex scenarios and/or for analyzing and interpreting reflections coming from deep layers, more advanced processing procedures have to be used. An up-to-date introduction to data-processing methods for GPR can be found in [2]; moreover, a special issue on advanced GPR signal-processing techniques was recently published by the Elsevier journal *Signal Processing* [5].

After the processing, GPR results are interpreted by recognizing reflection and diffraction patterns on the radargrams and by determining their position. Discrete buried objects with circular cross section typically appear as hyperbolic reflections in the raw data, with hyperbola prongs projected downward like an inverted "V" shape.

Subsurface layers appear as continuous reflectors on the radargrams. More in general, reflection and diffraction patterns in the raw GPR profiles may have different shapes, which do not resemble the true shape or orientation of the scatterers. Various factors, including the innate design of the survey equipment and the complexity of electromagnetic propagation in the scenario, can disguise complex structures recorded on GPR profiles. For simple scenarios, such as regular structures hosting widely spaced targets, raw reflection patterns can be adequately interpreted by trained operators. However, for complex scenarios such as natural subsoil with stones and rocks, ancient bridges, or any structures hosting closely spaced targets, it may be convenient to aid the interpretation by using an electromagnetic simulator or by applying to the radargrams suitable imaging and inversion algorithms.

The state of the art in the fields of electromagnetic modeling, imaging, and inversion techniques for GPR was composed of Part III of [3]; those areas of research are progressing fast, and for up-to-date information, we recommend the reader to check the proceedings of the most recent editions of the biennial events International Conference on Ground Penetrating Radar and International workshop on advanced Ground Penetrating Radar.

Risk factors related to on-site hazards should always be assessed prior to commencing a fieldwork. Sometimes, people are worried about the effects of the electromagnetic radiation emitted by the radar, whereas they do not think about other environmental factors that can be dangerous for the operators—such as working outdoor at high temperatures, in the traffic, or in a polluted area. The average signal power transmitted by small commercial GPR systems is in the order of 1 mW, about three orders of magnitude less than the power transmitted by a mobile phone. Thus, the levels of electromagnetic radiation produced by commercial GPR transmitters do not have to be considered as a risk to the health of the operators doing the measurements. It is instead very important to be aware of accidents that may happen during field survey operations. Care needs to be taken to prevent injury to the crew, especially when traffic or construction vehicles are moving around the site. Safe traffic control is required to avoid accidents during surveys on roads or bridges. Special care should be taken when conducting surveys on inclined slopes or at height, to prevent injuries from falling or rolling. Working underground or in confined spaces, such as inside a small tunnel, may require special procedures. Operations in confined spaces with high methane levels may present a risk from gas explosions or poisoning from noxious gases. Useful recommendations for planning safe GPR surveys and dealing with contingencies are given in [6].

For a more efficient inspection of transport infrastructures, GPR can be complemented with other NDT methods. For example, when roads, highways, airport pavements, harbor areas, and railway tracks are investigated, the falling weight deflectometer (FWD) can provide useful information concerning the pavement structural capacity, for overlay design and to determine if a pavement is being overloaded. The infrared camera and 3-D scanner can be used in the assessment of damaged pavement areas and to sense different pavement temperatures during asphalt compaction. Electric resistivity tomography (ERT), gravimetry, seismic propagation, spectral analysis of surface waves, and plate load tests can be combined with GPR for a better railway track prospection. Resistivity, ultrasound testing, impulse response measurement, impact echo, laser scanning, photogrammetry, and thermography are examples of techniques that can be utilized in the inspection of bridges and tunnels. Each NDT method provides different information and may be more or less useful depending on the physical properties of the construction materials of the structure under test. An in-depth introduction to nondestructive methods complementary to GPR can be found in [7].

An exceptional contribution to the GPR technology and methodology has been given, during the last 4 years, by European Cooperation in Science and Technology (COST, www.cost.eu) Action TU1208 *Civil Engineering Applications of Ground Penetrating Radar* (www.GPRadar.eu).

COST is the longest-running European framework supporting cooperation among scientists and researchers across Europe and beyond. Founded in 1971, it is currently integrated in the Horizon 2020 program. It contributes to reducing fragmentation in European research investment, building the European Research Area (ERA), and opening it to worldwide cooperation. It also aims at constituting a “bridge” toward the scientific communities of emerging countries, increasing the mobility of researchers across Europe, and fostering the establishment of excellence in key scientific domains. Gender balance, support to early-career investigators, and inclusiveness are strategic priorities of the COST program. COST does not fund research itself, but provides support for activities carried out within Actions: these are bottom-up science and technology networks, centered around nationally funded research projects, with a 4-year (or, exceptionally, slightly longer) duration and a minimum participation of five countries.

The Actions are active through a range of networking tools, such as meetings, workshops, conferences, training schools, short-term scientific missions, and dissemination activities; they are open to researchers and experts from universities, public and private research institutions, nongovernmental organizations, and industry and small- and medium-sized enterprises. By creating open spaces where people and ideas can grow, COST truly fosters the birth of new ideas and unlocks the full potential of science.

COST Action TU1208 started in April 2013 and is coming to an end in October 2017. It involves more than 300 experts from 150 partner institutes in 28 COST countries (Austria, Belgium, Croatia, Czech Republic, Denmark, Estonia, Finland, France, former Yugoslav Republic of Macedonia, Germany, Greece, Ireland, Italy, Latvia, Malta, the Netherlands, Norway, Poland, Portugal, Romania, Serbia, Slovakia, Slovenia, Spain, Sweden, Switzerland, Turkey, and the United Kingdom), a COST cooperating state (Israel), 6 COST near-neighbor countries (Albania, Armenia, Egypt, Jordan, Russia, and Ukraine), and 6 COST international partner countries (Australia, Colombia, Hong Kong Special Administrative Region of the People's Republic of China, the Philippines, Rwanda, and the United States). University researchers, software developers, civil and electronic engineers, archeologists, geophysics experts, NDT equipment designers and manufacturers, and end users from private companies and public agencies are participating in the Action.

The scientific structure of TU1208 is shown in Fig. 2 and includes four working groups (WGs), in which research activities cover all areas of the GPR technology, methodology, and applications. GPR surveying of transport infrastructure, which is the subject of the present chapter, is mainly dealt with in WG2. However, all WGs cooperate and interact, such that the contribution of the other WGs has been essential for the advancement of the GPR technique in this area of application.

WG1 focuses on the development of novel GPR instrumentation. Within this WG, novel equipment dedicated to the inspection of transport infrastructure has been designed, realized, and tested. Members from Finland have been developing a series of prototypes for road investigations, which are particularly effective in estimating the permittivity and thickness of near-surface and thin pavement layers [8,9]. Members from Norway proposed a new step-frequency GPR system with a ground-coupled array, for the investigation of roads and bridge decks; this system has 22 channels and uses boomerang monopole antennas ([10]; the paper includes useful examples for a better understanding of the advantages and drawbacks of

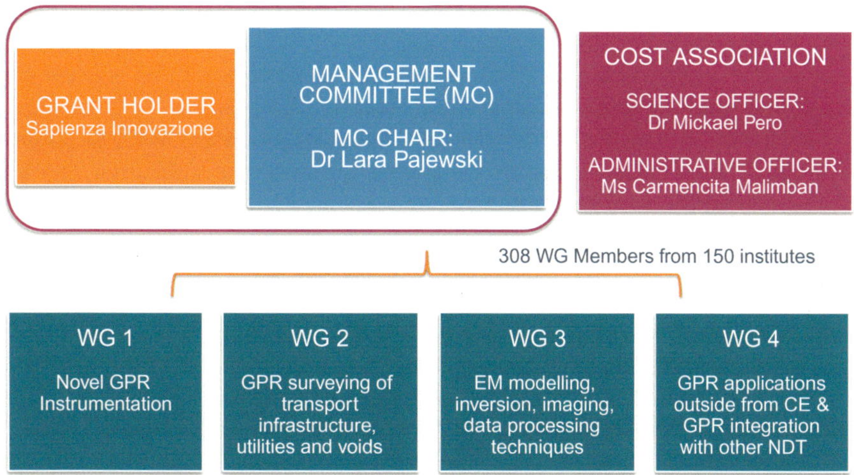

FIG. 2

Scientific structure of COST Action TU1208. The activities are coordinated by a management committee (MC), with 2–4 representatives per COST country, and by the Action Chair. Funds are administrated by the Grant Holder institution. Research activities are carried out within four working groups (WGs). The scientific and administrative support of COST Association has been crucial.

air- and ground-coupled antennas). The cooperation between members from Belgium and Italy has led to the development and testing of a new GPR system for tomographic reconstruction of vertical structures, which may be applied to bridge pillars and columns [11,12]. Members from Italy, Malta, and Norway cooperated at the improvement and testing of a reconfigurable step-frequency prototype, initially developed in Italy; this system allows reconfiguring the length of antenna arms (i.e., the operating frequency band), the power radiated at different frequencies, and the integration time of the received harmonic tones; it may be used for local pavement and bridge-deck inspections, although so far it has been mainly employed for geologic and archeological investigations [13–17]. A team of members from Italy developed a low-cost system integrating the potentialities of GPR and IMU, for accurate positioning [18].

WG2 focuses on the use of GPR in civil engineering. In particular, this WG produced the already mentioned recommendations for a safe use of GPR [6]. Moreover, it has developed guidelines for the nondestructive inspection of flexible pavements using GPR (with the main purposes to assess layer thicknesses and to detect and classify road damages; innovative applications, such as moisture-content and compaction evaluation, are dealt with). Furthermore, WG2 has produced guidelines for utility detection in urban areas using GPR and guidelines for the evaluation of concrete structures by using GPR (including bridges and tunnels). Among the most interesting outcomes of WG2, it is also worth mentioning a catalog of test sites, where GPR equipment and procedures can be verified and tested. The guidelines and catalog of test sites will be available in open access by September 2017 on the website of the Action. Additionally, WG2 developed a wide series of investigations and case studies: the most interesting ones, relevant to the use of GPR on transport infrastructures, are shortly resumed in Sections 2–5 of the present chapter.

WG3 studies electromagnetic forward and inverse methods for the solution of near-field scattering problems by buried structures, imaging techniques, and data-processing algorithms. Among the main activities and results of WG3 relevant to the topic of this chapter, the release of electromagnetic simulators has to be mentioned [19,20]. In particular, members from the United Kingdom developed a new and open-source version of the electromagnetic simulator gprMax [21]. Members from Italy developed a freeware CAD for 2-D gprMax models [22]. More information about those tools is given in Section 6 of the present chapter.

WG4 deals with the applications of GPR outside from the civil-engineering area and with the combined use of GPR and complementary NDT methods. The most interesting case studies produced by members of this WG and relevant to the inspection of transport infrastructures are briefly resumed in Sections 2–5 of the present chapter.

It is worth reporting that all WGs have been active in training activities (16 training schools were organized in 4 years) and they produced an educational package for teaching GPR in the university, which will be available in open access on the website of the Action as of September 2017. Educational material concerned with the use of GPR on transport infrastructure is included in the package.

The present chapter continues with Section 2, which describes more in detail the use of GPR for the inspection of roads, highways, and airport runways. Section 3 is devoted to GPR assessment of railways. Section 4 deals with GPR exploration of bridges. Section 5 focuses on the use of GPR in tunnels. Sections 3–5 all have the same structure: Firstly, the possible objectives of a GPR survey on the relevant infrastructure are listed; secondly, the investigation methodology is presented, and recommendations concerning the possibility to combine GPR with complementary NDT approaches are given; lastly, an example is presented, and an overview of the most interesting research activities carried out within COST Action TU1208 is included, with useful tables including information about the various studies and relevant references to documentation where more information can be found. Section 6 highlights the importance and usefulness of electromagnetic modeling as a tool for GPR data interpretation; in such section, recommendations for accurate modeling of GPR scenarios are provided, and the main software tools developed in COST Action TU1208 are presented. The chapter is concluded with Section 7, where final remarks are drawn. A rich bibliography completes the chapter, where several references are listed, for further readings.

2 ROADS, HIGHWAYS AND AIRPORT RUNWAYS

The use of GPR on pavements started in the 1990s. In the same period, the first air-coupled antennas were developed, dedicated to performing measurements on roads at traffic speed [3].

It is common knowledge that a pavement consists of several layers of bound and unbound materials placed over a natural soil, which is usually called subgrade. Each pavement layer can be considered as a mixture of small particles (aggregates) embedded in a homogeneous matrix. The aggregate size is small compared with the wavelength of the electromagnetic field emitted by the GPR; therefore, in the framework of a GPR pavement inspection by using GPR, each layer can be assumed as a homogeneous dielectric lossy material, with an effective complex relative permittivity.

As is well known, the so-called flexible pavements are constituted by asphalt layers on a subbase; semiflexible or semirigid pavements are made of asphalt layers on hydraulically bound layers, and rigid pavements are made of concrete and may be reinforced with steel bars, which behave as reflecting and scattering targets when they are illuminated by electromagnetic waves. There also are composite pavements, with asphalt on top and concrete below. All those pavements can be surveyed with GPR and pavements for pedestrian use, which are often made of paving blocks.

2.1 OBJECTIVES AND METHODOLOGY

In the beginning, GPR was mainly used to assess the thicknesses of pavement layers, and this still is the most common GPR application in pavement engineering. The theory behind this measurement is very simple. If the real part of the relative permittivity of the first layer is ε_r, an electromagnetic wave propagates in the layer with a velocity $v = c/\sqrt{\varepsilon_r}$, where c is the speed of light in a vacuum (3×10^8 m/s). When the relative permittivity (or the propagation velocity) is known, by measuring the two-way travel time τ to the interface between the first and second layers, it is possible to estimate the thickness of the first layer as $d = v\,\tau/2$ (the two-way travel time is the time for an electromagnetic wave to travel from the transmitting antenna to the interface and return back to the receiving antenna). Analogously, by using suitable formulas, the thicknesses of deeper layers can be estimated.

Cores are very important for calibrating GPR data. Reasonably good accuracy can be achieved in the estimation of pavement thicknesses, within 5% of real thicknesses. The accuracy improves if GPR data are calibrated with a higher number of cores.

A lot of research has focused on estimating the relative permittivity values of materials used to realize asphalt and rigid pavement layers [23,24]. The most commonly used methods are the following:

(i) Using data taken from the literature, for customary pavement materials—this is the least accurate method, mainly because the presence of moisture greatly affects the permittivity, but may be effective for a preliminary interpretation.

(ii) Coring, that is, drilling a core in a selected area of the pavement—from the extracted pavement sample, it is possible to accurately measure the thicknesses and calculate the permittivity values of the pavement layers; then, those values of the permittivity can be used for interpreting GPR data acquired in other positions and estimating the layer thicknesses.

(iii) Common midpoint (CMP) method—by using ground-coupled antennas with separate transmitting and receiving modules, various measurements are carried out by positioning the antennas at various distances from a central point.

(iv) Wide-angle reflection and refraction (WARR) method—by using ground-coupled antennas with separate transmitting and receiving modules, one antenna is kept in a fixed position, and various measurements are carried out by putting the other antenna at various distances.

(v) Reflection coefficient method—by using air-coupled antennas and a metal plate, the velocity is estimated by comparing the amplitude of the waves reflected by the pavement with the amplitude of the waves reflected by a metal plate positioned on the pavement surface.

For network-level assessment, usually, only the estimation of the first-layer thickness is required. For this purpose, air-coupled antennas with central frequencies from 1000 to 2000 MHz are commonly used (Fig. 3); the choice mainly depends on the pavement thickness of the network (i.e., required resolution). The effective penetration depth of a 1000 MHz antennas is around 1 m (can be more or less, depending on the pavement materials and conditions), with the ability to resolve the thickness of

FIG. 3

The apparatus for pavement assessment available at the National Laboratory for Civil Engineering (Laboratório Nacional de Engenharia Civil (LNEC)), in Portugal: air-coupled antennas and survey wheel are well visible in the picture.

the first layer down to a couple of centimeters. With a 2000 MHz antenna, the depth of penetration is less, and the resolution is better. The additional use of ground-coupled antennas with lower central frequencies, such as 500 MHz, is useful for measuring the thicknesses of both the pavement subbase and subgrade and/or for assessing in a more reliable way the wearing-course thickness. Measurements are normally performed along the external wheel path of the road, and usually, at least 1–4 scans per meter are acquired in Europe. The use of multichannel systems is not yet common, because they are expensive.

For road-level assessment, the procedure and antenna configuration are the same. However, as the investigation area is more limited, data can be collected on a few longitudinal profiles (instead of only one), and the scan density is usually higher, that is, 10–20 scans per meter. For rehabilitation design purposes, GPR tests are performed along a grid on the pavement area under study.

Nowadays, the use of GPR for layer thickness assessment is a firmly established application in several European countries and in the United States, and it is regularly included in pavement management strategies. However, in several other European countries, GPR is still used only for research purposes.

Another well-established application of GPR in pavement engineering is the detection of cracks and voids. The capability of a GPR to detect voids and cracks is a function of their dimensions and of the operating frequency; in the frequency range used by commercial systems conceived for civil-engineering applications, the size of the smallest detectable void ranges from 1 to 2.5 cm. In certain conditions, it is also possible to roughly evaluate the size of the voids. And it is possible to easily discriminate between air-filled and water-filled voids, by looking at the algebraic sign of the reflection originating from the void (the reflection coefficient is negative for water-filled void and positive otherwise).

Currently, GPR is also used for estimating asphalt density [25], for detecting defects and deteriorations in flexible pavements [19,20], and for assessing water content at different levels within the pavement structure [26,27]. These applications already are at advanced level, but more research is needed; indeed, they are not yet offered/required in a systematic way, in pavement management. Such investigations are mostly carried out for quality control during new pavement construction or overlaying and sometimes to monitor and support the maintenance planning of existing roads [24]. The estimation of asphalt density and water content is based on empirical formulas, which take into account how the air content and water content affect the relative permittivity of the material. Concerning pavement defects, it is rather easy to distinguish between point damages (singularities inside a homogeneous layer, such as cracks), short damages (small singularities at the interface between two layers, with a size comparable with the layer thicknesses, such as pumping), and long damages (singularities at the interface between two layers, wider than the layer thickness, such as depressions); further experiments and electromagnetic simulations will allow a more advanced radargram interpretation and a more accurate discrimination between different kinds of damages and to understand their causes.

The position of GPR traces is traditionally referenced by the odometer. However, when the acquisition is done outdoor such as in road inspections, it is mostly suitable to use also a global navigation satellite system (GNSS) to georeferenced data and to allow a more precise interpretation of the data. The following accuracy levels can typically be achieved, when GNSS technology is used (there also are a number of other solutions available):

(i) Standard uncorrected accuracy (typically 2–8 m)—this accuracy level is not recommended in road inspections.
(ii) Increased accuracy (typically from 0.5 to 2 m)—this accuracy level can be adequate in large-scale road investigations and can be achieved by using a differential (code-based) GNSS (DGNSS). Higher accuracy is required for the location of reinforcement in rigid pavement.
(iii) High-accuracy real-time kinematic (RTK) GNSS (typical from 0.01 to 0.05 m)—this accuracy level can be achieved in real time by receiving corrections (carrier-based) from a reference station (or from a network of reference stations) or else by storing the carrier-phase measurements and postprocessing the data. This is the preferred accuracy and is becoming more available at lower costs.

Prior to any operation on public roads and in order to avoid accidents during GPR surveys, it is recommended to arrange traffic controllers and use all the necessary speed control signs and barriers. At night, reflective barriers should be used, and the operators should wear appropriate reflective clothing; flashing lights and reflective tape can be attached to the antennas. For continuous surveys along busy roads, guard cars with flashing signs may be necessary, to protect the operators and equipment. Further safety recommendations can be found in [6].

2.2 COMBINATION OF GPR WITH COMPLEMENTARY NONDESTRUCTIVE TESTING TECHNIQUES

For the structural evaluation of pavement, load tests are customarily performed. These are often done by using an FWD [28]. In order to set up the pavement model for FWD interpretation, it is important to know the thickness of the pavement layers. Hence, the availability of GPR technique brought a significant improvement on the back-calculation process. In fact, GPR enables to estimate the thicknesses not only of the first asphalt or concrete layers but also of the unbounded material layers, namely, the base and subbase; this reduces significantly the uncertainly of the back-calculation process. Moreover, as GPR assesses the layer thickness continuously along the pavement, an accurate identification of changes in the pavement structure is possible based on GPR results. The identification of interface debonding between layers is also possible. This information is of primary importance for a reliable calculation of the elastic moduli of the layers.

Other NDT tests carried out in road engineering are thermography and 3-D scanner acquisitions. The results of those techniques can complement GPR data and allow a more accurate study of damaged areas. Thermography is also useful for the sensing of different pavement temperatures during asphalt compaction.

2.3 EXAMPLE AND TU1208 RESEARCH ACTIVITIES

2.3.1 Example: Detection of areas with subgrade settlements on a highway

The application of GPR to pavement assessment is crucial to determine subgrade settlements that are not visible from the surface, which cause continuous pavement deterioration, due to permanent deformation over the weak subgrade. Generally, when a surface settlement appears at the pavement surface, maintenance is performed by adding asphalt, in order to restore the design layout. If information about subgrade condition is not available and in case of weak subgrade, such overlaying and repairing only add weight over a soft base; thus, instead of solving the cause of settlement, the problem is increased, and this results in more frequent repairs needed in the future.

In Fig. 4, an example is presented. On a section of a highway, GPR tests were carried out, and we detected a subgrade settlement. Test pits were performed (in the position indicated by P9), and the observed asphalt thickness was 0.40 m, instead of 0.20 m as per design specifications, due to several resurfacing operations performed on the highway.

GPR measurements enabled not only to assess the layer thicknesses of the pavement but also to accurately locate the areas with settlements. This enabled the possibility to adopt proper rehabilitation measures, by reinforcing the subgrade layer and reconstructing the pavement on the damaged locations.

2.3.2 Research activities carried out within COST Action TU1208

As already mentioned in the Introduction, WG2 of COST Action TU1208 focuses on the use of GPR in civil engineering. WG2 is in turn organized in projects, that is, working teams dealing with specific civil-engineering applications of GPR; of course, all projects cooperate with other projects and WGs. Project 2.1 studies the use of GPR for the assessment of transport infrastructures, including roads, highways, and airport runways, which are the subject of the present section of the chapter.

A significant outcome of Project 2.1 will be the already mentioned guidelines for nondestructive inspection of flexible pavements using GPR, with the main purposes to assess layer thicknesses and to detect and classify road damages; moisture-content and compaction evaluation will also be considered. Moreover, several innovative studies have been carried out within Project 2.1, and the most interesting ones are described in the following subsection. A wide series of experimental investigations have been also performed, where GPR measurements were carried out at test site, road/highway/runway, or road network level: information about some case studies

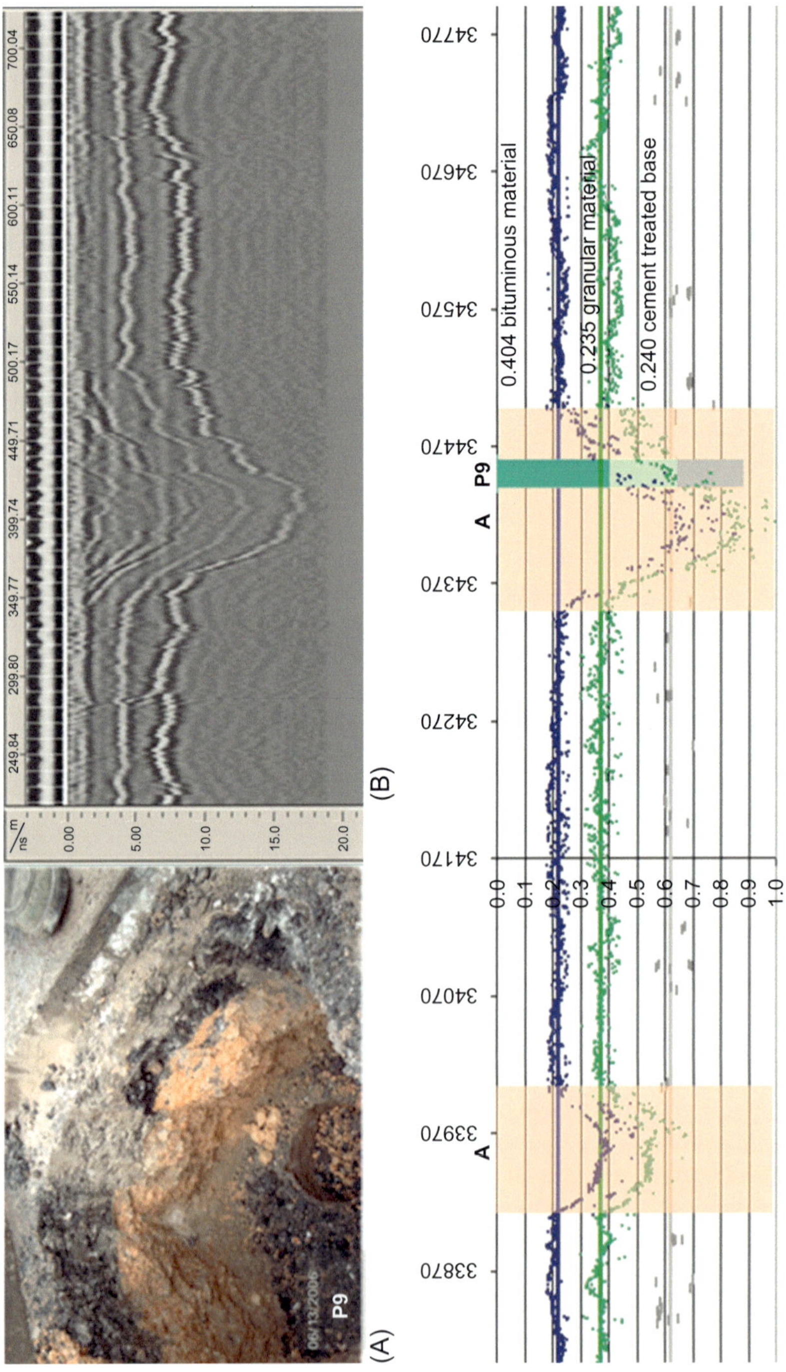

FIG. 4

Subgrade settlements detected with GPR: (A) test pit, (B) radargram, and (C) estimated layer thicknesses after processing.

is resumed in Table 1, whereas a more exhaustive list of case studies and further information can be found on the website of the Action.

Some researchers dealt with detecting damages in pavements. For example, in [30], the relationship between excavation works (and subsequent pavement reconditioning) and pavement damages was investigated. In [33], the possibility to predict rutting evolution in flexible pavement, by carrying out GPR surveys every 6 months and exploiting prediction models available in the literature, was studied; preliminary results were presented. Rodés et al. [26], studied the inner conditions and properties of pavements by analyzing the frequency spectrum of GPR signals. Several tests were carried out on a section of a highway, in two different stages of its service life. Results highlighted a good correlation between the spectrum shape of the recorded traces and the pavement conditions. In [34], and [19,20], the electromagnetic simulator gprMax [21] was used to model different types of damages in three different flexible pavements. Results show the high potential of GPR in distinguishing between different deteriorations and understanding their causes. A catalog of synthetic radargrams might be developed, which would represent a useful tool for aiding data interpretation during GPR road surveys (such a catalog might be combined with photo catalogs commonly used during visual inspection of roads).

A research team focused on developing a new technique for velocity estimation, to be employed when air-coupled antennas are used and the CMP method is not applicable [23]. In particular, a modified CMP method was proposed, accounting for the refraction at the air-pavement surface. Synthetic experiments were executed to demonstrate that the traditional method resulted in a persistent overestimation whereas the modified method improved the results significantly. Two field experiments were conducted on an existing road, to test and validate the new method. In order to improve the initial results, which were highly affected by the roughness of the surface, a further modification was proposed, and in-line data were incorporated in the procedure, which allowed obtaining a more accurate estimation of depth and velocity.

Plati et al. [25], studied the compactability of hot-mix asphalt by using GPR. In particular, they conducted a series of laboratory experiments and field tests, where they assessed the asphalt density during the compaction process. Three different electromagnetic-mixing-theory-based density models were used and compared. Additionally, different compaction procedures were used, and their effects on the estimated asphalt density were analyzed. The results obtained show that GPR is an effective and efficient tool for improving the asphalt mix compaction process and the assessment of in situ density.

Tosti et al. [33], used GPR to study the clayey soil properties in unsaturated conditions (clayey soils are often present in pavements). Several soils were tested, in different moisture conditions, from dry to saturated. Measurements were carried out for various levels of water content, by using a vector network analyzer spanning the 1–3 GHz frequency range and an impulse GPR with ground-coupled antennas and a central frequency of 500 MHz. Promising results were obtained, concerning the influence of clay in load-bearing layers and subgrade soils and its impact on the electromagnetic response of pavement layers.

Table 1 Examples of Applications of GPR on Roads, Highways, and Runways—and Related Laboratory Studies—Within COST Action TU1208

References	Technology	Methodology	Place	Road Type	Purpose of the Study	Major Findings and Remarks	Involved Countries
29	RIS 99-MF impulse GPR, with 600 and 1600 MHz antennas (IDS) Prima 100 LFWD (Carl Bro Pavement Consultants, Kolding)	GPR survey carried out in November 2013, acquisition of 11 longitudinal profiles (30 m long) and 76 transversal profiles (4 m long); trace spacing, 2.4 cm. Distance between adjacent GPR profiles, 40 cm; total number of recorded traces, 26,415 LFWD measurements carried out in October 2013, in 836 positions	Rome (Italy)	Test site (4 × 30 m), with a flexible pavement structure	To develop a new method for inferring mechanical properties of roads from GPR data	A semiempirical model for predicting pavement strength properties from GPR data was developed and calibrated. A good agreement was obtained between the values of Young's modulus measured by the LFWD and those predicted by the proposed model. Subsequently, the model was further tested and validated on Italian and Spanish roads (such results are in the process of being published)	Italy
30	GSSI SIR-20 impulse GPR, with 400 MHz (model 5103A) ground-coupled antenna	Acquisition of multiple longitudinal or transversal (depending on the direction of cracks) profiles; distance between adjacent profiles, 1 m; trace spacing, 0.5 cm. Data processed with ReflexW	Switzerland	Three road sections in Switzerland	To investigate the relationship between excavation works for the maintenance or realization of subsurface infrastructure and pavement damages occurring thereafter	Although the statistical basis is limited, results suggest that excavation works followed by pavement reconditioning cause the presence of a higher number of cracks, as expected	Switzerland
31	Impulse GPR (UTSI), with ground-coupled 1000 MHz antennas, eight channels Step-frequency GPR (3-D radar), with 11 channels, antennas at a small distance from the ground, central frequency of 1500 MHz	Survey carried out in 2013, during IWAGPR. The impulse GPR was pushed by an operator; distance between adjacent profiles, 12 cm; trace spacing, 1.5 cm; data processed with ReflexW. The step-frequency GPR was pulled by a vehicle; distance between adjacent profiles, 7.5 cm; trace spacing, 2.5 cm; data processed with examiner	Nantes (France)	IFSTTAR accelerated pavement testing facility, with different defects intentionally incorporated at the interface between two asphalt layers	To test two commercial multichannel systems and investigate the usefulness of 3-D images	A 3-D view of the pavement structure was reconstructed, with both systems. It is apparent that a 3-D image strongly facilitates the detection, localization, and identification of pavement anomalies while revealing their size, shape, and/or direction	France, Norway, the United Kingdom

26	Impulse GPR with 800MHz air-coupled antenna, mounted on a vehicle FWD Coring	First GPR survey carried out in 2004; second GPR survey carried out in 2013. Distance between antenna and road surface, 25cm; trace spacing, 25cm	Barcelona (Spain)	Highway (2.5km-long section), flexible pavement, constructed and opened to service in 2004	To investigate what kind of information can be deducted by looking at GPR data in the spectral domain (given that, in road surveys, GPR data are usually examined in the time domain)	In the spectral domain, signal peaks have lower amplitude and occur at lower frequencies when road layers are thicker. Different regions of the spectrum can be associated with reflections occurring at different interfaces. With higher water content, the bandwidth is narrower, and peaks occur at lower frequencies. There are more peaks when the road porosity is higher	Spain
32	Impulse GPR with two 1800MHz air-coupled antennas (GSSI), mounted on a vehicle Coring	Acquisition of 44 longitudinal or transversal profiles; trace spacing, 20cm in the first survey; 5cm in the second survey Data processed with Radan	NA	Concrete prestressed runway pavement	To detect the position of reinforcement in the concrete slab of an airport runway, for evaluating the bearing capacity of the runway	This is an example of application. Based on GPR data calibrated with cores, longitudinal and transversal reinforcements were located in the concrete slab	Portugal, Spain
23	Geoscope GS3F step-frequency GPR with VX1213 antenna, with seven channels (3-D radar). Bandwidth: 200–3000MHz	Distance between antenna and road surface, 15cm Trace spacing: 5cm A synthetic dataset was also created, by using the free software MEEP	(1) Wavre (Belgium) (2) Lovendegem (Belgium)	(1) BRRC test site, with four pavement sections made of different materials and with different layer thicknesses (2) Grassland	To develop a new technique for velocity estimation, to be employed when air-coupled antennas are used and the CMP method is not applicable	The proposed method can be used with small-offset air-coupled antenna configurations. By taking into account the refraction at the air-pavement interface and the roughness of the soil, it is possible to estimate depth and propagation velocity with higher accuracy than with traditional approaches	Belgium
33	RIS Hi-Pave HR1 impulse GPR with 1000MHz antenna (IDS)	Two surveys carried out on the same road sections, with 6months between the first and the second survey Data processed with GRED 3-D	Rieti (Italy)	Rural roads (three road sections, total length 160km)	To predict rutting evolution in flexible pavement, at road network level	By using prediction curves available in the literature and by exploiting the results of the surveys, rutting prediction curves were inferred for the considered roads. The accuracy of the prediction will be verified by performing again GPR surveys on the same roads in the future	Italy, the United Kingdom

The effectiveness of multichannel GPR systems and the usefulness of 3-D images of roads were investigated in Simonin et al. [31], and [35]. In Simonin et al. [31], an impulse GPR and a step-frequency GPR, with 8 and 11 channels, respectively, were tested and compared on the accelerated pavement testing facility of the French institute of science and technology for transport, spatial planning, development and networks (IFSTTAR). In [35], a noise-modulated GPR developed in Australia was tested; this system is specifically conceived for road assessment and has 8 transmitters and 20 receivers housed within 4 adjacent ground-coupled pods, each of which contains 2 transmitters and 5 receivers; by pairing different transmitters and receivers, a series of virtual antennas are achieved across the lane width, arranged into eight partially overlapping WARR groups. The major conclusions of those experimental works are that multichannel systems, in addition to enabling faster inspection of wide areas, allow constructing much more easily a 3-D view of the investigated structure (and horizontal slices); such view strongly facilitates the detection, localization, and identification of anomalies. Moreover, the use of multichannel systems improves the accuracy of thickness estimation, especially in the case of variable pavement conditions.

Further TU1208 studies regarded the combination of GPR with complementary techniques. For example, GPR was combined with load tests (FWD) for structural pavement evaluation and quality control during construction [24,36]. Solla et al. [37,38], studied the use of GPR and thermography for the detection of pavement cracking.

In Tosti et al. [29], the new challenge of inferring mechanical properties of road pavement and materials from their electromagnetic characteristics was investigated. An impulse GPR, with ground-coupled antennas, was used over a test site with a flexible pavement structure; Young's modulus of elasticity was measured in more than 800 positions by using an LFWD. A semiempirical model for predicting the pavement strength properties was developed; the model was calibrated by comparing the elastic modulus measured by using the LFWD and the electromagnetic response of the pavement structure. A good agreement between measured and predicted values was found, thereby showing good promises for large-scale mechanical inspections of pavements using GPR. Subsequent studies aimed at further testing and improving the model; measurements were carried out on a 45 km-long section of a rural Italian road, in the Rieti district, and on 90 km of Spanish roads, in the districts of Madrid and Guadalajara. The obtained results are promising, and they are in the process of being published; the proposed approach still needs to be refined and further investigated.

In Uddin et al. [39], an overview on the use of GPR in the United States for the evaluation of pavement thicknesses and cracking can be found. In the United States, GPR systems for road investigations are commonly mounted on a van and equipped with air-coupled horn antennas; surveys are carried out at traffic speed, without traffic interruption or separate control. The trace spacing varies from 30 cm to 3 m (depending on the survey purpose and required resolution), which is larger than the spacing commonly used in Europe. GPR power levels are lower than in Europe; hence, scanning in the field is more sensitive to noise interference from passing vehicles, radio transmissions, radio-frequency signals near airports, smartphones, and other facilities producing electromagnetic signals.

In Benedetto et al. [40], a review on GPR signal processing for road applications is provided.

Looking at Table 1, it can be observed that, in COST Action TU1208, experimental investigations on road pavements have been carried by using both impulse and step-frequency systems, with ground- and/or air-coupled antennas in the 200–3000 MHz range. The trace spacing along a profile ranges from 0.5 to 25 cm (4–200 scans per meter, that is, more traces are acquired than are usually done in professional works), depending on the employed system/antenna and on the purpose of the study. Comparisons between different radar systems are rarely carried out. For road applications, slice scan and 3-D data view are not common, though there is awareness that they allow a more accurate characterization of the road structure and anomalies. Most studies involved scientists from research-intensive countries. Notwithstanding all studies listed in Table 1 contemplate experimental activities, industry involvement is observed only in [31].

3 RAILWAYS

At European level, the main challenges for railway networks are the increased train traffic and traffic loads, in a scenario where the network has been in service for several decades, already. Due to economical restrictions, it is important to adopt maintenance and rehabilitation solutions that guarantee an extended track life cycle, by promoting the adoption of timing solutions and by preventing severe deteriorations that normally require total rehabilitation, which is expensive and time-consuming. Due to the increased traffic, it is also important to be able to monitor and maintain the existing tracks without significant traffic interruption.

Note that more traffic leads to less maintenance time availability and that heavier loads wear more down the track. Maintenance planning is a crucial factor for an effective and cost-efficient management of a railway network, and to plan a proper maintenance, it is essential to have at disposal a realistic evaluation of the track condition. Specially designed vehicles exist, for inspection operations, and they are regularly used in several countries to assess the track geometry, the rail profile, and other parameters. Maintenance works are planned when the track condition reaches a predefined level, called "intervention level" [41]. Generally, such works consist in adding ballast through tamping [42]: this is an operation that corrects the track geometry but does not solve the deterioration causes, which are mainly due to substructure settlements.

GPR started being applied for track diagnosis in the last 15 years [28,43], and it is mostly employed for the assessment of ballast layer thickness [44] (same as in road investigations, it is mainly used for estimating the thicknesses of pavement layers). GPR can also estimate the fouling level of ballast and the moisture content [45–48]. Water-content estimation is very important because the presence of water causes a variation of ballast permittivity; hence, it affects the estimation of the thickness of the ballast layer. Fouling evaluation is crucial as well, because the proper operation of a ballast track bed may be severely undermined by the presence of fine-grained material filling the voids between the ballast aggregates and affecting the original strength mechanisms.

Ballast thickness assessment by means of GPR is widely accepted and is applied on a regular basis in several European countries, whereas fouling and moisture-content evaluations are rather new applications and require further investigations.

By now, it is apparent that, by using a GPR, the causes of track deteriorations can be better understood and their evolution can be studied; hence, track maintenance can be more effectively planned over time and more appropriately designed in terms of type of interventions (e.g., it is possible to decide whether tamping or ballast cleaning is required and it is possible to evaluate the depth of ballast cleaning needed) [49].

3.1 OBJECTIVES AND METHODOLOGY

Currently, the use of GPR results combined with track geometry measurements enables a realistic and efficient rehabilitation of the track, as it allows the detection and proper localization of (i) zones where the thickness of the ballast layer is below the regulation limit, (ii) substructure settlements, (iii) sections with fouled ballast, (iv) sections with accumulated moisture, and (v) areas with the lack of ballast between sleepers. As already mentioned, GPR provides useful information about the fouling conditions and moisture content, but more research is needed to improve the effectiveness of the technique in those areas.

The use of GPR for railway assessment presents several challenges. Obviously, there is a practical difficulty of dragging the antennas over the track surface, due to the sleeper geometry. Another problem is the presence of strong reflections into radargrams, due to the presence of metal rails. To address the first issue, air-coupled antennas are used and, also, ground-coupled antennas specifically developed for railway assessment [47]; see Fig. 5. To reduce the influence of rail reflections, generally, three antennas are used: An antenna is positioned between the rails, and two antennas

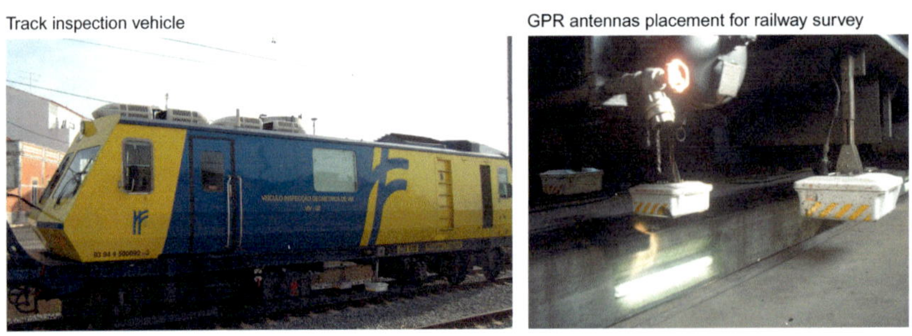

FIG. 5

Track inspection with GPR: track inspection vehicle, on the left, and suspended ground-coupled antennas placed laterally and between the rails, on the right.

are placed laterally. In this way, by means of suitable processing algorithms, it is possible to filter the rail reflections [44].

Usually, 500 MHz antennas are employed for thickness measurement, whereas 2000 MHz antennas are more suitable for fouling evaluation and moisture detection because of the higher-resolution images that they provide.

The processing of GPR data is different when the thickness of ballast layer or fouling levels have to be estimated [45,50]. Dedicated software tools have been developed by GPR manufacturers, for the processing of GPR data collected over railways. Time-domain processing is more developed and is the most suitable for thickness assessment; recently, frequency-domain approaches are being developed for ballast fouling evaluation. Future developments of GPR processing techniques are expected, not only to better identify the railway properties and conditions but also to automatize the whole activity, given the significant length of railway tracks to be normally inspected [44,51].

3.2 COMBINATION OF GPR WITH COMPLEMENTARY NONDESTRUCTIVE TESTING TECHNIQUES

The joint use of GPR with complementary NDT approaches is useful in all civil-engineering works, for a more in-depth evaluation of the structure under study; this is true in the field of railway assessment, as well.

Geophysical NDT methods, such as electric tomography, gravimetry, seismic propagation, spectral analysis of surface waves, and load tests performed by means of plate loads or LFWD, can be used with GPR for a better track prospection [7,49,52,53]. Most geophysical methods present the drawback of performing confined measurements, as opposed to GPR that can provide continuous data; moreover, they are affected by the rail and sleeper presence (same as GPR).

Measurements along the track, for assessing its geometric properties, are very useful, and they are regularly performed by railway administrations to classify the track quality and plan maintenance campaigns. Track geometry measurements are carried out at traffic speed, and they can be performed simultaneously to GPR evaluations, in the same positions. As already mentioned, a joint interpretation of track geometry measurements and GPR data enables a precise diagnostic of the track.

3.3 EXAMPLE AND TU1208 RESEARCH ACTIVITIES

3.3.1 Example: Studying the evolution of the track geometry and identifying zones with different ballast conditions

An example of track geometry evolution, due to subgrade settlement, is presented in Fig. 6A; the graph was built based on a series of radargrams collected on the same railway section every 6 months. One of those radargrams is shown in Fig. 6B. In this case, information about the railway section was inferred by looking at GPR data in the time domain. More details about this case study can be found in the open-access paper [44].

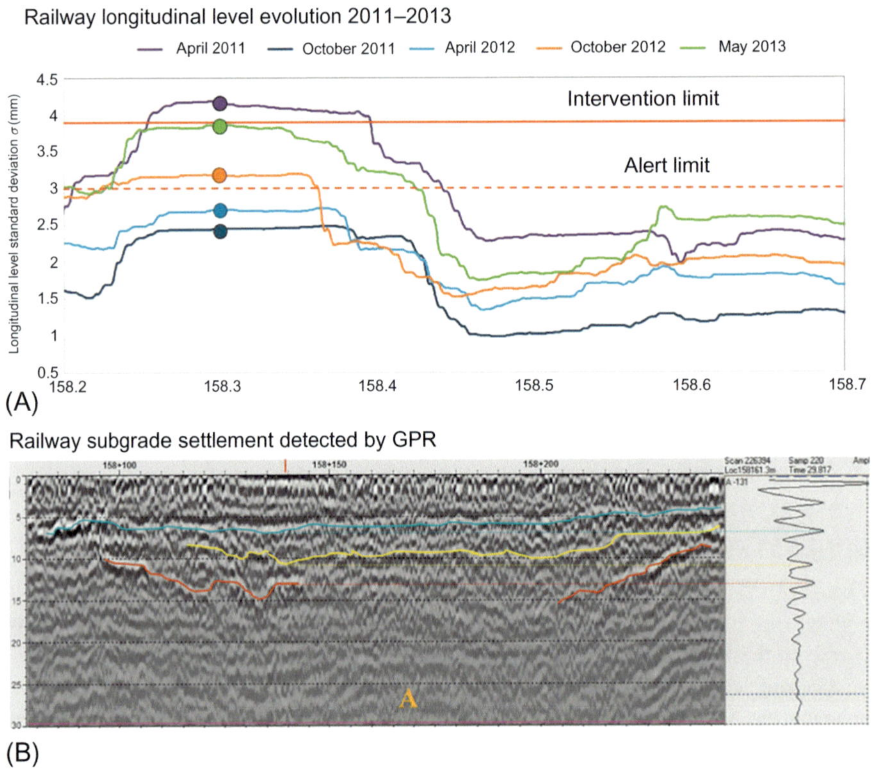

FIG. 6

(A) Longitudinal level evolution due to subgrade settlement, inferred from a series of GPR radargrams collected every 6 months on the same railway section. (B) An example of GPR radargram used for producing the graph in (A).

It is also possible to deduce interesting parameters by looking at data in the frequency domain; *inter alia*, patterns can be identified in the spectrum of the GPR signal, corresponding to deteriorated zones that are not evident in the time domain. To illustrate this concept, a second example is presented in Fig. 7: here, the analysis of the spectrum of the GPR signal allowed to better identify zones with different ballast conditions, probably due to different levels of fouling.

3.3.2 Research activities carried out within COST Action TU1208

The use of GPR for the assessment of railways is a topic dealt with in WG2, Project 2.1, of COST Action TU1208. Several studies have been carried out during the last 4 years, which have contributed to raising the level of knowledge in this area of application of the GPR technique. The performed investigations include experimental activities carried out in the laboratory or in situ and numerical simulations of electromagnetic propagation and scattering in ballast. The most interesting activities are resumed in the following (see also Table 2).

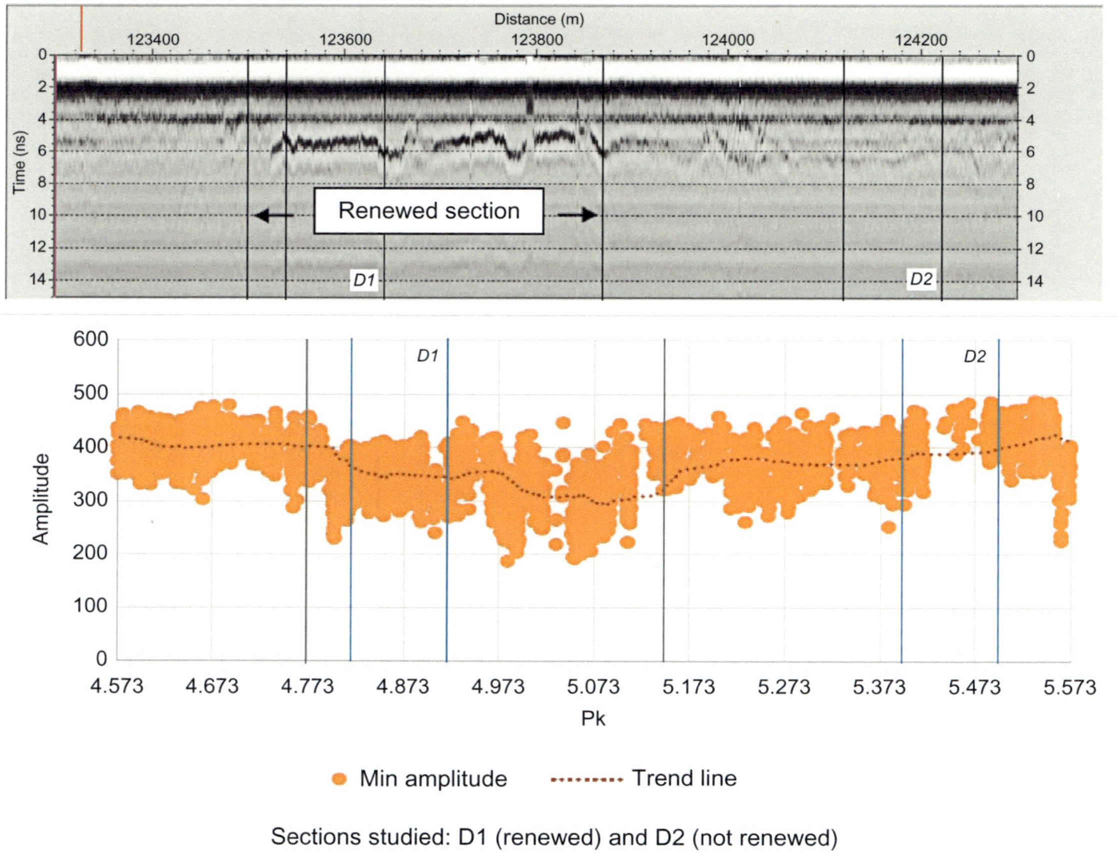

FIG. 7

Radargram and corresponding variation of the minimum amplitude in the fast Fourier transform of the GPR signal. The central frequency was 1700 MHz. The measurements were taken along a track section with different ballast conditions.

Table 2 Examples of Experimental Investigations Carried Out Within COST Action TU1208 in the Laboratory (on Ballast Samples) and on Railway Sections

References	Technology	Methodology	Place	Railway Type	Purpose of the Study	Major Findings and Remarks	Involved Countries
47	Impulse GPR systems SIR 20 and SIR 10H (GSSI), with (1) air-coupled 400 MHz antenna, (2) ground-coupled 500 MHz antenna, (3) ground-coupled 900 MHz antenna, (4) air-coupled 1000 MHz antenna, (5) air-coupled 1800 MHz antenna	The GPR signal was recorded over clean ballast, silt soil, and fouled ballast (five fouling levels were considered; for each fouling level, four different levels of water content were considered). For air-coupled antennas, the distance between antenna and sample under test was 30 cm. Measurements were repeated after rotating the antennas of 90°	Lisbon (Portugal)	Test sites in the laboratory are as follows: (1) 110 × 94 (base) × 40 cm (height) clean granite ballast, in a plastic box (2) 53 × 41 (base) × 30 cm (height) box silt soil, in a plastic box (3) 75 × 50 (base) × 30–40 cm (height) fouled ballast, in a wooden box All boxes with aluminum sheets at the bottom	To measure the effective permittivity of ballast, with different fouling and water-content levels	The ballast permittivity increases with both fouling and water-content levels; ballast has a frequency-dispersive behavior. Numerical modeling can be useful to generalize the results of this work	Italy, Portugal
44	Impulse GPR with air-coupled 400 MHz antennas	In order to detect critical areas, the standard deviation of the geometry measurements was calculated over a 200 m mobile window; the window was shifted with a 0.25 m step. Critical areas were identified by analyzing GPR data, as well. As a subsequent step, GPR data were analyzed more in depth in the critical areas, for a deeper understanding of the causes of defects	Portugal	A 400 km-long railway section, in service	To combine GPR data and track geometry measurements, for an accurate evaluation of railway conditions and for a deeper understanding of the causes of defects	The use of GPR allows to better identify track trenches with problems and also to understand the causes of the deteriorations, so that maintenance can be more effectively planned	Italy, Portugal
40	Impulse GPR system, with air-coupled antennas having center frequency of 1000 MHz (RIS Hi-Pave HR1 1000 by IDS) and 2000 MHz (RIS Hi-Pave HR1 2000 and 2000 NA, by IDS)	The GPR signal was recorded over ballast with four different levels of fouling. Two-dimensional FDTD simulations of the scenario were performed by using gprMax and E^2GPR	Rome (Italy)	Test sites in the laboratory are as follows: 147 × 147 (base) × 48 cm (height) fouled ballast, in a methacrylate tank, above a 2 × 2 m copper sheet Ballast aggregate size between 31.5 mm and 50 mm	To measure the effective permittivity of ballast, with different fouling levels	The ballast permittivity increases with fouling level. The best results are obtained with the 1000 MHz antenna. The usefulness of numerical modeling is shown	Italy, the United Kingdom

The influence of fouling level and water content on the GPR response was assessed in the laboratory by [47]. In this study, two impulse systems and five different ground-coupled and air-coupled antennas were used. Measurements were performed under controlled conditions: Boxes of ballast with five different fouling levels were prepared, and the GPR response was recorded over them; moreover, for each fouling level, four different levels of water content were considered. In this way, it was possible to understand how the effective permittivity of ballast varies, within the considered ranges of fouling and water-content levels. It was also observed that ballast has a frequency-dispersive behavior. Note that, in real railways, fouling and water content are not independent phenomena: when the ballast is more fouled, the water content in the ballast layer is higher.

More recently, laboratory studies were performed by [46]. The purpose of the experiments was, again, to characterize the ballast electromagnetic properties for different levels of fouling. Moreover, the test scenario was modeled numerically by using the open-source simulator gprMax [21] and the freeware tool E^2GPR [22]. Ballast was stored in a large square-based methacrylate tank and progressively polluted with a fine-grained silt soil. Four different levels of fouling were considered. An impulse GPR system, equipped with three different air-coupled antennas, was used. The permittivity of the ballast system under investigation was estimated from GPR data, and results were validated by using the volumetric mixing approach (available within the literature) and via a comparison with simulation results. Some experimental aspects of this research activity were presented in [54], whereas the electromagnetic simulation was described in detail in [55].

The GPR data collected with 2000 MHz antennas during the experiments presented in [46], were further analyzed in [56]. In [57], new data were presented, which were obtained by using 600 and 1600 MHz ground-coupled antennas.

Joint interpretation of GPR data and track geometry measurements was performed in [44]. The experimental activity was carried out on a 400 km-long railway section. The purpose of the work was to develop an improved procedure for the combination of GPR data and track geometry measurements, for a more accurate evaluation of railway conditions and for a better understanding of the causes of defects. The proposed procedure includes three main phases: detection of critical areas based on track geometry measurements and GPR data, accurate analysis of GPR data in the critical areas, and suggestions for maintenance planning. In order to detect critical areas, a new method was proposed for the analysis of track geometry measurements, consisting in calculating the standard deviation of the measured parameters over a 200 m mobile window; it was suggested to shift the window with a 0.25 m step. In the commonly employed approach, instead, the standard deviation of the measured parameters is calculated over 200 m nonoverlapping windows. Moreover, critical areas were identified by analyzing GPR data, as well. Then, GPR data were analyzed more in depth in the critical areas, in order to understand the causes of defects. More experimental investigations are needed, but the case study presented in [44], demonstrates the effectiveness of the GPR technique for the identification of track trenches with problems and for a better estimation of the causes of damages.

Looking at Table 2, it can be observed that, in COST Action TU1208, experimental investigations on railway ballast have been performed by using impulse systems with ground- and/or air-coupled antennas in the 400–2000 MHz range. Comparisons between different radar systems and operation frequencies have been often made. Slice scan and 3-D data view have not been used. In this field of application of the GPR technique, research activities are being carried out only by two research teams: One of them involves researchers from Sapienza University of Rome (Italy) and the National Laboratory of Civil Engineering of Portugal; the other team involves researchers from Roma Tre University (Italy), Sapienza University of Rome, and University of West London (the United Kingdom). Recently, some activities are starting in the University of Pardubice (Czech Republic). The publication of more case studies carried out on real railways is desirable, and the involvement of industry in this area of research needs to be fostered. The realistic electromagnetic modeling of railway scenarios is envisaged to be further developed, because of its high potential in supporting the interpretation of datasets and in understanding the electromagnetic phenomena taking place inside the structure under test.

4 BRIDGES

Bridges are engineering structures of primary importance, as they connect separated lands to improve economic and social development. Many bridge structures have experienced structural degradation because of increases in traffic loads and intense vibrations, which are significantly higher than those estimated for their original design. Therefore, bridges require special attention and frequent diagnosis of their changing structural condition, to help with their management and conservation. The use of NDT methods is highly recommended for bridge monitoring, to prevent potential structural damage that may result from more invasive testing techniques. GPR is quite commonly used in bridge inspection because it is capable to produce an overall internal image of the structure in a relatively quick time.

4.1 OBJECTIVES AND METHODOLOGY

This subsection starts with a brief overview of the main GPR applications on the condition monitoring of both concrete and masonry bridge structures. Subsequently, the inspection methodology is presented.

GPR diagnosis of concrete bridges includes the following:

- condition evaluation of a bridge deck [58,59]
- thicknesses of bridge deck [60]
- position of reinforcing bars and tendon ducts [61]

- diameter of reinforcement [62]
- concrete quality (e.g., porosity) [63]
- moisture content and chloride content (corrosion) [64,65]
- debonding and delamination or cracking [66].

GPR applications on masonry bridges are the following:

- detection of hidden arches and/or ancient profiles [67]
- building materials and evidences of restoration or reconstruction activities [37,68]
- moisture content [69]
- cavities and cracking in masonry [70]
- thicknesses of masonry (pavement, ring stone of arches, etc.) [71]
- ring separation [72]
- bridge foundations [69,73].

As mentioned in the Introduction, an important parameter controlling the depth range of the GPR method is the operation frequency. The antenna central frequency shall be carefully chosen, keeping an appropriate balance between penetration depth and resolution: A low-frequency antenna gives deeper signal penetration but poorer resolution; a higher-frequency antenna gives better resolution but shallower penetration. The 250 and 500 MHz central frequencies are the most suitable, for obtaining an overall image of a bridge structure: they allow reaching the bridge foundations while analyzing the shallower filling. When better resolution is required (e.g., for the analysis of rebar condition in reinforced concrete bridges), higher frequencies are used, in the 1000–2500 MHz frequency range.

Both air-coupled and ground-coupled antenna configurations are used in bridge surveying. The antennas are most commonly mounted on a van or in a trolley, with minimal traffic interruption, and they are connected to an external global positioning system (GPS) for trace tagging or to a distance measurement indicator (DMI) to control the trace interval and to measure the traveled distance. The ground-coupled antennas operate suspended at 5–10 cm over the surface, whereas the air-coupled antennas operate suspended 40–50 cm over the surface (suspended antennas allow gathering data at higher speed). Apart from their height of operation, the air-coupled antennas usually are horns working in the 1000–2000 MHz range, mainly for thickness measurements and rebar assessment; the ground-coupled antennas work in a wider frequency range, from 100 to 2000 MHz, for various purposes.

For large survey areas and, particularly, to inspect utilities and rebar in concrete bridges, the use of array multichannel systems and 3-D visualization are recommended, as they allow surveying large areas in less time. However, when a more detailed analysis is required (e.g., water-content or concrete quality evaluation), single antennas can be used. In the case of masonry arch bridges, single antennas are most commonly employed, because of the complexity of the bridge structure and heterogeneity of building materials; to measure the profile lengths and to control the distance interval, the GPR system is usually mounted on a survey cart with an odometer wheel. Some masonry arch bridges have intact ancient flagstone surfaces composed of large irregular blocks, which make measurements difficult as the continuous movement of the survey wheel is often interrupted. Moreover, for masonry bridges, it is recommended to record more than one profile because of the heterogeneity in filling (having at disposal only one profile may lead to inappropriate data interpretation).

A large variety of processing filters can be applied to GPR data collected over bridges, to reduce clutter and any unwanted noise. The objective is to correct the downshifting of the radar section due to the air-ground interface and to amplify the received signal, as well as to remove both low- and high-frequency noise. The most-used filters are time-zero correction, dewow (temporal filtering), background removal (spatial filtering), gain, band-pass filtering, and migration. Additionally, ancient bridges usually present a double-slope or steep profile, and topographic corrections are therefore applied to improve the accuracy in imaging subsurface features.

The use of advanced 3-D processing techniques has benefited the visualization and reconstruction of the occupied underground space, particularly in the case of piping network and rebar configuration.

More advanced and sophisticated processing includes the use of image processing (e.g., Hilbert transform) and algorithms for automated detection.

4.2 COMBINATION WITH COMPLEMENTARY NONDESTRUCTING TESTING TECHNIQUES

The combination of GPR with complementary NDT techniques is very useful in the assessment of bridge stability. Various NDT techniques can be applied to the inspection of bridges, such as sonic/ultrasonic, electromagnetic, and electric methods and infrared thermography. Each particular NDT method provides different information owing to the physical properties of the construction materials used. In [74], a review can be found of NDT methods commonly used for the assessment of concrete and masonry structures; although the paper is rather old, it is still interesting: the parameters that can be measured and the advantages and limitations of each method are described.

For the inspection of concrete bridges, the combined use of GPR and other NDT methods has earned the interest of researchers on the assessment of water and chloride contents in concrete and rebar corrosion, as well as concrete quality and delamination. Breysse et al. [75], presented different case studies combining complementary NDT techniques. A combination of GPR and ERT methods was used for the assessment of water content in a concrete bridge (Lille, Nord, France); the resulting data showed that an increase of water content decreases the wave amplitude (GPR) and decreases the resistivity (ERT) values. Another study was performed to evaluate the concrete quality of the Empalot Bridge (near Toulouse, France) by the combination of GPR and infrared thermography (IRT), and delamination was confirmed. In [76], an overview is presented on different NDT methods for quality assessment of concrete slabs with tendon ducts and other reinforcement, including ultrasounds, GPR, and impact echo (IE). Villain et al. [77], employed a combination of GPR, capacitive, and IE measurements to evaluate water and chloride contents and to analyze porosity. Alani et al. [78], presented an integrated health assessment on the Pentagon Road Bridge (Chatham, Kent, the United Kingdom), which is a reinforced concrete bridge, by the use of GPR, interferometry (IBIS-S technology), and accelerometer. All the NDT data produced were integrated in a finite-element method (FEM) model to assess the dynamic behavior of the bridge and to measure deterioration. The GPR allowed for the position of rebar and other reinforcements underneath the bridge deck, as well as damages such as moisture problems, whereas excessive deflection (IBIS-S) was related to inner defects (cracks or delamination).

With respect to the evaluation of stonemasonry arch bridges, [79], combined GPR, sonic, and conductivity to investigate inhomogeneity and layering within the bridge and to identify moisture-drainage problems. Flint et al. [80], combined GPR, seismic, and ERT methods to evaluate changes in the internal condition of a masonry viaduct. The resulting data allowed mapping the existence of cavities and moisture into the structure. Orbán and Gutermann [81], presented a combination of GPR, IRT, and sonic methods to investigate hidden characteristics such as cavities, moisture, and inhomogeneity in filling. Pérez-Gracia et al. [82], combined the GPR method with vibration modes to model the dynamic behavior of a bridge structure. Fauchard et al. [83], combined two complementary geophysical methods, GPR and ERT, to obtain noteworthy information such as cavities and moisture. Bergamo et al. [84], combined the use of GPR, seismic, and IRT methods to create the model of the bridge structure to be simulated using integral FEM analysis in order to provide the structural behavior of the structure. Russo [85], employed GPR and sonic methods to analyze the structural behavior of the Rialto Bridge in Venice (Italy). Stavroulaki et al. [86], presented a combined approach of GPR and terrestrial laser scanning (TLS) to elaborate data in a FEM analysis. The complex external geometry of the structure is provided through TLS, while the GPR gives information of its internal composition.

An interesting study was carried out on the Roman Bridge of Lugo (in Galicia, Spain), which is a masonry arch bridge that has suffered several reconstruction and restoration works from many eras. The bridge was assessed by using GPR in combination with light detection and ranging (LiDAR) equipment [68]. In particular, static and mobile LiDAR instruments were used for the evaluation and documentation of the external masonry and GPR for its internal inspection. The most important results are related to the creation of a well-documented 3-D bridge model and identification of different building materials in both external and internal stonework (granitic or slate materials), as well as inner modifications over its lifetime (e.g., a former double-slope profile from the Roman structure).

4.3 EXAMPLE AND TU1208 RESEARCH ACTIVITIES

4.3.1 Example: Interpretation of real data—The case studies of the Roman Bibei bridge and the medieval Traba bridge

Figure 8 shows data acquired with antennas of 500 and 250 MHz central frequencies. The purpose was to reach the bridge foundations while analyzing the shallower filling for a complete analysis of the structure. The 500 MHz data provided higher resolution, which allowed identifying the reflections produced by the fill-stone interface, in addition to the reflections produced by the arch-air interface and the layering distribution above the arches. The knowledge of these arch-ring thicknesses represents an important parameter for a mechanical analysis, because arches are structural elements that are crucial to the stability and durability of bridges. On the other hand, the 250 MHz data provided higher penetration, which allowed identifying the water level at the air-water interface and, subsequently, the heights of the arches.

An illustration of the usual heterogeneity in filling is presented in Fig. 9, which shows two longitudinal parallel profiles collected in the Traba bridge, in Spain, with a 250 MHz antenna. The occurrence of restoration activities in the stonework of the bridge is apparent from the external appearance of the bridge; the first and third arches from the left margin upstream are barrel arches, whereas the other two are gothic arches. These differences are due to different times in the construction process: The gothic arches are from the original medieval bridge, and the barrel ones come from a later restoration. The GPR data obtained showed differences for the detection of the third arch (around 15–16 m, in Fig. 9). Moreover, it was possible to distinguish a reflection pattern in form of consecutive small hyperbolas over the first arch (from 6 to 9 m, in Fig. 9), which was certainly produced by the existence of reinforced concrete. However, this reflection pattern is not present in the upper radargram so that this reinforcement probably does not occupy both ways of the pavement.

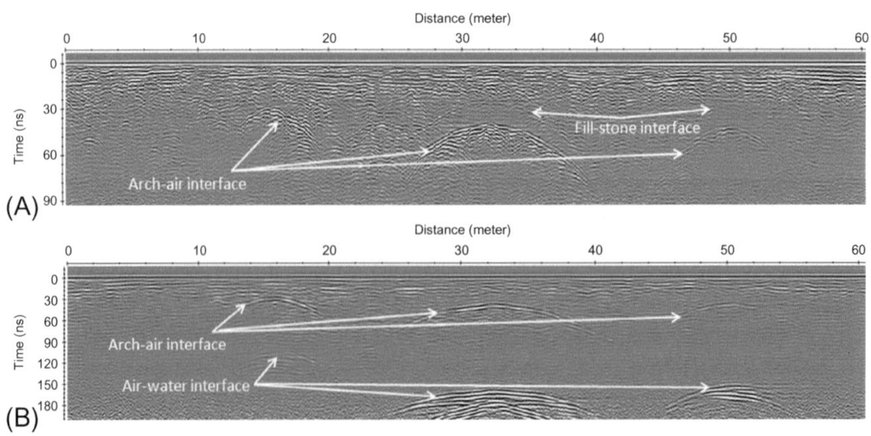

FIG. 8

500 MHz (A) and 250 MHz (B) radargrams obtained on the Roman Bibei bridge, in Spain.

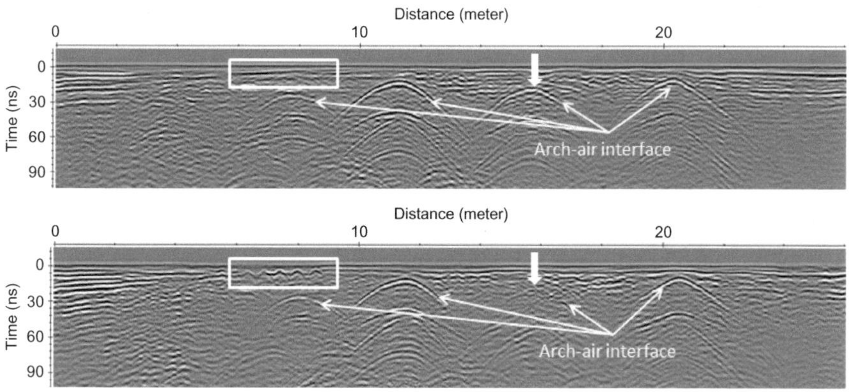

FIG. 9

250 MHz data recorded along two longitudinal parallel profiles, on the medieval Traba bridge (Spain); differences between the profiles highlight the importance of collecting more than one profile on masonry bridges, for a more complete and correct data interpretation.

4.3.2 Research activities carried out within COST Action TU1208

The use of GPR for the assessment of bridges is a topic dealt with in WG2, Project 2.1, of COST Action TU1208. The investigations performed within the Action include experimental activities carried out on various ancient and modern bridges; in many cases, the inspected bridges were also modeled and simulated. The already mentioned guidelines for the use of GPR on concrete structures will include recommendations for the assessment of bridges. In the following of this subsection, a selection of TU1208 case studies is presented (see also Table 3).

Several studies regarded the evaluation of historical stonemasonry arch bridges.

For example, Alani and Banks [87], combined GPR with 3-D laser scanner for assessing the Old Bridge of Aylesford (in Kent, England), which was built in the 12th century. As a result of this activity, the thicknesses of different layers of the deck structure were measured, the original stone base of the bridge was identified, structural ties were detected, and numerous defects or nonstructural features were mapped.

A team of Spanish members has used GPR and complementary NDT approaches to elaborate modeling methods for the electromagnetic simulation of complex bridge scenarios [89,90]. In Solla et al. [90], a very interesting and advanced study is presented, where the Lubiáns bridge (in Galicia, Spain) was investigated. This is a medieval masonry arch bridge dating back to the 15th century that has suffered several reconstructions using different construction materials. A GPR survey was carried out, with the main objectives to classify the different filling existents while defining the distribution of materials in the interior of the bridge structure and to detect damaged zones where restoration and maintenance are recommended. Given that the heterogeneity in masonry complicates the analysis of the signals detected by the

Table 3 Examples of Experimental Investigations Carried Out Within COST Action TU1208 on StoneMasonry Arch and Concrete Bridges

References	Technology	Methodology	Place	Bridge Type	Purpose of the Study	Major Findings and Remarks	Involved Countries
87	Impulse GPR (IDS RIS Hi-BrigHT), eight channels (10 cm spacing); central frequency of the antennas, 2000 MHz Leica P20 3-D laser scanner	The bridge deck was fully surveyed with GPR, by performing four equally spaced scans (the footprint of each scan being 80 cm wide). To break data into manageable files, the survey was broken into three zones. Three-dimensional views of data and horizontal slices were used for data analysis and interpretation	Old Bridge Aylesford (the United Kingdom)	Masonry arch bridge; the older part is from the twelfth century	To perform a GPR survey of the asphalt surface of the bridge, in order to determine the following parameters: the thickness of the upper asphalt layer, the total depth from the bridge surface to the historic stonework, the uniformity of the upper asphalt layer, and its total depth	The positions of different layers of the deck structure were established. The thicknesses are not uniform and vary across the surface of the bridge. Structural ties (anchors) have been detected and localized. Locations of numerous defects and nonstructural features were identified. A FEM model of the bridge was realized	The United Kingdom
88	Impulse GPR (GSSI SIR20), with 4205 horn antennas (2500 MHz central frequency)	Distance between profiles, 0.5 m; trace spacing, 2 cm. Two orthogonal antenna orientations used Data processed with ReflexW	A bridge on the National Road (Nr) 13 near Splügen, Grisons (Switzerland)	Concrete bridge, with a 79 m-long and 10 m-wide deck	To inspect, by using GPR, a bridge deck is designated for demolition, looking for chloride contamination. To compare GPR results with the results of a destructive testing (probing)	The comparison between GPR results and destructive testing demonstrated that the nondestructive approach detected all contaminated areas. Few areas that were not contaminated caused false alarms	Switzerland
10	Step-frequency GPR (Geoscope by 3-D radar) with commercial air-coupled and prototypal ground-coupled antenna arrays. The air-coupled array has 21 channels, and the ground-coupled array has 22 channels (7.5 cm spacing); both use boomerang monopole elements and cover the frequency band 200–3000 MHz	The air-coupled antenna was 40 cm from the ground. Trace spacing, 2.5 cm	Norway	Concrete bridge with asphalt layer on top	To test a new ground-coupled array and compare results with those of a commercial air-coupled array using the same antenna elements	The thickness of the asphalt layer was estimated as varying from 7 to 9 cm, with both antenna arrays. The ground-coupled array provided a clearer image of each rebar inside the bridge deck; moreover, it could reach a higher penetration depth	Norway

89	Impulse GPR (RAMAX CU-II), with 250 and 500 MHz antennas. Three-dimensional laser scanner. Photogrammetry	GPR trace spacing, 5 cm for the 250 MHz antenna and 2 cm for the 500 MHz antenna; 2 longitudinal profiles acquired, with a distance between profiles of 1 m	Carracedo bridge and Monforte bridge (Spain)	Stonemasonry arch bridges	To assess the internal condition of the bridges. To show the effectiveness of electromagnetic modeling as a supporting tool for the interpretation of experimental data	Valuable previously unknown information about the bridges was revealed, such as the hidden structural details, the geometry of ancient profiles, the existence of voids, and the evidence of restoration and reconstruction tasks. Nevertheless, it was not possible to obtain clear information about the presence of cracks or moisture within the bridges. Electromagnetic modeling significantly helped data interpretation	Spain
90	Impulse GPR (RAMAX CU-II), with a 500 MHz antenna Leica TCR 1102 total station Nikon D200 with Nikkor 20 mm f/2.8D lens IR camera NEC TH9260 with uncooled focal plane array	GPR trace spacing, 5 cm; 2 longitudinal profiles acquired, with 620 traces per profile; distance between profiles, 1 m. GPR data were processed with ReflexW. gprMax was used to create a realistic model of the bridge and calculate synthetic data that were compared with experimental data	Lubiáns bridge (Spain)	Medieval masonry arch bridge from the fifteenth century	To perform a GPR survey in order to define the internal condition of the bridge. To create an electromagnetic model of the bridge, for improving the interpretation of experimental data	The internal state of the structure was successfully defined. The combination of photogrammetric and thermographic data, jointly with the development of custom algorithms, was used to create a realistic electromagnetic model of the bridge that was simulated with gprMax. Synthetic data significantly aided the interpretation of experimental data. Such an advanced and multidisciplinary approach allowed determining the composition and distribution of different materials in the interior of the bridge	Spain

GPR and their interpretation, numerical modeling was used to assist such interpretation. The combination of photogrammetric and thermographic data, together with the application of the finite-difference time-domain simulator gprMax, allowed creating large-scale and rather realistic models. The synthetic data obtained from the simulation were compared with the experimental data, and this allowed understanding the composition and distribution of the filling in the interior of the bridge. In Solla and Nieto [89], the usefulness of electromagnetic modeling in the interpretation of experimental radargrams is shown; here, the Carracedo bridge and Monforte bridge (in Galicia, Spain) are studied.

The GPR inspections of concrete bridges carried out in the framework of COST Action TU1208 include a bridge located on the National Road (Nr) 13 near Splügen, Grisons, in Switzerland [88]. The bridge was designated for demolition; hence, it represented an opportunity for performing tests with GPR and comparing the results of the nondestructive inspection with the results of destructive (probing) tests. The main purpose was to detect and map areas contaminated by chlorides in the reinforced concrete bridge structure. The chloride content was evaluated by using the "quotient method," based on the assumption that chlorides and moisture in concrete increase the electric conductivity and, subsequently, damping. Thus, increased quotients between concrete surface and top of the rebar reflection amplitude allowed for the recognition of areas of possible contamination. The nondestructive approach successfully detected all contaminated areas, plus a few false alarms.

Another concrete bridge was inspected in [10], in Norway. The purpose of the study was to test a new prototype of a ground-coupled array and compare results with those of a commercial air-coupled array. Both antenna arrays provided similar results concerning the thickness of the asphalt layer covering the concrete bridge; however, the ground-coupled array provided a clearer image of reinforcing elements inside the bridge deck and could reach a higher penetration depth.

In Part II of Ref. [3], a chapter is devoted to presenting a compilation of published works in the frame of the applications of GPR for the diagnostics of transport infrastructures, including bridges. Concrete and stonemasonry bridges are considered; information about equipment and methods used is included, as well as an overview of the existent recommendations for bridge inspection (the most important ones being the American standard ASTM D6087-08 *Standard Test Method for Evaluating Asphalt-Covered Concrete Bridge Deck Using Ground Penetrating Radar*, the Great Britain DMRB 3.1.7 document "Advice notes on the nondestructive testing of highway structures—advice note 3.5 BA 86/2006: Ground Penetrating Radar," the Mara Nord Project "Recommendations for guidelines for the use of GPR in bridge deck surveys," the American SHRP 2 *Nondestructive Testing to Identify Concrete Bridge Deck Deterioration*, the German document B10 "Recommendation for nondestructive testing of civil engineering structures by GPR," and the German BASt-report B55 "Examination of GPR in combination with magnetic techniques for the determination of moisture and salinity of concrete bridge decks with asphalt cover").

In Part IV of Ref. [3], another chapter is devoted to presenting a compilation of published works in the frame of the GPR applications in combination with complementary NDT methods for the evaluation of transport infrastructures, including bridges. Both concrete and masonry bridges are dealt with.

Looking at Table 3, it can be observed that, in COST Action TU1208, experimental investigations on bridges have been carried by using both impulse and step-frequency systems, with ground- and/or air-coupled antennas in the 200–3000 MHz range. The trace spacing along a profile ranges from 2 to 5 cm (20–50 scans per meter), depending on the employed system/antenna and on the purpose of the study. Multiple parallel profiles (from 2 to 32) are always acquired over the bridge; the distance between parallel profiles ranges from 7.5 cm to 1 m. Comparisons between different radar systems are rarely carried out, but GPR is often and successfully combined with complementary NDT techniques. Slice scan and 3-D data views are commonly used. It is noted that all TU1208 studies involved scientists from research-intensive countries. Notwithstanding all studies listed in Table 1 contemplate experimental activities, industry involvement is observed only in Ref. [10]. It is worth mentioning that, in addition to the few masonry arch bridges mentioned in this subsection, the Spanish team coordinated by Solla investigated around 40 historical bridges located in the Galician territory, obtaining a huge amount of information about their inner structure and conditions, which will be crucial for planning effective conservation and strengthening activities.

5 TUNNELS

Tunnels have become key infrastructures in modern transport development as they improve the transportation capabilities and increase road/highway/railway efficiency. Nevertheless, they require routine diagnostics to assess their structural condition and extend their life cycle. Nondestructive testing techniques have proved to be valuable methods for tunneling inspections because of the need to protect the tunnel integrity while evaluating and to provide time-saver standards.

5.1 OBJECTIVES AND METHODOLOGY

GPR has been established as one of the most recommended NDT methods for routine subsurface inspections such as tunnel walls. The technique has proved its suitability for providing high-resolution imaging of the interior of the tunnels, which reveals internal structural information with respect to tunnel quality management and detection.

Different types of tunnel lining can be surveyed by GPR, such as unreinforced concrete lining, reinforced concrete lining, shotcrete lining with sprayed concrete, and brick lining. Shotcrete containing steel fibers cannot be inspected because fibers generate random electromagnetic scattering.

The main GPR applications in the field of tunneling inspection are the following:

- estimation of thicknesses in lining and backfill grouting [91]
- rebar location and corrosion assessment [92]
- location of reinforcement in lining [93]
- detection and location of discontinuities, cavities, and grouted space in lining, between concrete and rock face or between external and inner lining layers [94,95]
- moisture-content assessment [96]

Further GPR compliance checks with designed structural details include the assessment of homogeneity of the lining materials.

The main issues of GPR tunnel surveying are related to the geometry of the tunnel and to the difficult accessibility. Usually, traffic disruption is necessary, and it is not easy to fit the antenna systems on tunnel walls or roofs; operators have to pay attention to the presence of cables and conduits running along the tunnel, which can cause obstruction to the GPR.

The GPR systems used to inspect roadway tunnels typically employ air-coupled antennas mounted on suitable vehicles and work in a frequency range of 1000–2000 MHz. For accurate thickness measurements (such as concrete cover thickness, shotcrete layer thickness, or grouting/rock contact thickness) and for locating reinforcement in lining, ground-coupled handheld antennas are recommended; the working frequency needs to be selected based on the size of the sought target and ranges from 400 to 2000 MHz; when a deeper analysis is required, to detect internal defects such as cavities, lower frequencies (250–500 MHz) are necessary.

When monostatic/bistatic systems are used, the survey is normally carried out longitudinally, along the length of the tunnel. If antennas are air-coupled, the survey is carried out at high speed (30 km/h or even faster), so that traffic does not have to be completely interrupted. More longitudinal profiles at different clock times can be acquired, by crossing the tunnel several times. Ground-coupled antennas are normally used for analyzing selected zones of interest, not the entire tunnel. Multichannel systems for tunnel inspections use air-coupled or ground-coupled arrays of antennas; they provide excellent results, but their popularity is still limited, because they entered recently in the market and are much more expensive than single-channel systems.

The connection with external GPS devices for data referencing is obviously limited when surveying the interior of a tunnel structure. Hence, the profile lengths and the distance (trace) interval are controlled by using an odometer wheel attached to the GPR.

Similar data-processing algorithms can be used, as for the analysis of GPR data acquired in bridges (see Section 4).

In some cases, the GPR equipment may require safety inspection from regulatory staff, to ensure that it complies with local safety standards, especially with regard to gas explosion risk, which is higher inside tunnels.

5.2 COMBINATION WITH COMPLEMENTARY NONDESTRUCTIVE TESTING TECHNIQUES

Various works are available in the literature, where the GPR method was combined with other NDT approaches for a more complete assessment of tunnel condition and stability. Karlovšek et al. [94], presented an overview of different NDT methods applied to the study of tunnel integrity in bored tunnels, including the advantages and limitations of each method. Wimsatt et al. [97] published a review of structural health monitoring techniques and NDT methods for high-speed mapping of voids, for the estimation of debonding, delamination, and moisture content, and for the condition assessment of tunnel lining; the techniques considered in [97] include GPR, electrical resistivity measurement, ultrasound testing, impulse response and impact echo (IE) diagnostics, laser scanning, photogrammetry, and thermography.

Abraham and Dérobert [98], presented a combination of GPR with seismic refraction for the assessment of the fired walls of the Mont Blanc Tunnel. The seismic method was used to obtain the mechanical characteristics of the concrete, while the GPR method allowed for the location of reinforcing in concrete. This combination allowed classifying the degree of deterioration in concrete.

Cardarelli et al. [99], presented the evaluation of a tunnel constructed in 1938–40 in the Central Apennines using integrated geophysical methods. The objectives of this work were to investigate the discontinuities in the rock, the elastic characteristics of the rock, the quality of the concrete-rock contact, and the extent of the cave-in zone and loosened rock around the structure. Seismic surveys (seismic refraction and seismic tomography) were carried out in addition to GPR surveys with frequencies of 200, 225, and 450 MHz. The GPR data defined the quality of the concrete-rock contact and revealed some discontinuities in the zone, whereas the seismic methods provided the distribution of the mean elastic properties of the rock.

Lehmann [100], presented the application of GPR and IE methods for the assessment of single-shell tunnel lining. GPR data were useful to locate reinforcement and to evaluate the concrete cover thickness. The IE method was employed to detect cracks and cavities and to distinguish interfaces in subsurface (e.g., concrete/grouting interface).

Xie and Zeng [101], combined the GPR method with 3-D laser scanning to evaluate the shield tunnel condition on a double-deck highway tunnel in Shanghai (China). The GPR method was used to measure thicknesses and to detect cavities behind lining, while 3-D laser scanning was applied to test the surface condition of the tunnel liner.

A major work concerned with GPR survey of tunnels is the SHRP-2 American report entitled *Mapping voids, debonding, delaminations, moisture, and other defects behind or within tunnel linings*. This document was produced in 2013 and can be found on the web; it studies six different NDT methods, including GPR, by comparing their effectiveness in detecting deteriorations and in measuring thicknesses. The main conclusions of the report are that air-coupled and ground-coupled GPR systems are very effective; the air-coupled systems are recommended for detecting areas of high moisture or low density (due to high content of air voids), whereas ground-coupled GPR systems are recommended for detecting defects at different cover depths within or behind the tunnel linings. The report recommend to first collect thermal images of the tunnel and air-coupled GPR data; then, selected areas can be studied with more detail by using a ground-coupled GPR along with ultrasonic tomography, ultrasonic echo, or portable seismic property analyzers. It is clear that GPR plays a central role in tunnel inspections. Note that this sequence of operations is applicable and likely effective also in other types of infrastructures.

5.3 RESEARCH ACTIVITIES CARRIED OUT WITHIN COST ACTION TU1208

The use of GPR for the assessment of tunnels is a topic dealt with in WG2, Project 2.1, of COST Action TU1208. The investigations performed within the Action include experimental activities carried out on a few tunnels. The already mentioned guidelines for the use of GPR on concrete structures will include recommendations for the assessment of tunnels. In the following of this subsection, TU1208 case studies concerned with GPR inspection of tunnels are shortly resumed.

The Medway Tunnel (North Kent, the United Kingdom) was studied by Alani and Banks [92]; the GPR survey was carried out by using a 2000 MHz antenna in order to analyze structural details of the tunnel roof at immersion joints, estimate the thickness of covering concrete at joints, estimate the depth and location of reinforcement, and determine the position of an Omega Seal and Rockwall layers. The results obtained allowed to determine the construction materials used (this information was previously unclear) and the actual structure in comparison with the design drawings; it was also possible to find the existence of a previously unknown feature, below the road surface. Additionally, Alani and Banks [102], carried out a second GPR survey to assess possible underground construction details of the road on a section of the Medway Tunnel; this time, four different frequencies were used including 200, 600, 900, and 2000 MHz. The results showed the existence of a previously unknown feature below the road surface.

Other members of COST Action TU1208 have employed the GPR technique for investigating an excavated high-speed railway tunnel (the Seiró tunnel in Ourense, Spain), during its early stages of construction [103]. The GPR survey was carried out by using a 1000 MHz antenna frequency. The results provided information about the thickness of the shotcrete layer, the positioning of steel ribs on coating, the presence of cavities and delamination in lining, and moisture problems.

Another interesting TU1208 case study was presented by Núñez-Nieto et al. [104], and was concerned with the evaluation and characterization of an ancient underground concrete tunnel in Marín (Galicia, Spain). The main objective of this work was to create a volumetric reconstruction of the tunneling space and the whole framework of lateral galleries. Three-dimensional methodologies and imaging techniques were employed. Moreover, problems of corrosion were detected over a sector of the structure, which were analyzed with the aid of finite-difference time-domain modeling.

In Part II of Ref. [3], as already mentioned, a chapter is devoted to presenting a compilation of published works in the frame of the applications of GPR for the diagnostics of transport infrastructures, including tunnels. The equipment and methods used and an overview to the existent recommendations for tunneling inspection are described. In the same context, a chapter in Part IV deals with the combined use of GPR and complementary NDT methods in the evaluation of tunnel structures.

6 ELECTROMAGNETIC MODELING AS A TOOL FOR GPR DATA INTERPRETATION

Understanding how heterogeneous environments reflect and scatter electromagnetic fields is a challenging problem. Consequently, the interpretation of GPR data is often difficult, and considerable experience gained through extensive experimentation is needed. Extracting quantitative information from traces and radargrams is even more complicated than performing a qualitative analysis.

To advance our level of knowledge about GPR and enable a more advanced data interpretation, it is very important to have at disposal and learn to use an accurate and robust simulator, which can be employed for implementing and executing numerical models of GPR scenarios.

In order to predict correctly the GPR response from a particular scenario, electromagnetic simulators solve Maxwell's equations subject to the physical and geometric properties of the considered problem and to its initial conditions. Several analytic-numerical methods have been developed for this purpose; one of the most versatile is the finite-difference time-domain (FDTD) technique [21], which is based on a discretization of both space and time continua.

6.1 OBJECTIVES AND METHODOLOGY

The usefulness of electromagnetic modeling as a tool supporting the interpretation of complex datasets has been already mentioned in the previous sections of the present chapter. Furthermore, electromagnetic modeling allows studying and understanding the limitations and capabilities of GPR in a specific environment, prior to designing and performing a survey. An electromagnetic simulator can also be used to design new antennas or characterize existing ones, as well as to produce synthetic radargrams with the purposes of testing data-processing, imaging, and inversion algorithms. Additionally, it is worth mentioning that a fast and accurate forward solver may be used as part of an inverse solver.

As far as the simulation methodology is concerned, this depends on the theoretical principles and numerical methods underlying the chosen simulator. The most commonly used simulator in the GPR field is gprMax (see Subsection 6.2), which implements the FDTD technique.

In the FDTD technique, objects with continuous boundaries are represented using a staircase approximation (the space is discretized in cells), and obviously, the smaller the spatial discretization steps, the more accurate the model.

Concerning time discretization, to obtain a correct solution to the electromagnetic problem, it is necessary to satisfy a stability condition, known as the CFL condition after the initials of Courant, Friedrichs, and Lewy:

$$\Delta t \leq \frac{1}{c\sqrt{\frac{1}{\Delta x^2}+\frac{1}{\Delta y^2}+\frac{1}{\Delta z^2}}} \quad (1)$$

where, in a three-dimensional space, Δx, Δy, and Δz are the spatial discretization steps; Δt is the time discretization step; and c is the speed velocity of light in a vacuum. The stability condition in a two-dimensional space (for simplified models) can be easily found by letting $\Delta z \rightarrow \infty$ in Eq. (1). From this equation, it is apparent that Δt cannot be chosen independently from the spatial discretization steps, being bounded by the values of Δx, Δy, and Δz.

Due to the limited amount of storage and finite processing capacity of computers, a good compromise between simulation accuracy and execution time has always to be achieved, when choosing the spatial and time discretization steps. This has to be done according to the frequency content of the pulse emitted by the transmitter and to the size of targets.

As far as the frequency content of the emitted pulse is concerned, in order to keep the errors associated with numerical induced dispersion at a minimum, the following rule of thumb has to be satisfied: the spatial discretization step has to be at least 10 times smaller than the smallest wavelength of the propagating electromagnetic fields. First, it has to be noted that such wavelength is not associated with the center frequency of the pulse emitted by the transmitter, as the pulse contains a significant amount of energy at higher frequencies than the center one: three to four times the center frequency of the pulse usually gives a good estimation for the highest frequency that should be taken into account in the simulations. Second, it has to be kept in mind that the wavelength corresponding to the highest considered frequency f has a different length in different media: in particular, in a medium with relative permittivity ε_r, the wavelength is $\lambda = \lambda_0/\sqrt{\varepsilon_r}$, being $\lambda_0 = c/f$ the wavelength in a vacuum (having assumed that the medium has a relative permeability $\mu_r = 1$). Hence, the abovementioned rule of thumb becomes as follows: the spatial discretization step has to be at least 10 times smaller than the wavelength corresponding to the highest considered frequency, calculated in the medium with the highest relative permittivity.

Regarding the size of targets, the staircase approximation of continuous objects cannot be too coarse; for example, a sphere should be modeled by discretizing its radius through at least 10 cells, and the discretization step has to be chosen so that the smallest features are properly modeled.

Another crucial point in the FDTD approach is that the modeled scenario has a finite size. An approximation has to be made in order to represent the whole region of interest in the model but limit the computational space. To this aim, absorbing boundary condition (ABC) is usually adopted. The ABCs are substantially virtual layers, which are added all around the simulated scenario and absorb electromagnetic waves impinging on them; they have to be applied at a sufficient distance from all sources and targets of the model, thus allowing to simulate an unbounded space. A good rule is to keep all sources and targets of the model at least 15 cells away from the ABCs, to guarantee reliable results; better results are obtained: the more cells there are between all sources/targets and ABCs. Furthermore, a lower or higher number of ABC layers can be used; normally, 8–10 layers give good results if the ABCs are properly implemented in the simulator.

In order to reduce the computational resources required and run simulations in a reasonable time, the physical structure of the GPR transmitting and receiving antennas is often not included in the model (i.e., the focus is on modeling the environment). In such cases, the transmitting antenna is represented through an idealized source: a Hertzian dipole (in a 3-D model) or a line of current (in a 2-D model); the receiving antenna is not represented, and the electric field impinging on it is calculated [19,34,38,46,55,69,89,90].

Electromagnetic simulations of GPR scenarios including realistic models of the antennas are not yet common. Accurate models of GPR antennas have been only occasionally developed during the past two decades, both for antennas designed for research purposes [105–108] and, more recently, within COST Action TU1208 and beyond, for commercial antennas [109–112]. Rarely, realistic models of GPR antennas have been combined with realistic models of complex environments [113]. Running realistic models of GPR scenarios including antenna still is a challenging task nowadays, notwithstanding computing power that is increasing dramatically and becoming more accessible.

Further information about the use of FDTD for GPR modeling can be found in the manual of gprMax [21], available at www.gprMax.com. It is always crucial to carefully read the manual of a software tool, before using it. This may seem a trivial recommendation, and yet, we have observed that manuals are rarely read; many works in the GPR literature present and interpret wrong synthetic results, coming from simulations set incorrectly because the users of the simulator do not have a solid electromagnetic background and did not take the time to properly read the simulator instructions.

6.2 ELECTROMAGNETIC SIMULATORS DEVELOPED IN COST ACTION TU1208

In the framework of COST Action TU1208, electromagnetic simulators based on the FDTD technique and on the boundary element method (BEM) have been developed.

In Section 6.2.1, we provide basic information about the new open-source version of the FDTD software gprMax, which includes advanced features for the modeling and simulation of realistic scenarios; Section 6.2.2 is about E^2GPR, a freeware package conceived to ease the use of gprMax. Those tools have been developed in the United Kingdom and in Italy, respectively, and they have been widely tested by Action members in several countries.

Further simulators, implementing FDTD and BEM approaches, have been developed by members from Ukraine and Croatia, respectively; they are not presented here for space reasons.

6.2.1 The new open-source version of gprMax

The software gprMax was originally developed in 1996, and it was written in C; it was available on the web for free download, for both research and commercial use. Recently, the software has been rewritten in Python with performance-critical parts implemented in Cython and released under the GNU General Public License [21]. The new version of gprMax is a contribution to COST Action TU1208 and was officially released by the authors during a training school that the Action organized in Thessaloniki, Greece, in November 2015.

In the process of rewriting the software, several important changes have been made to improve its efficiency, speed, and usability. For instance, the perfectly matched layer boundary conditions have been improved. Moreover, all geometric and physical properties defining the scenario can now be parameterized. Furthermore, new advanced features have been added to the simulator:

- A library of antennas that can be included in the models.
- It is possible to build heterogeneous media using fractals and objects with rough surfaces.
- Anisotropic media can be defined. This allows materials such as wood and fiber-reinforced concrete, which are often inspected with GPR, to be accurately modeled.
- Media with arbitrary frequency-dispersive properties can be defined. This paves the way to the use of gprMax in new areas, such as the modeling of human tissues.
- Optimization of parameters based on Taguchi's method can be performed. This feature can be useful to optimize material properties based on experimental data or to design new antennas.

Much more information about the new version of gprMax and an active forum dedicated to gprMax users can be found at www.gprMax.com.

6.2.2 The freeware tool E^2GPR

E^2GPR stands for "Edit your geometry, Execute GprMax2D and Plot the Results!" This freeware tool is conceived to ease the use of gprMax for the modeling and simulation of 2-D scenarios. In particular, the main purposes of E^2GPR are the following:

- Assist in the creation, modification, and analysis of 2-D gprMax models, through computer-aided design.
- Facilitate parallel and/or distributed computing with gprMax, on multicore machines or networks of computers.
- Automatically plot radargrams generated by gprMax.

gprMax is command-line driven and does not feature a graphical user interface (GUI). This implies that, for complicated scenarios, creating or modifying a model can be a not straightforward work. Moreover, the geometry of a model cannot be checked in real time while editing it. E^2GPR helps overcoming these limitations and assists the user in the creation of gprMax 2-D models. It is written in HTML and JavaScript; hence, it runs on any device. A new scenario can be created from scratch, or an existing input file can be loaded and modified. Electromagnetic properties of different media can be defined. Basic shapes are available, and arbitrary shapes can be hand-drawn by using any pointing device. Objects can be copied and pasted, deleted, shifted, and resized. It is possible to load a background picture and draw the model by tracing it. For the shift of transmitting and receiving antennas, a fixed step can be defined, or an arbitrary path can be hand-drawn, to move the antennas on irregular curved and rough interfaces.

In the new version of gprMax, the FDTD loops (which are the most computationally intensive parts of the code) have been parallelized by using OpenMP. All CPU cores available on a machine can be exploited to perform calculations. A message-passing interface has been employed to implement a task farm, which can be utilized to distribute a series of models to different cores as independent tasks. This is very useful when a B-scan is required: each A-scan composing the B-scan can be task-farmed as an independent model. E^2GPR facilitates the use of this feature (the user can set all parameters through an intuitive GUI and monitor how the calculation is advancing, through reader-friendly progress bars).

Finally, E^2GPR includes a plotter. This script automatically plots A- and B-scans generated by gprMax, to offer a first and easy view of the results.

E^2GPR will be available for free download, for both research and commercial purposes, on the website of COST Action TU1208 (www.GPRadar.eu) as of September 2017.

6.3 EXAMPLE OF APPLICATION

In Sections 2–5, several works were mentioned where electromagnetic modeling was successfully used to facilitate the interpretation of GPR data collected over civil-engineering structures. In this subsection, we wish to present a new example.

The Lubiáns bridge (Galicia, Spain) is a medieval bridge built in the 15th century that has suffered several restorations throughout its lifetime. The most important intervention was the reconstruction of its third arch from the left margin downstream (Fig. 10). This arch was rebuilt from ruin after a great flood in 1984 using different construction materials than the original ones and breaking the continuity of its medieval pavement composed of granitic flagstones.

A GPR survey was carried out over the rebuilt arch, aiming to assess the most probable composition and distribution of the construction materials used for reconstruction. The survey was conducted with a RAMAC CU-II system. A 500 MHz antenna was used, and the survey parameters selected were 5 cm trace interval with

FIG. 10

General view of the Lubiáns bridge, medieval pathway, and rebuilt arch.

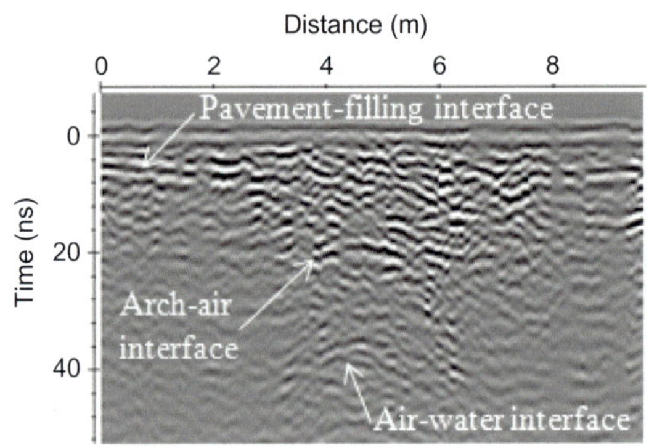

FIGS. 11

500 MHz data and interpretation of the main reflections produced.

a 90 ns total time window and 620 samples per trace. The GPR data were acquired along the longitudinal direction of the bridge.

Figure 11 presents the experimental GPR data obtained, showing the reflection produced by the arch-air interface and a continuous reflection, in the left part of the radargram around 5–7 ns, which certainly corresponds to the pavement-filling interface. Moreover, it is possible to observe the interruption in the continuity of the pavement over the arch, due to reconstruction. The analysis and interpretation of the field data are complicated because of the heterogeneity in filling, which provides complex reflection patterns characterized by the occurrence of multiple reflections and diffraction events. To assist field data interpretation, electromagnetic modeling was therefore elaborated based on the FDTD algorithm.

The synthetic model was elaborated using the Matlab software along with gprMax, on the basis of the orthothermogram generated from the external face of

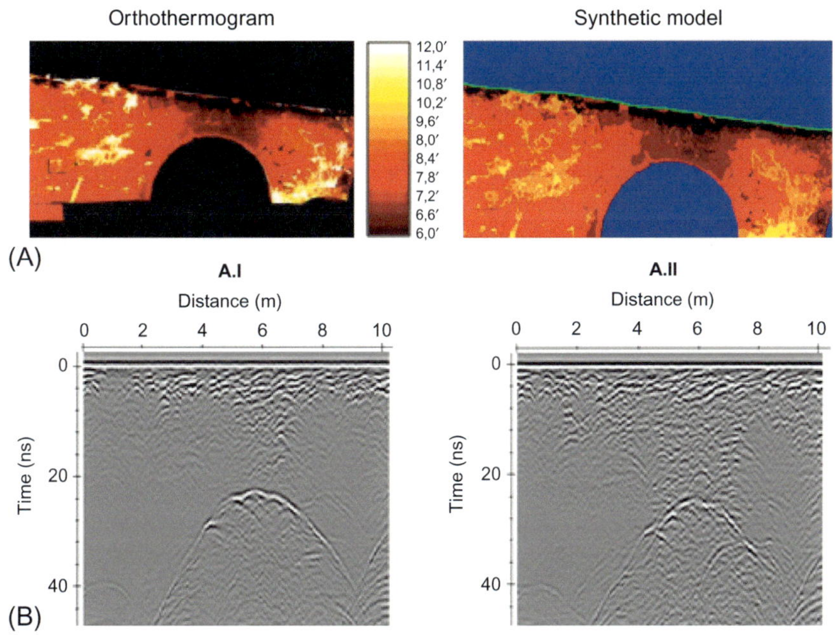

FIG. 12

Orthothermogram of the rebuilt arch and synthetic model used for the simulation (A) and synthetic radargrams produced by using gprMax (B).

the bridge (Fig. 12A). Although the orthothermogram gives information for the first 2–5 cm of the stonework, coinciding with the external masonry, the emissivity values obtained allow zoning different regions attending to a different composition of construction materials. The composition and distribution of these materials in the interior of the bridge, outside the range of thermography, can be adjusted from the field GPR data produced. This approach provides large-scale and realistic models to be compared with real data.

The model was created with a small spatial step (grid cell size), equal to 6 mm; the electromagnetic source was a line of current emitting a 500 MHz Gaussian pulse. The trace interval and the total time window were the same used during field data acquisition.

Once the model was built, different simulations were performed in gprMax by assuming different assumptions of material composition in the rebuilt zone. Fig. 12B shows two examples of simulated radargrams.

The first assumption (A.I) was to consider a layered layout in pavement, composed of stone and cement. Pareto on/off distributions was used, considering a composition of 60% granitic flagstones and 40% cement, with an average length of 60 cm and thicknesses of 5 and 3 cm for stone and cement, respectively. Heterogeneity within each layer was also simulated that was randomly chosen using a uniform distribution. The dielectric constant for stone ranged from 6 to 9, whereas for cement this constant varied from 11 to 14. The rebuilt arch and its upper filling were simulated considering a granitic masonry with an average length of 20 cm modeled with a uniform distribution by considering a dielectric constant that ranged from 6 to 9. Finally, the original masonry was assumed as granitic ashlars with dielectric constant between 6 and 7.5. The conductivity value for all the media was 0.001 S/m. Fig. 12B, left panel, presents the synthetic radargram obtained, where the presence of the reflection produced by the arch-air interface is well visible. However, these results did not provide a good agreement with those of the field data (Fig. 11), which showed more layering in the pavement and greater heterogeneity in the rebuilt zone close to the arch. More adjustments were therefore needed to configure the Pareto on/off layout.

The second assumption (A.II) consisted of considering the pavement layout composed of 60% granitic flagstones and 40% cement, but the average length was set to 80 cm and thicknesses of 7 and 3 cm for stone and cement, respectively. In the case of the rebuilt arch and its surrounding filling, the composition consisted of 60% granitic flagstones and 40% cement, with average lengths of 60 and 50 cm and thicknesses of 3 and 6 cm, for stone and cement, respectively. The ranges of dielectric constant to simulate heterogeneity in cement, stone, and granitic ashlars were the same assumed in the first assumption. Fig. 12(b), right panel, presents the synthetic radargram obtained for this second assumption. A good agreement between the synthetic data and the field data was found for the reflections produced by pavement, in form of flat continuous reflections along the whole reflection profile. Furthermore, regarding the filling over the arch, a similar reflection pattern was obtained where several continuous strong reflections are visible corresponding to the different layers of masonry, filling, and cement used for simulation.

The modeling strategies assumed and the synthetic data produced demonstrate the capabilities of electromagnetic modeling as an effective interpretational tool, which can be used in the analysis of complex and heterogeneous structures such as stonemasonry bridges. The results presented here allow a good recreation of the most probable construction techniques and materials used during the reconstruction of the arch. Identifying the materials involved in masonry and their distribution are important tasks, given their influence on the stability and durability of this type of bridge structures.

7 CONCLUSIONS AND TRENDS

The goal of this chapter was to provide an overview about the use of ground-penetrating radar (GPR) for the nondestructive testing (NDT) of transport infrastructures.

After a general introduction to the topic of the chapter (Section 1), we dedicated particular attention to roads, highways, and airport runways (Section 2); railways (Section 3); bridges (Section 4); and tunnels (Section 5). For each application, we listed the possible objectives of a GPR survey, gave practical recommendations about the methodology of application, suggested complementary noninvasive techniques that can be combined with GPR, discussed recent or new examples, and resumed the relevant research activities carried out within COST Action TU1208 *Civil Engineering Applications of Ground Penetrating Radar* (running from April 2013 to October 2017).

Throughout Sections 1–5, we highlighted the importance of electromagnetic modeling as a useful and effective tool for GPR data interpretation. In Section 6, we dealt with this subject more in detail. In particular, we provided recommendations for a correct implementation and execution of electromagnetic models by using the finite-difference time-domain (FDTD) technique; we resumed the main features of two software tools developed within COST Action TU1208 (*viz.*, the new and open-source version of the well-known simulator gprMax and the freeware package E^2GPR conceived to ease the use of gprMax); and we presented a new example of application, where electromagnetic modeling was exploited for interpreting data collected on the medieval Lubiáns bridge in Galicia, Spain.

Some observations can be made, based on the information presented in this chapter.

Firstly, it is apparent that GPR is a powerful, accurate, efficient, and increasingly used imaging method, assuming a crucial importance in the monitoring and maintenance of transport infrastructures. In this area and more in general in the context of civil-engineering works, some GPR applications are well established; further applications are still under study.

Secondly, we wish to highlight that GPR surveys require engagement of skilled, trained, and experienced professionals; using properly the GPR technique is not easy, and experimental data are not at all straightforward to be interpreted. To obtain the best results, GPR must be applied by experts who are familiar with the physical principles of the method and its limitations. More training in professional education programs on NDT and GPR is needed. For a wider and more effective use of GPR, it is also important to standardize GPR data acquisition, processing, and interpretation procedures, in each particular civil-engineering task and beyond. In this regard, the open-access publication of the TU1208 guidelines for the use of GPR in civil engineering and the TU1208 educational package for teaching GPR in the university will be of great use.

Thirdly, it can be observed that the GPR technology is still evolving, with much future potential residing, for example, in the development of new reconfigurable impulse and step-frequency systems, less expensive multichannel systems, and systems with multiple transmitting and receiving antennas fully exploiting the potential offered by antenna arrays.

Furthermore, it is noted that GPR signal-processing approaches are still being improved and extensively discussed. Imaging and full-waveform inversion techniques are at a rather infant stage; their progress will ease the interpretation of radargrams, allow removing antenna effects from data, and enable the estimation of new quantitative information useful for civil-engineering purposes. The advancement and wider spread of electromagnetic simulators is envisaged and will yield huge benefits.

Finally, this chapter underlines that GPR should not be used alone but rather as one of many techniques of a multidisciplinary NDT toolbox. In this area, several advancements are expected: For example, there is space for optimizing the combination of different approaches in the various applications, and there is a huge potential in the development of innovative methods for a joint processing and interpretation of data measured by different instruments.

For up-to-date information on the use of GPR on transport infrastructure and beyond, we recommend the reader to check the proceedings of the most recent editions of the International Conference on Ground Penetrating Radar and International Workshop on Advanced Ground Penetrating Radar. We also hope that the website of COST Action TU1208 will continue being updated and enriched in the upcoming years, turning into a platform of exchange for sustaining the dynamic in the GPR field and for creating new opportunities.

ACKNOWLEDGMENTS

This work is a contribution to European Cooperation in Science and Technology (COST; www.cost.eu) Action TU1208 *Civil Engineering Applications of Ground Penetrating Radar* (www.GPRadar.eu). The authors thank COST for funding and supporting COST Action TU1208.

REFERENCES

[1] D.J. Daniels, Ground Penetrating Radar, John Wiley & Sons Inc., 2005.

[2] R. Persico, Introduction to Ground Penetrating Radar: Inverse Scattering and Data Processing, John Wiley & Sons, Inc., 2014.

[3] A. Benedetto, L. Pajewski, Civil Engineering Applications of Ground Penetrating Radar, Springer Transactions in Civil and Environmental Engineering, Springer International, 2015. e-book ISBN: 978–3–319-04813-0; hardcover book ISBN: 978–3–319-04812-3, https://doi.org/10.1007/978-3-319-04813-0.

[4] W. Wai-Lok Lai, X. Dérobert, P. Annan, A Review of Ground Penetrating Radar Application in Civil Engineering: A 30-Year Journey from Locating and Testing to Imaging and Diagnosis, NDT & E Int. (2017), https://doi.org/10.1016/j.ndteint.2017.04.002.

[5] N. Economou, F. Benedetto, M. Bano, A. Tzanis, J. Nyquist, K.-J. Sandmeier, N. Cassidy, Advanced ground penetrating radar signal processing techniques—Editorial, Signal Process. 132 (2017) 197–200, https://doi.org/10.1016/j.sigpro.2016.07.032.

[6] R. Persico, A. Provenzano, C. Trela, M. Sato, K. Takahashi, S. Arcone, S. Koppenjan, L.G. Stolarczyk, E.C. Utsi, S. Ebihara, K. Wada, E. Pettinelli, L. Pajewski, Recommendations for the safety of people and instruments in ground-penetrating radar and near-surface geophysical prospecting, in: European Association of Geoscientists and Engineers, EAGE Publications bv, 2015.

[7] Riveiro, B., and Solla, M., "Non-Destructive Techniques for the Evaluation of Structures and Infrastructure" CRC Press, Structures and Infrastructures Series: ISSN: 1747-7735 vol. 11, https://doi.org/10.1201/b19024-1 (2016).

[8] E. Huuskonen-Snicker, P. Eskelinen, T. Pellinen, M.-K. Olkkonen, A new microwave asphalt radar rover for thin surface civil engineering applications, Frequenz: J. RF-Eng. Telecom. 69 (7–8) (2015) 377–381, https://doi.org/10.1515/freq-2015-0034.

[9] M.-K. Olkkonen, P. Eskelinen, E. Huuskonen-Snicker, T. Pellinen, P. Olmos Martinez, in: A new microwave asphalt radar, Proceedings of the International RADAR Conference, Lille, France, 2014, pp. 13–17, https://doi.org/10.1109/RADAR.2014.7060386.

[10] E. Eide, P.A. Våland, J. Sala, Ground-Coupled Antenna Array for Step-Frequency GPR, in: Proceedings of the 15th International Conference on Ground Penetrating Radar (GPR 2014), Brussels, Belgium, 30 June – 4 July, 2014, pp. 756–761, https://doi.org/10.1109/ICGPR.2014.6970527.

[11] A. Fedeli, J. Ježová, S. Lambot, M. Pastorino, A. Randazzo, L. Pajewski, in: Tomographic reconstruction of structures using a novel GPR system, Geophysical Research Abstracts, European Geosciences Union (EGU) General Assembly 2017, 23–28 April, 2017. Vienna, Austria, article ID EGU2017–18265.

[12] J. Jezova, S. Lambot, A. Fedeli, A. Randazzo, in: Ground-penetrating radar for tree trunk investigation, Proc. 9th International Workshop on Advanced Ground Penetrating Radar (IWAGPR 2017), 28–30 June, Edinburgh, United Kingdom, 2017, pp. 1–6.

[13] L. Matera, M. Ciminale, E. Eide, J. Sala, R. Persico, A. Anderson Stamnes, in: Applications of conventional and non-conventional GPR systems in the Viking remains of Torget, Proc. 8th International Workshop on Advanced Ground Penetrating Radar (IWAGPR 2015), 7-10 July 2015, Florence, Italy, 2015, pp. 1–4, https://doi.org/10.1109/IWAGPR.2015.7292631.

[14] L. Pajewski, R. Persico, S. Chicarella, V. Ferrara, F. Frezza, F. Troiani, in: Ground Penetrating Radar prototypes developed in COST Action TU1208, Proceedings of 24th International Conference on Software, Telecommunications and Computer Networks (SoftCOM), 22–24 September 2016, Split, Croatia, 2016, pp. 1–5.

[15] R. Persico, D. Dei, F. Parrini, L. Matera, Mitigation of narrowband interferences by means of a reconfigurable stepped frequency GPR system, Radio Sci. 51 (8) (2016) 1322–1331, https://doi.org/10.1002/2016RS005986.

[16] R. Persico, L. Pajewski, S. D'Amico, V. Pérez-Gracia, in: Ground-Penetrating Radar prospection at the Jesuits' church in Valletta, Malta, 22nd European Meeting of Environmental and Engineering Geophysics - Near Surface Geoscience 2016, 4–8 September 2016, Barcelona, Spain, 2016, pp. 1–4, https://doi.org/10.3997/2214-4609.201602031 (cited in the text as *Persico et al., 2016–2*).

[17] R. Persico, G. Leucci, Interference mitigation achieved with a reconfigurable stepped-frequency GPR system, Remote Sensing 8 (11) (November 2016). Article No. 926, 11 pp. https://doi.org/10.3390/rs8110926.

[18] V. Ferrara, A. Pietrelli, S. Chicarella, L. Pajewski, GPR/GPS/IMU system as buried objects locator, Measurement (2017), https://doi.org/10.1016/j.measurement.2017.05.014.

[19] L. Pajewski, A. Giannopoulos, D. Pirrone, C. Warren, S. Antonijevic, V. Doric, D. Poljak, in: Development of electromagnetic simulators for Ground Penetrating Radar, Conference of the Applied Computational Electromagnetics Society, Florence, Italy, 26–30 March 2017, 2017. https://doi.org/10.23919/ROPACES.2017.7916393.

[20] A. Benedetto, F. Tosti, L. Pajewski, F. D'Amico, W. Kusayanagi, FDTD simulation of the GPR signal for effective inspection of pavement damages, in: Proc. 15th International Conference on Ground Penetrating Radar (GPR 2014), Brussels, Belgium, 30 June–4 July 2014, pp. 513–518, https://doi.org/10.1109/ICGPR.2014.6970477.

[21] C. Warren, A. Giannopoulos, I. Giannakis, gprMax: Open source software to simulate electromagnetic wave propagation for ground penetrating radar, Comput. Phys. Commun. 209 (2016) 163–170, https://doi.org/10.1016/j.cpc.2016.08.020.

[22] D. Pirrone, L. Pajewski, in: E^{2}GPR – Edit your geometry, Execute GprMax2D and Plot the Results! Proc. IEEE 15th Mediterranean Microwave Symposium, 30 November-2 December 2015, Lecce, Italy, 2015, pp. 1–4, https://doi.org/10.1109/MMS.2015.7375414.

[23] J. De Pue, M. Van Meirvenne, W.M. Cornelis, Accounting for surface refraction in velocity semblance analysis with air-coupled GPR, IEEE J. Selected Topics Appl. Earth Obser. Remote Sens. (JSTARS) 9 (1) (2016) 60–73.

[24] V. Marecos, M. Solla, S. Fontul, V. Antunes, Assessing the pavement subgrade by combining different non-destructive methods, Constr. Build. Mater. 135 (2017) 76–85, https://doi.org/10.1016/j.conbuildmat.2017.01.003.

[25] C. Plati, P. Georgiou, A. Loizos, A Comprehensive Approach for the Assessment of in situ Pavement Density Using GPR technique, EAGE Near Surface Geophy. (NSG) 14 (2) (2016) 117–126, https://doi.org/10.3997/1873-0604.2015043.

[26] J. Pedret Rodés, V. Pérez-Gracia, A. Martínez-Reguero, Evaluation of the GPR frequency spectra in asphalt pavement assessment, in: Construction and Building Materials, vol. 96, Elsevier, 2015, pp. 181–188, https://doi.org/10.1016/j.conbuildmat.2015.08.017.

[27] F. Tosti, A. Benedetto, L. Bianchini Ciampoli, S. Lambot, C. Patriarca, E.C. Slob, GPR analysis of clayey soil behaviour in unsaturated conditions for pavement engineering and geoscience applications, EAGE Near Surface Geophys. (NSG) 14 (2) (2016) 127–144, https://doi.org/10.3997/1873-0604.2016011 (cited in the chapter as *Tosti et al., 2016–2*).

[28] S. Fontul, Structural Evaluation of Flexible Pavements Using Non-destructive Tests, PhD Thesis, University of Coimbra, Portugal, 2004.

[29] F. Tosti, S. Adabi, L. Pajewski, G. Schettini, A. Benedetto, in: Large-scale analysis of dielectric and mechanical properties of pavement using GPR and LFWD, Proc. 15th International Conference on Ground Penetrating Radar (GPR 2014), Brussels, Belgium, 30 June – 4 July, 2014, pp. 868–873, https://doi.org/10.1109/ICGPR.2014.6970551.

[30] J. Hugenschmidt, A. Herlyn, Damages in pavements caused by previous excavation work? in: Proc. 15th International Conference on Ground Penetrating Radar (GPR 2014), Brussels, Belgium, 30 June – 4 July, 2014, pp. 852–856, https://doi.org/10.1109/ICGPR.2014.6970548.

[31] J.-M. Simonin, V. Balthazart, P. Hornych, X. Dérobert, E. Thibaut, J. Sala, V. Utsi, in: Case study of detection of artificial defects in an experimental pavement structure using 3D GPR systems, Proc. 15th International Conference on Ground Penetrating Radar (GPR 2014), Brussels, Belgium, 30 June – 4 July, 2014, pp. 847–851, https://doi.org/10.1109/ICGPR.2014.6970547.

[32] V. Marecos, S. Fontul, M.L. Antunes, M. Solla, in: Assessment of a concrete prestressed runway pavement with ground penetrating radar, Proc. 8th International Workshop on Advanced Ground Penetrating Radar (IWAGPR), 7–10 July, Florence, Italy, 2015, pp. 1–4, https://doi.org/10.1109/IWAGPR.2015.7292635.

[33] F. Tosti, A. Benedetto, L. Bianchini Ciampoli, A. Calvi, F. D'Amico, in: Prediction of rutting evolution in flexible pavement life cycle at the road network scale using an air-launched Ground-Penetrating Radar system, Proc. 16th International Conference on Ground Penetrating Radar (GPR 2016), Hong Kong, 13–16 June, 2016, pp. 1–7, https://doi.org/10.1109/ICGPR.2016.7572604.

[34] A. Benedetto, F. Tosti, L. Pajewski, F. D'Amico, W. Kusayanagi, in: FDTD Simulation of the GPR Signal for Effective Inspection of Pavement Damages, Proc. 15th International Conference on Ground Penetrating Radar (GPR 2014), Brussels, Belgium, 30 June – 4 July, 2014, pp. 513–518, https://doi.org/10.1109/ICGPR.2014.6970477.
[35] W. Muller, in: Self-correcting pavement layer depth estimates using 3D multi-offset Ground Penetrating Radar (GPR), Proc. 15th International Conference on Ground Penetrating Radar (GPR 2014), Brussels, Belgium, 30 June – 4 July, 2014, pp. 887–892, https://doi.org/10.1109/ICGPR.2014.6970554.
[36] J. Domitrović, T. Rukavina, Application of GPR and FWD in assessing pavement bearing capacity, Romanian Journal of Transport Infrastructure 2 (2) (2013) 11–21, ISSN 2286–2218 ISSN-L 2286–2218.
[37] M. Solla, S. Lagüela, H. González-Jorge, P. Arias, Approach to identify cracking in asphalt pavement using GPR and infrared thermographic methods: Preliminary findings, Elsevier NDT & E Int. 62 (2014) 55–65, https://doi.org/10.1016/j.ndteint.2013.11.006.
[38] M. Solla, B. Riveiro, H. Lorenzo, J. Armesto, Ancient stone bridge surveying by ground-penetrating radar and numerical modeling methods, J. Bridg. Eng. 19 (1) (2014) 110–119, https://doi.org/10.1061/(ASCE)BE.1943-5592.0000497#sthash.Pf0oOuLT.dpuf (cited in the text as *Solla et al., 2014-2*).
[39] W. Uddin, in: A Synthesis Study of GPR Applications for Evaluation of Pavement Thickness and Cracking, Proc. 15th International Conference on Ground Penetrating Radar (GPR 2014), Brussels, Belgium, 30 June – 4 July, 2014, pp. 925–930, https://doi.org/10.1109/ICGPR.2014.6970561.
[40] A. Benedetto, F. Tosti, L. Bianchini Ciampoli, F. D'Amico, An overview of ground penetrating radar signal processing techniques for road inspections, Signal Processing (Elsevier) 132 (2017) 201–209, https://doi.org/10.1016/j.sigpro.2016.05.016.
[41] EN 13848-5:2008, Track geometry quality – Geometric quality levels, 2008.
[42] C. Esveld, Modern Railway Track, MRT-Productions, Zaltbommel, 2001.
[43] A. Loizos, M. Silvast, S. Dimitrellou, Railway trackbed assessment using the GPR technique, Adv. Char. Pave. Soil Eng. Mater. 1 (2007) 1817–1826.
[44] S. Fontul, E. Fortunato, F. De Chiara, R. Burrinha, M. Balderais, Railways Track Characterization Using Ground Penetrating Radar, Procedia Eng.: Adv. Transport. Geotech. III 143 (2016) 1193–1200, https://doi.org/10.1016/j.proeng.2016.06.120.
[45] I.L. Al-Qadi, W. Xie, D.L. Jones, R. Roberts, Development of a time–frequency approach to quantify railroad ballast fouling condition using ultra-wide band ground-penetrating radar data, Int. J. Pavement Eng. 11 (4) (2010) 269–279.
[46] A. Benedetto, F. Tosti, L. Bianchini Ciampoli, A. Calvi, M.G. Brancadoro, A.M. Alani, Railway ballast condition assessment using ground-penetrating radar—An experimental, numerical simulation and modelling development, in: Construction and Building Materials, vol. 140, 2017, pp. 508–520, https://doi.org/10.1016/j.conbuildmat.2017.02.110 (cited in the chapter as *Benedetto et al., 2017–2*).
[47] F. De Chiara, S. Fontul, E. Fortunato, GPR laboratory tests for railways materials dielectric properties assessment, Remote Sens. 6 (10) (2014) 9712–9728, https://doi.org/10.3390/rs6109712.
[48] R. Roberts, J. Rudy, I.L. Al Qadi, E. Tutumluer, J. Boyle, in: Railroad ballast fouling detection using ground penetrating radar–a new approach based on scattering from voids, Ninth European Conference on NDT, 2006.

[49] E. Fortunato, A. Pinelo, M. Matos Fernandes, Characterization of the fouled ballast layer in the substructure of a 19th century railway track under renewal, Soils Found. 50 (1) (2010) 55–62.

[50] O. Attoh-Okine, Big Data and Differential Privacy: Analysis Strategies for Railway Track Engineering, (2017). ISBN: 9781119229063.

[51] W. Shao, A. Bouzerdoum, S. Phung, L. Su, B. Indraratna, C. Rujikiatkamjorn, Automatic classification of ground-penetrating-radar signals for railway-ballast assessment, IEEE Trans. Geosci. Remote Sens. 49 (10) (2011) 3961–3972.

[52] E. Berggren, Railway Track Stiffness: Dynamic Measurements and Evaluation for Efficient Maintenance, University West, 2009.

[53] S. Fontul, E. Fortunato, F. De Chiara, in: B.H.V. Topping, Y. Tsompanakis (Eds.), Non-Destructive Tests for Railway Infrastructure Stiffness Evaluation, Stirlingshire, UK, 2011.

[54] F. Tosti, A. Benedetto, A. Calvi, L. Bianchini Ciampoli, in: Laboratory Investigations for the Electromagnetic Characterization of Railway Ballast through GPR, Proc. 16th International Conference on Ground Penetrating Radar (GPR 2016), Hong Kong, 13–16 June, 2016, pp. 1–7, https://doi.org/10.1109/ICGPR.2016.7572605 (cited in the chapter as *Tosti et al., 2016–3*).

[55] A. Benedetto, F. Tosti, L. Bianchini Ciampoli, L. Pajewski, D. Pirrone, A. Umiliaco, M. G. Brancadoro, in: A simulation based approach for railway applications using GPR, Proc. 16th International Conference on Ground Penetrating Radar (GPR 2016), Hong Kong, 13–16 June, 2016, pp. 1–6, https://doi.org/10.1109/ICGPR.2016.7572609.

[56] L. Bianchini Ciampoli, F. Tosti, M.G. Brancadoro, F. D'Amico, A.M. Alani, A. Benedetto, A spectral analysis of ground-penetrating radar data for the assessment of the railway ballast geometric properties, NDT & E Int. (Elsevier) 90 (2017) 39–47, https://doi.org/10.1016/j.ndteint.2017.05.005.

[57] F. Benedetto, F. Tosti, A.M. Alani, An Entropy-Based Analysis of GPR Data for the Assessment of Railway Ballast Conditions, IEEE Trans. Geosci. Remote Sens. 55 (7) (2017) 3900–3908, https://doi.org/10.1109/TGRS.2017.2683507 (cited in the chapter as *Benedetto et al., 2017–3*).

[58] A.M. Alani, M. Aboutalebi, G. Kilic, Applications of ground penetrating radar (GPR) in bridge deck monitoring and assessment, J. Appl. Geophys. 97 (2013) 45–54, https://doi.org/10.1016/j.jappgeo.2013.04.009.

[59] R. Parrillo, R. Roberts, A. Haggan, Bridge Deck Condition Assessment using ground penetrating radar, in: ECNDT Geophysical Survey Systems, 2006. North Salem, NH, USA.

[60] A. Van der Wielen, L. Courard, F. Nguyen, Static detection of thin layers into concrete with ground penetrating radar, Restorat. Build. Monum. 18 (3/4) (2012) 247–254, https://doi.org/10.1515/rbm-2012-6532.

[61] J. Hugenschmidt, A. Kalogeropoulos, F. Soldovieri, G. Prisco, Processing strategies for high-resolution GPR concrete inspections, NDT&E Int. 43 (2010) 147–155, https://doi.org/10.1016/j.ndteint.2010.02.002.

[62] P. Kaur, K.J. Dana, F.A. Romero, N. Gucunski, Automated GPR rebar analysis for robotic bridge deck evaluation, IEEE Trans. Cybernet. 46 (10) (2016) 2265–2276, https://doi.org/10.1109/TCYB.2015.2474747.

[63] X. Dérobert, J.F. Lataste, J.P. Balayssac, S. Laurens, Evaluation of chloride contamination in concrete using electromagnetic non-destructive testing methods, NDT&E Int. 89 (2017) 19–29, https://doi.org/10.1016/j.ndteint.2017.03.006.

[64] J. Hugenschmidt, R. Loser, Detection of chlorides and moisture in concrete structures with ground penetrating radar, Mater. Struct. 41 (4) (2008) 785–792, https://doi.org/10.1617/s11527-007-9282-5.

[65] G. Klysz, J.P. Balayssac, Determination of volumetric water content of concrete using ground-penetrating radar, Cem. Concr. Res. 37 (2007) 1164–1171, https://doi.org/10.1016/j.cemconres.2007.04.010.

[66] S. Cardimona, B. Willeford, J. Wenzlick, N. Anderson, in: Investigation of bridge decks utilizing ground penetrating radar, Proc. International Conference on the Application of Geophysical Technologies to Planning, Design, Construction and Maintenance of Transportation Facilities, St. Louis, USA, 2000.

[67] M. Solla, H. Lorenzo, F.I. Rial, A. Novo, GPR assessment of the medieval arch bridge of san Antón, Galicia, Spain, Archaeol. Prospect. 17 (2010) 223–232, https://doi.org/10.1002/arp.390.

[68] I. Puente, M. Solla, H. González-Jorge, P. Arias, NDT documentation and evaluation of the roman bridge of lugo using GPR and mobile and static LiDAR, J. Perfor. Construct. Facil. (2013). 2015.29, 06014004-1-5, https://doi.org/10.1061/(ASCE)CF.1943-5509.0000531.

[69] M. Solla, H. Lorenzo, F.I. Rial, A. Novo, Ground-penetrating radar for the structural evaluation of masonry bridges: Results and interpretational tools, Constr. Build. Mater. 29 (2012) 458–465, https://doi.org/10.1016/j.conbuildmat.2011.10.001.

[70] M. Solla, H. Lorenzo, B. Riveiro, F.I. Rial, Non-destructive methodologies in the assessment of the masonry arch bridge of Traba, Spain, Eng. Fail. Anal. 18 (3) (2011) 828–835, https://doi.org/10.1016/j.engfailanal.2010.12.009.

[71] B. Riveiro, M. Solla, I. de Arteaga, P. Arias, P. Morer, A novel approach to evaluate masonry arch stability on the basis of limit analysis theory and non-destructive geometric characterization, Autom. Constr. 31 (2013) 140–148, https://doi.org/10.1016/j.autcon.2012.11.035.

[72] N. Diamanti, A. Giannopoulos, M. Forde, Numerical modelling and experimental verification of GPR to investigate ring separation in brick masonry arch bridges, NDT&E Int. 41 (5) (2008) 354–363, https://doi.org/10.1016/j.ndteint.2008.01.006.

[73] V. Pérez-Gracia, Radar de subsuelo. Evaluación para aplicaciones en arqueología y en patrimonio histórico-artístico, PhD Thesis, Universidad de Cataluña, 2001.

[74] D.M. McCann, M.C. Forde, Review of NDT methods in the assessment of concrete and masonry structures, NDT&E Int. 34 (2001) 71–84, https://doi.org/10.1016/S0963-8695(00)00032-3.

[75] D. Breysse, G. Klysz, X. Déboret, C. Sirieix, J.F. Lataste, How to combine several non-destructive techniques for a better assessment of concrete structures, Cem. Concr. Res. 38 (2008) 783–793, https://doi.org/10.1016/j.cemconres.2008.01.016.

[76] H. Wiggenhauser, in: Advanced NDT methods for quality assurance of concrete structures, Proc. Non-Destructive Testing in Civil Engineering (NDTCE'09), 30 June – 3 July, Nantes, France, 2009, , pp. 1–12.

[77] G. Villain, Z.M. Sbartaï, X. Dérobert, V. Garnier, J.P. Balayssac, Durability diagnosis of a concrete structure in a tidal zone by combining NDT methods: Laboratory tests and case study, Constr. Build. Mater. 37 (2012) 893–903, https://doi.org/10.1016/j.conbuildmat.2012.03.014.

[78] A.M. Alani, M. Aboutalebi, G. Kilic, Integrated health assessment strategy using NDT for reinforced concrete bridges, NDT&E Int. 61 (2014) 80–94, https://doi.org/10.1016/j.ndteint.2013.10.001.

[79] C. Colla, P.C. Das, D. McCann, M. Forde, Sonic, electromagnetic and impulse radar investigation of stone masonry bridges, NDT&E Int. 30 (4) (1997) 249–254, https://doi.org/10.1016/S0963-8695(96)00067-9.

[80] R.C. Flint, P.D. Jackson, D.M. McCann, Geophysical imaging inside masonry structures, NDT&E Int. 32 (1999) 469–479, https://doi.org/10.1016/S0963-8695(99)00024-9.

[81] Z. Orbán, M. Gutermann, Assessment of masonry arch railway bridges using nondestructive in-situ testing methods, Eng. Struct. 31 (10) (2009) 2287–2298, https://doi.org/10.1016/j.engstruct.2009.04.008.

[82] V. Pérez-Gracia, D. Di Capua, O. Caselles, F. Rial, H. Lorenzo, R. González-Drigo, J. Armesto, Characterization of a Romanesque bridge in Galicia (Spain), Int. J. Architec. Heritage 5 (3) (2011) 251–263, https://doi.org/10.1080/15583050903560249.

[83] C. Fauchard, R. Antoine, F. Bretar, J. Lacogne, Y. Fargier, C. Maisonnave, P. Guilbert Marjerie, P.F. Thérain, J.P. Dupont, M. Pierrot-Deseilligny, Assessment of an ancient bridge combining geophysical and advanced photogrammetric methods: Application to the Pont De coq, France, J. Appl. Geophys. 98 (2013) 100–112, https://doi.org/10.1016/j.jappgeo.2013.08.009.

[84] O. Bergamo, G. Campione, S. Donadello, G. Russo, In-situ NDT testing procedure as an integral part of failure analysis of historical masonry arch bridges, Eng. Fail. Anal. 57 (2015) 31–55, https://doi.org/10.1016/j.engfailanal.2015.07.019.

[85] S. Russo, Integrated assessment of monumental structures through ambient vibrations and ND tests: The case of rialto bridge, J. Cult. Herit. 19 (2016) 402–414, https://doi.org/10.1016/j.culher.2016.01.008.

[86] M.E. Stavroulaki, B. Riveiro, G.A. Drosopoulos, M. Solla, P. Koutsianitis, G. E. Stavroulakis, Modelling and strength evaluation of masonry bridges using terrestrial photogrammetry and finite elements, Adv. Eng. Softw. 101 (2016) 136–148, https://doi.org/10.1016/j.advengsoft.2015.12.007.

[87] A.M. Alani, K. Banks, in: Applications of Non-destructive methods (GPR and 3D Laser Scanner) in Historic Masonry Arch Bridge Assessment, Geophysical Research Abstracts, European Geosciences Union (EGU) General Assembly 2014, 27 April–2 May 2014, Vienna, Austria, article ID EGU2014–2727-1, 2014.

[88] J. Hugenschmidt, F. Wenk, E. Brühwiler, in: GPR chloride inspection of a RC bridge deck slab followed by an examination of the results, Proc. 15th International Conference on Ground Penetrating Radar, Brussels, Belgium, 30 June – 4 July, 2014, pp. 456–460, https://doi.org/10.1109/ICGPR.2014.6970465.

[89] M. Solla, X. Nieto, in: FDTD modelling of the GPR signal based on data obtained from other NDT methods: an approach for more exhaustive interpretation of field data, Proceedings of the Second General Meeting, COST Action TU1208, Vienna, Austria, April–May 2014, Aracne Editrice, 2015, , pp. 179–189.

[90] M. Solla, R. Asorey-Cacheda, X. Núñez-Nieto, B. Conde-Carnero, Evaluation of historical bridges through recreation of GPR models with the FDTD algorithm, Constr. Build. Mater. 77 (2016) 19–27, https://doi.org/10.1016/j.ndteint.2015.09.003.

[91] H. Liu, X. Xie, M. Sato, in: Accurate thickness estimation of a backfill grouting layer behind shield tunnel lining by CMP measurement using GPR, Proc. 14th International Conference on Ground Penetrating Radar, Shanghai, China, 4–8 June, 2012.

[92] A.M. Alani, K. Banks, in: Applications of ground penetrating radar in the medway tunnel—Inspection of structural joints, Proc. 15th International Conference on Ground Penetrating Radar, Brussels, Belgium, 30 June – 4 July, 2014, pp. 461–464, https://doi.org/10.1109/ICGPR.2014.6970466 (cited in the chapter as *Alani and Banks, 2014–2*).

[93] G. Parkinson, C. Ékes, Ground penetrating radar evaluation of concrete tunnel linings, in: Proc. 12th International Conference on Ground Penetrating Radar, Birmingham, UK, 2008.

[94] J. Karlovšek, A. Scheuerman, D.J. Willimas, in: Investigation of voids and cavities in Bored Tunnels using GPR, Proc. 14th International Conference on Ground Penetrating Radar, Shanghai, China, 4–8 June, 2012.

[95] X. Lei, Z. Hui-lin, S. Zhen, T. Si-hao, L. Guo-quing, Z. Jian, GPR evaluation of the Damaoshan highway tunnel: A case study, NDT&E Int. 59 (2013) 68–76, https://doi.org/10.1016/j.ndteint.2013.05.004.

[96] I. Saricicek, A. Seren, in: Imaging of deteriorations and concrete quality of the tunnels (Zinaga, Torull) with ground penetrating radar, Proc. 15th International Conference on Ground Penetrating Radar, Brussels, Belgium, 30 June–4 July, 2014.

[97] A. Wimsatt, T. Scullion, S. Hurlebaus, D. Zollinger, K. Wieghaus, P. Shokouhi, H. Wiggenhauser, T. Saarenketo, F. Tonon, S. Nazarian, in: High-speed nondestructive testing methods for mapping voids, debonding, delaminations, moisture and other defects behind or within tunnel linings, Proc. Structural Materials Technology Conference, 2010.

[98] O. Abraham, X. Dérobert, Non-destructive testing of fired tunnel walls: The Mont-Blanc tunnel case study, NDT&E Int. 36 (2003) 411–418, https://doi.org/10.1016/S0963-8695(03)00034-3.

[99] E. Cardarelli, C. Marrone, L. Orlando, Evaluation of tunnel stability using integrated geophysical methods, J. Appl. Geophys. 52 (2003) 93–102, https://doi.org/10.1016/S0926-9851(02)00242-2.

[100] F. Lehmann, in: Practical application of non-destructive test methods at a single-shell tunnel lining, Proc. 17th fib PhD Symposium in Stuttgart, Germany, 11–13 September, 2008.

[101] X. Xie, C. Zeng, in: Non-destructive evaluation of shield tunnel condition using GPR and 3D laser scanning, Proc. 14th International Conference on Ground Penetrating Radar, Shanghai, China, 4–8, 2012.

[102] A.M. Alani, K. Banks, in: Medway Tunnel Road Pavement Survey Using Different Frequency GPR Antenna Systems—A Case Study, Geophysical Research Abstracts, European Geosciences Union (EGU) General Assembly 2015, 12–17 April 2015, Vienna, Austria, article ID EGU2015-15509, 2015.

[103] F.J. Prego, M. Solla, X. Núñez-Nieto, P. Arias, Assessing the applicability of ground-penetrating radar to quality control in tunneling construction, J. Constr. Eng. Manage. 142 (5) (2016). 06015006. 10/1061/(ASCE)CO.1943-7862.0001095.

[104] X. Núñez-Nieto, M. Solla, A. Novo, H. Lorenzo, Three-dimensional ground-penetrating radar methodologies for the characterization and volumetric reconstruction of underground tunneling, Constr. Build. Mater. 71 (2014) 551–560, https://doi.org/10.1016/j.conbuildmat.2014.08.083.

[105] A. Ahmed, Y. Zhang, D. Burns, D. Huston, T. Xia, Design of UWB antenna for air-coupled impulse ground-penetrating radar, IEEE Geosci. Remote Sens. Lett. 13 (1) (2016) 92–96.

[106] L. Gurel, U. Oguz, Three-dimensional FDTD modeling of a ground-penetrating radar, IEEE Trans. Geosci. Remote Sens. 38 (4) (2000) 1513–1521.

[107] G. Klysz, X. Ferrieres, J. Balayssac, S. Laurens, Simulation of direct wave propagation by numerical FDTD for a GPR coupled antenna, NDT & E Int. 39 (4) (2006) 338–347.

[108] S. Lambot, E.C. Slob, I. van den Bosch, B. Stockbroeckx, M. Vanclooster, Modeling of ground-penetrating radar for accurate characterization of subsurface electric properties, IEEE Trans. Geosci. Remote Sens. 42 (11) (2004) 2555–2568.

[109] N. Diamanti, A.P. Annan, Characterizing the energy distribution around GPR antennas, J. Appl. Geophys. 99 (2013) 83–90.

[110] C. Warren, Numerical modelling of high-frequency ground-penetrating radar antennas, PhD Thesis, The University of Edinburgh, 2009 (available for download at http://hdl.handle.net/1842/4074).

[111] C. Warren, A. Giannopoulos, Creating FDTD models of commercial GPR antennas using Taguchi's optimisation method, Geophysics 76 (2) (2011) G37–G47.

[112] C. Warren, A. Giannopoulos, Experimental and Modeled Performance of a Ground Penetrating Radar Antenna in Lossy Dielectrics, IEEE Journal of Selected Topics in Applied Earth Observations and Remote Sensing 9 (1) (2016) 29–36 (cited in the chapter as *Warren et al., 2016–2*).

[113] I. Giannakis, A. Giannopoulos, C. Warren, A realistic FDTD numerical modeling framework of ground penetrating radar for landmine detection, IEEE J. Selected Topics Appl. Earth Observ. Remote Sens. 9 (1) (2016) 37–51.

CHAPTER

THz imaging and data processing: State of the art and perspective

11

Ilaria Catapano, Francesco Soldovieri

Institute for Electromagnetic Sensing of the Environment, National Research Council of Italy, Naples, Italy

CHAPTER OUTLINE

1 INTRODUCTION

Ranging between microwaves and far infrared, Terahertz waves ($1\,\text{THz} = 10^{12}$ Hz) represent a new frontier of sensing technologies [1], whose development is growing up significantly thanks to the recent advances concerning instrumentations for effective generation and detection of THz signals [2].

Innovation in Near-Surface Geophysics. https://doi.org/10.1016/B978-0-12-812429-1.00011-8

In this framework, progress is motivated by the attractive features characterizing radiations belonging to the electromagnetic spectral range from 0.3 to 10 THz. THz waves are, indeed, capable of penetrating dry, nonpolar, and nonmetallic materials, which are typically opaque at visible/infrared frequencies. Moreover, they allow the excitation of vibrational mode in a range of organic and inorganic materials, which is not possible by using microwaves [3]. As a further advantage, THz waves exhibit a low photon energy and accordingly are classified as nonionizing radiations, like microwave, millimeter, and infrared waves.

Therefore, THz waves allow characterization of the inner features of manmade objects, with a submillimetric spatial resolution. Moreover, they offer the possibility of recognizing the adopted materials, while assuring negligible long-term risks to the molecular stability of the exposed objects. These features motivate the current interest about the use of THz sensing as a nondestructive and contactless technique able to perform diagnostic and monitoring surveys.

Thanks to their peculiarities, THz waves deserve interest in many applicative fields [4]. Specifically, the ability to propagate through a wide variety of materials, among which plastics, papers, ceramics, wood, and fabrics, motivates the adoption of THz technology as a high-resolution, contactless, and noninvasive diagnostic tool. As such, THz sensing is of interest to perform nondestructive imaging within security applications [5], quality control [6–9], medical diagnosis [10], pharmaceutics [11], art conservation [12], and vegetation [13]. In addition, the capability of exciting vibrational modes makes THz signals suitable to perform spectroscopic analysis [14–16]. Material identification is, indeed, possible by taking into account the specific absorption spectra or spectral fingerprint that many substances exhibit according to their chemical composition and crystalline structures. Therefore, THz waves deserve huge attention for the identification of drugs and explosives [17,18], to distinguish paints having the same colors but different chemical composition [19], and to discriminate benign and malignant tissues [20].

Thanks to their remarkable potentialities and massive applicative interest, nowadays, THz technologies represent the new frontier for diagnostics and sensing, and this chapter aims at providing an overview of the currently available measurement setups and imaging techniques. In addition, the potentialities offered by THz imaging are discussed by means of experimental results concerning two different applicative scenarios. The first one deals with artwork characterization, and specifically, results of a survey carried out on a fresco fragment are presented in order to provide an example of the THz technology capability of detecting overlapping layers made by different materials. The second example aims at assessing THz imaging potentialities in the frame of agrofood industry, and surveys carried out on adulterated and genuine chocolate tablets are reported.

The chapter is organized as follows. Section 2 reviews briefly the available THz measurement technologies, while Section 3 summarizes the commonly adopted imaging methodologies. Section 4 deals with denoising filtering procedures and assess their ability to enhance imaging. Section 5 presents THz surveys, while conclusions and perspectives are traced in Section 6.

2 THz MEASUREMENT SETUPS AND CONFIGURATIONS

The measurement setup is the core element of all sensing technologies because its features determine the amount and quality of the acquired data and, thus, performance of data processing approaches, so affecting reliability and accuracy of the overall chain of the diagnostic analysis.

Generally, THz systems work in transmission and/or reflection mode. In the first measurement configuration, the emitter and receiver are located at two opposite sides of the object under test, and the transmitted signal is measured. In the second configuration, the THz probes are located at the same side, and the signal reflected by the investigated object is acquired. Moreover, THz systems can be grouped into pulsed systems and continuous wave systems, whose main characteristics are briefly summarized in the following.

2.1 THz PULSED SYSTEMS

Commonly known as THz time-domain spectroscopy (THz TDS), THz pulsed systems adopt a femtosecond laser to excite a photoconductive antenna and a coherent detector. This latter can be a photoconductive antenna similar to the emitter or can exploit an electrooptical conversion to transfer the information contained into the THz beam to an optical probe beam recorded by a visible or near-infrared detector [2,21]. The object under test is scanned point by point through a focused beam, and at each pixel, the time evolution of the THz electric field due to a single incident pulse is measured. Precisely, the time-domain traces of the transmitted or reflected THz waveform, within the observation time window, are stored.

THz TDS systems allow a high characterization of the interaction between wave and material. As said above, at each pixel of the sample under test, they gather a time-dependent waveform. Hence, by computing, the Fourier transform, one obtains both amplitude and phase spectrum of the THz field. Since these latter are related to several physical features of the material through which the propagation occurs, object characterization is possible. For instance, the distribution of the complex permittivity of the sample can be retrieved.

Currently, THz TDS systems are the most well established and widespread used measurement devices. This is mainly due to the progress of laser technology as a new-generation tool for the development of flexible, compact, and portable commercial systems [22–24]. These systems are characterized by optical-fiber-coupled source and detector probes and do not require complex optical alignments. Hence, they can be easily reconfigured to scan objects under transmission and reflection modalities.

Although there are advantages, THz TDS systems have some drawbacks. First of all, typically, they cover a limited frequency band, usually ranging from 0.1 to 3 THz. This limits the spatial resolution, and often, these systems are not successful to perform spectroscopy, because several materials exhibit specific fingerprints at frequencies higher than 3 THz. Another limitation is that THz TDS systems generate pulses whose average power is less than a dozen of microwatts and require high dynamic range sensors, which have to be capable of measuring low-power signals. In this framework, to improve the signal-to-noise ratio (SNR), the stored field is usually the average of several measured waveforms. In addition, since no efficient 2D detectors are currently available, a mechanic scanner is necessary to cover the object surface of interest. Consequently, the measurement time may be very long and not suitable for the integration of THz surveys in mass production processes.

It is worth mentioning that quantum cascade laser (QCL) technology offers a significant improvement in terms of performances of THz TDS system. Indeed, QCL has emerged recently as compact, high-power, spectrally pure sources for THz waves in the 1.2–5.2 THz range [25]. Therefore, they make possible to relax constrains in terms of the required dynamic range or equivalently about the tolerable degree of attenuation of the signal propagating through the object. At the same time, they increase the ability of THz TDS systems to perform spectroscopic analysis.

2.2 CONTINUOUS WAVE THz SYSTEMS

Continuous wave (CW) THz systems mainly gather information about the average intensity of the THz field. Phase information is, in principle, possible, but this is complex and time-consuming. Hence, THz CW systems are used in those cases wherein a partial object characterization is sufficient, which is achievable by taking into account only the attenuation of the signal amplitude occurring when the wave travels through the object [26].

With respect to THz TDS systems, THz CW systems are able to work at higher frequencies than 3–5 THz, and thus, they are the most suitable to perform spectroscopy. As further positive feature, they generate signals whose average power is some milliwatts: thus, they allow an improvement of the SNR. Accordingly, for each pixel of the object, it is sufficient to collect a single waveform, that is, the stored datum is not obtained by computing the average on a certain number of waveforms, and this affects positively the data collection stage, which is not time-consuming.

3 THz IMAGING

The first 2D THz imaging dates back in 1995 [27], when Hu and Nuss used a THz radiation, generated by using subpicosecond optical pulses, to scan test objects and collect the transmitted waveforms.

However, the main development of 2D and 3D THz imaging occurred at about the beginning of this millennium thanks to Zhang and coworkers, who developed a fast electrooptical detector, capable of working at video rate [28].

Starting from the beginning of the 21st century, THz imaging has experienced a rapid expansion, and several imaging methodologies to obtain 2D and 3D characterizations have been developed and tested.

Among the approaches devoted to perform 2D imaging, it is worth mentioning that THz false-color images and techniques are based on pixel-by-pixel representation of (a) the amplitude of a pulse peak or of the spectrum at a specific frequency and (b) the amplitude or power integration over a specific time windows or over a certain frequency range [29]. Conversely, THz tomography groups all those approaches capable of providing cross-sectional images and thus performing a 3D characterization of the object [30].

This section provides a brief and not exhaustive review of tomographic techniques. It is worth remembering that the word tomography derives from the Greek words "tomos" and "graphia" and means "describing by slices."

3.1 THz DIFFRACTION TOMOGRAPHY

THz diffraction tomography aims at reconstructing the refractive index of the investigated object by exploiting the relation existing among diffracted THz waves and electromagnetic parameters of materials. This relationship is described by Maxwell's equations and depends on the relative position of the object with respect to THz transmitting and receiving probes.

More precisely, diffraction tomography exploits a linearized model to describe the interaction between THz wave and material, which is obtained by introducing approximations commonly exploited in electromagnetic sensing, like Born or Rytov approximations. Hence, the imaging is performed by applying the Fourier diffraction theorem [31].

The first implementation of THz diffraction tomography was reported in Ref. [32], wherein an imaging approach based on the Born approximation was proposed and exploited to characterize a polyethylene cylinder.

Despite its potentialities, THz diffraction tomography has a limited applicability due to the complexity of the required measurement setup, which must allow acquisition of 2D images of the diffracted THz field over a wide range of angles. This involves also a time-consuming data acquisition step, which is not suitable for many applicative scenarios. Another limitation is the power of the probing signal, which constrains the investigation depth significantly, due to material attenuation. Indeed, the very low SNR hinders the imaging, and poor reconstruction capabilities are experienced.

3.2 THz TOMOSYNTHESIS

Tomosynthesis allows reconstruction of few slices of the investigated object without requiring measurement in all directions. Tomographic images are obtained, indeed, by moving THz emitter along a linear trajectory in such a way to cover a limited angle of view, which range from −50 to 50 degrees usually. Then, THz waveforms transmitted through the object are measured on a detector plane opposed to the emitter plane. Accordingly, for each point of view, a radiography of the object is obtained, similarly to the standard tomography, and the imaging approaches developed for the standard tomography can be exploited.

Tomosynthesis is often used for medical imaging [33] and can be considered a tomography with a reduced amount of data. This implies a restriction on the imaging performance that can be in principle achieved but allows a not trivial simplification of the measurement setup.

3.3 THz TIME OF FLIGHT

Time-of-flight (TOF) THz imaging provides a 3D reconstruction by exploiting data collected in reflection mode and was used for the first time by Mittleman and coworkers to produce a 3D image of a floppy disk [34].

The object is probed by a pulse signal, and the reflected waveform is collected as a time-dependent function in a certain observation time window. The time behavior of the reflected signal provides information about the inner features of the object along the wave propagation path, and the achievable spatial resolution along this direction depends on the width of the probing THz pulse. Specifically, it improves when pulse time-duration decreases.

By moving emitter and receiver along a straight trajectory and plotting all together the acquired waveforms, a space-time image, referred as THz radargram, is obtained. The radargram is a two-dimensional image, wherein the horizontal axis is the space axis and represents the measurement line, while the vertical axis is the time axis and represents the time of flight, that is, the time T that the waveform employs to propagate from the emitter to an electromagnetic discontinuity and to go back to the receiver. The time of flight T is related to the distance d between THz probes and detected discontinuities by

$$T = 2d/v \tag{1}$$

v being the electromagnetic wave propagation velocity into the object.

Therefore, a THz radargram provides a cross-sectional representation of the inner features of the object, which gives information about position and thickness of possibly inner layers.

A 3D characterization is obtained by collecting radargrams along parallel profiles and visualizing all together the stored data. In addition, since amplitude and phase of the reflected signals are available by means of the Fourier transform, the spatial distribution of the refractive index can be retrieved in principle.

3.4 THz HOLOGRAPHY

THz holography is derived from radar and optical holography [35]. It requires the knowledge of amplitude and phase of the interference pattern between the field produced by the probed object and a reference waveform. Hence, holograms are obtained by exploiting the Fourier optics theory and provide an accurate 3D view of the object. On the other hand, this is possible only if the separation between scattering centers is larger than the width of the pulse. Moreover, the estimation of the refractive index is not possible.

4 ENHANCED THz IMAGING

Whatever the imaging methodology is exploited, one of the main limitations of THz imaging is related to the signal-to-noise ratio, which is particularly critical when THz TDS systems are exploited. In this framework, data filtering procedures provide some benefits.

This section deals with two filtering procedures, which are commonly applied in electromagnetic sensing, and compare them by considering an applicative example referred to nondestructive testing of composite materials.

4.1 FOURIER FILTER

The Fourier filter method exploits the concept of the actual frequency bandwidth of the data and can be implemented as described in the following.

First of all, all the time-dependent THz waveforms gathered along a line or a planar scanning surface are arranged into a vector, and the direct fast Fourier transform (FFT) is used to obtain their spectra. Then, the effective frequency range of the data is fixed as the range where the amplitude of spectra, as normalized to their maximum value and plotted in logarithmic scale, are no lower than a prefixed threshold (usually, 3 dB). Finally, the selected portion of the spectrum of each THz signal is transformed in time domain by means of the inverse FFT (IFFT).

Therefore, let $s(t)$ be the single waveform, that is, the raw THz data measured at the generic measurement point; the filtered signal is given as

$$\hat{s}(t) = \mathrm{IFFT}(\Pi(\mathrm{FFT}(s(t)))) \tag{2}$$

where Π denotes the band-pass filter designed to select the effective spectrum of the signal. It is worth observing that the restriction of the spectra at the effective frequency range of the signal affects the range resolution, that is, the depth resolution when a normal incidence is assumed for the probing signal. According to the diffraction theory, the range resolution is, indeed, $\Delta R = v/(2B)$, where v is the wave propagation velocity in the probed medium and B is the considered bandwidth [36]. This procedure is widespread used, as in the case of THz imaging of majolica tiles in Ref. [37].

4.2 SINGULAR VALUE DECOMPOSITION FILTER

This procedure exploits the concept of singular-value decomposition (SVD) of the data [38]. The items described below summarize the procedure:

(1) Data are organized into a matrix $\mathbf{A}$, whose rows and columns represent the N measurement points and the time samples M discretizing the observation time window, respectively.

(2) SVD of the data matrix $\mathbf{A}$ is computed, that is, $\mathbf{A}$ is represented as $\mathbf{A} = \mathbf{USV}^T$, $\mathbf{U}$ and $\mathbf{V}$ being orthogonal matrixes having size $N \times N$ and $M \times M$, respectively, while S is the $N \times M$ diagonal matrix whose elements are the singular values of $\mathbf{A}$ arranged in a decreasing order.

(3) Normalized singular values $\hat{\sigma}_i = \sigma_i / \sigma_1$ for $i = 1, \ldots, \min\{M,N\}$, that is, the normalized spectrum of $\mathbf{A}$, is plotted,

(4) Index where the singular spectrum changes its slope and starts to go to zero slowly is chosen as the truncation threshold P.

(5) Filtered data matrix $\tilde{\mathbf{A}}$ is obtained as

$$\tilde{\mathbf{A}} = \sum_{n=1}^{P} \sigma_i \mathbf{u}_i \mathbf{v}_i^T \quad (3)$$

Since $\mathbf{A}$ can be regarded as an image of surface and inner features of the inspected object, the SVD indicates that the object image, that is, the data matrix $\mathbf{A}$, can be represented as the weighted sum of the base images given by $\mathbf{u}_i \mathbf{v}_i^T$ for $i = 1, \ldots, \min\{M,N\}$. These base images are $N \times M$ matrices and are weighted by their corresponding singular value σ_i. Due to the decreasing behavior of the singular values, one can expect the base images associated with the first singular values that represent the most significant part of the data matrix, whereas those related to the smallest singular values are referring to less important information that can be overwhelmed by noise. Therefore, the first P base images account for the useful data, while the remaining ones mostly account for noise.

4.3 A COMPARATIVE EXAMPLE

In order to assess the effectiveness of the above described denoising filters and compare their performances, we present results referred to a composite material covered by an icephobic coating. This object was designed and produced by the Italian Aerospace Research Centre (CIRA) and investigated by means of the fiber-coupled Terahertz time-domain (FiCO) system [24] available at the Institute for Electromagnetic Sensing of the Environment-National Research Council of Italy (IREA-CNR) [39].

The adopted THz system is made up of three main components: (a) the laser source, (b) the base unit, and (c) fiber optic coupled transmitting and receiving probes reconfigurable in transmission and normal reflection mode. The system collects signals into a 100 ps observation time window, which can be moved along a time scan range of 1 ns according to the path length between transmitter and receiver. The waveform acquisition speed can be up to 500 Hz, and the maximum dynamic range (DNR) is 30 dB, while the typical DNR is 20 dB. The effective frequency range is from about 40 GHz to 3 THz.

We analyzed a $50 \times 50 \times 4$ mm specimen in normal reflection mode. Herein, normal reflection mode means that we performed data acquisition by using a specific holder for both emitter and receiver, which, by exploiting suitable mirrors, allows the emission of a probing signal impinging on the object with a null incidence angle and the collection of the corresponding reflected field. This holder is on an automatic positioning system allowing the survey of objects whose maximum extent can be 150×150 mm with a spatial offset not lower than 120 μm. The measurement setup is shown in Fig. 1A, while the object under test is pictured in Fig. 1B.

In the case herein considered, the measurement area was 54×54 mm wide, and the spatial offset between the acquisition points was 0.15 mm along both x- and y-axes. Hence, $N = 361 \times 361$ waveforms were measured. The observation time window was about 50 ps wide, and the trace was sampled by using $M = 804$ points.

Fig. 2 compares the waveform collected at the center of the specimen (i.e., raw data) and its filtered versions obtained by applying FFT and SVD filtering procedures. Specifically, the gray solid line represents the raw signal, the black dashed line is the FFT filtered signal, and the *black dotted line* denotes the SVD filtered signal. Fig. 2 corroborates that raw data are sufficient to discriminate the icephobic

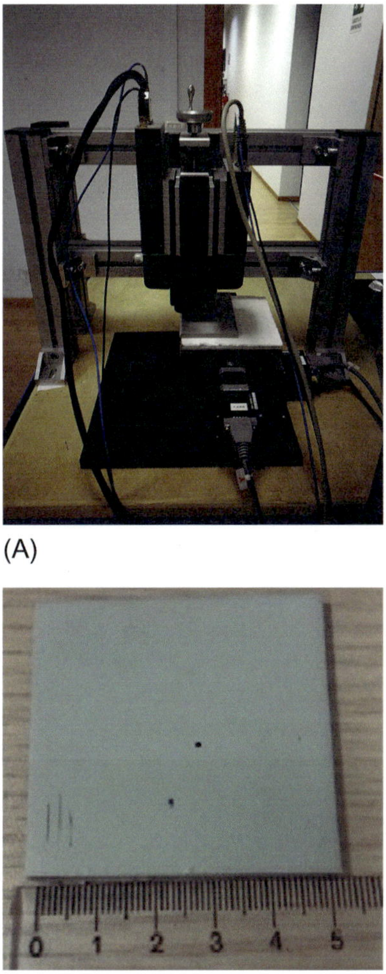

(A)

(B)

FIG. 1

Image of the THz survey: (A) picture of the normal reflection measurement setup and (B) picture of the investigated specimen.

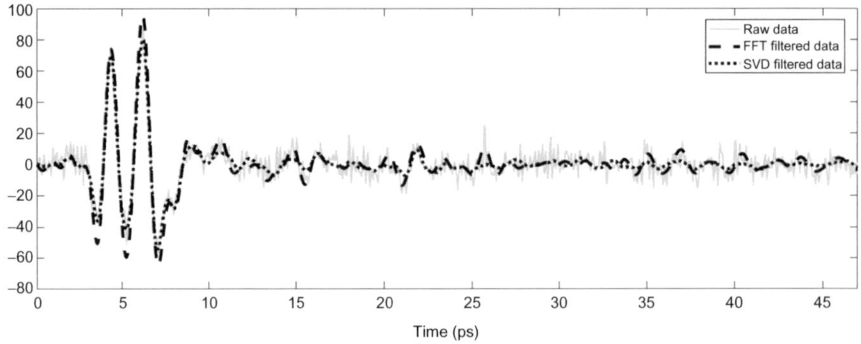

FIG. 2

Comparison of raw and filtered waveforms referring to the central point of the specimen. The *gray solid line* represents raw measured signal, the *black dashed line* is the FFT filtered waveform, and the *black dotted line* refers to the SVD filtered waveform.

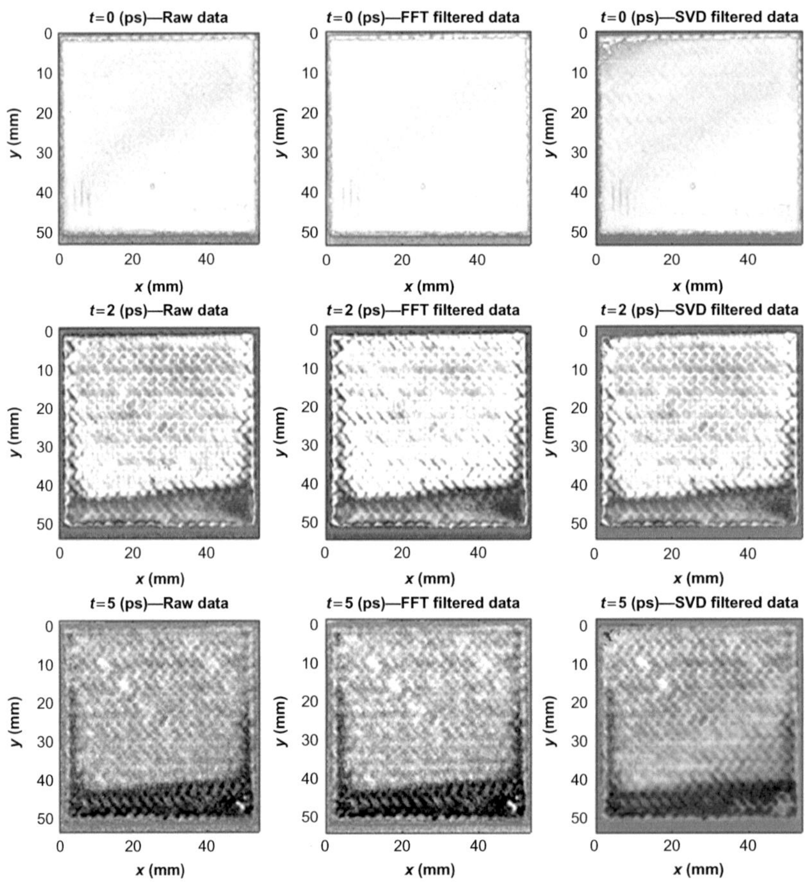

FIG. 3

THz images provided by raw and filtered data for increasing times. The images were obtained by applying topography correction.

(Continued)

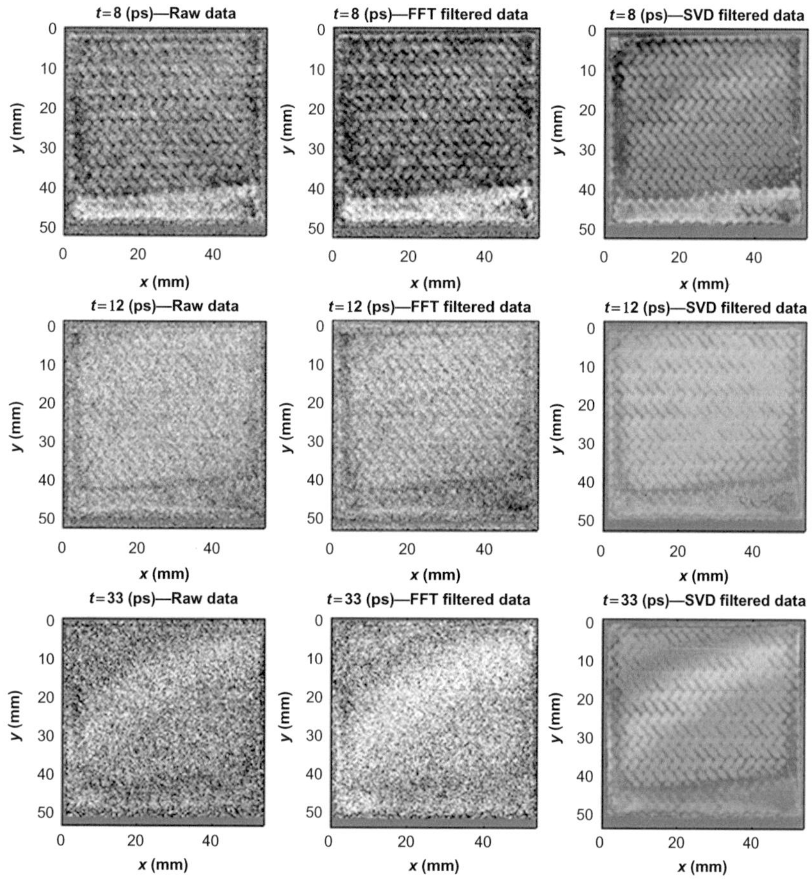

FIG. 3, CONT'D

coating (first peak) and the underling composite material (second peak), while both the filtering procedures are useful to properly image the inner texture of the specimen and possibly hidden defects.

This appears clear by taking into account Fig. 3, which shows slices at increasing time instants referred to both raw and filtered data. These images were obtained by applying a topography correction procedure [37], which performs an alignment of the first reflection by setting the zero time, that is, $t = 0\,\text{ps}$, at the beginning of the ascending front of the first peak occurring in each one of the measured waveforms.

Fig. 3 corroborates that both FFT and SVD filters allow an improved visualization of inner specimen texture; FFT filter introduces a slight loss of spatial resolution (f.i., see images at $t = 2\,\text{ps}$), while SVD filter allows a better noise rejection and provides more clean images, especially for increasing observation time instants.

5 TEST CASES

This section aims at assessing the potentialities offered by THz technology as a diagnostic tool capable of characterizing inner features of the investigated object. In particular, results concerning surveys carried on a fresco fragment and chocolate pralines are presented. The surveys were carried out at IREA-CNR by using the FiCO system described in the Section 4.3. The data were gathered in normal reflection mode by using a specific holder hosting both emitter and receiver. Hence, the SVD filtering procedure was applied in order to improve the imaging capabilities.

5.1 THz FOR ARTWORK CHARACTERIZATION

The first considered case study concerns the use of THz technology to perform nondestructive survey in the frame of cultural heritage. This is one of the most relevant applicative sectors, wherein THz waves allow the gaining of information useful to improve knowledge about the design technique adopted by the artist and to detect possible damages affecting the conservation state of precious artworks [12].

Fig. 4A shows the investigated fragment. In this survey, the measurement area was 40 × 25 mm wide and was scanned with a spatial offset of 0.5 mm in both the measurement directions. The data collection stage required about 10 min.

The false-color image obtained by plotting the maximum of the stored waveform at each measurement point is in Fig. 4B, while Fig. 4C shows the filtered radargrams with a spatial offset of 1 mm along the y direction.

Based on Fig. 4B, it is possible to infer that the fragment is characterized by a not uniform reflectivity. Moreover, Fig. 4C reveals the presence of different interfaces. In particular, one can recognize five different interfaces classified as it follows:

- I: air-decorative paint interface
- II: decorative paint-preparation paint interface

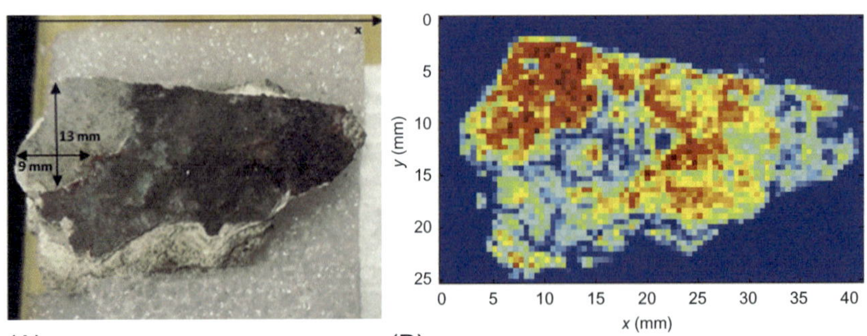

(A) (B)

FIG. 4

THz survey of a fresco fragment: (A) picture of the fragment under test, (B) THz false-color image, and

(Continued)

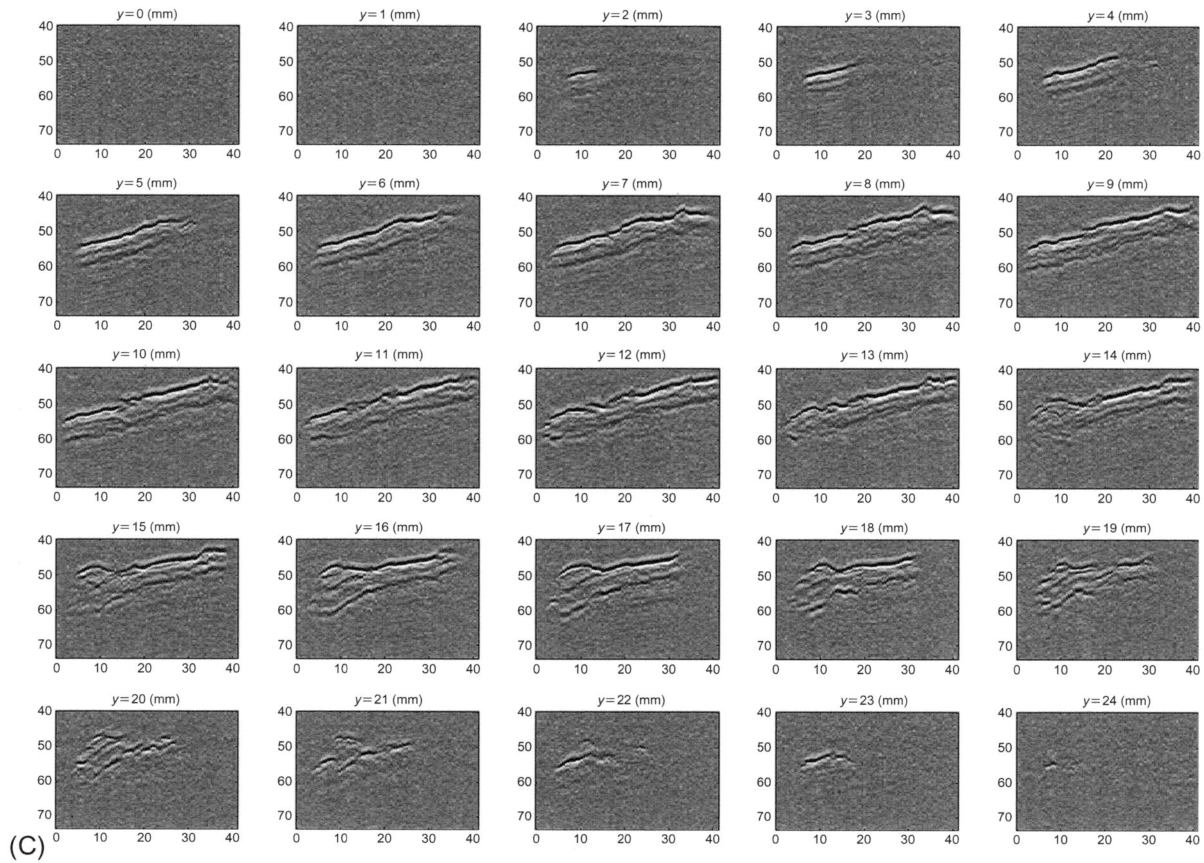

FIG. 4, CONT'D

(C) radargrams at y ranging from 0 to 24 mm, with 1 mm offset.

- III: preparation paint-stucco
- IV: stucco-mortar interface
- V: air-preparation paint interface

This classification is made evident in Figs. 5A–E, which show the radargram corresponding to $y = 8$ mm and the waveforms collected at $x = 10$, 19, 25, and 30 mm. Moreover, Fig. 5 shows that, in certain fragment points, a delamination phenomenon occurs, between decorative and preparation paint layers and between preparation paint and stucco.

5.2 THz FOR FOOD QUALITY CONTROL

The second example deals with food quality control, an applicative field where THz technology is deserving of huge attention thanks to its ability to detect and discriminate, based on their different refractive indexes, undesired foreign objects, f.i. plastic, glass, stone, or metallic pieces [9,13,40,41].

As an example, we consider the survey of dark chocolate tablets, produced by a famous Italian chocolate factory. More in detail, we investigated an uncontaminated praline and a second one wherein a foreign object was added in the bottom side

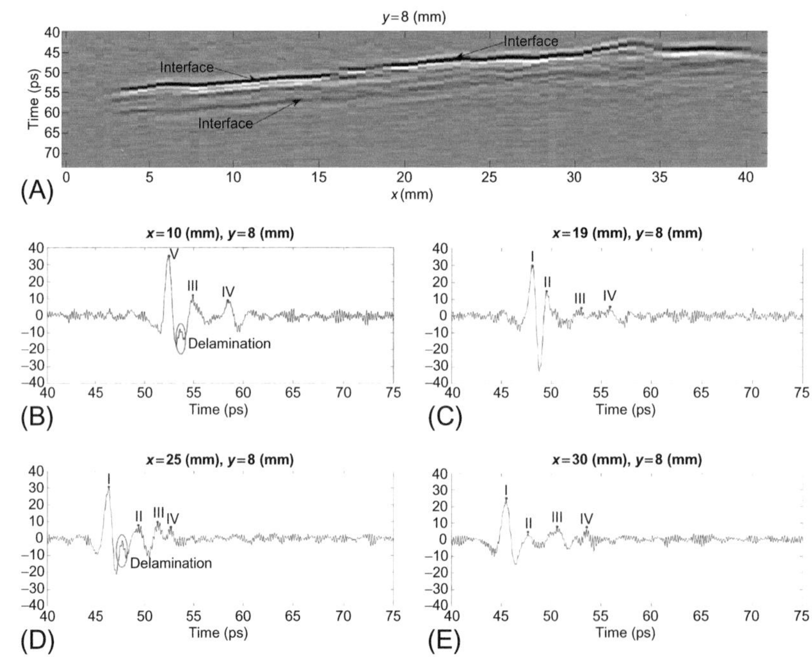

FIG. 5

Details of THz survey concerning fresco fragment: (A) radargrams at $y = 8$ mm and (B)–(E) waveforms referred to $x = 10$, 19, 25, and 30 mm and $y = 8$ mm.

without changing the praline structure in a visible way. The foreign object is a small calcareous stone whose cross section is about circular with diameter equal to 4 mm and whose height was about 1 mm.

The thickness of the chocolate tablet is 5 mm, and the foreign object was located at a depth of about 3.5 mm from the upper surface of the praline where the logo of the factory is printed.

The data were collected by probing the upper side of the praline, the measurement area was 23 × 23 mm wide, and the spatial offset was 0.2 mm along both the scanning directions.

Fig. 6 compares the reconstructed slices of the two pralines referred to increasing and constant depths. In particular, Fig. 6A and B shows the slices at $z = 0$ mm and at

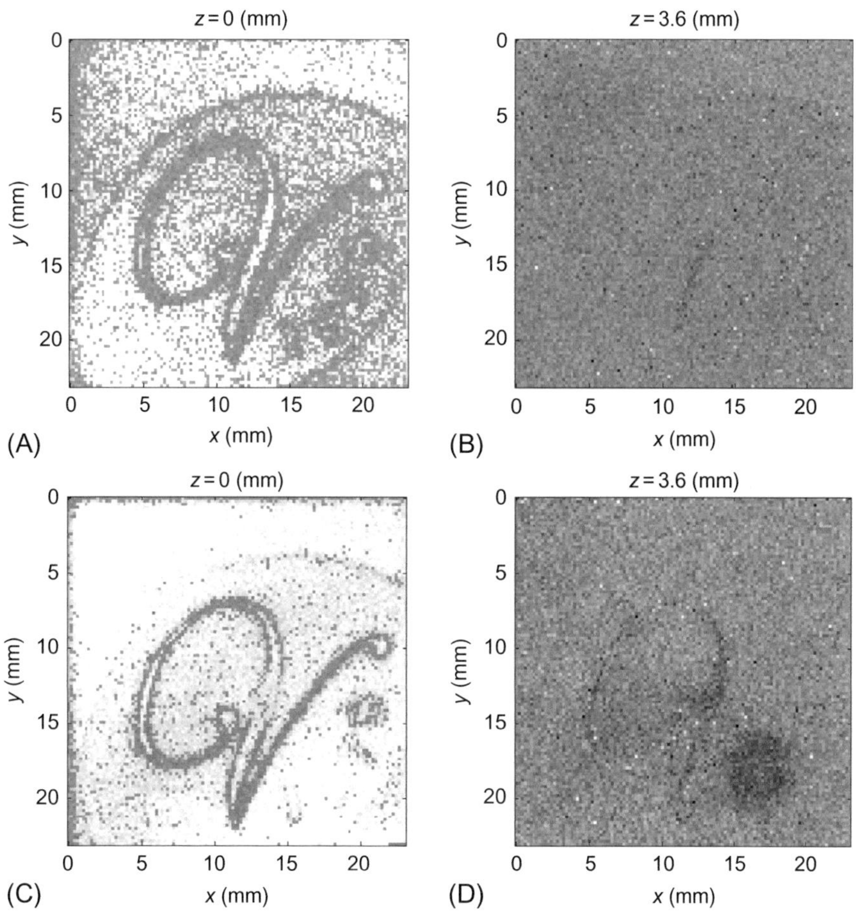

FIG. 6

THz survey of uncontaminated and not chocolate tablets: (A) and (B) slices referred to the uncontaminated tablet at $z = 0$ mm and $z = 3.6$ mm, and (C) and (D) slices referred to the contaminated tablet at $z = 0$ mm and $z = 3.6$ mm.

$z = 3.6\,\text{mm}$ referred to the uncontaminated praline, while Fig. 6C and D shows the slices at the same depths but corresponding to the contaminated chocolate tablet. It is worth mentioning that the conversion time depth was made by assuming the relative permittivity of chocolate equal to 3 as stated in Ref. [42].

Based on the images in Fig. 6, one can state that THz waves allow us not only to image the praline surface properly but also to detect and localize hidden objects that may be dangerous for human life. In Fig. 6D, indeed, an anomaly appears in the bottom right corner at x and y ranging from 15 to 20 mm, while this does not happen in Fig. 6B.

6 CONCLUSIONS AND PERSPECTIVES

This chapter has provided an overview of the current state of the art of THz technology by taking into account both the available measurement systems and the proposed imaging methodologies. Moreover, the usefulness of filtering procedures has been taken into account, and two test cases have been summarized in order to stress the potentialities offered by THz waves as a noninvasive, contactless, and safe diagnostic tool.

Although, it is the newest among the electromagnetic sensing techniques, THz technology has experienced a rapid and fast growth in the last few years thanks to the advancements made in terms of effective and stable detectors. Accordingly, today, its sensing capabilities are well known, and its use begins to attract interest in several applicative contexts.

However, despite the fact that state-of-the-art instrumentations allow satisfactory performance, constant efforts are addressed toward a more and more competitive and widespread use of THz technology. In this framework, research activities regard mainly the following:

(i) The design of systems capable of gathering large amount of data in a wide frequency range while keeping fast the scanning time and acceptable the SNR
(ii) The development of effective data processing strategies able to reconstruct reliable and accurate information about the investigated object

In order to give a contribution to the first necessity, systems exploiting array of detectors, like THz scanners, are appearing on the market [43]. Moreover, systems able to gather simultaneously data in reflection and transmission mode represent a possible future technological improvement, since they allow an enhancement of the information achievable on the objects under test, if their features are compliant with transmission measurements.

Moreover, in order to improve the achievable imaging capabilities in terms of localization and morphological characterization of the detected anomalies, focusing approaches, like those proposed in THz diffraction tomography, attract huge interest. They allow, indeed, a proper imaging of hidden targets whose size is equal or less than the minimum probing wavelength. In this framework, examples assessing the improvements offered by an imaging approach based on the first-order Born approximation are provided in Ref. [44].

Other progresses are dictated mainly by the applicative context of interest. For instance, in the frame of cultural heritage diagnosis, one of the main necessities is the availability of compact, lightweight, and robust systems that can be easily moved in the place wherein the object to be investigated is located. Conversely, in industrial applications, most important requirements are fast data acquisition time and automatic detection of anomalies, since they are crucial to assure the integration of THz technology within the productive process.

REFERENCES

[1] P.H. Siegel, Terahertz technology, IEEE Trans. Microwave Theory Tech. 50 (3) (2002) 910–928.

[2] Y.-S. Lee, Principles of Terahertz Science and Technology, Springer, Berlin, 2009.

[3] J.B. Baxter, G.W. Guglietta, Terahertz spectroscopy, Anal. Chem. 83 (12) (2011) 4342–4368.

[4] D. Mittleman, Sensing With THz Radiation, Springer, Berlin, 2003.

[5] M.C. Kemp, P.F. Taday, B.E. Cole, J.A. Cluff, A.J. Fitzgerald, W.R. Tribe, Security applications of terahertz technology, Proc. SPIE 5070, Terahertz for Military and Security Applications, July 29, 2003, p. 9.

[6] T. Yasui, et al., Terahertz paintmeter for noncontact monitoring of thickness and drying progress in paint film, Appl. Opt. 48 (33) (2005) 6541–6546.

[7] C. Jordens, M. Koch, Detection of foreign bodies in chocolate with pulsed terahertz spectroscopy, Opt. Eng. 47 (3) (2008) 037003.

[8] E.K. Rahani, et al., Mechanical damage detection in polymer tiles by THz radiation, IEEE Sensors J. 11 (8) (2011) 1720–1725.

[9] K. Wang, D.-W. Sun, H. Pu, Emerging non-destructive terahertz spectroscopic imaging technique: principle and applications in the agri-food industry, Trends Food Sci. Technol. 67 (2017) 93–105.

[10] K. Humphreys, et al., Medical applications of terahertz imaging: a review of current technology and potential applications in biomedical engineering, 26th Annual International Conference of the IEEE Engineering in Medicine and Biology Society, 2004 (IEMBS'04), 2004.

[11] Y.-C. Shen, Terahertz pulsed spectroscopy and imaging in pharmaceutic setting—a review, J. Pharm. Pharmacol. 47 (1–2) (2011) 48–60.

[12] K. Fukunaga, THz Technology Applied to Cultural Heritage in Practice, Cultural Heritage Science, Springer, Japan, 2016.

[13] R. Gente, M. Koch, Monitoring leaf water content with THz and sub-THz waves, Plant Methods (2015) 9.

[14] M. Hangyo, M. Tani, T. Nagashima, Terahertz time-domain spectroscopy of solids: a review, Int. J. Infrared Millimeter Waves 26 (12) (2005) 1661–1690.

[15] J.A. Zeitler, P.F. Taday, D.A. Newnham, M. Pepper, K.C. Gordon, T. Rades, Terahertz pulsed spectroscopy and imaging in the pharmaceutical setting—a review, J. Pharm. Pharmacol. 59 (2007) 209–223.

[16] A.G. Davies, A.D. Burnet, W. Fan, E.H. Linfield, J.E. Cunningham, Terahertz spectroscopy of explosives and drugs, Mater. Today 11 (2008) 18–26.

[17] K. Kawase, Y. Ogawa, Y. Watanabe, H. Inoue, Non-destructive terahertz imaging of illicit drugs using spectral fingerprints, Opt. Express 11 (2003) 2549–2554.

[18] Y.C. Shen, T. Lo, P.F. Taday, B.E. Cole, W.R. Tribe, M.C. Kemp, Detection and identification of explosives using terahertz pulsed spectroscopic imaging, Appl. Phys. Lett. 86 (2005) 241116.

[19] K. Fukunaga, M. Picollo, Terahertz spectroscopy applied to the analysis of artists' materials, Appl. Phys. A: Mater. Sci. Process. 100 (2010) 591–597.

[20] A.J. Fitzgerlad, V.P. Wallace, M. Jimenez-Linan, L. Bobrow, R.J. Pye, A.D. Purushotham, D.D. Arnone, Terahertz pulsed imaging of human breast tumors, Radiology 239 (2006) 533–540.

[21] K.E. Peiponen, A. Zeitler, M. Kuwata-Gonokami, Terahertz Spectroscopy and Imaging, vol. 171, Springer-Verlag, Berlin, Heidelberg, 2012.

[22] http://lunainc.com/thz/products/t-ray5000.

[23] http://www.menlosystems.com/en/products/thz-time-domain-solutions/all-fiber-coupled-terahertz-spectrometer/.

[24] https://www.crisel-instruments.it/phisical-science-elettroottica/thz/terahertz-time-domain-solution/.

[25] C.W.I. Chan, Q. Hu, J.L. Reno, Ground state terahertz quantum cascade lasers, Appl. Phys. Lett. 101 (2013) 151108.

[26] N. Karpowicz, H. Zhong, C. Zhang, K.I. Lin, J.S. Hwang, J. Xu, X.C. Zhang, Compact continuous-wave subterahertz system for inspection applications, Appl. Phys. Lett. 86 (2005) 54105 (4pp).

[27] B.B. Hu, M.C. Nuss, Imaging with THz waves, Opt. Lett. 20 (1995) 1716–1719.

[28] Z.P. Jiang, X.G. Xu, X.C. Zhang, Improvement of terahertz imaging with a dynamic subtraction technique, Appl. Opt. 39 (17) (2000) 2982–2987.

[29] W.L. Chan, J. Deibel, D.M. Mittleman, Imaging with terahertz radiation, Rep. Prog. Phys. 70 (2007) 1325–1379.

[30] V.P. Wallace, et al., Three-dimensional imaging of optically opaque materials using nonionizing terahertz radiation, J. Opt. Soc. Am. A Opt. Image Sci. Vis. 25 (12) (2008) 3120–3133.

[31] M. Bertero, P. Boccacci, Introduction to Inverse Problems in Imaging, IOP Publishing Ltd., London, 1998.

[32] B. Ferguson, et al., T-ray computed tomography, Opt. Lett. 27 (15) (2002) 1312–1314.

[33] L.T. Niklason, et al., Digital tomosynthesis in breast imaging, Radiology 205 (2) (1997) 399–406.

[34] D.M. Mittleman, et al., T-ray tomography, Opt. Lett. 22 (12) (1997) 904–906.

[35] P. Matoba, et al., Real-time three-dimensional object reconstruction by use of a phase-encoded digital hologram, Appl. Opt. 41 (29) (2002) 6187–6192.

[36] M. Born, E. Wolf, Principles of Optics, Cambridge University Press, Cambridge, 1999.

[37] I. Catapano, A. Affinito, L. Guerriero, B. Bisceglia, F. Soldovieri, Majolica imaging with THz waves: preliminary result, Appl. Phys. A: Mater. Sci. Process. 122 (2016) 11.

[38] M. Bertero, Linear inverse and ill-posed problems, Adv. Electron. Electron. Phys. 45 (1989) 1–120.

[39] I. Catapano, F. Soldovieri, L. Mazzola, C. Toscano, THz imaging as a method to detect defects of aeronautical coatings, J. Infrared Millim. Terahertz Waves 38 (2017) 14.

[40] A.A. Gowen, C. O'Sullivan, C.P. O'Donnell, Terahertz time domain spectroscopy and imaging: emerging techniques for food process monitoring and quality control, Trends Food Sci. Technol. 25 (1) (2012) 40–46.
[41] Y.-K. Lee, et al., Detection of foreign bodies in foods using continuous wave terahertz imaging, J. Food Prot. 75 (1) (2012) 179–183.
[42] Endress, Hauser, Messtechnik GmbH Co. KG, Relative Dielectric Constant *εr* (*dk* value) of Liquids and Solid Material*s*, Weil am Rhein, 2000.
[43] http://terasense.com/applications/terahertz-food-inspection/.
[44] R. Dahlbäck, T. Rubæk, M. Persson, J. Stake, A system for THz imaging of low-contrast targets using the born approximation, IEEE Trans. Terahertz Sci. Technol. 2 (3) (2012) 361–370.

CHAPTER

Ambient noise techniques to study near-surface in particular geological conditions: a brief review

12

D'Amico Sebastiano*, Panzera Francesco†, Martino Salvatore‡, Iannucci Roberto‡, Paciello Antonella§, Lombardo Giuseppe†, Galea Pauline*, Farrugia Daniela*

Department of Geosciences, University of Malta, Malta Department of Biological Geological and Environmental Sciences, University of Catania, Catania, Italy† Department of Earth Sciences, University of Rome "La Sapienza", Italy‡ ENEA, Rome, Italy§*

CHAPTER OUTLINE

1 INTRODUCTION

Generally, the term *site effects* refers to the amplification of seismic waves due to some particular geologic conditions [1–6]. Site effects occur as a result of several physical phenomena such as multiple reflections, diffraction, focusing, and resonance, to which the incoming wave front is subjected. This is a consequence of the various mechanical properties of terrains, the presence of heterogeneities and discontinuities, and the geometry of shallower layers and the existence of topographic irregularities both in the basement and in the surface. Several different methodological approaches are commonly adopted to quantitatively assess the local seismic response and to study peculiar geologic conditions such as mud volcanoes, faults, and landslides. The site response can be evaluated through various approaches each

Innovation in Near-Surface Geophysics. https://doi.org/10.1016/B978-0-12-812429-1.00012-X

of them having specific advantages and/or drawbacks well known in literature (see Ref. [7]). The experimental methods allow us to evaluate the local seismic response using the records of seismic signals that can be generated by earthquakes, artificial seismic sources, or ambient noise. Ambient noise has, in recent years, become widely used for site amplification studies. Its use appears opportune for significant reductions in field data acquisition time and costs. The evaluation of site response using this technique is largely adopted since it requires only one mobile seismic station with no additional measurements at rock sites for comparison. In the early 1950s, a method to estimate the dynamic characteristics of surface layers was introduced. Nogoshi and Igarashi [8] introduced the horizontal-to-vertical spectral ratio (HVSR) method, also known as the H/V method that later started to be referred to as the Nakamura technique. The scope of the HVSR method is to get the amplification of the S waves due to the contrast of impedance between geologic layers by recording ambient noise vibrations [9]. Several advantages of this method include the fact that it is low cost, it is simple to use and apply, it can be used at various site conditions, and it does not have any negative impact on the environment.

Ambient noise recordings are analyzed in several time series [10] and transformed in the frequency domain where the horizontal Fourier amplitude spectra are averaged (usually using the geometric mean) and divided by the Fourier amplitude spectrum of the vertical component. The curves for each window are then averaged to get the final HVSR curve. There is an agreement on the fact that the HVSR method gives an accurate measure of the S-wave resonance frequency f_0, which is related to the S-wave velocity of the sediment layer versus and the thickness h by

$$h = Vs/(4 \times f_0)$$

We would like to stress that the technique of the H/V method on its own is not sufficient to characterize the complexity of site effects and in particular the absolute values of seismic amplification [10]. The main use and recommended application of such technique are to map the fundamental period of the site and help to constrain some of the geologic and geotechnical parameters to be used for numerical computations [10].

A list of practical guidelines can be found in the SESAME Working Group [10], technical report that mainly focuses on measurements of ambient noise in free field. The main guidelines can be summarized as follows:

1. Recording parameters: The instrument sensor should be exactly leveled, and the gain level is fixed at the maximum possible without signal saturation.
2. Measurement spacing: A large spacing implies good results for obtaining the fundamental frequency and also getting a clear view of how the main geologic interfaces that are dipping (e.g., in mapping a sediment deposit or alluvial valley) is composed. However, denser set of measurements can be necessary in particular geologic and/or geomorphological conditions.
3. Sensor setting: The sensor should be placed horizontally and fixed tightly to the ground.

4. Nearby structures: Tall buildings, trees, and other similar structures should be avoided since oscillations of the structures may affect the readings due to low-frequency perturbations in the ground.
5. Weather conditions: Avoid measurement during windy days since this can cause large perturbations at low frequency. Heavy rain should also be avoided, but slight rain does not influence the results.
6. Disturbances: Recording near construction machines, pumps, generators and any other type of monochromatic noise sources should be avoided. If there is transient noise such as footsteps and passing vehicles, the recording duration is increased so that the transient time windows could be eliminated, but still, there would be enough windows for the required analysis.
7. Recording duration: For a reliable measurement, the condition $f_0 > 10/l_w$ must be satisfied where f_0 is the fundamental frequency and l_w is the window length. This condition says that there are at least 10 cycles in each window for the fundamental frequency. A large number of windows and cycles are needed. It is suggested that the total number of significant cycles $n_c = (l_w)(n_w)(f_0)$, where n_w is the number of windows, should be larger than 200.

2 AMBIENT VIBRATION MEASUREMENTS IN MUD VOLCANO AREAS

Hydrothermal systems occur in areas having high heat fluxes both on continents, near convergent plate boundaries, and on the ocean floor, near the mid-ocean ridges. Their formation requires the existence of three important components: fluids, heat, and permeability through rocks so that fluids can circulate. Mud volcanoes are observed and studied in different geologic settings in active and passive margins and more specifically in basins with high sedimentation rates where organic-rich sediments are deposited [11–13]. Regions characterized by sedimentary volcanism are typical targets for hydrocarbon exploration as they are often associated with oil and gas reservoirs. At the same time, mud volcanoes represent a potential natural hazard due to their sudden and violent eruptions (e.g., [14]). The development of mud volcanoes follows the presence in the subsurface of (a) clay-bearing strata that can be buoyant relative to the surrounding units and (b) overpressured fluids that facilitate the growth of diapirs through sedimentary rocks and structural and/or lithologic discontinuities [15,16]. Sedimentary-hosted hydrothermal systems are hybrid structures sharing the features of hydrothermal systems and mud volcanoes. These hybrid system phenomena involve the intrusion of magmatic bodies and migration of hydrothermal fluids in fertile and organic-rich sedimentary basins. The high temperatures trigger metamorphic reactions that generate production of CO_2 and CH_4 and ultimately accelerate the maturation processes of the organic matter existing in the sediments [17–20]. The overpressure results in hydrothermal vent systems that ultimately rise to the surface (e.g., Ref. [21]). In this framework, ambient vibration methods, which are common tools used for site effect investigations aiming at assessing the thickness of the sediments, were tested. The approach used is based on the ratio of the horizontal-to-vertical components of ground motion (HVSR) that generally produces recognizable peaks that correspond to the fundamental frequency of the site. Experimental data spectral ratio peaks usually fit quite well the resonance frequency of the theoretical curves derived from modeling the response of the layer package, so that HVSR contains valuable information about the underlying structure.

Here, we present a review of data collected at the mud volcanoes and sedimentary-hosted hydrothermal system sites in order to investigate the depths of the main discontinuities and of the hypothesized hydrocarbon reservoirs. Galgaro et al. [22] investigated the Euganean thermal basin, located in the Eastern Po plain (Italy) (see location in Fig. 1A). This area produces the largest amount of

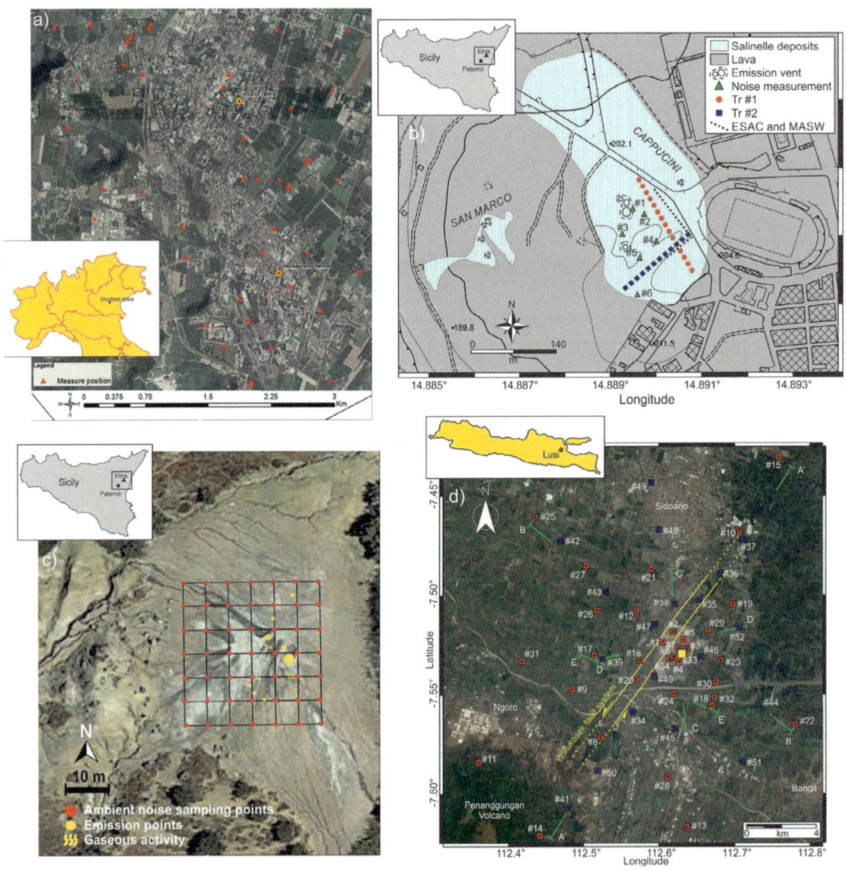

FIG. 1

Locations of ambient vibration measurement points used during the experiments performed by Galgaro et al. [22] (A), Panzera et al. [23] (B), Imposa et al. [24] (C), and Panzera et al. [25] (D). The pictures clearly show that the space distribution of the recording sites is a function of the chosen target.

geothermal energy in NE Italy [26]. The Euganean thermal district, about 23 km^2 in size, is located at the eastern edge of the Veneto region volcanic district. Two distinct geothermal subunits have commonly been distinguished within the Euganean basin: the Abano-Montegrotto unit and the Battaglia-Galzignano unit [27]. The authors investigated the Abano-Montegrotto subunit thermal aquifers related to the first Eocene eruptive Euganean activity, characterized by the emplacement of laccoliths within the fractured carbonate rocks [28,29]. They correlated the results from 50 ambient vibration measurements with information coming from remote sensing and thermal well data. In particular, the HVSR method was used for characterizing the thermal basin, giving information about the thickness of the soft sedimentary cover, and it was also employed as a detector of the presence of either smooth or more abrupt changes in the mechanical properties between covers and bedrock. This technique appeared also to be efficient preliminary estimator of the bedrock fracturing, therefore permitting to correlate the features of HVSR peaks to the thermal property of the explored area (Fig. 2). Another microtremor survey, aiming at obtaining information on the subsoil structure in mud volcano areas, was accomplished in the lower southwestern flank of Mt. Etna (Sicily, Italy) by Panzera et al. [23]. This experiment was carried out in the Salinelle of Paternò mud volcano area (Fig. 1B), emitting waters characterized by an abundant free gaseous phase and showing typical features of waters

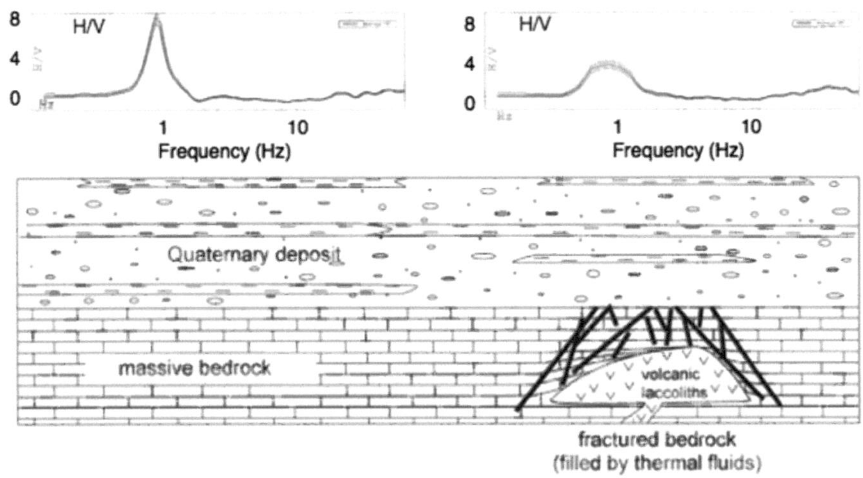

FIG. 2

Schematic interpretation of HVSR obtained in the Euganean basin. Left panel shows an example of HVSR that can be used for estimating the thickness of quaternary deposits, whereas the right panel depicts an example of a broad peak HVSR whose presence can be associated with the existence of a fractured bedrock.

Modified from A. Galgaro, J. Boaga, M. Rocca, 2013. HVSR technique as tool for thermal-basin characterization: a field example in N-E Italy. DOI 10.1007/s12665-013-2838-5.

accompanying the hydrocarbon reservoirs [30]. In the study area, three important stratigraphic discontinuities were identified through spectral ratio techniques after processing the ambient vibration data. The synthetic HVSR obtained by 1-D modeling of the area, using both literature information and the results coming from extended spatial autocorrelation (ESAC) and multichannel analysis of surface wave (MASW) investigations, accurately reproduces the shape of the experimental spectral ratios. This evidence supports the hypothesis of the presence of three discontinuities, linked to HVSR peaks at frequencies 0.1–0.3, 0.6–1.0, and 4.0 Hz, roughly located at depths of 10, 200, and 1000 m, respectively. The theoretical model also corroborates the hypothesis of a low-velocity zone, located at about 700–800 m depth, that is reasonably interpreted as a hydrothermal reservoir (Fig. 3). Imposa et al. [24] carried out an ambient vibration survey in the area of San Biagio Salinelle (Fig. 1C), located a few kilometers to the SE of the Salinelle of Paternò mud volcanoes, aimed at identifying the probable ascent path of the emitted products. Seismic ambient noise was recorded at the nodes of a specially designed grid, and subsequently, the V_s values were obtained from an active seismic survey. A digital elevation model (DEM) of the area was obtained by a topographic survey, carried out with the global navigation satellite system (GNSS) technique, in real-time kinematic mode. Through this experiment, the authors provided an accurate and detailed 3-D subsurface model showing the shallower feeding system of the investigated mud volcano (Fig. 4). More recently, Panzera et al. [25] by means of horizontal-to-vertical spectral ratio (HVSR) proposed insights about the geometry of the shallow layers around the Lusi hydrothermal clastic geysering system [31] (Fig. 1D). Lusi is a sedimentary-hosted hydrothermal system located in Eastern Java, Indonesia (see inset map in Fig. 1A), at about 40 km to the NE of the Arjuno-Welirang volcanic complex and seated along the Watukosek fault system (WFS) (yellow lines in Fig. 1B) [32]. The HVSRs identified the presence of a low-frequency peak at about 0.4–0.6 Hz that shifts to values of 0.3 Hz in the proximity of the Lusi vent. The authors located a resonant layer, having a thickness ranging between 130 and 340 m, linked to the alluvial deposits overlying the Pucangan formation. They also observed an increase in thickness, close to the Lusi crater, highlighting the presence of a "depocenter" that could be interpreted as the subsidence caused by withdrawal of mud and fluids from depth (Fig. 5). In conclusion, it seems worthwhile to point out that up to the present time, the only information about hydrothermal system areas and reservoir location has been mainly inferred from geochemical, electric, and gravimetric investigations. The above-described geophysical approaches, based on HVSR methods, shed further light on the subsoil structure of the investigated areas. The results of the ambient vibrations approach show indeed different frequency ranges from which it is possible to deduce information about the subsurface structure and the location of conduits for hydrothermal materials. Moreover, the preliminary sketches obtained constitute fast and simple methods that enable the planning of more specific and detailed investigations leading to a comprehensive model of the investigated areas.

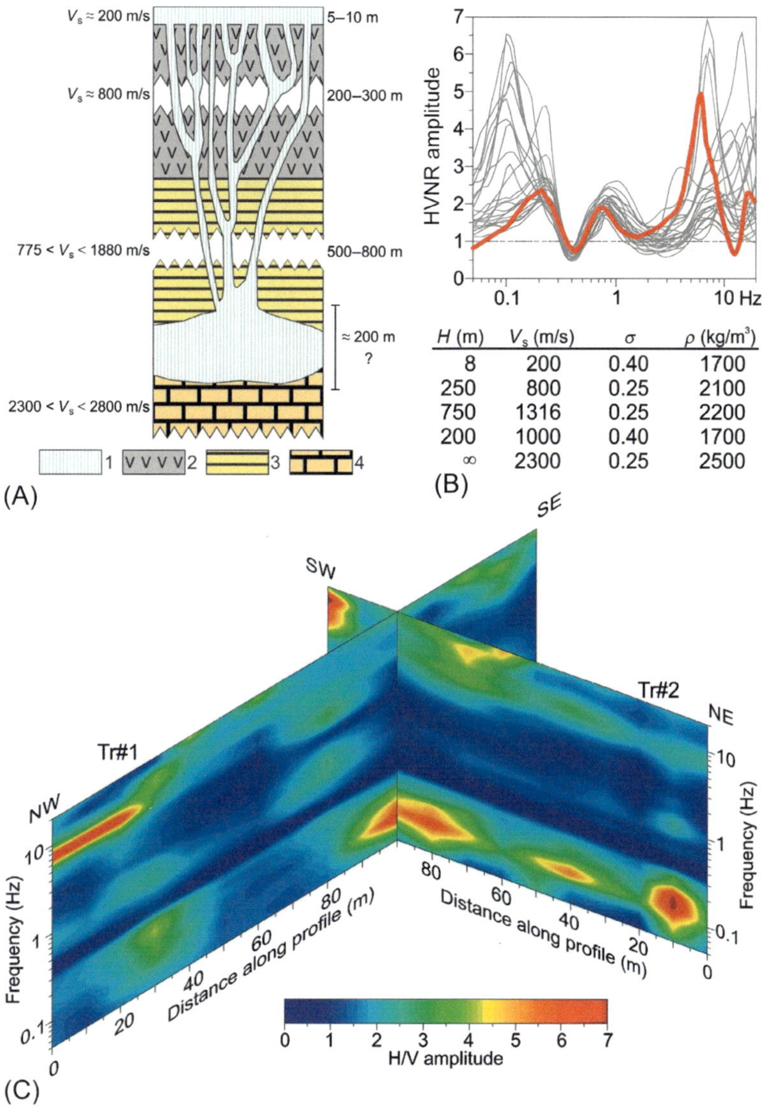

FIG. 3

(A) Thickness of the main lithotypes and corresponding S-wave velocity as used by Panzera et al. [102] for the 1-D modeling. (1) Salinelle deposits, (2) lava, (3) clay, and (4) limestone. (B) Comparison between the results of 1-D model *(red line)* and experimental HVSRs. (C) Two-dimensional diagram obtained by combining all the ambient noise measurements performed along the profiles (Tr#1 and Tr#2) shown in Fig. 1B, as a function of distance (*x*-axis) and frequency (*y*-axis).

(modified from F. Panzera, S. Sicali, G. Lombardo, S. Imposa, S. Gresta, S. D'Amico, A microtremor survey to define the subsoil structure in a mud volcanoes area: the case study of Salinelle (Mt. Etna, Italy), Environ. Earth Sci. 75 (2016) 1140, https://doi.org/10.1007/s12665-016-5974-x)

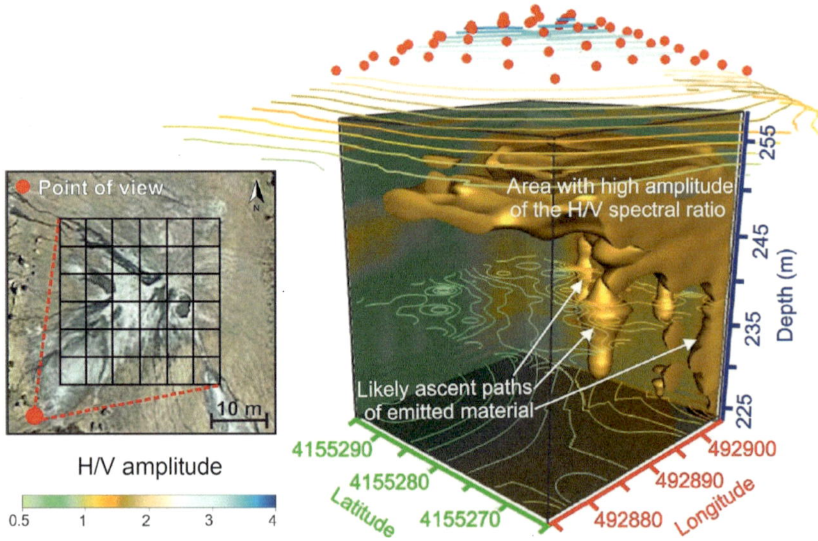

FIG. 4

Three-dimensional subsoil model of the San Biagio Salinelle area, obtained by inverting HVSR information, and related 3-D DEM of the surface *(contour lines)*. The *red dots* indicate the ambient vibration measurement points, whereas the box on the left shows the perspective point of view.

Modified from S. Imposa, S. Grassi, G. De Guidi, F. Battaglia, G. Lanaia, S. Scudero, 3D Subsoil model of the San Biagio 'Salinelle' mud volcanoes (Belpasso, Sicily) derived from geophysical surveys, Surv. Geophys. 37 (2016) 1117, https://doi.org/10.1007/s10712-016-9380-4.

3 AMBIENT VIBRATION MEASUREMENTS IN FAULT ZONES

It is well known that rocks in fault areas have elastic properties that generate local amplifications of the horizontal ground motion during earthquakes. Fault zones are generally characterized by a highly fractured low-velocity belt (damage zone) that may be hundreds of meters wide, bounded by higher-velocity material (host rock) that can extend for some kilometers ([33–35] and references therein). Such a geometric setting and impedance contrast are theoretically similar to the well-known situations widely studied in engineering geology and seismology, when soft sediments overly stiff rock. In the presently depicted case, the discontinuity is almost vertically oriented, and as described by Irikura and Kawanaka [36], it has, in principle, the potential to produce local amplification of ground motion [37–40] and to support the development of fault-zone-trapped waves (e.g., Refs. [3,41]).

In this section, we present results of studies about local seismic response in the vicinity of fault zones (e.g., Ref. [42]) performed by several authors. There are a large number of papers that describe propagation properties of fault-guided waves, and most of them are based on theoretical and experimental approaches dealing with almost pure strike-slip faults (see Refs. [33,43]). In the Calaveras fault, California,

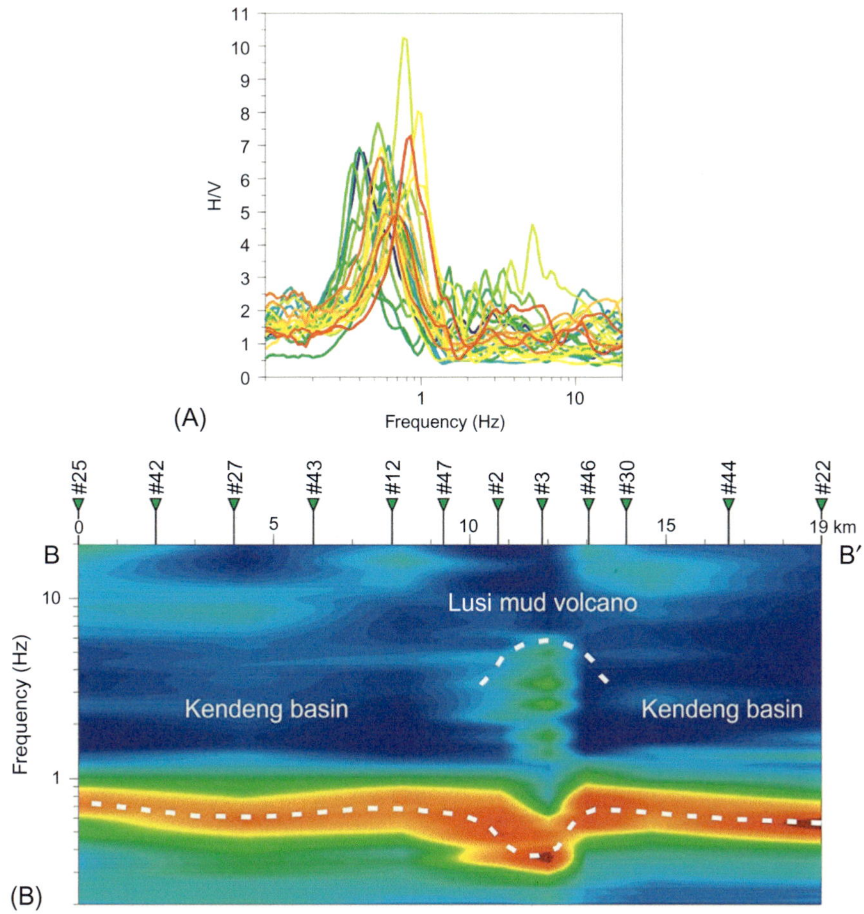

FIG. 5

(A) Fundamental frequency distribution of the HVSRs within the investigated area. (B) Two-dimensional diagram obtained by combining all the ambient noise measurements along the transect BB′ (see location in Fig. 1) as a function of distance (*x*-axis) and frequency (*y*-axis). *Modified from F. Panzera, S. D'Amico, M. Lupi, G. Mauri, K. Karyono, A. Mazzini, Lusi hydrothermal structure inferred through ambient vibration measurements, Mar. Pet. Geol. (2017). https://doi.org/10.1016/j.marpetgeo.2017.06.017.*

Boore et al. [44] noticed that directional effects observed during earthquakes were occurring parallel to the fault strike and not perpendicularly as would be expected for a strike-slip mechanism of faulting. In Italy, Rovelli et al. [45], investigating the Nocera Umbra fault, observed a shallow trapping zone (1–1.5 km) with evidence of measurable amplification effects and pronounced polarization of the ground motion in a direction parallel to the fault strike. On the other hand, Cultrera et al. [38], Calderoni et al. [37], and Pischiutta et al. [46] observed evidence of ground-motion amplification in the fault-zone environments and strong directional effects with high angle to the fault strike, therefore indicating a mechanism different than trapped waves. Similar studies, performed by Cara et al. [47], Imposa et al. [2], Lombardo and Rigano [48], Rigano et al. [49], Di Giulio et al. [50], Pischiutta et al. [46,51,52], and Panzera et al. [4,5,23,53], documented the presence, besides amplification effects, of a systematic horizontal ground-motion polarization near faults. These directional effects, never coincident with the strike of the tectonic structures, were observed both during local and regional earthquakes and using ambient vibration measurements, therefore suggesting the use of microtremors for investigating ground-motion polarization properties along and across the main tectonic structures. The faults characterizing the Etna volcano area (Fig. 6) were the main test sites for the abovementioned studies. Observations of Imposa et al. [2] and Lombardo and Rigano [48] pointed out the presence of significant amplitude increase in the HVSRs, obtained by recording ambient vibrations nearby the shear zone of Tremestieri and Ragalna faults (see locations in Fig. 6). Rigano et al. [49] obtained similar results by comparing ambient vibration data with local and regional earthquakes near the Tremestieri, Pernicana, and Acicatena faults (see locations in Fig. 6). In all these fault areas, rotated spectral ratios showed pronounced HVSR peaks with strong

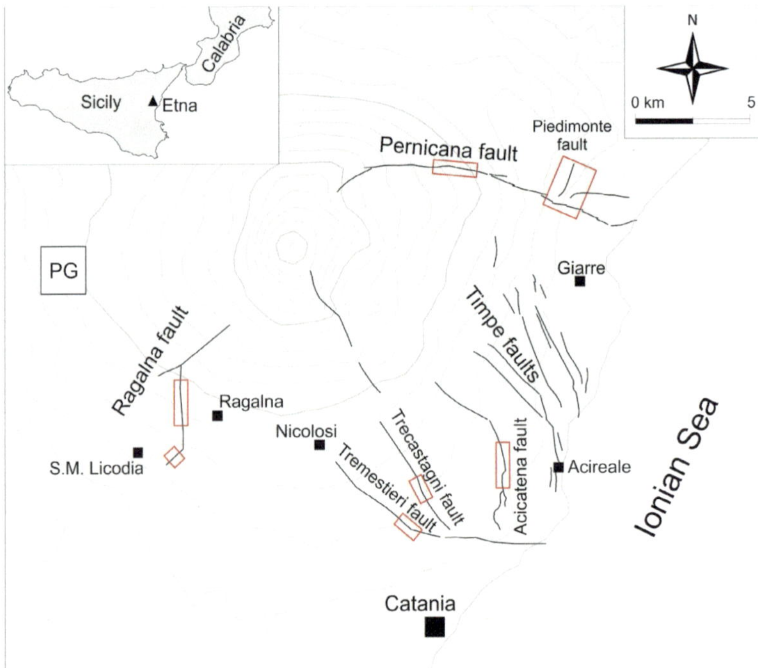

FIG. 6

Location of the main tectonic structures in the Etnean area; Piano dei Grilli (PG) and other rectangles indicate the areas where measurements were performed.

directionality. It is interesting to observe that in all the investigated sites, the ground-motion polarization azimuths were never coincident with the strike of the investigated structures (Fig. 7). Tests were also performed by Di Giulio et al. [50] who mapped azimuth-dependent horizontal-to-vertical spectral ratios along and across the Pernicana fault. The authors found strong directional resonances of horizontal components of motion around 1.0 Hz when measurements sample the most damaged part of the fault zone. This directional effect is not time-dependent, at least at a seasonal scale (it has been proved that seasonal variations [54] can affect the recording both in terms of amplitude and detected presence of the fundamental picks, especially at lower frequencies). They also used frequency-wave-number techniques, applied to array data, in order to investigate the nature of ambient vibrations and verified that the observed polarizations were not affected by features of local sources, such as the volcanic tremor. By analyzing several faults, Pischiutta et al. [46,52] found a near orthogonal relation between the horizontal polarization and the orientation of the main fracture field produced by the fault kinematics. They modeled the

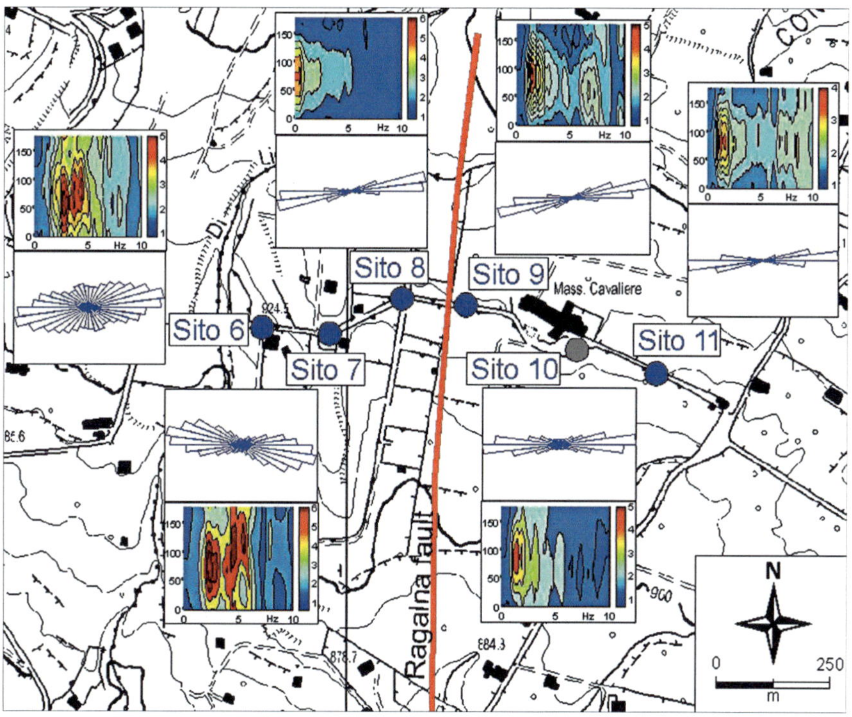

FIG. 7

Examples of directional resonances and polarization plots obtained from ambient noise measurements performed in the Ragalna fault.

Modified from R. Rigano, F. Cara, G. Lombardo, A. Rovelli, Evidence of ground motion polarization on fault zones of Mount Etna volcano, J. Geophys. Res. 113 (2008) B10306, https://doi.org/10.1029/2007JB005574.

expected brittle deformation pattern in the fault damage zone using a combination of numeric and analytic tools and interpreted the directional amplification as produced by stiffness anisotropy, with higher stiffness parallel to the dominant sets of fractures. The Ragalna fault system (see locations in Fig. 6) was then investigated by Panzera et al. [5] by comparing directional amplification effects experimentally observed, with the direction and the modeled type of fractures that are expected in these fault zones. Theoretical results are in good agreement with the field observations of the strike of the fractures, and the resulting directional effects appear to have an orthogonal relationship with these (Fig. 8). Moreover, the authors applied the time-frequency (TF) polarization analysis exploited by Burjánek et al. [55] in order to provide a direct estimate of the polarization parameters (ellipticity, azimuth, and dip). Following the outcomes of the abovementioned results, Panzera et al. [23] investigated the Piedimonte fault system and the surrounding area (see location in Fig. 6). The authors observed the existence of two main fault systems, namely, the NE-SW Piedimonte and the WNW-ESE Pernicana faults. Their results identified prevailing polarization effects almost perpendicular with respect to the faults striking WNW-ESE, confirming the existence of orthogonal relationships. The local extensional deformation field, related to the shear zone of the Pernicana fault system, therefore appears to lead to a local stress orientation that characterizes the seismic response of this portion of the volcano. The role played by anisotropy on the strong directional effects in fault zones was demonstrated by Pischiutta et al. [51] and Panzera et al. [53] by comparing shear-wave anisotropy, inferred from shear-wave splitting, and directional site effects (Fig. 9). They found, in Val d'Agri area and South Iceland, respectively, an orthogonal relation between the horizontal polarization and the fast shear-wave direction, suggesting that wave-field polarization and fast velocity directions in a fractured medium originate from the same source mechanism. A broad depiction of local seismic effects linked to the presence of various fault systems characterizing the eastern flank of the Etnean area, using ambient noise vibrations, was summarized by Panzera et al. [4]. The authors performed 135 measurements using a grid (spaced 1.0×1.0 km), with the recording sites located at its nodes. The contour map obtained by interpolating the frequency classes assigned to each measurement highlights the important role played by the fracture fields linked to the main fault systems present in the area. The results of both directional seismic site effects and polarization diagrams of the horizontal components of motion highlighted indeed the existence of areas with strong directional effects that appear almost perpendicular to the trend of the main fault systems (Fig. 10). To observe how directional effects can change at increasing distance from the fault lines, measurements were performed in areas significantly distant from the studied tectonic structures such as the Piano dei Grilli area (PG in Fig. 1) and the Augusta area (southeastern Sicily). The results (Fig. 11) show that ground-motion amplification directions, coming from both rotated spectral ratios and polarization diagrams, tend to become randomly distributed and/or uniformly scattered at increasing distance from the fault lines. The outcomes of the above studies allow us to make the following observations:

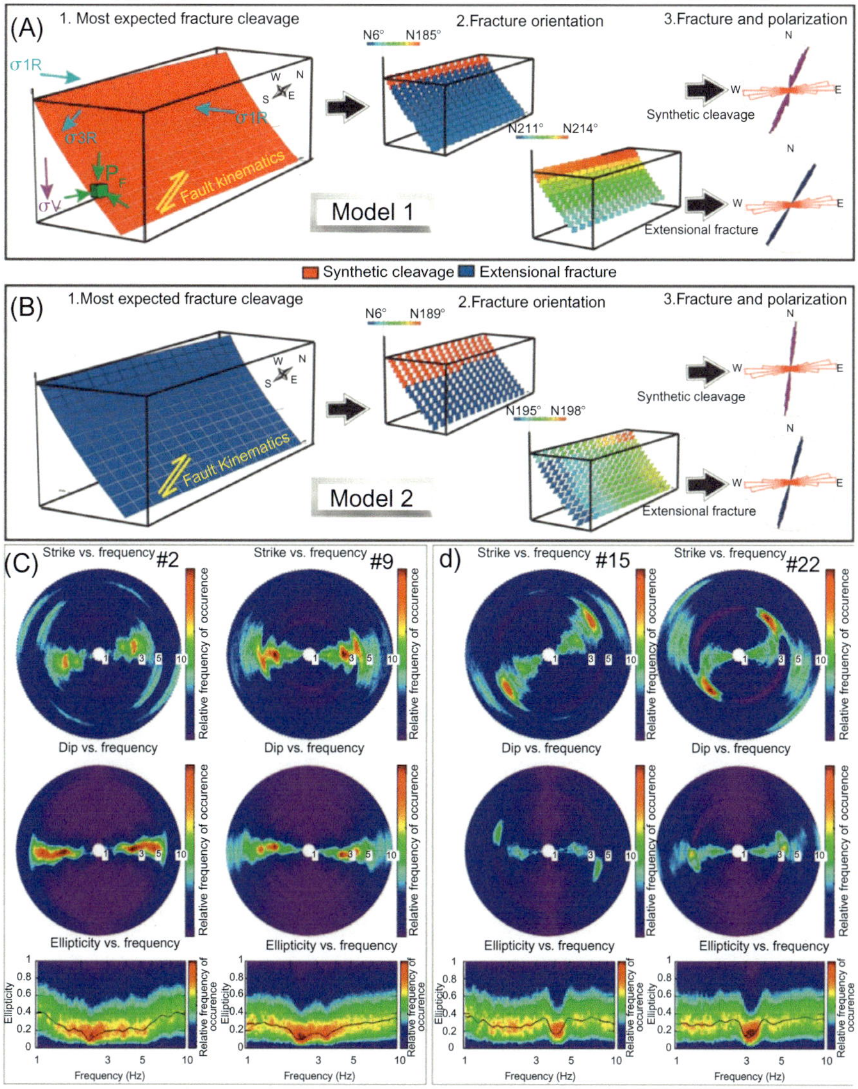

FIG. 8

(A and B) Results of brittle deformation modeling performed in the Ragalna fault system; model 1 uses as stress components the regional stress, overburden, fluid pressure PF, and kinematic stress; model 2 uses only the kinematic stress. The fault surface is discretized through a grid of cells whose color depends on the predominant fracture type (*red* for synthetic cleavage and *blue* for extensional fracture). The fault kinematics is assumed to be oblique right-lateral strike-slip *(yellow arrows)*. The *cyan*, *violet*, and *green arrows* in (A) represent the orientation of the regional stresses, the overburden, and the fluid isotropic pressure, respectively. The expected orientation of synthetic cleavage and extensional fractures is drawn using a 3-D plane visualization projected on the fault surface (the *color bars* represent the plane strike). The wind-rose diagrams show the expected orientation of synthetic cleavage and extensional fractures with the observed polarization across the investigated fault (*red* rose over imposed diagram). (C and D) Examples of polar plots and trends of the azimuth, dip, and ellipticity in selected sites of the Ragalna structures, obtained by using time-frequency polarization analysis.

Modified from F. Panzera, M. Pischiutta, G. Lombardo, C. Monaco, A. Rovelli, Wavefield polarization in fault zones of the western flank of Mt. Etna: observations and fracture orientation modelling, Pure Appl. Geophys. 171 (11) (2014) 3083–3097, https://doi.org/10.1007/s00024-014-0831-x.

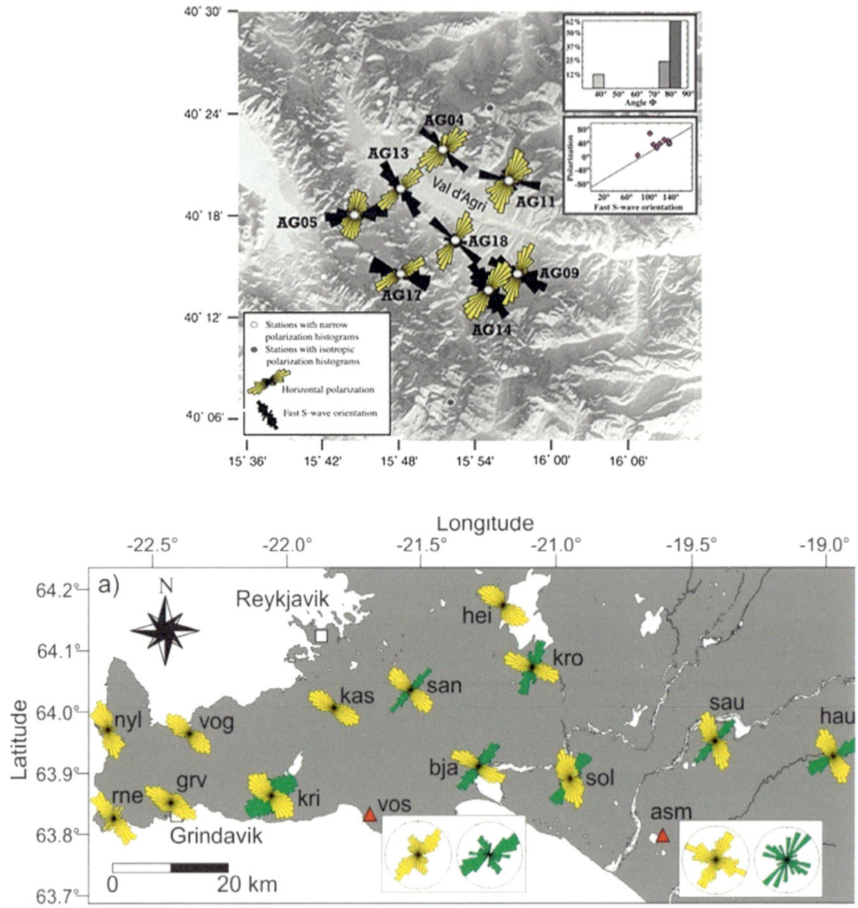

FIG. 9

(A) Horizontal polarization (*yellow* rose diagrams) and fast anisotropic direction (*black* rose diagrams) at stations in the Val d'Agri area (modified from M. Pischiutta, M. Pastori, L. Improta, F. Salvini, A. Rovelli, Orthogonal relation between wavefield polarization and fast S-wave direction in the Val d'Agri region: an integrating method to investigate rock anisotropy, J. Geophys. Res. 119 (2014) 1–13, doi:10.1002/2013JB010077); (B) map of polarization azimuths (*yellow* rose diagrams) and fast anisotropic direction (*green* rose diagrams) at selected seismic stations in Iceland.

(modified from F. Panzera, B. Halldorsson, K. Vogfjörðc, Directional effects of tectonic fractures on ground motion site amplification from earthquake and ambient noise data: a case study in South Iceland, Soil Dyn. Earthq. Eng. 97 (2017) 143–154)

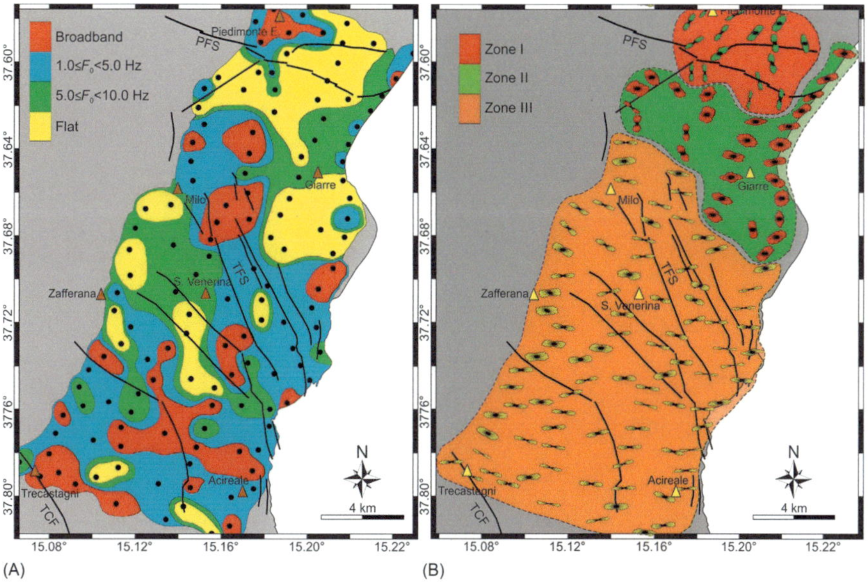

FIG. 10

Contour map bounding sites having different spectral ratio features from ambient noise measurements performed in the lower eastern flank of Mt. Etna (A) and polarization diagrams for each measurement site, obtained by gathering together *rose* diagrams showing similar features and/or prevailing azimuths (B).

Modified from F. Panzera, G. Lombardo, E. Longo, H. Langer, S. Branca, R. Azzaro, V. Cicala, F. Trimarchi, Exploratory seismic site response surveys in a complex geologic area: a case study from Mt. Etna volcano (southern Italy), Nat Hazards 86 (2017) S385–S399. doi: 10.1007/s11069-016-2517-4.

- In the neighborhood of fault areas, the presence of a damage zone implies the existence of ground-motion amplifications and persistent directional effects of the horizontal component of motion, set into evidence by both earthquake and ambient noise records that are observed several hundred-meter distance from the fault line.
- The directional site effects and the polarization angles observed for all the investigated structures are never (or not always) parallel to the fault strike, casting into doubt a simple explanation in terms of fault-trapped waves. To attempt a possible explanation for this recurrent ground-motion property, we postulate as an interpretative hint the existence of a tight relationship with the expected cleavage patterns [6].
- The directional resonance and the polarization analysis demonstrate that fault effects appear concentrated in the frequency range 1.0–6.0 Hz. The stability of this frequency interval, observed both in the western and in the eastern flanks of the Etna volcano, encourages us to affirm that it is a possible marker for observing site effects in fault zones, at least in the Etnean area.

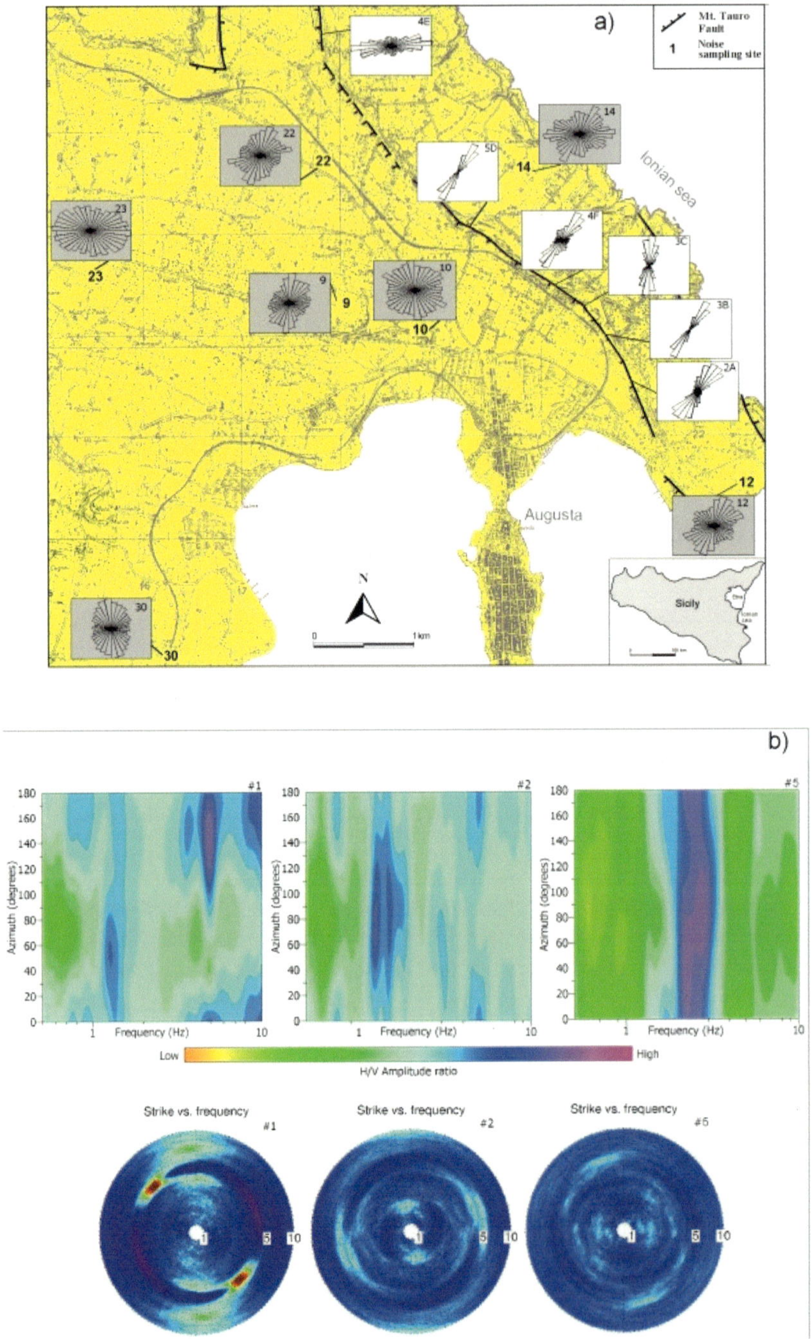

FIG. 11

Ambient noise polarization plots in the Augusta area (A) and examples of directional resonances and horizontal polarization azimuths (B) obtained in the Piano dei Grilli area (PG in Fig. 1).

- Directions coming from both rotated spectral ratios and polarization diagrams tend to become randomly distributed and/or uniformly scattered when noise measurements are performed in areas significantly far from fault zones.
- Correlation between fast shear-wave velocities and directional amplification effects nearby the fault zones underlines the prevalent role played by the anisotropy linked to the existence of oriented fractures.

Finally, it seems worth noting that such findings emphasize the importance of delineating the fault damage zones in city planning and, as observed in the Etnean area where faults cross villages, indicate the importance of accounting for such effects in seismic code provisions.

4 LOCAL SEISMIC RESPONSE IN LANDSLIDE INVOLVED SLOPES

Site conditions responsible for local seismic amplification (tectonic features, stratigraphic conditions, and morphology) can also be regarded as responsible for a significant reduction of slope stability due to possible interactions between seismic input, local geologic setting, slope geometry, geotechnical properties, and preexisting landslide mass.

Depending on the geologic setting of landslide slopes, the role of interactions between seismic waves and landslide masses on gravity-induced instabilities can be demonstrated through noise measurements and seismic arrays.

In this regard, two different geologic contexts for landslides can be identified (according to Varnes [56]): (i) jointed rock masses, involved in falls, toppling, slides, or lateral spreading, and (ii) soils or weak rocks involved in translational or rototranslational landslides, which can be characterized by unique or complex mechanisms, the latter being responsible for initiating landslide submasses.

As regard rock landslides, recent studies highlighted the role of jointing that causes a differential mobility of rock blocks, evidenced by superficial wave polarization [55,57], and contributes to their detachment and falling. In several cases, it was demonstrated that failure events on rock slopes are preceded by microseismic events that can be recorded on micro- to nanoseismic devices [58–60]. Such recordings may be able to indicate increasing weakening of the rock mass and changes in eigenmodes of vibration of the rock blocks before failure.

As regard earth landslides, several studies [61–64] have shown that physical interactions between seismic waves and landslide masses are very complex and result in several often simultaneous phenomena: (i) stratigraphic resonance due to the impedance contrast between the landslide mass and its substratum, (ii) basin-like amplification effects related to the geologic setting of the slope and to the landside mass, and (iii) seismic wave polarization within the landslide mass.

In the last decade, different case studies [61,65–68] have demonstrated that the aforementioned interactions can occur, conditioning the slope stability and, as a consequence, the potential of the landslide to be (re)activated. Moreover, several authors analyzed such effects through instrumental measurements [69–78]. When these site-dependent phenomena occur, the resulting effect on the earthquake-induced landslide displacements can be generally referred to as a "self-excitation" process [66], which implies an active role of the landslide mass during the dynamic shaking. Such an effect consists in modifying the seismic wave properties (i.e., PGA, frequency content, and polarity) and in favoring the landslide reactivation occurrence; the self-excitation process can also explain far-field earthquake reactivation of large landslides [79].

From a theoretical point of view [64,68,80], the maximum earthquake-induced displacements are expected when the resonant period related to the landslide mass thickness (T_s) is equal to the characteristic period [81] of the seismic input (T_m), a single parameter used to characterize the frequency content of an earthquake ground motion. A characteristic period ratio T_s/T_m was introduced to describe such a condition, that is, maximum displacements are theoretically expected for a T_s/T_m ratio equal to 1. Moreover, the role of the 2-D geometry (i.e., slope angle and landslide mass length) also needs to be taken into consideration for a better prediction of earthquake-induced landslide displacements. In particular, the landslide mass mobility is theoretically favored by a characteristic period of the seismic input (T_m) that is double with respect to the period (T_l) associated to the length of the landslide mass itself (i.e., for a characteristic period ratio T_l/T_m equal to 0.5).

Ambient noise measurements can significantly contribute to demonstrate both local seismic response in landslide areas and polarization effects due to rock mass jointing, in case of rock masses, or scattered superficial waves within the landslide mass, in case of earth landslides. In this last case, the seismic wave scattering due to the landslide mass over the slope induces a 2-D basin-like effect that is intensified with increasing earthquake intensity [80]. It is worth stressing that ambient noise does not allow to quantify amplification inside landslide masses and/or slopes involved in gravity-induced instabilities; for this purpose, standard spectral ratios (SSRs) computed from reference to receiver stations should be obtained [82], and numerical modeling should be performed to evaluate the effect of earthquake intensity, proximity, and/or spectral content in modifying the local seismic amplification.

In the following, some case studies are presented that exemplify the efficiency of ambient noise techniques for demonstrating local seismic effects in both rock and earth landslide slopes.

Geophysical techniques were applied to study the seismic response of unstable rock slopes in (i) Cala Rossa bay (Sicily, Italy), (ii) Selmun Promontory (Malta), and (iii) Xemxija bay (Malta).

Cala Rossa bay (Fig. 12) is located in the eastern part of Favignana island (Sicily, Italy), off the northwest coast of Sicily (Italy). Its geologic succession [83,84] is characterized by a 20m plateau of porous calcarenites (Lower Pleistocene) overlying a high-plasticity clay (Pliocene) 5 – 10m thick, in turn overlying a nonoutcropping massive calcarenite (Lower-Middle Miocene). All these formations show

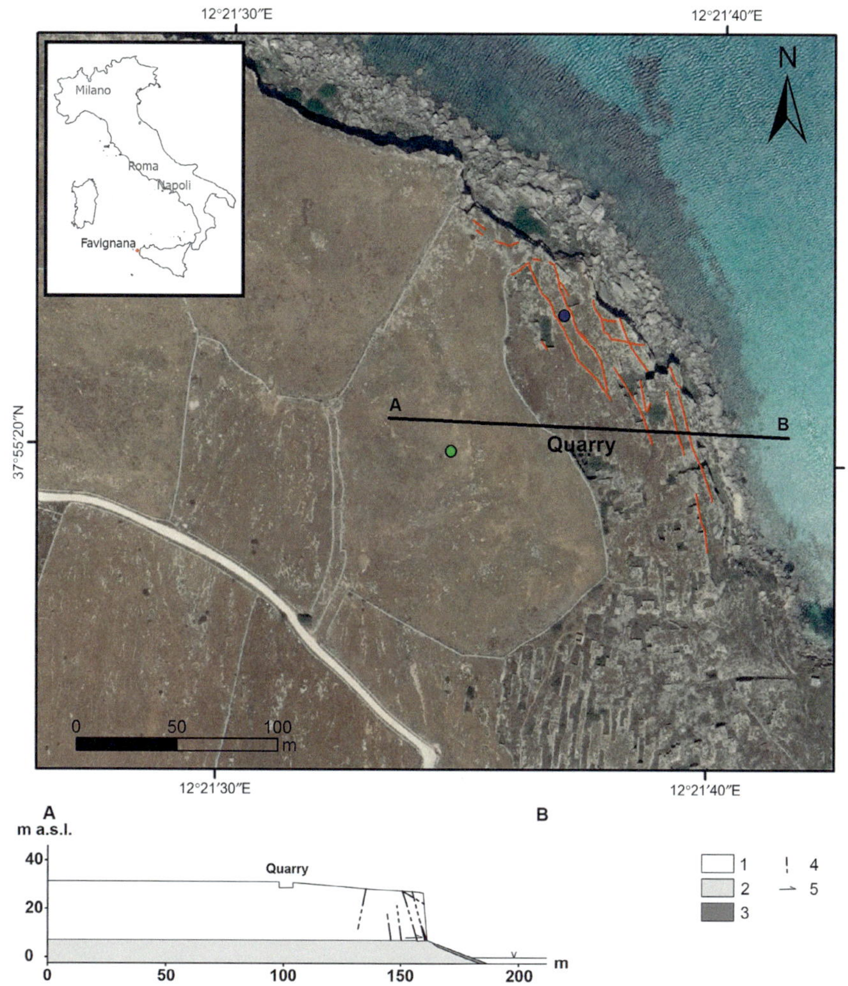

FIG. 12

Top: Satellite view of Cala Rossa bay and position of Favignana island in Italy; the *red lines* indicate the open fractures, the *green dot* a seismic measurement in the stable area, and the *blue dot* a seismic measurement in the unstable area. Bottom: Geologic cross section: (1) porous calcarenites (Lower Pleistocene), (2) high-plasticity clay (Pliocene), (3) debris slope deposit (Quaternary), (4) open fractures (*dashed line* for supposed part), and (5) direction of mass movement.

subhorizontal strata. Since the Roman Age and until the last century, intense quarrying activity took place on Favignana island that significantly changed the morphology of the eastern part of the island. This mining activity produced an extensive network of open quarries, underground quarries, and tunnels, locally named "Pirrere," hosted into the Pleistocene calcarenites.

Selmun Promontory and Xemxija are both located in the northwestern part of the island of Malta (Figs. 13 and 14), in the middle of the Mediterranean Sea, about 100km south from Sicily. The geology of Malta is characterized by marine sedimentary rock formations, generally having subhorizontal or slightly dipping (<5°) strata. The exposed stratigraphic succession of the Selmun Promontory (Fig. 13) shows [85,86], from top to bottom, 20–30m of upper coralline limestone (UCL), pale gray and orange fossiliferous coarse-grained limestone (Upper Miocene); 30m of blue clay (BC), a soft pelagic blue or greenish-gray marl and limey clay (Middle Miocene); and globigerina limestone (GL), a soft yellowish fine-grained limestone (Lower Miocene). The Għajn Ħadid Tower, the first of the 13 watchtowers built in 1658 by Grand Master Martin de Redin, stands precisely on the Selmun Promontory. Historical documents testify a significant damaging of the tower during an earthquake that struck the Maltese archipelago on October 12, 1856.

Xemxija (Fig. 14) shows the same sedimentary rock succession, except that the Globigerina Limestone does not outcrop in this area. In the three case studies, the local geologic setting favors landslide processes. In fact, the overposition of a stiff rock on a plastic clay induces a lateral spreading phenomenon [87] associated with falls, slides, and/or topples of different-size rock blocks. The landslide process shapes a typical landscape with a stable plateau, corresponding to the stiff limestone, bordered by unstable cliff slopes. The gravity-induced instability processes in the three selected sites were widely documented [1,88–91].

In each case study, ambient noise was recorded deploying several single stations in dense arrays, distributed between the unstable edge zones and the stable inland zones: 47 stations in Cala Rossa bay, 116 stations on the Selmun Promontory, and 45 stations in Xemxija. The results obtained from the geophysical investigations (Fig. 15) showed an absence of peaks in the HVSR(f) function and no evidence of polarization features in the particle motion for the inland measurements at Cala Rossa bay. On the contrary, the stations on the unstable edge show several peaks in the HVSR(f) function [92], and particularly, for a frequency range between 6.0 and 8.0Hz, the polarization analysis shows a particle motion that is almost linear and clearly polarized in the same direction of the dip direction of the main joints.

For the two case studies on Malta, the results (Figs. 16 and 17) for the inland zones show a peak at about 1.5Hz that is typical of the northwestern sector of Malta and is related to the stratigraphic succession [93,94] composed of UCL, BC, and GL. The polarization analysis does not show polarization of the particle motion in these regions. At the Selmun Promontory, the measurements carried out on the unstable plateau edge zones show several peaks in the HVSR(f) functions, in addition to the site resonance at 1.5Hz. Polarization analysis clearly shows that the particle motion for the peak at 3.5Hz has marked features of linearity and azimuthal polarization in a direction perpendicular to the main joint direction. Also at Xemxija, the HVSR(f) functions contain the stratigraphic peak (1.5Hz) and several other peaks, among which the one at 25Hz shows marked

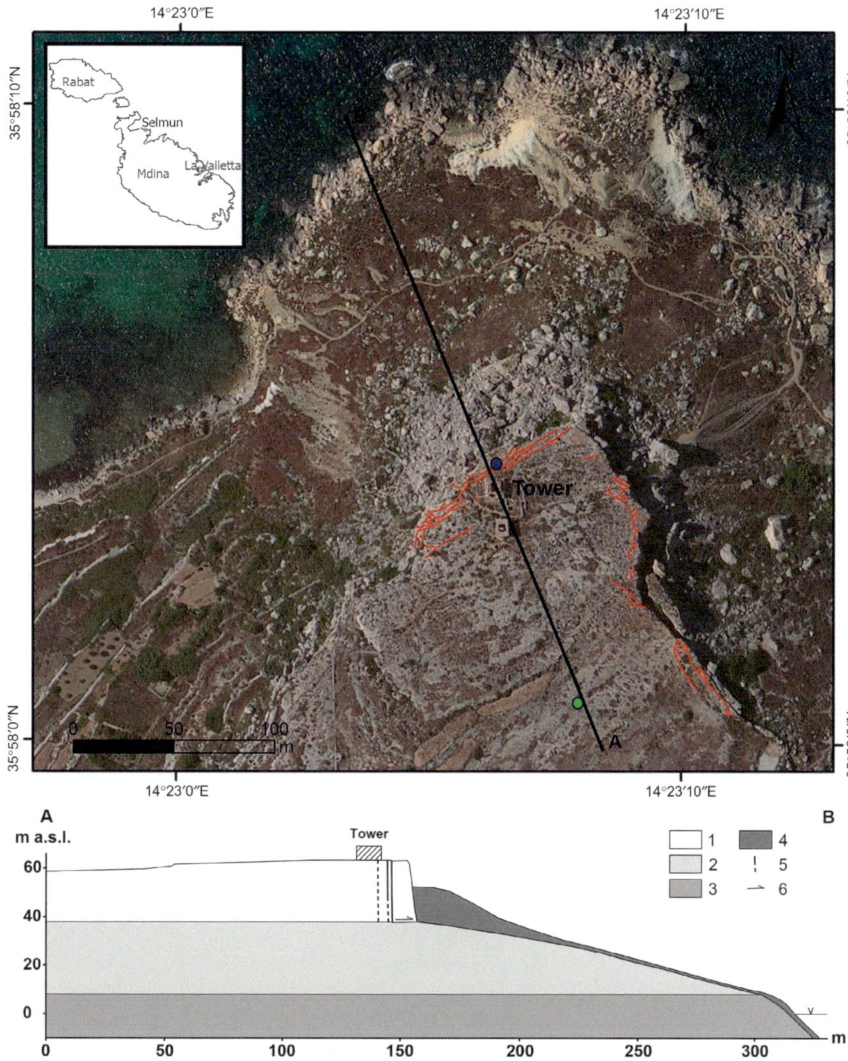

FIG. 13

Top: Satellite view of Selmun Promontory and its position in Malta; the *red lines* indicate the open fractures, the *green dot* a seismic measurement in the stable area, and the *blue dot* a seismic measurement in the unstable area. Bottom: Geological cross section: (1) upper coralline limestone (Tortonian-Messinian), (2) blue clay (Serravallian), (3) globigerina limestone (Aquitanian-Langhian), (4) debris slope deposit (Quaternary), (5) open fractures (*dashed line* for supposed part), and (6) direction of mass movement.

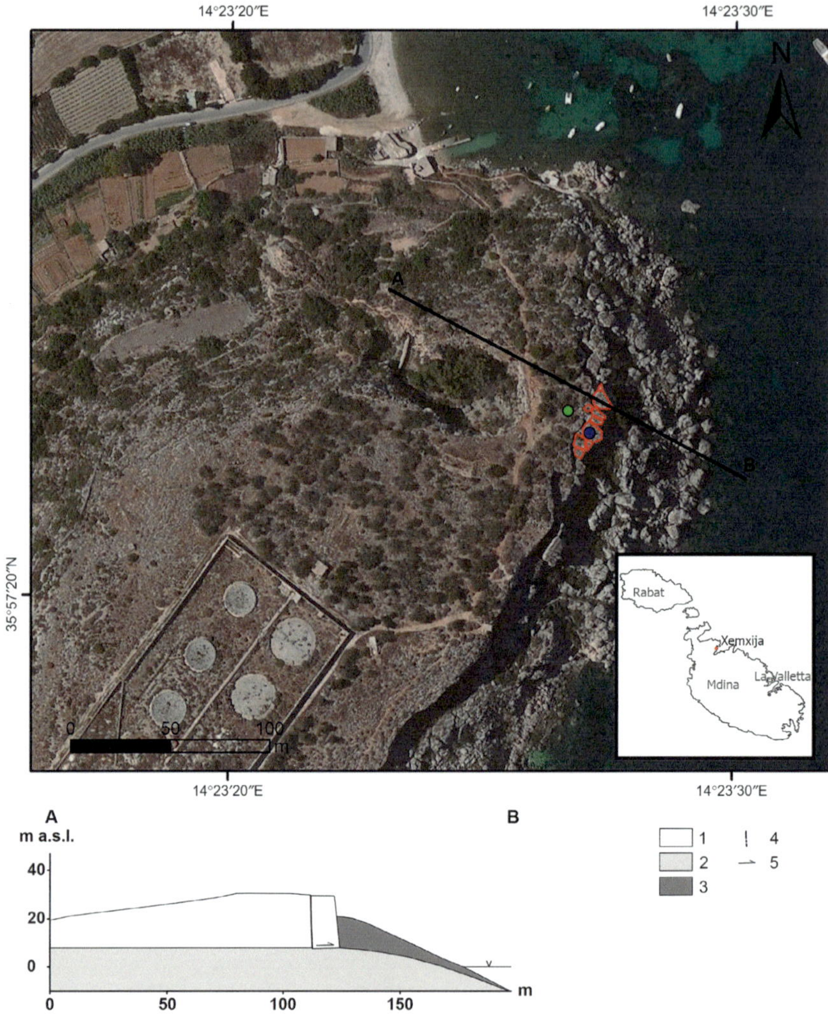

FIG. 14

Top: Satellite view of Xemxija and its position in Malta; the *red lines* indicate the open fractures, the *green dot* a seismic measurement in the stable area, and the *blue dot* a seismic measurement in the unstable area. Bottom: Geological cross section: (1) upper coralline limestone (Tortonian-Messinian), (2) blue clay (Serravallian), (3) debris slope deposit (Quaternary), (4) open fractures (*dashed line* for supposed part), and (5) direction of mass movement.

features of linearity and azimuthal polarization of the particle motion by polarization analysis.

In conclusion, for the three case studies of rock landslides, the seismic response in the stable zones is related to stratigraphic effects only, that is, to the juxtaposition of the lithotechnical units ascribable to the local geologic succession, and this is

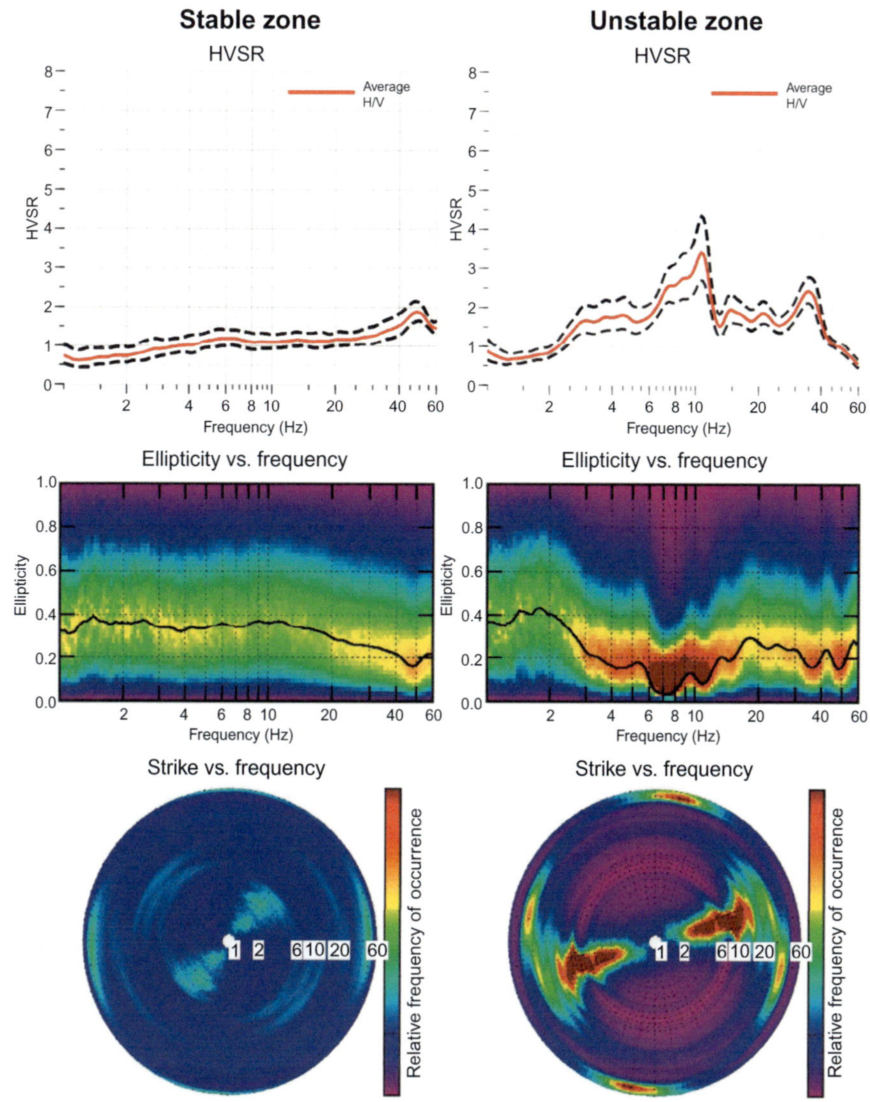

FIG. 15

Examples of ambient noise analysis results in the stable and the unstable zones of Cala Rossa bay (see Fig. 12 for station locations), from top to bottom: HVSR(f) function, ellipticity diagram, and polar strike plot.

consistent with the absence of polarization effects. On the other hand, the results obtained for the unstable zones show several peaks in the HVSR(f) functions and a marked linearity and a strong azimuthal polarization of the particle motion in the average dip direction of the main joint set. Ambient noise measurements were carried out in three different landslide slopes that show evidence of local seismic

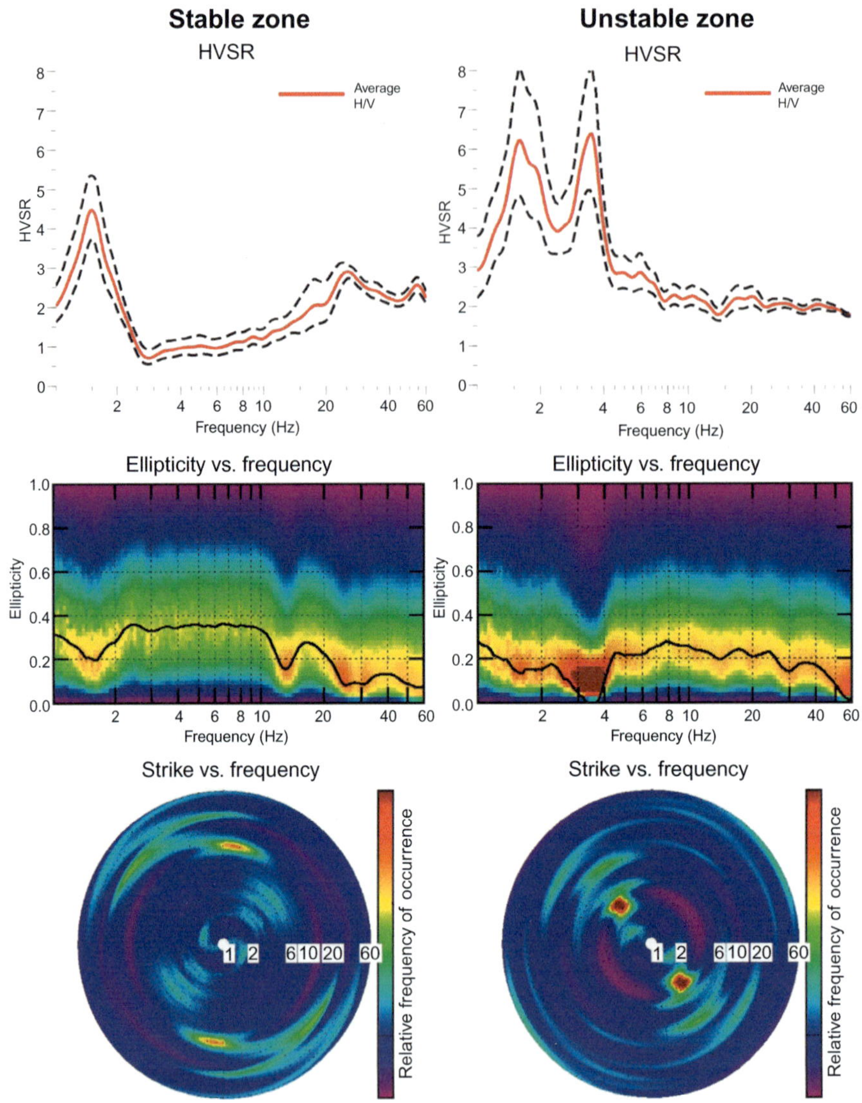

FIG. 16

Examples of ambient noise analysis results in the stable and the unstable zones of the Selmun Promontory (see Fig. 13 for station locations), from top to bottom: HVSR(f) function, ellipticity diagram, and polar strike plot.

response due to the existing landslide masses: (i) Cerda (Sicily, Italy), (ii) Salcito (Molise, Italy), and (iii) Diezma (Andalusia, Spain).

The Cerda landslide (Fig. 18) mainly involves clay shales of the Argille Varicolori Formation (Cretaceous-Eocene) and marls and calcarenites of the Polizzi Formation (Eocene-Oligocene). The Argille Varicolori Formation overthrusts Triassic

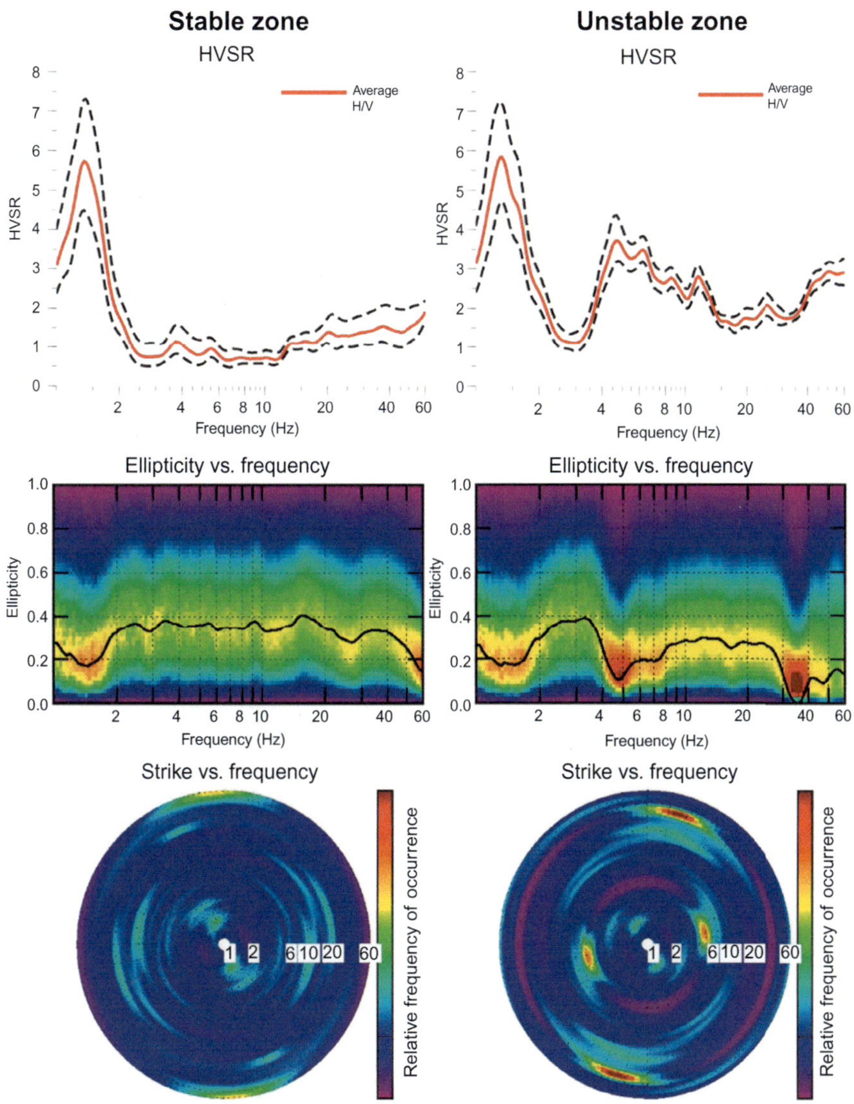

FIG. 17

Examples of ambient noise analysis results in the stable and the unstable zones of Xemxija (see Fig. 14 for station locations), from top to bottom: HVSR(f) function, ellipticity diagram, and polar strike plot.

marls and limestones (Cerda-Roccapalumba Formation), widely outcropping in a tectonic window SW of the Cerda landslide [95]. The thrust limits the right side of the landslide mass, corresponding to the creek that bounds the landslide mass southward. The same thrust is offset by numerous tear faults, one of them partly corresponding to the main scarp of the landslide [96].

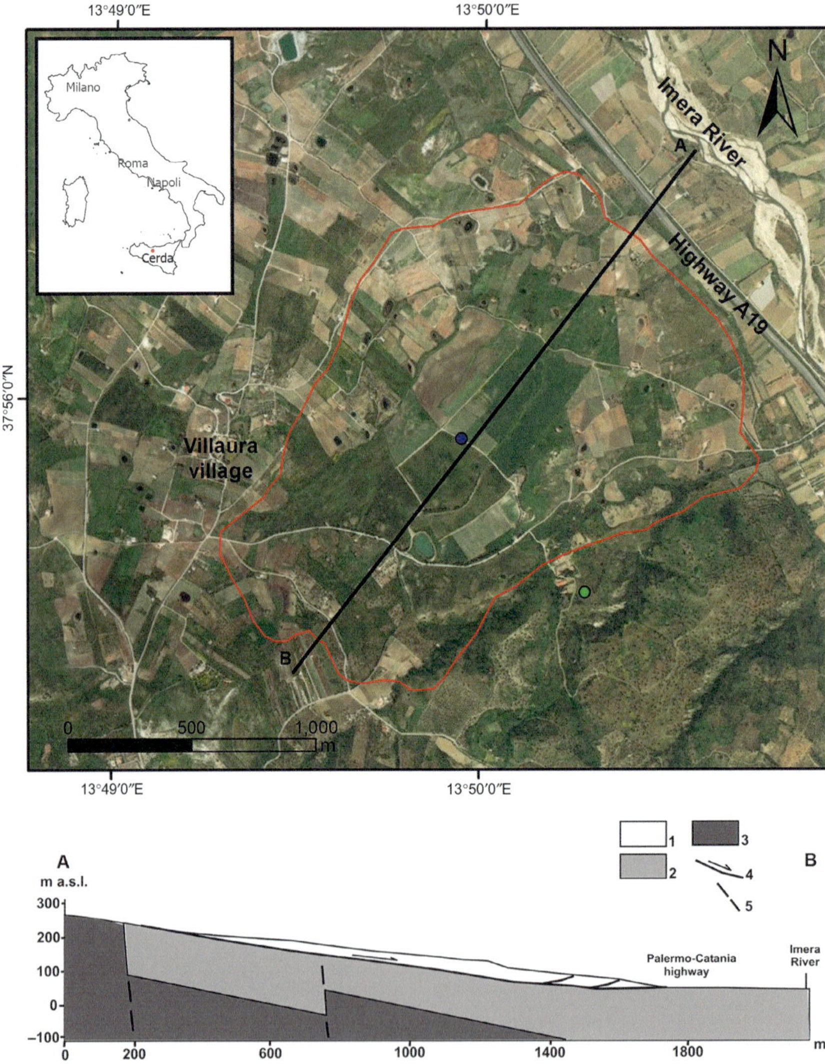

FIG. 18

Top: Satellite view of the Cerda landslide and its position in Italy; the *red lines* indicate the landslide boundaries, the *green dot* a seismic measurement in the stable area, and the *blue dot* a seismic measurement in the unstable area. Bottom: Geological cross section: (1) landslide debris, (2) scaly clays (Argille Varicolori Formation, Cretaceous-Eocene), (3) marls and calcarenites (Roccapalumba Formation, Triassic), (4) sliding surface, and (5) fault.

The Cerda landslide (Sicily, Italy) was reactivated by the September 6, 2002 Palermo earthquake (M_w 5.9), with epicenter about 50km away [66], a distance significantly higher with respect to the corelations between maximum landslide epicentral distance and earthquake magnitude by Keefer [97]. The landslide caused damage to farm houses, roads, and aqueducts, close to the homonymous village, and involved about 40 Mm^3 of clay shales; the first ground cracks due to the landslide movement formed about 30min after the main shock. The landslide extends about 1.5km and is characterized by a typical translational mechanism with failure zones up to 50m below ground level. An S-wave velocity of about 400m/s was measured in the landslide mass.

In the Cerda landslide area, the obtained HVSRs (Fig. 19) demonstrate a resonance in the range 0.5–1Hz close to the crown, while resonances in the range 2.5–4.5Hz are observed all along the landslide perimeter [66].

The resonance observed at lower frequency (<2Hz) in the landslide slope can be related to the local geologic setting of the landslide area (i.e., to the thick scaly clay layer above the local seismic bedrock). This effect can increase the amplitude of the long-period waves that are more likely to interact with the kilometer-extended masses of the landslide. A temporary velocimetric array was also installed to record possible small-magnitude events [69]. The recorded earthquake accelerograms were FFT transformed to obtain the receiver functions [98] in the related stations and spectral ratios to a reference station [82]. The results are consistent with those obtained from the HVSR from ambient noise.

Polarization analysis does not show evidence of polarization of the particle motion in both the stable and the landslide areas; this is consistent with a main stratigraphic effect for the local seismic amplification in the landslide slope, that is, due to impedance contrast between landslide mass and substratum. The Salcito landslide (Fig. 20) is located in the Molise section of the Apennine chain, where main structural stratigraphic units of the orogenic belt are tectonically juxtaposed: the Sannio pelagic basin unit and the Molise pelagic basin unit [99]. The Salcito landslide developed within the Argille Varicolori Formation of the Sannio unit. This formation (Oligo-Miocene) is composed of fissured clay shales, locally interlayered with intensely sheared arenaceous and marly-limestone beds. The upper part of the Sannio unit, outcropping south of the Salcito village, consists of reddish marls and massive calcarenites, dislodged by an NW-SE multithrust system and by an NE-SW tear-fault system. The Early-Middle Pliocene compressive tectonics produced an EW-trending thrust of the Sannio unit over the Molise unit. Starting from the Late Pliocene, compressive tectonics has evolved into strike-slip faulting; as a consequence, the thrust front was shifted northward by a NE-SW left-lateral tear fault. With regard to hydrogeological features, the marly-calcarenitic members of the Molise unit (Tufillo Formation) contain an aquifer that is hydraulically confined by the Argille Varicolori aquitard, which belongs to both the Molise and Sannio units. Along the slope affected by the landslide, a shallow aquifer drains into the Argille Varicolori of the Sannio unit. The aquifer lies within the first 15 m below ground level (b.g.l.); the groundwater table is located about 1 m b.g.l. [70]. The main crown in the detachment zone of

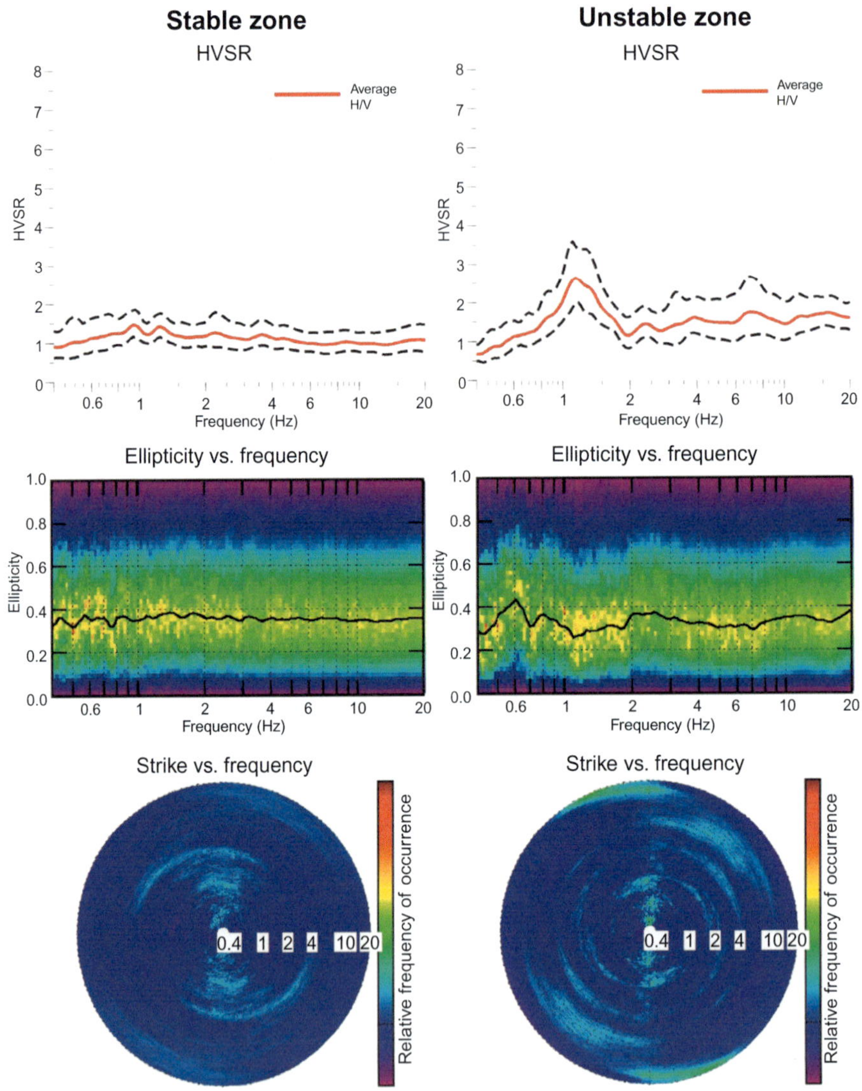

FIG. 19

Examples of ambient noise analysis results in the stable and the unstable zones of the Cerda landslide (see Fig. 18 for station locations), from top to bottom: HVSR(f) function, ellipticity diagram, and polar strike plot.

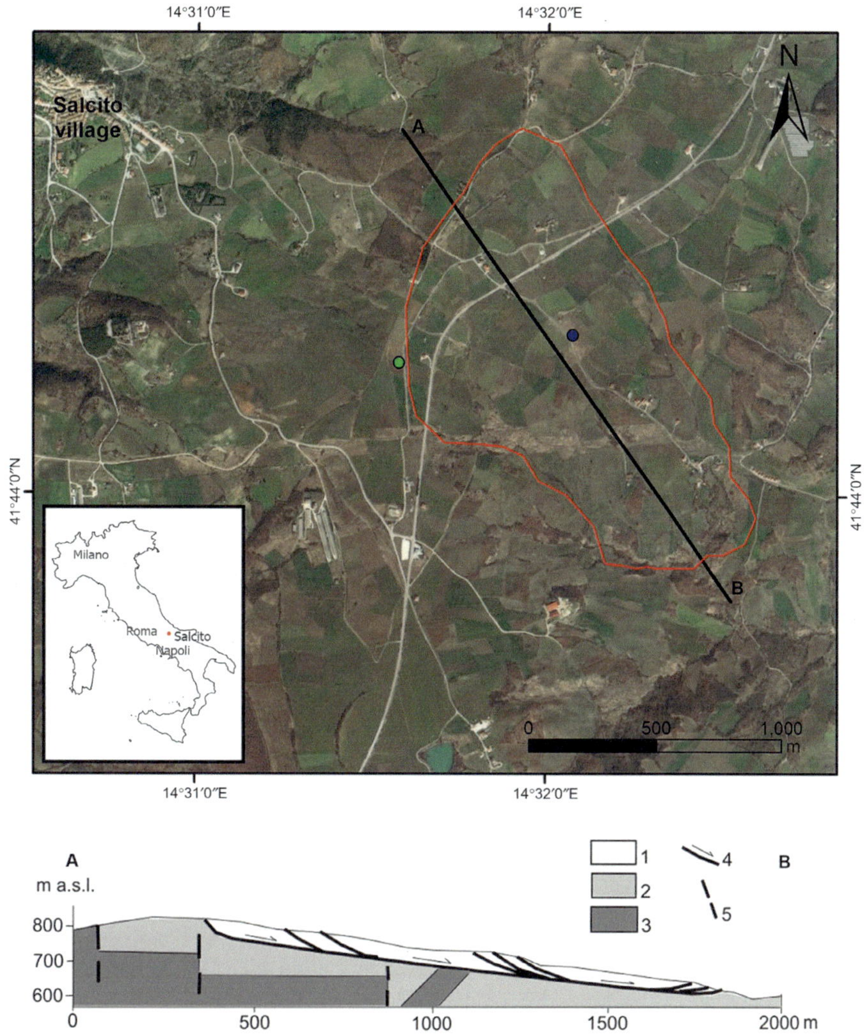

FIG. 20

Top: Satellite view of the Salcito landslide and its position in Italy; the *red lines* indicate the landslide boundaries, the *green dot* a seismic measurement in the stable area, and the *blue dot* a seismic measurement in the unstable area. Bottom: Geologic cross section: (1) landslide debris, (2) scaly clays (Sannio unit, Oligocene-Miocene), (3) marls and calcarenites (Sannio unit, Oligocene-Miocene), (4) sliding surface, and (5) fault.

the landslide, observed after the seismically induced reactivation, lies close to one of the main strike-slip faults. The Salcito landslide, located close to the Salcito village [70], was reactivated on 2002 by the 31st of October Molise earthquake (M_w 5.7), that is, at an epicentral distance of about 50 km that is significantly higher than the maximum landslide epicentral distance expected by Keefer [97] corelations.

The earthquake caused a complex ground crack pattern and downslope displacements up to 2 m, damaging some farmhouses and lifelines. No rainfall occurred in the days preceding the landslide reactivation; nevertheless, the landslide mass was saturated owing to groundwater seepage.

The Salcito landslide shows very similar features to the Cerda landslide as it involves a volume in the order of 30×10^6 m^3 of scaly clays; it is about 1.5 km, and it is characterized by a mainly translational mechanism with failure zones up to 50 m below ground level. The S-wave velocity measured in the landslide mass is about 400 m/s.

In the Salcito landslide area, the HVSRs (Fig. 21) show a resonant zone in the upper part of the slope, close to the left flank of the landslide; the main HVSR peaks correspond to frequency values in the range 1–2 Hz, while frequency values in the range 3.5–4.5 Hz are amplified within the landslide mass only [69]. These results show that after past activations, the landslides become more prone to seismically induced reactivations, as their softened masses can induce significant amplification effects at frequency values in the range 2–6 Hz, related to both the 1-D seismic resonance and the 2-D interactions mainly due to the superficial seismic waves. Such conditions are emphasized by a shallow debris layer over the landslide mass that is generated by weathering and remolding, also due to superficial earth slides and soil creep processes.

The polarization analysis does not show features of polarization in the particle motion for the measurement carried out in the stable area, while there is a polarization effect for frequencies in a range between 0.4 and 1 Hz in the station located in the landslide zone. The Diezma landslide (Fig. 22) is located along the tectonic contact between the internal (*Maláguide*) and *dorsal* domains of the Betic Cordillera chain [72]. The *Numidoide* Formation outcrops in the landslide slope, while Mesozoic to Eocene limestones of the *dorsal* domain outcrop at its top. Several authors have recognized high-plasticity levels in the smectite clay within the *Numidoide* deposits and associated them with the sliding surface of the Diezma landslide mass ([72] and references therein). The complex structural setting of the Diezma landslide is also the result of a main NE-SW fault system that can be identified in the landslide slope. In addition to this fault system, the geologic contacts are slightly vertically displaced, even at close distances; this fact has been interpreted as due to a set of fractures with small-scale vertical movement that dislodge the lateral continuity of the materials. In the Diezma landslide area, the local seismic response was studied [68,72] by ambient noise measurements at 21 sites. The results obtained in the sites located outside the landslide (Fig. 23) are characterized by flat HVSR functions with amplitudes below 2. On the other hand, all the measurements performed on the landslide mass exhibit a clear peak that occurs at frequencies of 4–5 Hz in the middle-low part of landslide mass, while the resonance frequency varies for sites located in the upper parts of the landslide. The amplitude of the HVSR peak is usually higher than 4, but it decreases toward the landslide flanks and toe, where a large rock fill exists. The low amplitude of the HVSRs and the corresponding lack of spectral

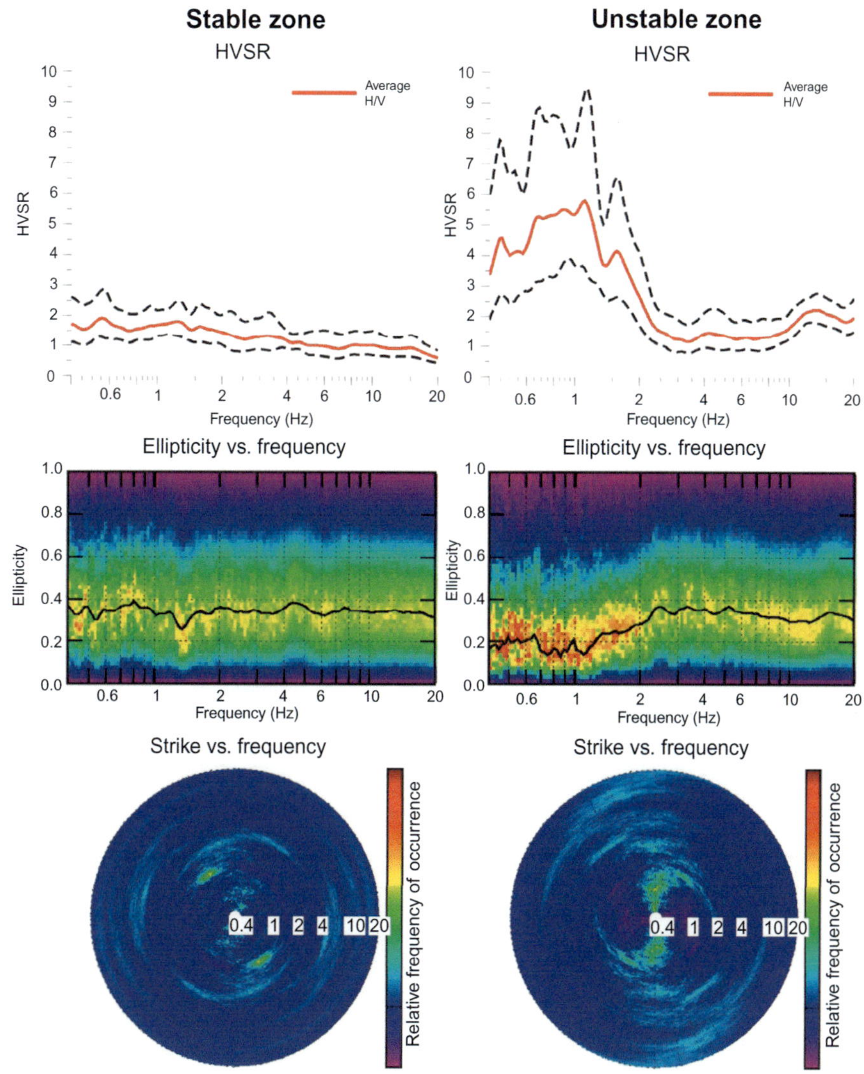

FIG. 21

Examples of ambient noise analysis results in the stable and the unstable zone of the Salcito landslide (see Fig. 20 for station locations), from top to bottom: HVSR(f) function, ellipticity diagram, and polar strike plot.

peaks at sites located outside the landslide mass are a consequence of the direct outcrop of the seismic bedrock. Features of polarized particle motion are not present for the measurement carried out both in the stable area and in the landslide zone. In conclusion, for the three case studies of earth landslide, the seismic response in the stable zones does not show any amplification effect, while the

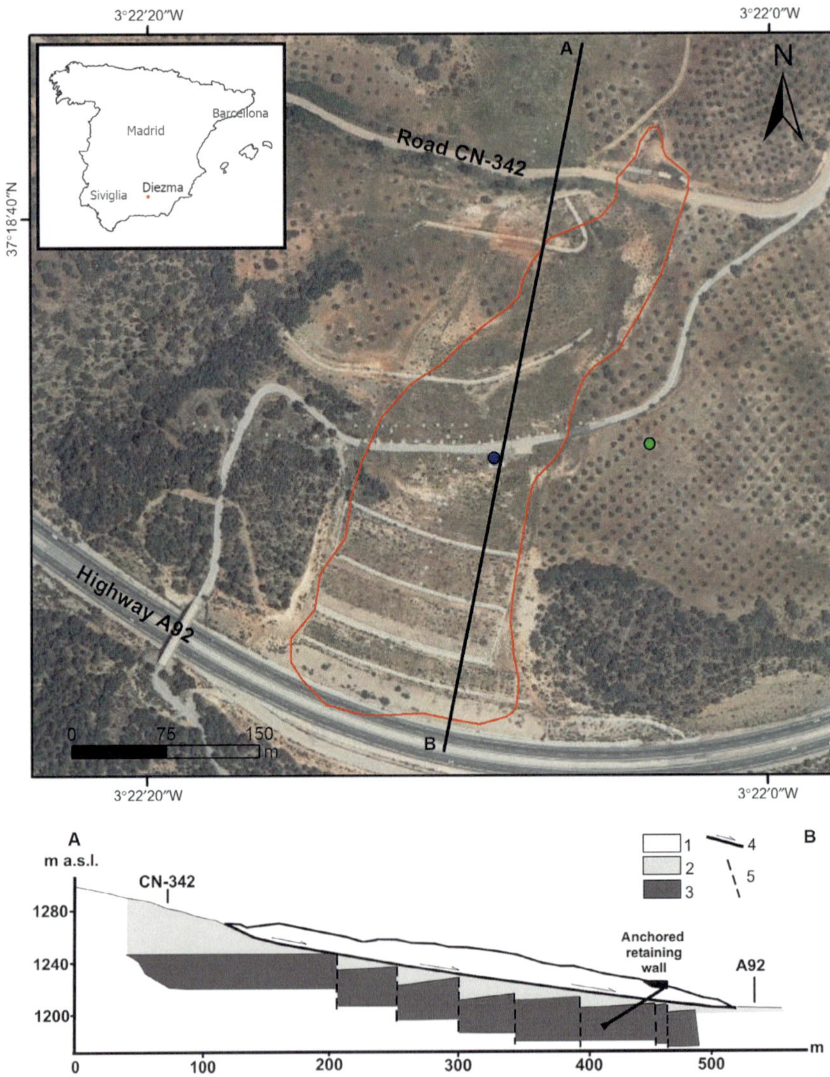

FIG. 22

Top: Satellite view of the Diezma landslide and its position in Spain; the *red lines* indicate the landslide boundaries, the *green dot* a seismic measurement in the stable area, and the *blue dot* a seismic measurement in the unstable area. Bottom: Geological cross section: (1) landslide debris, (2) silty clay with heterometric blocks of limestones of the Numidoide formation (Aquitanian-Burdigalian), (3) silts and clays of the Maláguide domain formations (Devonian-Triassic), (4) sliding surface, and (5) fault.

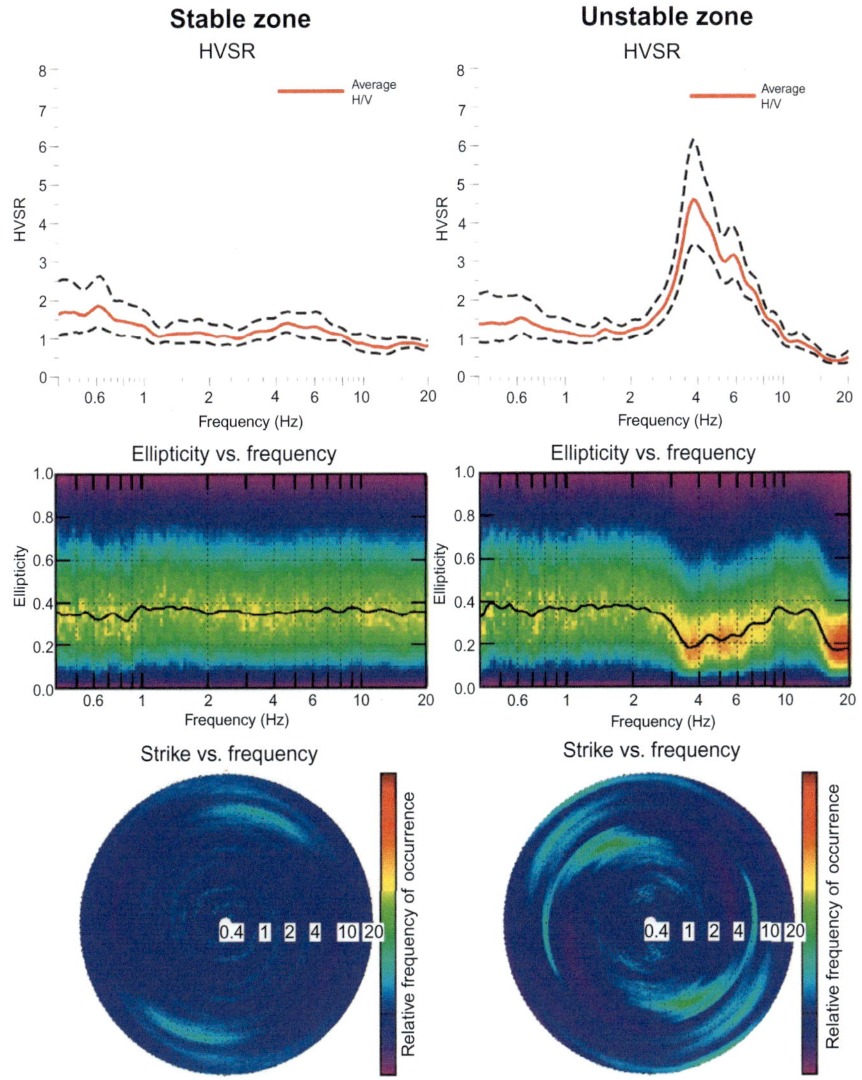

FIG. 23

Examples of ambient noise analysis results in the stable and the unstable zone of the Diezma landslide (see Fig. 22 for station locations), from top to bottom: HVSR(f) function, ellipticity diagram, and polar strike plot.

results obtained in the landslide mass show peaks in the HVSR(f) functions mainly related to the first stratigraphic resonance. In general, linear polarization effects are not typically found in this type of landslide; however, evidence of a slight polarization could be related to 2-D effects generated in a basin-like context where lateral boundary of the landslide mass could represent scatters for the superficial seismic waves.

5 AMBIENT NOISE MEASUREMENTS IN PRESENCE OF VELOCITY INVERSION

The investigation of local ground conditions is an important part of seismic hazard assessment (e.g., Refs. [6, 54, 100–103]). Soft sediments overlaying hard rock strata are responsible for amplifying the ground motion in case of a moderate-large seismic event. Ambient noise measurements, both with single station and array techniques, can surely contribute to the knowledge of the shear-wave velocity profile at a particular site that is an important tool toward the prediction of ground motion and hence in the prevention or mitigation of earthquake disasters.

In this section, we briefly describe the use of ambient noise measurement in case of the presence of shallow velocity inversion in the seismic velocity profile that can play a key role in the evaluation of site effects at this particular sites. The shallow velocity inversion is a common feature in the case of the Maltese archipelago (central Mediterranean; e.g., [93,104]) and other sites in Italy (e.g., Ref. [105]), and it may represent a problem for simplified zonation methods like Vs30. Di Giacomo et al. [105] showed that between the HVSR curves obtained by triggered noise and earthquake data recorded at the Venosa site (Italy) have significant differences. These have been investigated by using numerical simulations, and they have been attributed to the presence of an inversion in the shear-wave velocity profile. This is due to the fact that the sources of microtremor (seismic noise) are usually located at the surface and the wave field is dominated by surface waves in contrast with the seismic waves generated by tectonic events for which the wave front has a nearly vertical propagation. This study firstly demonstrated how ambient noise can be used as a proxy to evaluate amplification functions, but in the presence of a velocity inversion, it could lead to an underestimation of the amplification function of earthquakes.

Farrugia et al. [93] present a step toward a holistic seismic risk assessment that includes the implications on the site effects induced by shear-wave velocity inversion. This study justifies the use of the ESAC method [106], the joint inversion, and the genetic inversion algorithm [107–109] that have been shown to perform very well in resolving both the presence and the characteristics of a low-velocity layer in the stratigraphy. Farrugia et al. [93] demonstrated that Passive seismic surface wave measurements have been used to obtain 1-D, shear-wave velocity profiles at sites characterized by velocity inversion.

ACKNOWLEDGMENTS

Sebastiano D'Amico was partially supported by the research project funded by University of Malta, "Dynamic characteristics of active coastal spreading areas: evaluation of cliff instability" (GSCRP02-17).

REFERENCES

[1] R. Iannucci, S. Martino, F. Martorelli, L. Falconi, V. Verrubbi, in: M. Mikoš, N. Casagli, Y. Yin, K. Sassa (Eds.), Susceptibility to sea cliff failures at Cala Rossa Bay in Favignana Island (Italy), Advancing Culture of Living With Landslides. WLF 2017, vol. 4, Springer International Publishing AG, Cham, Switzerland, 2017, pp. 537–546, https://doi.org/10.1007/978-3-319-53485-5_63.

[2] S. Imposa, G. Coco, M. Corrao, Site effects close to structural lineaments in eastern Sicily (Italy), Eng. Geol. 72 (3) (2004) 331–341.

[3] Y.G. Li, K. Aki, D. Adams, A. Hasemi, Seismic guided waves in the fault zone of the Landers, California, earthquake of 1992, J. Geophys. Res. 99 (1994) 11705–11722.

[4] F. Panzera, G. Lombardo, E. Longo, H. Langer, S. Branca, R. Azzaro, V. Cicala, F. Trimarchi, Exploratory seismic site response surveys in a complex geologic area: A case study from Mt. Etna volcano (southern Italy), Nat Hazards 86 (2017) S385–S399, https://doi.org/10.1007/s11069-016-2517-4.

[5] F. Panzera, M. Pischiutta, G. Lombardo, C. Monaco, A. Rovelli, Wavefield polarization in fault zones of the western flank of Mt. Etna: observations and fracture orientation modelling, Pure Appl. Geophys. 171 (11) (2014) 3083–3097, https://doi.org/10.1007/s00024-014-0831-x.

[6] F. Panzera, G. Lombardo, S. Sicali, S. D'Amico, Surface geology and morphologic effects on seismic site response: the study case of Lampedusa, Italy, Phys. Chem. Earth (2016), https://doi.org/10.1016/j.pce.2016.08.006.

[7] K. Pitilakis, Chapter 5: Site effects, in: A. Ansal (Ed.), Recent Advances in Earthquake Geotechnical Engineering and Microzonation, Kluwer Academic Publishers, Netherlands, 2004, pp. 139–198.

[8] M. Nogoshi, T. Igarashi, On the propagation characteristics of microtremors, J. Seismol. Soc. Jpn. 23 (1970) 264–280.

[9] M. Mucciarelli, M.R. Gallipoli, Non-parametric analysis of a single seismometric recording to obtain building dynamic parameters, Ann. Geophys. 50 (2007) 259–266.

[10] SESAME Working Group, Guidelines for the implementation of the h/v spectral ratio tecnique on ambient vibration measurements, processing and interpretation, http://sesame-fp5.obs.ujf-grenoble.fr/Delivrables/Del-D23-HV_User_Guidelines.pdf, 2005.

[11] L.I. Dimitrov, Mud volcanoes—the most important pathway for degassing deeply buried sediments, Earth Sci. Rev. 59 (2002) 49–76.

[12] G. Etiope, Natural Gas Seepage. The Earth's Hydrocarbon Degassing, Springer International Publishing, Switzerland, 2015, p. 199. ISBN 978-3-319-14601-0 (eBook), https://doi.org/10.1007/978-3-319-14601-0.

[13] A. Kopf, Significance of mud volcanism, Rev. Geophys. 40 (2002) 1–52.

[14] Tinivella U., Giustiniani M., 2012. An overview of mud volcanoes associated to gas hydrate system, Updates in Volcanology—New Advances in Understanding Volcanic Systems, Dr. K. Nemeth (Ed.), InTech, UK, https://doi.org/10.5772/51270

[15] A.A. Feyzullayev, F.A. Kadirov, C.S. Aliyev, Mud volcano model resulting from geophysical and geochemical research, in: G. Martinelli, B. Panahi (Eds.), Mud Volcanoes. Geodynamics and Seismicity, NATO Science Series, Springer, Dordrecht, 2005, pp. 251–262.

[16] M. Yusifov, P.D. Rabinowitz, Classification of mud volcanoes in the South Caspian Basin, offshore Azerbaijan, Mar. Pet. Geol. 21 (2004) 965–975.

[17] G. Ciotoli, G. Etiope, F. Marra, F. Florindo, C. Giraudi, L. Ruggiero, Tiber delta CO_2-CH_4 degassing: a possible hybrid, tectonically active Sediment-Hosted Geothermal System near Rome, J. Geophys. Res. Solid Earth 121 (2016), https://doi.org/10.1002/2015JB012557.

[18] A. Mazzini, H. Svensen, G. Etiope, N. Onderdonk, D. Banks, Fluid origin, gas fluxes and plumbing system in the sediment-hosted Salton Sea Geothermal System (California, USA), J. Volcanol. Geotherm. Res. 205 (2011) 67–83.

[19] K.L. Vondamm, J.M. Edmond, C.I. Measures, B. Grant, Chemistry of submarine hydrothermal solutions at Guaymas Basin, Gulf of California, Geochim. Cosmochim. Acta 49 (1985) 2221–2237.

[20] P.F. Zarate-del Valle, B.R.T. Simoneit, Hydrothermal bitumen generated from sedimentary organic matter of rift lakes—Lake Chapala, Citala Rift, western Mexico, Appl. Geochem. 20 (2005) 2343–2350.

[21] H. Svensen, S. Planke, A. Malthe-Sørenssen, B. Jamtveit, R. Myklebust, T. Eidem, S. S. Rey, Release of methane from a volcanic basin as a mechanism for initial Eocene global warming, Nature 429 (2004) 542–545.

[22] A. Galgaro, J. Boaga, M. Rocca, HVSR technique as tool for thermal-basin characterization: a field example in N-E Italy, Environ. Earth Sci. 71 (10) (2013) 4433–4446, https://doi.org/10.1007/s12665-013-2838-5.

[23] F. Panzera, G. Lombardo, C. Monaco, New evidence of wavefield polarization on fault zone in the lower NE slope of Mt. Etna. Ital. J. Geosci. 135 (2) (2016) 250–260, https://doi.org/10.3301/IJG.2015.22.

[24] S. Imposa, S. Grassi, G. De Guidi, F. Battaglia, G. Lanaia, S. Scudero, 3D Subsoil model of the San Biagio 'Salinelle' mud volcanoes (Belpasso, Sicily) derived from geophysical surveys, Surv. Geophys. 37 (2016) 1117, https://doi.org/10.1007/s10712-016-9380-4.

[25] F. Panzera, S. D'Amico, M. Lupi, G. Mauri, K. Karyono, A. Mazzini, Lusi hydrothermal structure inferred through ambient vibration measurements, Mar. Pet. Geol. (2017), https://doi.org/10.1016/j.marpetgeo.2017.06.017.

[26] G.V. Dal Piaz, R. Antonelli, L. Bellucci, P. Fabbri, V. Iliceto, P. Noto, C. Panichi, R. Sedea, Relazione finale sulle ricerche sul Bacino Termale Euganeo (Final report of research contract on Thermal Euganean Basin), Università di Padova, Regione Veneto, 1994.

[27] R. Bellati, E. Di Lallo, R. Sedea, G. Piccoli, Carta Geologica dei Colli Euganei. Memorie degli Istituti di Geologia e Mineralogia dell'Universita' di Padova, Mem. Sci. Geol. 34 (1976) 101–129.

[28] R. Antonelli, P. Fabbri, V. Iliceto, C. Majorana, P. Previatello, B. Schreer, R. Sedea, The geothermal Euganean eld. A subsidence modelling approach, in: Proceedings of the World Geothermal Congress, vol. 2, IGA, World Geothermal Congress, Florence, 1995, pp. 1263–1268.

[29] Piccoli G, Bellati R, Binotti C, Lallo E, Sedea R, Dal Prà A, Cataldi R, Gatto G, Ghezzi G, Marchetti M, Bulgarelli G, Schiesaro G, Panichi C, Tongiorgi E, Baldi P, Ferrara GC, Medizza F, Norinelli A, De Vecchi G, Gregnanin A, Sbettega G, 1976. Il sistema idrotermale euganeo-berico e la geologia dei Colli Euganei (The Hydrothermal Euganean system and the geology of the Euganean hills). Mem. Ist. Geol. Miner. Padova, vol. 30, pp. 1–266 (in Italian).

[30] A. Aiuppa, R. Moretti, C. Federico, G. Giudice, S. Gurrieri, M. Liuzzo, P. Papale, H. Shinohara, M. Valenza, Forecasting Etna eruptions by real-time observation of volcanic gas composition, Geology 35 (2007) 1115–1118, https://doi.org/10.1130/G24149A.1.

[31] Karyono K., Obermann A., Mazzini A., Lupi M., Syafri I., Abdurrokhim A., Masturyono M., Hadi S., 2016. Monitoring and characterizing the geysering and seismic activity at the Lusi mud eruption site, East Java, Indonesia. Geophysical Research Abstracts, 18, EGU2016-367-6.

[32] A. Mazzini, G. Etiope, H. Svensen, A new hydrothermal scenario for the 2006 Lusi eruption, Indonesia. Insights from gas geochemistry, Earth Planet. Sci. Lett. 317 (2012) 305–318.

[33] Y. Ben-Zion, Z. Peng, D. Okaya, L. Seeber, J.G. Armbruster, N. Ozer, A.J. Michael, S. Baris, M. Aktar, A shallow fault-zone structure illuminated by trapped waves in the Karadere-Duzce branch of the North Anatolian Fault, western Turkey, Geophys. J. Int. 152 (2003) 699–717.

[34] Y. Ben-Zion, C.G. Sammis, Characterization of fault zones, Pure Appl. Geophys. 160 (2003) 677–715, https://doi.org/10.1007/PL00012554.

[35] Y. Ben-Zion, C.G. Sammis, Mechanics, structure and evolution of fault zones, Pure Appl. Geophys. 166 (10/11) (2009) 1533–1536, https://doi.org/10.1007/s00024-009-0509-y.

[36] K. Irikura, T. Kawanaka, Characteristics of microtremors on ground with discontinuous underground structure, Bull. Disaster Prev. Inst. Kyoto Univ. 30 (3) (1980) 81–96.

[37] G. Calderoni, A. Rovelli, R. Di Giovambattista, Large amplitude variations recorded by an on-fault seismological station during the L'Aquila earthquakes: evidence for a complex fault induced site effect, Geophys. Res. Lett. 37 (2010) L24305, https://doi.org/10.1029/2010GL045697.

[38] G. Cultrera, A. Rovelli, G. Mele, R. Azzara, A. Caserta, F. Marra, Azimuth-dependent amplification of weak and strong ground motions within a fault zone (Nocera Umbra, Central Italy), J. Geophys. Res. 108 (B3) (2003) 2156, https://doi.org/10.1029/2002JB001929.

[39] Z. Peng, Y. Ben-Zion, Temporal changes of shallow seismic velocity around the Karadere-Duzce Branch of the North Anatolian Fault and strong ground motion, Pure Appl. Geophys. 163 (2006) 567–600.

[40] L. Seeber, J.G. Armbruster, N. Ozer, M. Aktar, S. Baris, D. Okaya, Y. Ben-Zion, E. Field, The 1999 earthquake sequence along the North Anatolia transform at the juncture between the two main ruptures, in: A. Barka, O. Kazaci, S. Akyuz, E. Altunel (Eds.), The 1999 Izmit & Duzce Earthquakes: Preliminary Results, Istanbul Technical University, Istanbul, 2000, pp. 209–223.

[41] T. Mizuno, K. Nishigami, Deep structure of the Nojima fault, Southwest Japan, estimated from borehole observation of fault-zone trapped waves, Tectonophysics 417 (2006) 231–247.

[42] F. Villani, S. D'Amico, F. Panzera, M. Vassallo, G. Bozionelos, D. Farrugia, P. Galea, Shallow high-resolution geophysical investigation along the western segment of the Victoria Lines Fault (island of Malta), Tectonophysics 724–725 (2018) 220–233, https://doi.org/10.1016/j.tecto.2018.01.010.

[43] Y.G. Li, J.E. Vidale, K. Aki, F. Xu, Depth-dependent structure of the Landers fault zone using fault zone trapped waves generated by aftershocks, J. Geophys. Res. 105 (2000) 6237–6254.

[44] D.M. Boore, V.M. Graizer, J.C. Tinsley, A.F. Shaka, A study of possible ground-motion amplification at the Coyote Lake Dam, California, Bull. Seism. Soc. Am. 94 (4) (2004) 1327–1342, https://doi.org/10.1785/012003144l.

[45] A. Rovelli, A. Caserta, F. Marra, V. Ruggiero, Can seismic waves be trapped inside an inactive fault zone? The case study of Nocera Umbra, Central Italy, Bull. Seismol. Soc. Am. 92 (2002) 2217–2232.

[46] M. Pischiutta, F. Salvini, J. Fletcher, A. Rovelli, Y. Ben-Zion, Horizontal polarization of ground motion in the Hayward fault zone at Fremont, California: dominant fault-high-angle polarization and fault-induced cracks, Geophys. J. Int. 188 (3) (2012) 1255–1272.

[47] F. Cara, G. Di Giulio, A. Rovelli, A study on seismic noise variations at Colfiorito, Central Italy: implications for the use of H/V spectral ratios, Geophys. Res. Lett. 30 (18) (2003) 1972–1976.

[48] G. Lombardo, R. Rigano, Amplification of ground motion in fault and fracture zones: observations from the Tremestieri fault, Mt. Etna (Italy), J. Volcanol. Geotherm. Res. 153 (2006) 167–176.

[49] R. Rigano, F. Cara, G. Lombardo, A. Rovelli, Evidence of ground motion polarization on fault zones of Mount Etna volcano, J. Geophys. Res. 113 (2008) B10306, https://doi.org/10.1029/2007JB005574.

[50] G. Di Giulio, F. Cara, A. Rovelli, G. Lombardo, R. Rigano, Evidences for strong directional resonances in intensely deformed zones of the Pernicana fault, Mount Etna, Italy, J. Geophys. Res. 114 (2009) B10308, https://doi.org/10.1029/2009JB006393.

[51] M. Pischiutta, M. Pastori, L. Improta, F. Salvini, A. Rovelli, Orthogonal relation between wavefield polarization and fast S-wave direction in the Val d'Agri region: an integrating method to investigate rock anisotropy, J. Geophys. Res. 119 (2014) 1–13, https://doi.org/10.1002/2013JB010077.

[52] M. Pischiutta, A. Rovelli, F. Salvini, G. Di Giulio, Y. Benzion, Directional resonance variations across the pernicana fault, Mt. Etna, in relation to brittle deformation fields, Geophys. J. Int. 193 (2) (2013) 986–996, https://doi.org/10.1093/gji/ggt031.

[53] F. Panzera, B. Halldorsson, K. Vogfjörðc, Directional effects of tectonic fractures on ground motion site amplification from earthquake and ambient noise data: a case study in South Iceland, Soil Dyn. Earthq. Eng. 97 (2017) 143–154.

[54] E. Paolucci, D. Albarello, S. D'Amico, E. Lunedei, L. Martelli, M. Mucciarelli, D. Pileggi, A large scale ambient vibration survey in the area damaged by May-June 2012 seismic sequence in Emilia Romagna, Italy, Bull. Earthq. Eng. (2015), https://doi.org/10.1007/s10518-015-9767-5.

[55] J. Burjánek, J.R. Moore, F.X. Yugsi Molina, D. Fäh, Instrumental evidence of normal mode rock slope vibration, Geophys. J. Int. 188 (2) (2012) 559–569.

[56] D.J. Varnes, Slope movement types and processes, in: R.L. Schuster, R.J. Krizek (Eds.), Special Report 176 Landslides: Analysis and Control, Transportation Research Board, National Research Council, Washington, DC, 1978, pp. 11–33.

[57] J. Burjánek, G. Gassner-Stamm, V. Poggi, J.R. Moore, D. Fäh, Ambient vibration analysis of an unstable mountain slope, Geophys. J. Int. 180 (2) (2010) 820–828.

[58] J.-L. Got, P. Mourot, J. Grangeon, Pre-failure behaviour of an unstable limestone cliff from displacement and seismic data, Nat. Hazards Earth Syst. Sci. 10 (2010) 819–829.

[59] G. Senfaute, A. Duperret, J.A. Lawrence, Micro-seismic precursory cracks prior to rock-fall on coastal chalk cliffs: a case study at Mesnil-Val, Normandie, NW France, Nat. Hazards Earth Syst. Sci. 9 (2009) 1625–1641.

[60] T. Spillmann, H. Maurer, A.G. Green, B. Heincke, H. Willenberg, S. Husen, Microseismic investigation of an unstable mountain slope in the Swiss Alps, J. Geophys. Res. 112 (2007) B07301, https://doi.org/10.1029/2006JB004723.

[61] C. Bordeau, H.B. Havenith, Site effects modelling applied to the slope affected by the Suusamyr earthquake (Kyrgyzstan, 1992), Eng. Geol. 97 (2008) 126–145.

[62] G. Danneels, C. Bourdeau, I. Torgoev, H.B. Havenith, Geophysical investigation and dynamic modeling of unstable slopes: case-study of Kainama (Kyrgyzstan), Geophys. J. Int. 175 (1) (2008) 17–34.

[63] L. Lenti, S. Martino, The interaction of seismic waves with step-like slopes and its influence on landslide movements, Eng. Geol. 126 (2012) 19–36.

[64] L. Lenti, S. Martino, A parametric numerical study of the interaction between seismic waves and landslides for the evaluation of the susceptibility to seismically induced displacements, Bull. Seismol. Soc. Am. 103 (1) (2013) 33–56.

[65] C. Bourdeau, L. Lenti, S. Martino, O. Oguz, E. Yalcinkaya, P. Bigarrè, S. Coccia, Comprehensive analysis of the local seismic response in the complex Büyükçekmece landslide area (Turkey) by engineering-geological and numerical modelling, Eng. Geol. 218 (2017) 90–106.

[66] F. Bozzano, L. Lenti, S. Martino, A. Paciello, G. Scarascia Mugnozza, Evidences of landslide earthquake triggering due to self-excitation process, Int. J. Earth Sci. 100 (2011) 861–879.

[67] V. Del Gaudio, J. Wasowski, Directivity of slope dynamic response to seismic shaking, Geophys. Res. Lett. 34 (2007) L12301.

[68] S. Martino, L. Lenti, J. Delgado, J. Garrido, C. Lopez-Casado, Application of a characteristic periods-based (CPB) approach to estimate earthquake-induced displacements of landslides through dynamic numerical modelling, Geophys. J. Int. 206 (2016) 85–102.

[69] F. Bozzano, L. Lenti, S. Martino, A. Paciello, G. Scarascia Mugnozza, Self-excitation process due to local seismic amplification responsible for the reactivation of the Salcito landslide (Italy) on 31 October 2002, J. Geophys. Res. 113 (2008) B10312.

[70] F. Bozzano, S. Martino, G. Naso, A. Prestininzi, R.W. Romeo, G. Scarascia Mugnozza, The large Salcito landslide triggered by the 31st October 2002, Molise earthquake, Earthq. Spectra 20 (2) (2004) 1–11.

[71] V. Del Gaudio, J. Wasowski, Advances and problems in understanding the seismic response of potentially unstable slopes, Eng. Geol. 122 (2011) 73–83.

[72] J. Delgado, J. Garrido, L. Lenti, C. Lopez-Casado, S. Martino, F.J. Sierra, Unconventional pseudostatic stability analysis of the Diezma landslide (Granada, Spain) based on a high-resolution engineering-geological model, Eng. Geol. 184 (2014) 81–95.

[73] M. Gallipoli, V. Lapenna, P. Lorenzo, M. Mucciarelli, A. Perrone, S. Piscitelli, F. Sdao, Comparison of geological and geophysical prospecting techniques in the study of a landslide in southern Italy, Eur. J. Environ. Eng. Geophys. 4 (2000) 117–128.

[74] H.B. Havenith, A. Strom, D. Jongmans, K. Abdrakhmatov, D. Delvaux, P. Tréfois, Seismic triggering of landslides, part A: field evidence from the northern Tien Shan, Nat. Hazards Earth Syst. Sci. 3 (2003) 135–149.

[75] H.B. Havenith, M. Vanini, D. Jongmans, E. Faccioli, Initiation of earthquake-induced slope failure: influence of topographical and other site specific amplification effects, J. Seismol. 7 (2003) 397–412.

[76] L. Lenti, S. Martino, A. Paciello, A. Prestininzi, S. Rivellino, Recorded displacements in a landslide slope due to regional and teleseismic earthquakes, Geophys. J. Int. 201 (2015) 1335–1345.
[77] O. Méric, S. Garambois, J.P. Malet, H. Cadet, P. Guéguen, D. Jongmans, Seismic noise-based methods for soft-rock landslide characterization, Bull. Soc. Geol. Fr. 178 (2) (2007) 137–148.
[78] E. Yalcinkaya, H. Alp, O. Ozel, E. Gorgun, S. Martino, L. Lenti, P. Bigarrè, S. Coccia, Near-surface geophysical methods for investigating the Buyukcekmece landslide in Istanbul, Turkey, J. Appl. Geophys. 134 (2016) 23–35.
[79] J. Delgado, J. Garrido, C. López-Casado, S. Martino, J.A. Peláez, On far field occurrence of seismically induced landslides, Eng. Geol. 123 (2011) 204–213.
[80] S. Martino, Earthquake-induced reactivation of landslides: recent advances and future perspectives, in: S. D'Amico (Ed.), Earthquakes and Their Impact on Society, Springer International Publishing AG, Cham (Switzerland), 2015, pp. 291–322. ISBN: 978-3-319-21752-9, https://doi.org/10.1007/978-3-319-21753-6_10.
[81] E.M. Rathje, N.A. Abrahamson, J.D. Bray, Simplified frequency content estimates of earthquake ground motions, J. Geotech. Eng. ASCE 124 (2) (1998) 150–159.
[82] R.D. Borcherdt, Estimates of site-dependent response spectra for design (methodology and justification), Earthq. Spectra 10 (1994) 617–653.
[83] B. Abate, A. Incandela, P. Renda, Carta geologica delle isole di Favignana e Levanzo (Arcipelago delle Egadi), Scala 1:12.500, Dip. di Geologia e Geodesia, Università di Palermo, 1997.
[84] E. Tondi, A. Cilona, F. Agosta, A. Aydin, A. Rustichelli, P. Renda, G. Giunta, Growth processes, dimensional parameters and scaling relationships of two conjugate sets of compactive shear bands in porous carbonate grainstones, Favignana Island, Italy, J. Struct. Geol. 37 (2012) 53–64.
[85] H.P.T. Hyde, Geology of the Maltese Islands, Lux Press, Valletta (Malta), 1955. 135 pp.
[86] M. Pedley, M. Clarke, P. Galea, Limestone Isles in a Crystal Sea: The Geology of the Maltese Islands, Publishers Enterprises Group Ltd, San Gwann (Malta), 2002. 109 pp.
[87] A. Goudie, Encyclopedia of Geomorphology, Routledge, London, 2004. 1156 pp.
[88] L. Falconi, A. Peloso, C. Puglisi, A. Screpanti, A. Tati, V. Verrubbi, Rockfalls monitoring along eastern coastal cliffs of the Favignana island (Egadi, Sicily): preliminary remarks, in: G. Lollino, D. Giordan, C. Marunteanu, B. Christaras, I. Yoshinori, C. Margottini (Eds.), Engineering Geology for Society and Territory, vol. 8, Springer International Publishing AG, Cham (Switzerland), 2015, pp. 287–291, https://doi.org/10.1007/978-3-319-09408-3_50.
[89] G. Gigli, W. Frodella, F. Mugnai, D. Tapete, F. Cigna, R. Fanti, E. Intrieri, L. Lombardi, Instability mechanisms affecting cultural heritage sites in the Maltese Archipelago, Nat. Hazards Earth Syst. Sci. 12 (2012) 1883–1903.
[90] R. Iannucci, S. D'Amico, D. Farrugia, P. Galea, S. Martino, A. Paciello, in: Seismic measurements in unstable sea cliff areas, Atti del 34° Convegno Nazionale Gruppo Nazionale di Geofisica della Terra Solida. Tema 3: Geofisica Applicata. Trieste (TS), 17–19 November 2015, vol. 3, 2015, pp. 56–63.
[91] F. Panzera, S. D'Amico, A. Lotteri, P. Galea, G. Lombardo, Seismic site response of unstable steep slope using noise measurements: the case study of Xemxija Bay area, Malta, Nat. Hazards Earth Syst. Sci. 12 (11) (2012) 3421–3431.

[92] Y. Nakamura, A method for dynamic characteristics estimation of subsurface using microtremor on the ground surface, Q Rep RTRI 30 (1) (1989) 25–33.

[93] F. Farrugia, E. Paolucci, S. D'Amico, P. Galea, Inversion of surface-wave data for subsurface shear-wave velocity profiles characterised by a thick buried low-velocity layer, Geophys. J. Int. 206 (2016) 1221–1231, https://doi.org/10.1093/gji/ggw204.

[94] P. Galea, S. D'Amico, D. Farrugia, Dynamic characteristics of an active coastal spreading area using ambient noise measurements-Anchor Bay, Malta, Geophys. J. Int. 199 (2014) 1166–1175.

[95] B. Abate, P. Renda, M. Tramutoli, Schema geologico dei Monti di Termini Imerese e delle Madonie Occidentali (Sicilia), Mem. Soc. Geol. Ital. 41 (1988) 465–474.

[96] L. Bonci, F. Bozzano, S. Calcaterra, V. Eulilli, F. Ferri, P. Gambino, M.R. Manuel, S. Martino, G. Scarascia Mugnozza, in: Geological control on large seismically induced landslides: the case of Cerda (Southern Italy), Proc IX ISL, Rio De Janeiro, 2004, pp. 985–991.

[97] D.K. Keefer, Landslides caused by earthquakes, Geol. Soc. Am. Bull. 95 (1984) 406–421.

[98] E.H. Field, K. Jacob, A comparison and test of various site response estimation techniques, including three that are non reference-site dependent, Bull. Seismol. Soc. Am. 85 (1995) 1127–1143.

[99] Patacca E, Scandone P, Bellatalla M, Perilli N, Santini U (1992) La zona di giunzione tra l'arco appenninico meridionale nell'Abruzzo e nel Molise, Studi Geologici Camerti, spec. vol. (1991/2), CROP 11, 417–441.

[100] S. D'Amico, G. Lombardo, F. Panzera, Seismicity of the Mediterranean Region and mitigation of earthquake losses. Phys. Chem. Earth 63 (2013) 1–2, https://doi.org/10.1016/j.pce.2013.07.001.

[101] D. Fah, F. Kind, D. Giardini, Inversion of local S-wave velocity structures from average H/V ratios, and their use for the estimation of site-effects, J. Seismol. 7 (2003) 449–467.

[102] F. Panzera, S. Sicali, G. Lombardo, S. Imposa, S. Gresta, S. D'Amico, A microtremor survey to define the subsoil structure in a mud volcanoes area: the case study of Salinelle (Mt. Etna, Italy), Environ. Earth Sci. 75 (2016) 1140, https://doi.org/10.1007/s12665-016-5974-x.

[103] F. Panzera, S. D'Amico, G. Lombardo, E. Longo, Evaluation of building fundamental periods and effects of local geology on ground motion parameters in the Siracusa area, Italy, J. Seismol. 20 (2016) 1001–1019, https://doi.org/10.1007/s10950-016-9577-5.

[104] D. Farrugia, P. Galea, S. D'Amico, E. Paolucci, Sensitivity of ground motion parameters to local shear-wave velocity models: the case of buried low-velocity layers, Soil Dyn. Earthq. Eng. 100 (2017) 196–205, https://doi.org/10.1016/j.soildyn.2017.05.033.

[105] D. Di Giacomo, M.R. Gallipoli, M. Mucciarelli, S. Parolai, S.M. Richwalski, Analysis and modeling of HVSR in the presence of a velocity inversion: the case of Venosa, Italy, Bull. Seism. Soc. Am. 95 (6) (2015) 2364–2372, https://doi.org/10.1785/0120040242.

[106] M. Ohori, A. Nobata, K. Wakamatsu, A comparison of ESAC and FK methods of estimating phase velocity using arbitrarily shaped microtremor arrays, Bull. Seism. Soc. Am. 92 (2002) 2323–2332.

[107] S. Parolai, M. Picozzi, S.M. Richwalski, C. Milkereit, Joint inversion of phase velocity dispersion and H/V ratio curves from seismic noise recordings using a genetic

algorithm, considering higher modes, Geophys. Res. Lett. 32 (2005), https://doi.org/10.1029/2004GL021115.

[108] M. Picozzi, D. Albarello, Combining genetic and linearized algorithms for a two-step joint inversion of Rayleigh wave dispersion and H/V spectral ratio curves, Geophys. J. Int. 169 (2007) 189–200.

[109] H. Yamanaka, H. Ishida, Application of genetic algorithms to an inversion of surface-wave dispersion data, Bull. Seism. Soc. Am. 86 (1996) 436–444.

FURTHER READING

[110] R. Rigano, G. Lombardo, Effects of lithological features and tectonic structures in the evaluation of local seismic response: an example from Hyblean plateau (Eastern Sicily), Geol. Carpath. 56 (4) (2005) 297–306.

CHAPTER

13

Multimethodological approach to investigate urban and suburban archaeological sites

Piro Salvatore*, Papale Enrico*, Zamuner Daniela*, Kucukdemirci Melda†

Institute for Technologies Applied to Cultural Heritage ITABC CNR, Roma, Italy Engineering Geophysical Department, University of Istanbul, Istanbul, Turkey†*

CHAPTER OUTLINE

Innovation in Near-Surface Geophysics. https://doi.org/10.1016/B978-0-12-812429-1.00013-1

1 INTRODUCTION

Nondestructive ground surface geophysical prospecting methods are often used for the investigation of archeological sites, where a detailed physical and geometric reconstruction of hidden volumes is required prior to any excavation work.

Geophysical prospecting methods are considered valuable techniques for nondestructive detection of shallow-buried features of archeological interest. Usually, detection is possible because the physical property of the target features contrasts with the surrounding medium.

All methods measure the variations of single physical parameters; therefore, if these are used singularly, they could not permit a complete location and characterization of anomalous bodies.

The probability of a successful result rapidly increases if a multimethodological approach is adopted, according to the logic of objective complementarity of information and of global convergence toward a high-quality multiparametric imaging of the buried structures.

The representation of the static configuration of the bodies in the subsoil and of the space–time evolution of the interaction processes between targets and hosting materials has to be actually considered fundamental elements of primary knowledge in archeological prospecting.

Geophysical techniques provide raw data that must then be evaluated. The first step is to present the data in a form that can be understood by archeologists, either by constructing a model of the physical phenomena thought to be responsible and changing this until the measured data are accounted for within a minimum error or by using information that provides some degree of separation of the components of the measurements that are due to archeological sources from those of natural or modern origin.

The main effort in geophysical prospecting for archeology is therefore the integration of different, absolutely noninvasive techniques, especially if managed in view of an ultrahigh-resolution three-dimensional (3-D) tomographic representation mode.

Generally, the limited dimensions and depth of the archeological remains and the presence of structures made with the same material of the surrounding ground, as well as soil inhomogeneity and other environmental disturbances, can yield low signal-to-noise ratios (S/N). This makes it difficult to define the spatial position and extent of the buried structure.

However, due to (a) the often limited dimensions and depth of the archeological structures, (b) the presence of structures made with the same material of the hosting rocks, (c) the soil inhomogeneity, and (d) the environmental and anthropic disturbances, it can be rather difficult to individuate their position and extension because of the generally low values of the signal-to-noise ratio.

It is possible to contrast such a shortcoming: (a) by improving *data acquisition techniques* (high-resolution 2-D and 3-D acquisition), (b) by improving *data processing methods* (modeling and techniques to enhance S/N), and (c) by *integrating the results* obtained with different geophysical methods.

Some approaches to overcome this problem have been considered, during the last three decades [1–7]. They generally consist of improving data acquisition techniques (the recent technical instrumental evolution involves the application of the high-resolution 2-D and 3-D acquisition techniques) and by improving data processing methods (modeling and techniques to enhance S/N) [8–15]. Our recent approach, discussed in previous papers and applied here, is related to the integration of the results obtained with different geophysical methods [12,16–23].

There is no doubt that a multimethodological survey, which compares and contrasts magnetic, ground-penetrating radar (GPR), and resistive methods at the same site, could provide the opportunity to gather significant information. This is possible because each of these methods evaluates different physical properties of the target features with respect to the ground [17,24–27].

All classical geophysical methods, such as magnetic and resistivity, yield measurements that are average values over a given volume. In contrast, the GPR method produces three-dimensional values in the form of radar traces, which cannot be handled in the same way as most other data. Another aspect of combining and contrasting multiple method data is that a sampling spatial interval needs to be as low as possible. In studying archeological remains, the integrated survey must therefore be performed at a spatial sampling interval that is in accordance with the penetration depth of the single method and with the integration volume. These varying parameters include electrode distance in geoelectric methods and sensor height in magnetic surveys. As GPR reflections can be considered as nearly accurate (respect to the dimension of the searched bodies), the risk of spatial aliasing is much higher. Therefore, GPR spatial sampling needs to be compatible with the wavelength and appropriate to the dimensions of the target features in the ground.

Some techniques have already been developed and discussed in previous papers in the last three decades; Brizzolari et al. [2,9,28–30] and many others have presented overlaid geophysical maps obtained by multiple methods. Neubauer and Eder-Hinterleitner [27,31] proposed an arithmetic and/or color-coded combination of resistivity and magnetic surveys. Piro et al. [26] and Piro and Gabrielli [32] suggested quantitative methods of data integration by applying functions that give an indicator of source occurrence (ISO) based on the spatial correlation of three employed geophysical methods. Kvamme [25] introduced several graphical, discrete, and continuous integration methods to a dataset to have an improved understanding of the subsurface. Ogden et al. [33] furthermore applied some of these integration techniques to investigate a site in Italy and to test the effectiveness of these methods.

Typically, for archeological prospections, no more than two or three geophysical methods are employed to solve problems of detection contouring and the development of three-dimensional models of archeological sites [34]. The prospection techniques most frequently used in archeological sites are the GPR, fluxgate differential magnetic (FDM), and electric resistivity tomography (ERT). Since each method relies on measuring a specific physical parameter of the subsurface, it is well understood that multiple geophysical surveys could provide the opportunity to gather significant information about the investigated area [25,26,34].

The developments in geophysical data acquisition and analysis suggest that together with a multimethodological approach, the quantitative integration techniques can offer meaningful insights into archeological structures and for researchers to combine patterns through multivariate datasets that might otherwise be not evaluated.

Following the above outlined approach, we have integrated geophysical methods that measure the variations of potential field (magnetic gradiometric method) with active methods that measure the variations of physical properties due to the body's geometry and volume (GPR and ERT).

In the present chapter, some of these integration methodologies will be described, and it will be shown how these have been applied to four case studies in urban and suburban areas, three in Italy and one in Turkey. In particular, the integration techniques have been tested using all three or two of the mentioned methods, depending on the conditions of the investigated sites. The first site of interest is located in an area in the Aurunci Natural Regional Park (Italy) along the ancient Appian Way; the second site is located in an urban context in the city of Rome (Italy) near the Santa Balbina's church complex. In these two sites, it has been possible to acquire data employing only two of the three available geophysical methods (GPR and ERT). The third investigated site, where it has been possible to use all three methods (GPR, ERT, and FDM), is located in a suburban area in Cerveteri about 40 km far from the city of Rome (Italy). The fourth investigated site, where have been employed the magnetic gradiometer and GPR methods, is in the archeological site of Heraion Teichos (Turkey).

2 INTEGRATION METHODS

The techniques for integrating the geophysical datasets have been separated in two main groups: *qualitative* and *quantitative* methods. The qualitative methods are the most commonly used and give out purely graphical results; the quantitative techniques are instead more advanced and consist in a number of mathematical or statistical operations that yield new datasets that can be further analyzed. It is important to specify that all the integration operations have to be performed on planimetric bidimensional geophysical maps (slices) that correspond to the same depth for all the methods employed.

The *qualitative* integration methods used include the following:

- *Contour map overlays*: This consists in the simple overlaying of the geophysical maps represented as contour lines (as shown in Fig. 1). To obtain this type of representation, any kind of software to represent geophysical data can be used; in this case we employed Surfer 13 (Golden Software). This is just one of the various combinations of representation that can be used for two-dimensional overlays. This graphical integration method is particularly useful in discerning sharp differences in geophysical signatures and clearly defines linear archeological features [33]. Other overlaying methods that can be used for displaying continuous map data include gray or color scales, pseudo-three-dimensional views, and surface shadowing [25]. Many combinations using all the mentioned representation techniques are possible, though these maps become too complex for understanding if more than two or three dimensions are combined.
- *Red-green-blue (RGB) color composite*: In this visualization method, each of three datasets is assigned a primary color, and the images are then combined to create color mixes that span the full visual spectrum indicating the degree of

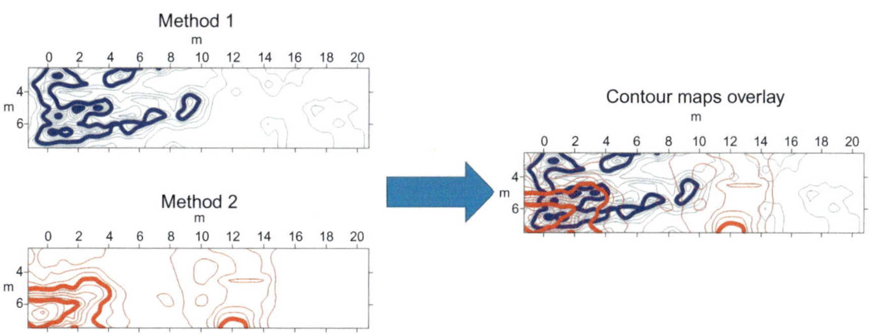

FIG. 1

Example of the contour map overlay integration method. On the left are the two methods represented separately. On the right is the result given by the overlay. In this case, a different color for the contour lines has been applied for the different methods in order to discern between the two in the final overlay map.

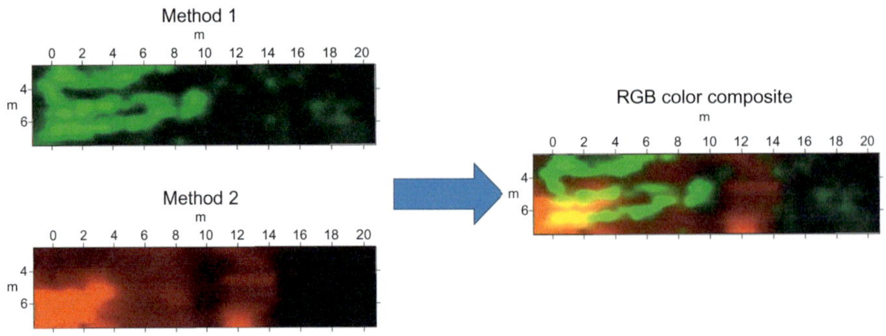

FIG. 2

Example of color composite integration. Each of the dataset is assigned a primary color, and the images are then combined.

overlapping of the anomalies detected by the single methods [25], as observed in Fig. 2. Color composites have been used initially for displaying satellite imagery; successively, this technique has been tested for integrating geophysical data [25,30,31,35]. In the case of Kvamme's [25] Army City data, he represented each of the geophysical techniques with the following colors: red for EM resistivity, green for EM conductivity, and blue for magnetic susceptibility, resulting in different structural features being represented simultaneously in the same image but at the same time provide a tool to easily discern the objects detected by the different geophysical methods. Moreover, in Kvamme's [25] work, some features were robustly indicated in all three datasets, resulting in a white color, and some features were instead represented by a black color indicating a low response for all three surveys. Knowledge of the RGB color model in fact enhances interpretation. A bright primary color indicates high values of the variable assigned to the respective variable. White indicates large values in all datasets, whereas black indicates the opposite.

The *quantitative* integration techniques are subdivided in two more groups that include the following:

- *Discrete* integrating techniques in which case data are divided into discrete classes and include the following:
    - *Binary sum*: In this case, a reclassification is performed in order to have binary data that are representative of the presence (1) and absence (0) of archeological features [25,33]. The reclassified datasets are then combined to have a Boolean union of the features detected by the methods employed that occurs wherever at least one of the data inputs is coded as 1. One advantage of this method is the immediate determination of the presence or absence of an anomaly in the archeological site as shown in Fig. 3.
    - *K-means cluster analysis*: Cluster analysis includes a series of advanced algorithms that define natural groupings in multivariate data such that each

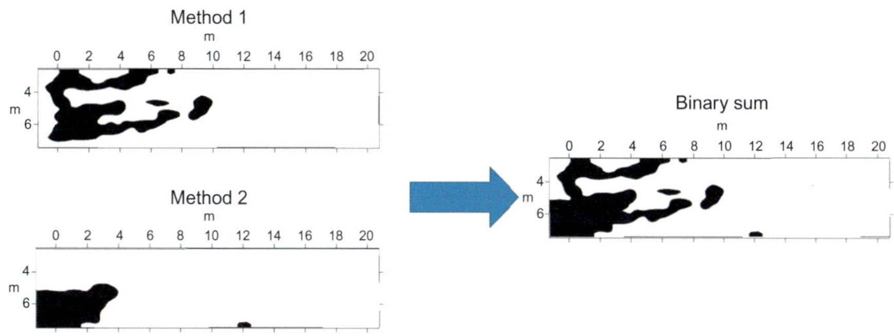

FIG. 3

Example of binary sum integration.

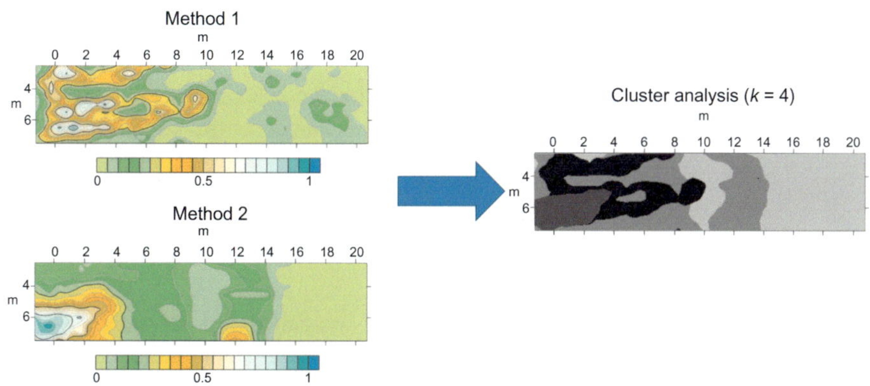

FIG. 4

Example of *k*-means cluster analysis. On the left, the normalized field data for two methods are represented. On the right, the result of cluster analysis integration method is shown.

one is more or less homogeneous and distinct from others [36]. This technique divides the group of values or attributes into a specified number of clusters as defined by the user [33]. The chosen number of classes defines the anomaly types that will then be represented in the final image as shown in Fig. 4. For example, two classes might represent "anomaly" versus "background"; three classes might represent positive anomaly, negative anomaly, and background; or a higher number of classes can differentiate between different features detected by the methods such as walls, floors, pipeline, streets, and background [25].

- *Continuous integrating methods* give new datasets from the mathematical or statistical combinations of the multiple geophysical datasets, which can be further analyzed. Robust and subtle anomalies may be expressed simultaneously, producing composite imagery with high information content. To perform quantitative integration of the results obtained with different geophysical methods, it is necessary, after the elaboration developed for each dataset, to make all obtained values independent of their original physical parameters and numerically comparable [26]. This is achieved by normalizing the processed data, resulting in data with a range between 0 and 1 as indicated by the function

$$F_{m_i}(x,y) = \frac{\left| f_{m_i}(x,y) - \tilde{f}_{m_i} \right|}{\max \left| f_{m_i}(x,y) - \tilde{f}_{m_i} \right|}$$

where $\tilde{f}_{m_i}$ represents the undisturbed value of the data function $f_{m_i}(x,y)$ or the response of a given method if no anomalous bodies were present underground. Thus, the function of $\left| f_{m_i}(x,y) - \tilde{f}_{m_i} \right|$ indicates the departure of the physical properties investigated by any given method at a specific depth [26].

This new dataset is then managed in order to obtain a grid of values of comparable sizes for all methods employed (with *Surfer*), and a few operations can be performed:

- *Data sum*: By summing the grids calculated for the different methods, it is possible to obtain an indicator of the spatial distribution of the anomaly's source occurrence detected by at least one method [26,32] by applying the following function:

$$\overline{F}(x,y) = \frac{1}{M} \sum_{m_i}^{M} F_{m_i}(x,y)$$

The sum of the normalized data will then include anomalies from all sources simultaneously, including those of large and small magnitude, as shown in Fig. 5. In fact, the function is equal to 1 only at those points where the methods give maximum anomaly pattern and is equal to 0 where all methods do not reveal geophysical anomalies. This method, in Piro et al. [26], gave good results especially in one of the investigated areas, where after the integration of GPR, magnetometry, and ERT methods, archeological features appeared clearly reconstructed as shape respect to the single methods analyzed individually that showed only portions of the archeological structures.

- *Data product*: In this case, the datasets are multiplied, which will give an indicator of the spatial distribution of the anomaly's source occurrence detected by all the methods used following the function:

$$F^*(x,y) = \prod_{m_i}^{M} F_{m_i}(x,y)$$

That means that only anomalies detected by all methods are shown on the final dataset and robust anomalies tend to be better contrasted and slightly exaggerated, whereas subtle ones are eliminated, since the function $F^*(x,y)$ is equal to 0 at those points where also only one of the methods does not reveal geophysical anomalies (Fig. 5).

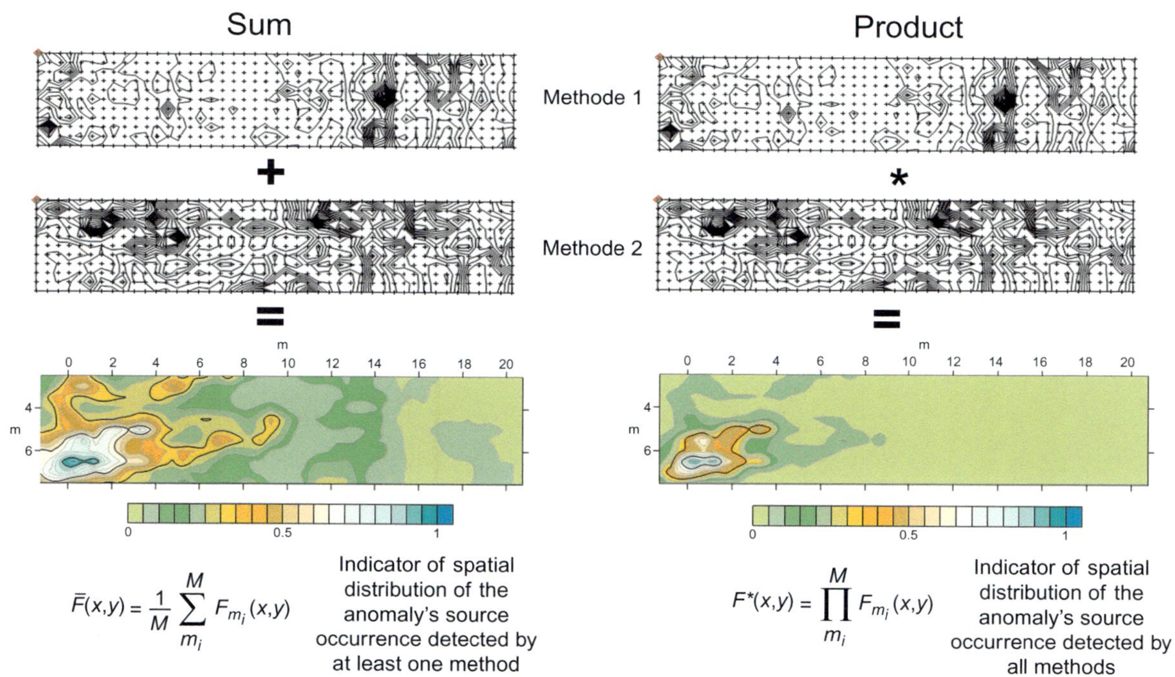

FIG. 5

Example of data sum (on the left) and data product (on the right) integration methods.

- *Principal component analysis*: Principal component analysis (PCA) is a mathematical procedure that transforms a number of correlated variables into a smaller number of uncorrelated variables called principal components, which let us reduce the dimension of large set of variables to a small set that still contains most of the information, as shown in Fig. 6. The first component accounts for a greatest amount of total variance between the observed variables. The second component represents the greatest amount of variance in the dataset that was not explained by the first component, with each new component accounting for gradually smaller and smaller amounts of variance (measures of how far a set of (random) numbers are spread out from their mean). In order to obtain a significant reduction in redundancy and therefore extract more meaningful information from this reclassification, input variables must be highly correlated. In geophysics, one might expect that this condition is satisfied, especially when using methods that are related to the same physical parameter. However, in Kvamme's analysis of Army City and Ogden's analysis of Portus data, the overall correlation between the input data variables was relatively low. In the first case, this method gave good results despite the low correlation between the variables, where the first and second components represented a combination of all the variables used. In the case of Ogden's Portus data, the applicability of this type of analysis was questioned since the three principal components only represented the single input variables.

3 ANCIENT APPIAN WAY SITE

3.1 SITES CHARACTERISTICS

The area, situated along the segment of the Ancient Appian Way, near the Gorge of St. Andrew (Gola di S. Andrea), is characterized by the presence of various archeological features such as the well-preserved old basalt paving of the road from the mid-Imperial period or the Fort of St. Andrew, which is a mixture of Roman, medieval, and modern fortification that closed the way to the gulf. Moreover, there is proof of the presence of an ancient temple dedicated to the God Apollo, which has been dismantled over the centuries in order to build other structures such as the Fort of St. Andrew. The subsurface is also characterized by the presence of a series of cisterns for water storage dating back to the third century BC.

Geologically, the sedimentary succession cropping out in the area belongs to the carbonate platform domain, and it consists of neritic carbonate deposits of Upper Triassic to Upper Cretaceous age [37–39].

3.2 GEOPHYSICAL SURVEYS

The area was investigated using the GPR and ERT.

GPR surveys were conducted during 2013 in a 14 m × 24 m area, using the SIR 3000 (GSSI) system equipped with a 400 MHz bistatic antenna with constant offset.

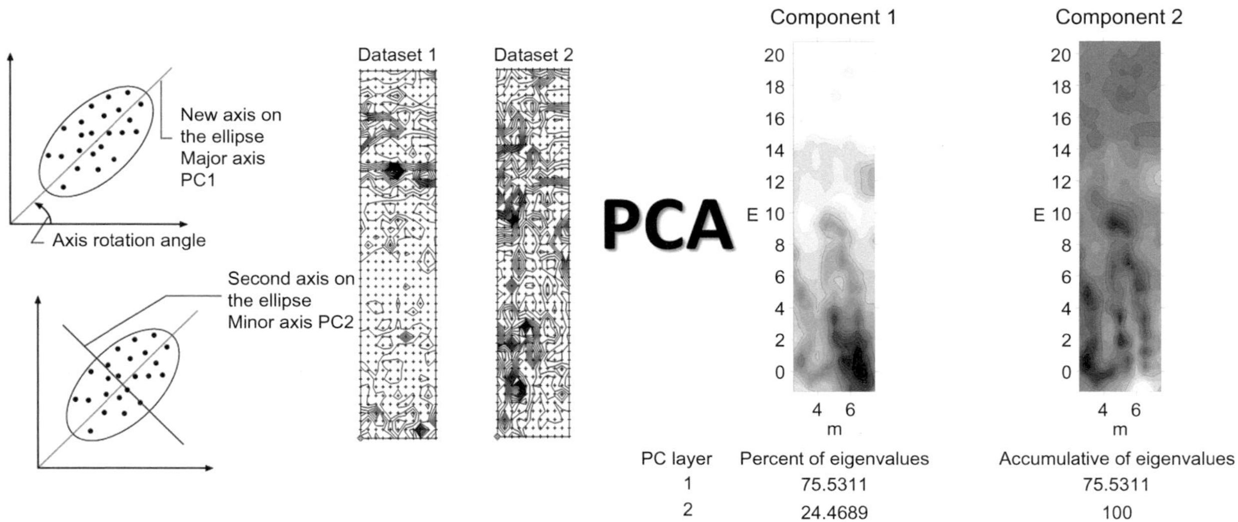

FIG. 6

PCA integration example for two dataset. In this case, the two datasets are highly correlated, and as can be seen from the result of the principal component analysis, the first component accounts for most of the total variance.

A total of 29 profiles were collected alternatively in forward and reverse directions employing the GSSI cart system equipped with odometer. The horizontal spacing between parallel profiles was 0.5 m, and along each profile markers were set every 1 m to provide spatial reference. All responses within 80 ns (two-way travel time) time window were recorded in the field as 16-bit data and 512 samples per radar scan.

An area of 5 × 23.5 m overlapping the surveyed GPR area was investigated by 11 parallel ERT profiles, distanced by 0.5 m, each consisting of a 0.5 m spaced 48 electrodes, dipole–dipole array. The measurements were taken employing an IRIS SYSCAL Junior Switch-72.

3.3 DATA PROCESSING AND RESULTS

3.3.1 GPR data

The raw reflection data acquired by GPR are a collection of many individual traces along the 2-D profiles within the acquisition grid.

Reflection data, collected along profiles with 0.50 m spacing, were processed using standard techniques [40–51].

The basic radargram signal processing steps included (i) postprocessing pulse regaining, (ii) DC drift removal, (iii) trace resampling along the profile (an averaged trace every 4 cm), (iv) band-pass filtering, (v) migration, and (vi) background filter.

With the aim of obtaining a planimetric map of all possible anomalous bodies in the investigated areas, the time-slice representation technique was applied, using all processed profiles. A total of 30 time slices were obtained for both sites with a 4 ns slicing interval. Time slices are calculated by creating two-dimensional horizontal contour maps of the averaged absolute value of the wave amplitude from a specified time value across parallel profiles and then placed in a three-dimensional space using x and y horizontal coordinates for the grid, with z being the two-way travel time (twt) [52]. All processing operations were performed employing the GPR-Slice v7 software [53]. From the two-way travel time, it is possible to estimate the depths of the slices by calculating the electromagnetic wave velocity from the shape of diffraction hyperbolas. An average velocity of 0.1 m/ns was calculated for both sites.

Fig. 7 illustrates some of the time slices generated by the software at depths of 0.6, 1.3, 1.7, and 2.7 m for the Appian Way site. Some anomalies are clearly visible, associated with the empty portions of the cisterns earlier described. These anomalies are clearly visible from a depth of about 1 m.

3.3.2 ERT data

In the suburban site along the Via Appia, the 11 ERT profiles were inverted using the software RES2DINV (Geotomo); all the inversions can be assumed to be quite reliable, giving a final RMS of about 1.5% after seven to nine iterations and a better representation of electric data. All the inverted electric sections display approximately the same features and are characterized by high-resistivity anomalies (over 10,000 Ωm) some of which are surely related to the presence of the empty sections of the cisterns. These inverted pseudosections were then interpolated using Voxler

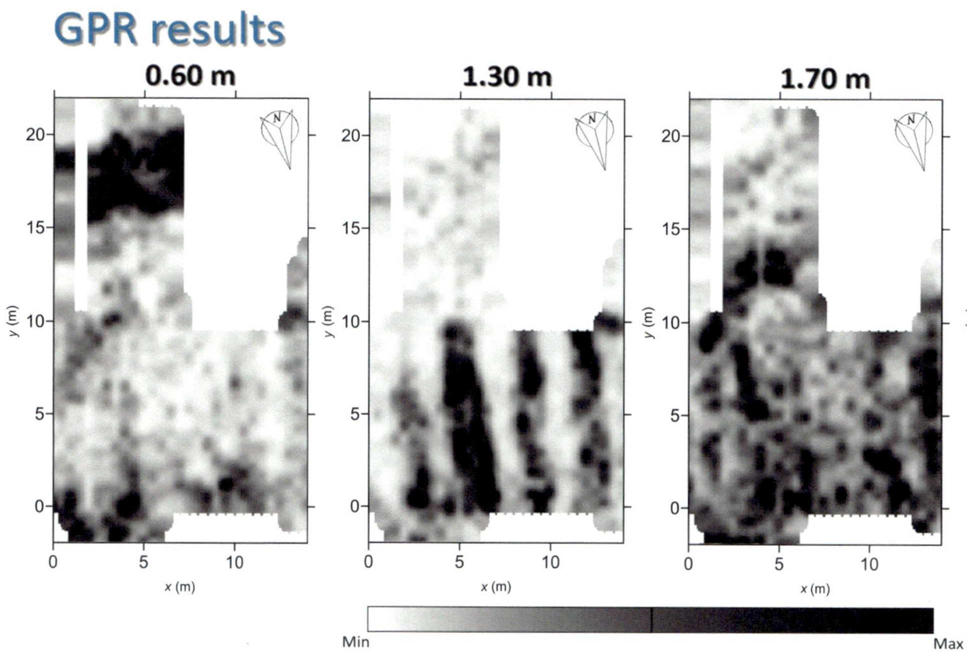

FIG. 7

GPR field data time slices. High-intensity anomalies indicate the presence of the empty sections of the cisterns in the subsurface.

(Golden Software) in order to obtain a three-dimensional model of the resistivity distribution of the subsurface, from which it was possible to extract slices corresponding to different depths, as shown in Fig. 8.

4 DATA INTEGRATION METHODS

4.1 QUALITATIVE INTEGRATION (CONTOUR MAPS OVERLAYS AND RGB)

This integration method consists in representing the maps obtained from the data acquired by the different methods in a single image.

For the Appian Way site, it can be seen that overall, the GPR and the ERT anomalies corresponding to the cisterns show good matching positions. The level of detail and the difference in size of the anomalies detected by the two methods is linked to their resolution. The ERT data anomalies tend to overestimate the real size of the cisterns, due to the method's lower resolution with respect to the resolution of GPR method. It can in fact be seen that the ERT anomaly related to the cistern is bigger in size, and it is therefore detected at a depth of 0.6 m, while the GPR method detects it at a deeper depth of 1.0 m (Fig. 9). The RGB color composite gives very good results since it shows in a single map the anomalies detected by the two methods while it is still possible to clearly discern between the anomalies detected by both methods and to highlight the common features.

4.2 DISCRETE QUANTITATIVE INTEGRATION (BINARY SUM AND CLUSTER ANALYSIS)

In the case of the binary combination, the data represent all features within the normalized range 0.4–1. In this case, the main features detected by both methods are represented, but the methodology relies on the arbitrary threshold chosen by the user in order to define the significant anomalies, while the subtler ones, using this method, are completely ignored. K-means cluster analysis was instead applied using $k = 4$ that highlights the main anomalies, which are given by the presence of the cisterns (Fig. 10).

4.3 CONTINUOUS QUANTITATIVE INTEGRATION (SUM, PRODUCT AND PRINCIPAL COMPONENT ANALYSIS)

The continuous quantitative integration techniques performed were the sum, the product, and the principal component analysis. As can be seen in Fig. 11, the sum is probably the best method, since it is the only method to include anomalies from all sources simultaneously, including those of large and small magnitude without leaving out any anomaly that could be also meaningful for the final interpretation. The product integration as stated previously highlights only the anomalies detected

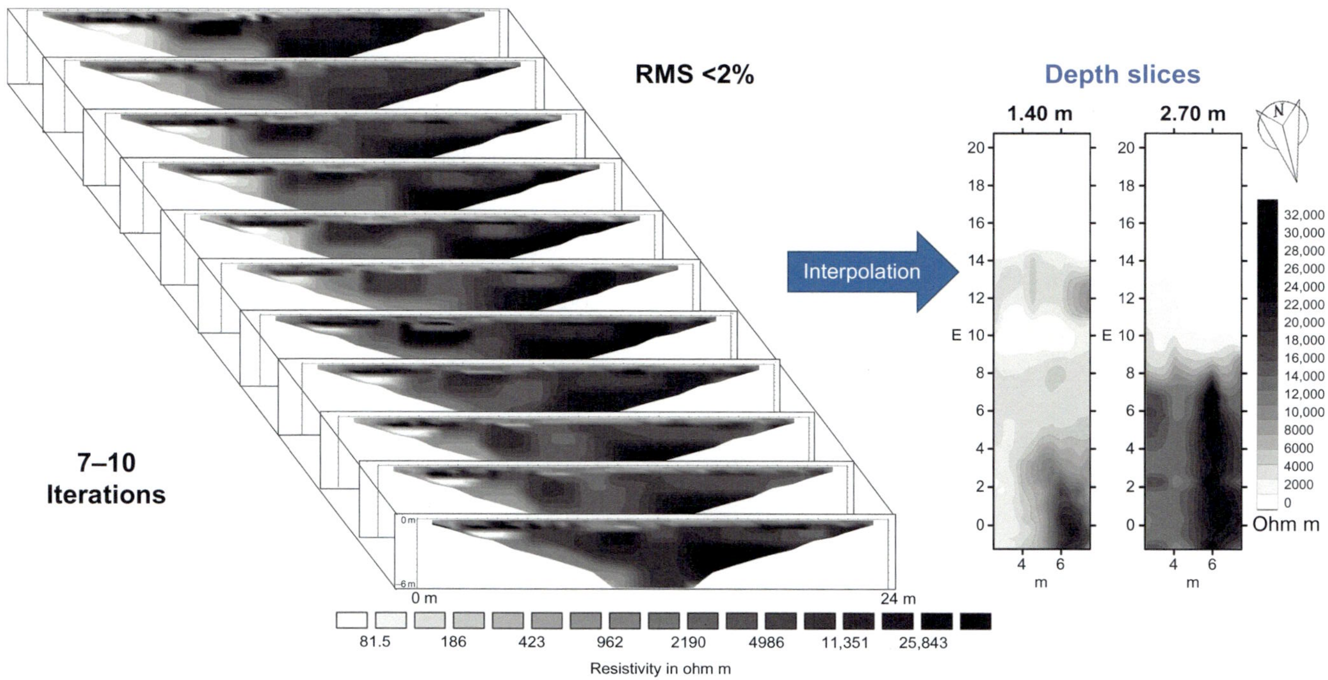

FIG. 8

Ancient Appian Way ERT pseudosections relative to the 11 profiles (left). Depth slices are obtained by interpolating these pseudosections. Strong anomalies are related to the empty sections of the cisterns.

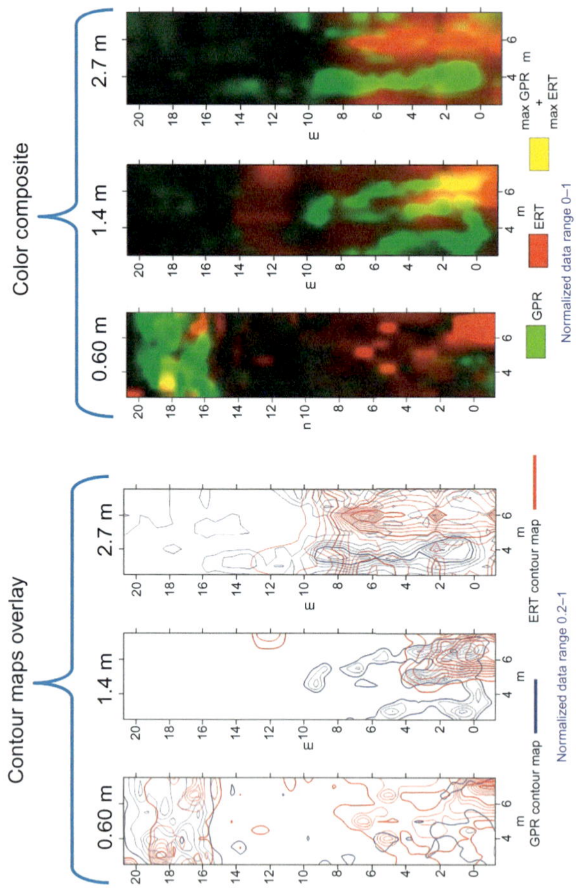

FIG. 9

Contour map overlay and RGB color composite integration for the Appian Way site.

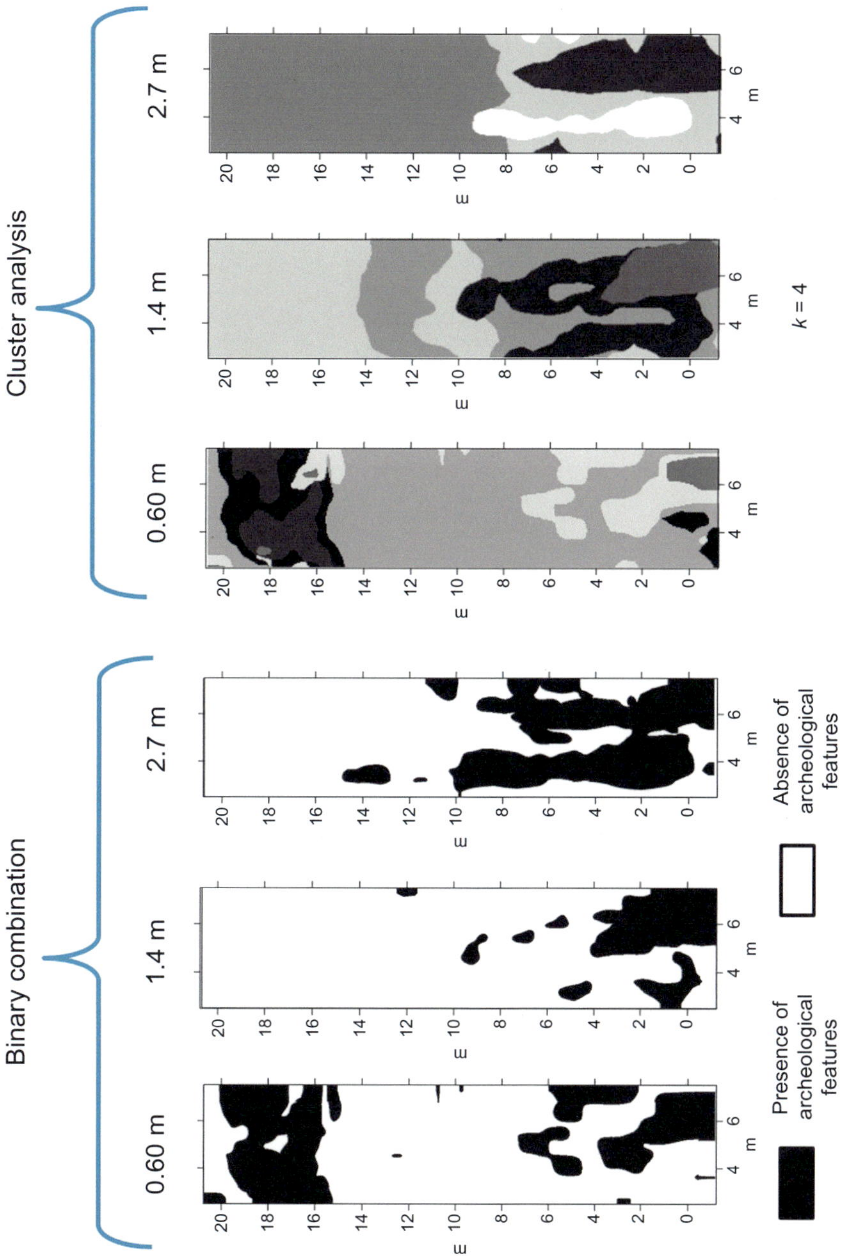

FIG. 10

Binary sum and *k*-means cluster analysis integration for the Appian Way site for depths of 0.60, 1.4, and 2.7 m.

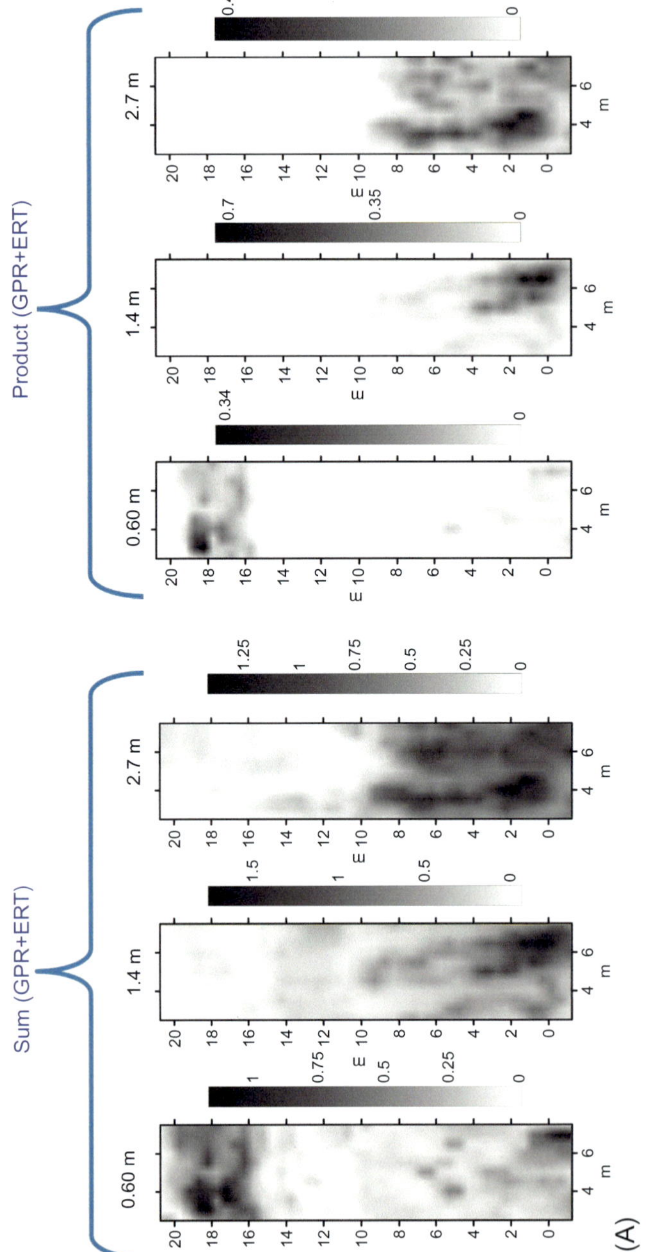

FIG. 11

Continuous quantitative integration applied to the Appian Way site. (A) The sum and the product of the datasets are represented.

(Continued)

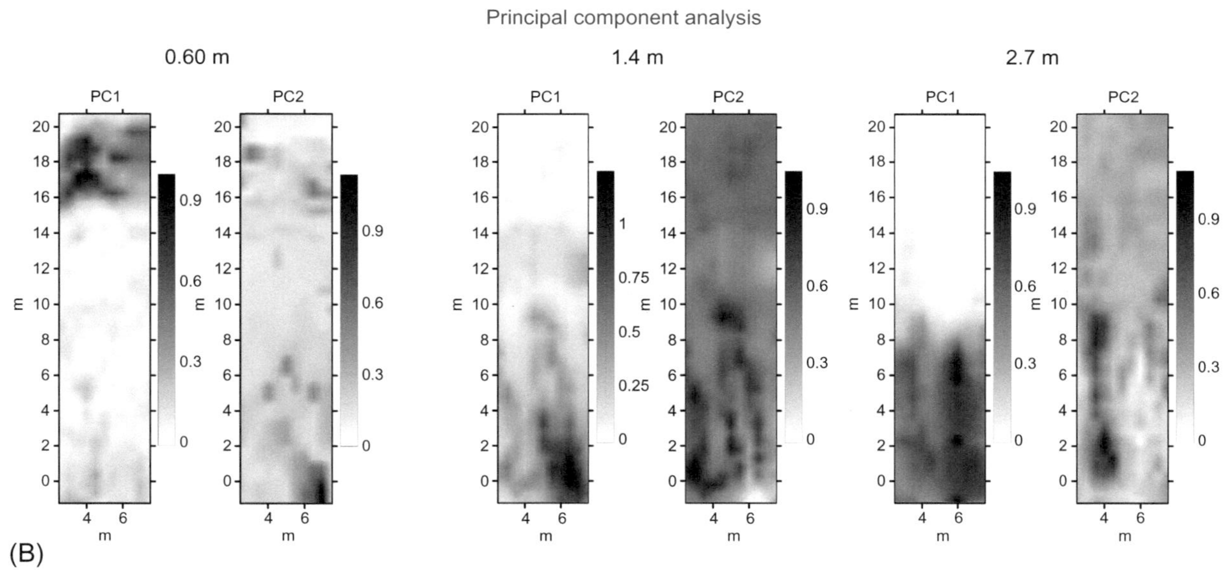

FIG. 11, CONT'D

(B) The two principal components for the three investigated depths are represented.

by both methods, so if some anomalies are only detected by one method, they will not be represented in the final map.

Regarding the principal component analysis, it is important, for it to work and properly eliminate redundant data, to have highly correlated data. In this case, we get a high level of correlation only for the depths of 1.4 and 2.7 m, where in fact both methods detected the same features. In this case, the correlation of the two datasets was exceptionally high, giving out a maximum value of r of 0.832. This was the highest value of correlation that was calculated for the datasets of all the case studies, since for the others, the maximum value of r achieved was never higher than 0.3. Although the high value of correlation, and its utility in this context, still remain questionable, this technique was applied to only two datasets, and it is generally recommended to use more methods to achieve better results with this type of analysis.

5 SANTA BALBINA SITE

5.1 SITE CHARACTERISTICS

The second investigated site is located in the central part of the city of Rome, near the Santa Balbina all' Aventino's Church. There is not much archeological information regarding this area; however, it is suggested that there may be the presence of structures associated with the Servian Wall (fourth century BC), and ancient quarries exploited for the extraction of tuff as construction material. Geologically, the area is characterized by the presence of the products related to the Alban Hills volcanic activity and more specifically by the "Villa Senni" formation deposits.

5.2 GEOPHYSICAL SURVEYS

The second site was also investigated using the GPR and ERT methods. For GPR surveys, the site was divided in two areas investigated in two separate times using the SIR 3000 (GSSI) system equipped with a 400 MHz bistatic antenna with constant offset. The first area of 50 m × 45 m was investigated during December 2013 by collecting 101 profiles with horizontal spacing of 0.5 m, and along each profile, markers were set every 1 m to provide spatial reference. The second area was investigated in May 2015 with an extent of 60 m × 35 m; a total of 121 profiles were collected with a horizontal spacing of 0.5 m. All profiles were collected alternatively in forward and reverse directions employing the GSSI cart system equipped with odometer. For both areas, all responses within 90 ns (two-way travel time) time window were recorded in the field as 16-bit data and 512 samples per radar scan.

The ERT survey was performed in an area overlapping the two GPR surveyed areas; in order to acquire 3-D data, a special array type was employed ("snake" geometry with a dipole-dipole configuration), consisting of 72 electrodes laying in a regular rectangular grid of 18 by 4 electrodes with a spacing of 2 m; 11 of these ERT grids were used in order to cover a total area of 72 m × 40 m. The measurements were taken employing an IRIS SYSCAL Junior Switch-72.

5.3 DATA PROCESSING AND RESULTS

5.3.1 GPR data

For the GPR, the same processing techniques used for the previous site were employed. In particular, the basic radargram signal processing steps included (i) postprocessing pulse regaining, (ii) DC drift removal, (iii) trace resampling along the profile (an averaged trace every 4 cm), (iv) band-pass filtering, (v) migration, and (vi) background filter. Also in this case, a total of 30 time slices were obtained with a 4 ns slicing interval, and an average velocity of 0.1 m/ns was calculated. In Fig. 12, some of the most significant time slices obtained after the processing are represented.

5.3.2 ERT data

Data were processed employing RES3DINV (Geotomo) software that generates directly the depth slices. The inversion process gave a final RMS of about 8% after seven iterations. Fig. 13 shows the depth slices produced by the software with the indicated corresponding depth. The highest resistivity anomalies ($\rho > 150\,\Omega\text{m}$) may be related to the presence of structures such as buildings, while sections of the investigated area with lower resistivity ($\rho < 50\,\Omega\text{m}$) can indicate the presence of portions of excavated terrain such as quarries, filled up by inhomogeneous material.

6 DATA INTEGRATION METHODS

6.1 QUALITATIVE INTEGRATION (CONTOUR MAPS OVERLAYS AND RGB)

As for the first case study, various integration techniques were applied to the data acquired on the field. In this case, as can be seen from the field data, the two methods detected completely different results. The GPR, given its high resolution, shows mainly the presence of groups of small-sized anomaly sources, while the ERT method gives out regular geometric anomalies. This situation is especially highlighted by the contour map overlay, where with the blue contour lines are represented the GPR anomalies, while in red are the ERT anomalies. With the RGB color composite, we get a clearer idea of the correlation between the detected anomalies for the two methods. The results for the qualitative integration are represented in Fig. 14.

6.2 DISCRETE QUANTITATIVE INTEGRATION (BINARY SUM AND CLUSTER ANALYSIS)

Also in this case for the binary combination, the data represent all features within the normalized range between 0.4 and 1, and the binary sum shows the areas where for that given range of values there is the presence or absence of anomalies. For the cluster analysis in this case, the calculations were performed using $k=5$ by which it is possible to highlight the main robust features detected by all methods (Fig. 15).

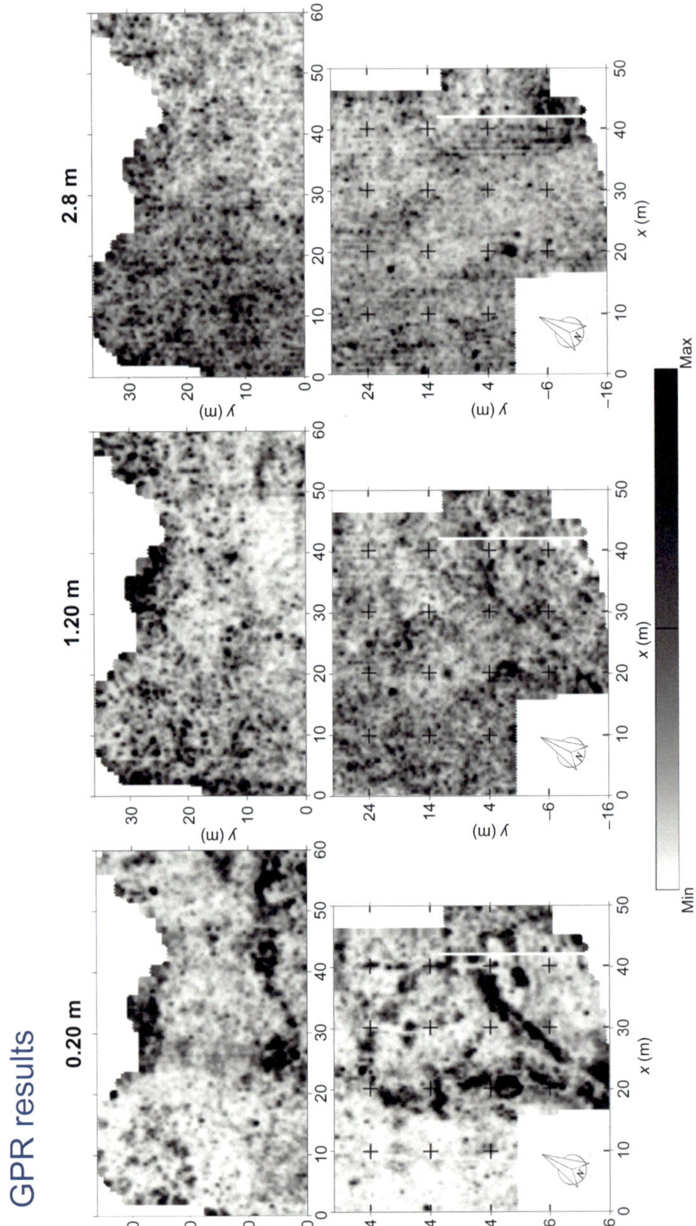

FIG. 12

GPR time slices at the indicated estimated depths. The anomalies at shallow depths may indicate the presence of structures. The diffused anomalies at deeper depths are related to the presence of collapsed structures.

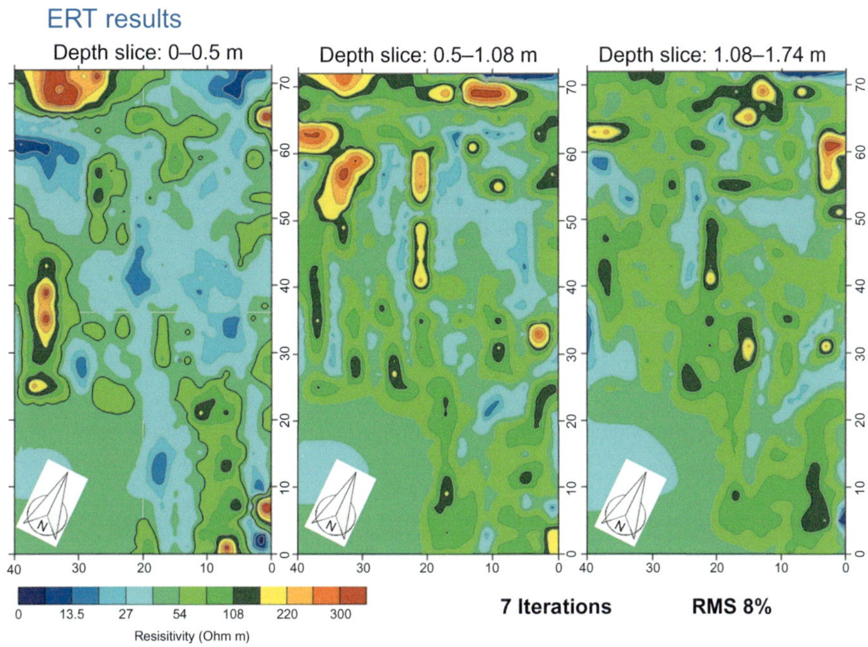

FIG. 13

Santa Balbina ERT depth slices obtained after 3-D inversion.

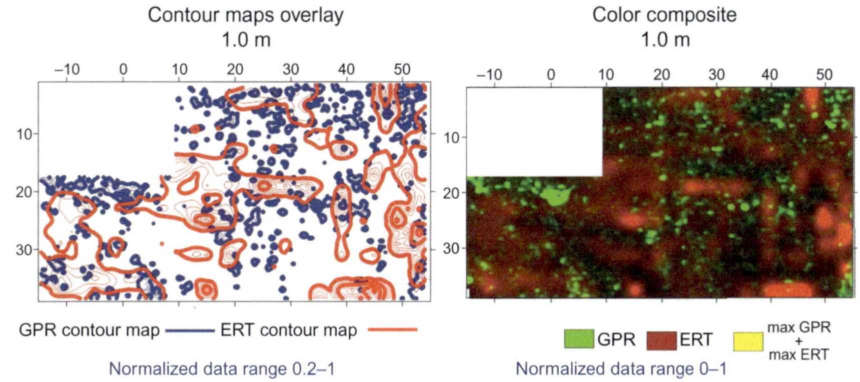

FIG. 14

Contour map overlay and RGB color composite integration for Santa Balbina case study.

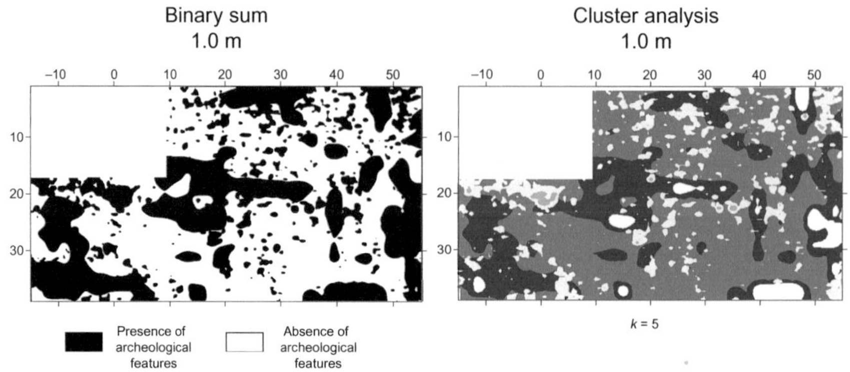

FIG. 15

Binary sum and *k*-means cluster analysis ($k=5$) integration for Santa Balbina case study.

6.3 CONTINUOUS QUANTITATIVE INTEGRATION (SUM, PRODUCT AND PRINCIPAL COMPONENT ANALYSIS)

As in the previous example, the best result for the continuous quantitative integration is given by the arithmetic sum of the two datasets as it points out all the anomalies detected by both GPR and ERT methods. The product image gives also an indication that no anomaly was strongly detected by both methods, highlighting the fact that the two methods have detected completely different kinds of structures. Regarding the principal component analysis, in this case, we have a very low correlation coefficient for the two datasets ($r=0.1$), and the usefulness of the methodology remains questioned since the two components are the two input variables, resistivity and GPR, respectively. The results given by these integration methods are represented in Fig. 16.

7 CERVETERI ARCHEOLOGICAL SITE

7.1 SITE CHARACTERISTICS

The selected area is located on the northwest urban plateau, outside the actual modern town of Cerveteri, between the Manganello creek and Vignali plateau, a place where from the iron age the archeological site has been founded. This area is characterized by the presence of a sanctuary (Etruscan sanctuary of Manganello) that represents the main development hub and urbanistic organization hub of the eastern part of the urban plateau, very close to the main part of the ancient acropolis of Cerveteri. The sacred area is delimited by *temenos* structures, which separate the sacred building from the service buildings. Between this sacred area and the eastern part of the ancient town, there is a portion of the plateau that can be considered as hinge between them. The surface archeological surveys have indicated that this area has been used

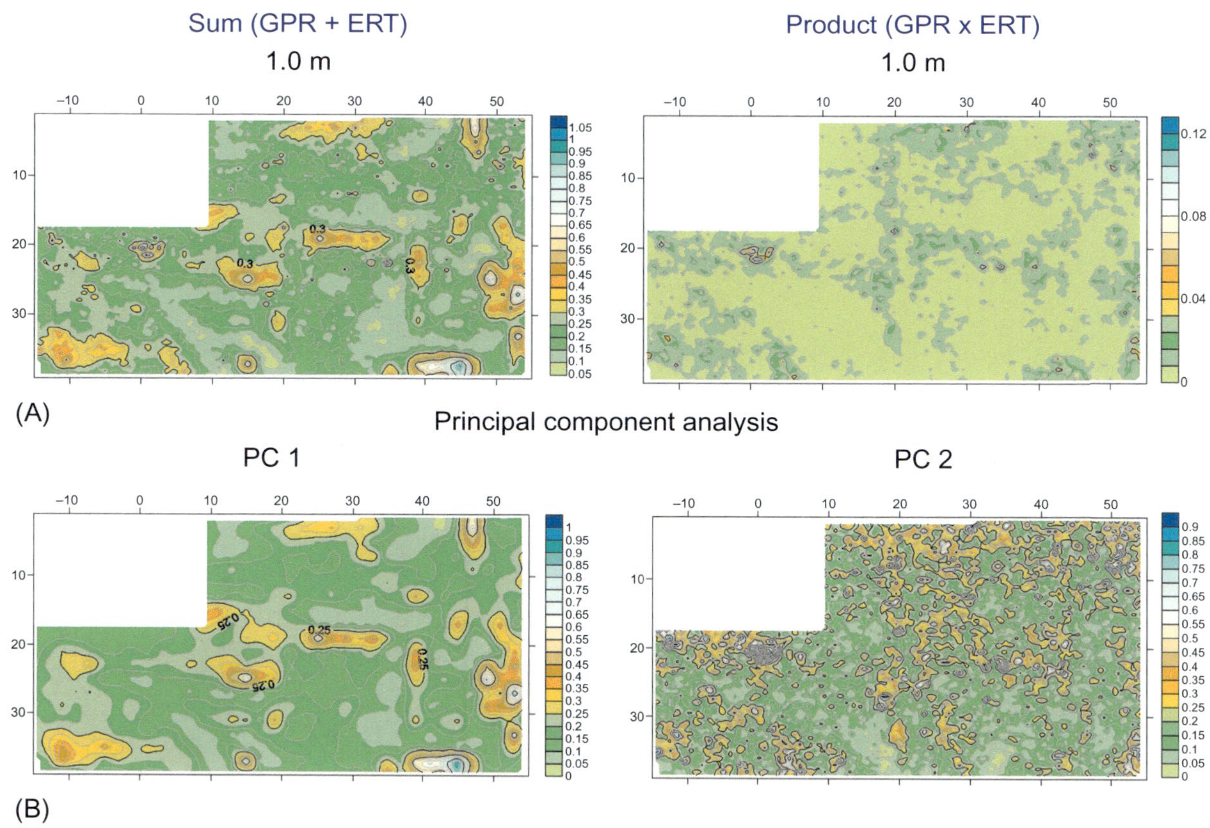

FIG. 16

Continuous quantitative integration applied to the Santa Balbina case study. (A) The sum and the product of the datasets are represented. (B) The two principal components for the depth of 1 m are represented.

from the Iron Age till roman empire time and is characterized by the presence of external fortified walls and hydraulic structures. To enhance the knowledge of this area, a first stage of integrated geophysical surveys has been made during 2014, employing GPR, gradiometric, and ERTs.

7.2 GEOPHYSICAL SURVEYS

The geophysical surveys were carried to detect buried structures as remains of buildings and portion of drainage tunnels. Surveys were carried out using the GPR, the FDM, and ERTs. For GPR surveys, measurements were collected, along parallel profiles, employing the SIR3000 (GSSI) system, equipped with a 400MHz (GSSI) bistatic antenna with constant offset. The horizontal spacing between parallel profiles at the site was 0.50m. In the investigated areas, a total of 260 adjacent profiles across the site were collected alternatively in forward and reverse directions, employing the GSSI cart system equipped with odometer. All radar reflections within the 75ns (twt) time window were recorded in the field as 16-bit data and 512 samples per radar scan.

Part of the area, overlapping those surveyed with GPR, has been investigated employing the differential magnetic method. In particular, the area was divided in three 30 × 30m squares where the gradient of the vertical component of the magnetic field was measured using a fluxgate gradiometer FM256 (GEOSCAN Research, the United Kingdom) along parallel profiles with a horizontal spacing of 1 m and with a sampling interval of 0.5 m.

The second stage, carried out at a later time, involved the use of the ERT method to detect deep unexplored drainage tunnels in the upper part of the archeological site. The measurements have been acquired in an area where the presence of these drainage tunnels has been hypothesized. A total of 21 parallel profiles, with a length of 48m, have been acquired, using an IRIS SYSCAL Junior Pro Switch-72; 48 electrodes with a spacing of 1 m for all the profiles was used, employing the Wenner-Schlumberger and dipole-dipole array configurations.

7.3 DATA PROCESSING AND RESULTS

7.3.1 GPR data

All the GPR profiles were processed with GPR-SLICE v7.0 ground-penetrating radar imaging software [53]. The basic radargram signal processing steps included (i) postprocessing pulse regaining, (ii) DC drift removal, (iii) data resampling, (iv) band-pass filtering, (v) migration, and (vi) background filter. With the aim of obtaining a planimetric vision of all possible anomalous bodies, the time-slice representation technique was applied using all processed profiles [44]. Time slices for a depth of about 0.80m are shown in Fig. 17.

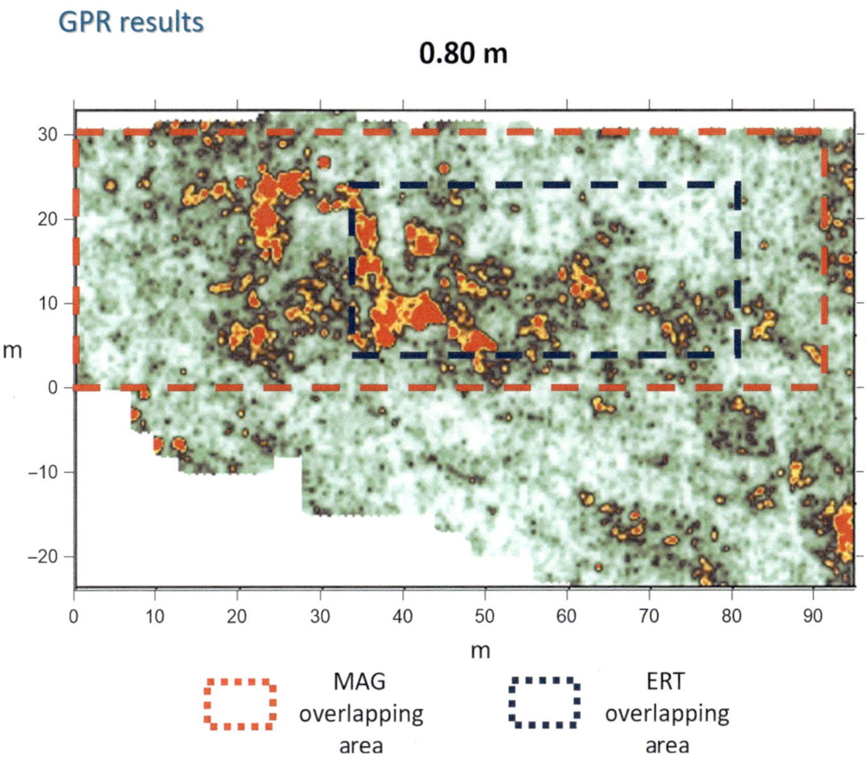

FIG. 17

GPR time slices for a depth of about 0.80m. The *dashed red and blue rectangles* represent, respectively, the area investigated with the magnetic and ERT methods.

7.3.2 Magnetic data

The magnetic data were processed with GEOPLOT 3.0 software (GEOSCAN Research). After despiking, filtering, and rearranging, the data were assembled in a contour of the vertical gradient of the total magnetic field as shown in Fig. 18. The contour map representing the results of the magnetic method is characterized by the presence of many dipole anomalies that can be related to the presence of a portion of walls.

The magnetic data are currently being processed using the bidimensional cross correlation technique in order to enhance the S/N ratio and to better define the spatial location and orientation of the possible targets [11]. This method is a measure of the similarity between the raw data and calculated synthetic anomalies.

7.3.3 ERT data

In the selected area, the 21 ERT profiles were inverted using the software RES2DINV (Geotomo); all the inversions can be assumed to be quite reliable, giving a final RMS of about 0.9%–1.0% after seven iterations and a better representation of

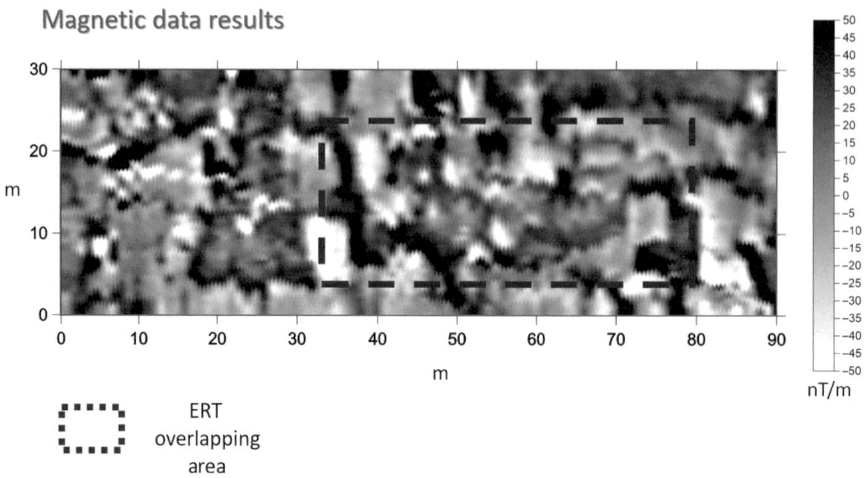

FIG. 18

Magnetic map of the investigated area at Cerveteri. Range from −50 to +50 nT/m. The *blue dashed rectangle* indicated the area investigated with the ERT method. This is the area where the integration methods have been applied on the three methods.

electric data. All the inverted electric sections display approximately the same features and are characterized by high-resistivity anomalies (in the range between 400 and 1000 Ωm) some of which are surely related to the presence of the empty sections of the searched drainage tunnels. These inverted pseudosections were then interpolated using Voxler (Golden Software) in order to obtain a three-dimensional model of the resistivity distribution of the subsurface, from which it was possible to extract slices corresponding to different depths. Fig. 19 shows an example of the obtained tomography after the inversion. This information, regarding the presence of the drainage tunnels in the investigated area, can enable us to determine the extension, depth, thickness, and electric characteristics of subsurface features including geologic units.

8 DATA INTEGRATION METHODS

8.1 QUALITATIVE INTEGRATION (CONTOUR MAPS OVERLAYS AND RGB)

For this case study, the integration techniques are performed using three datasets. The contour map overlay in this case becomes confused given to too many layers that overlapped, and it is difficult to identify specific features relative to the single

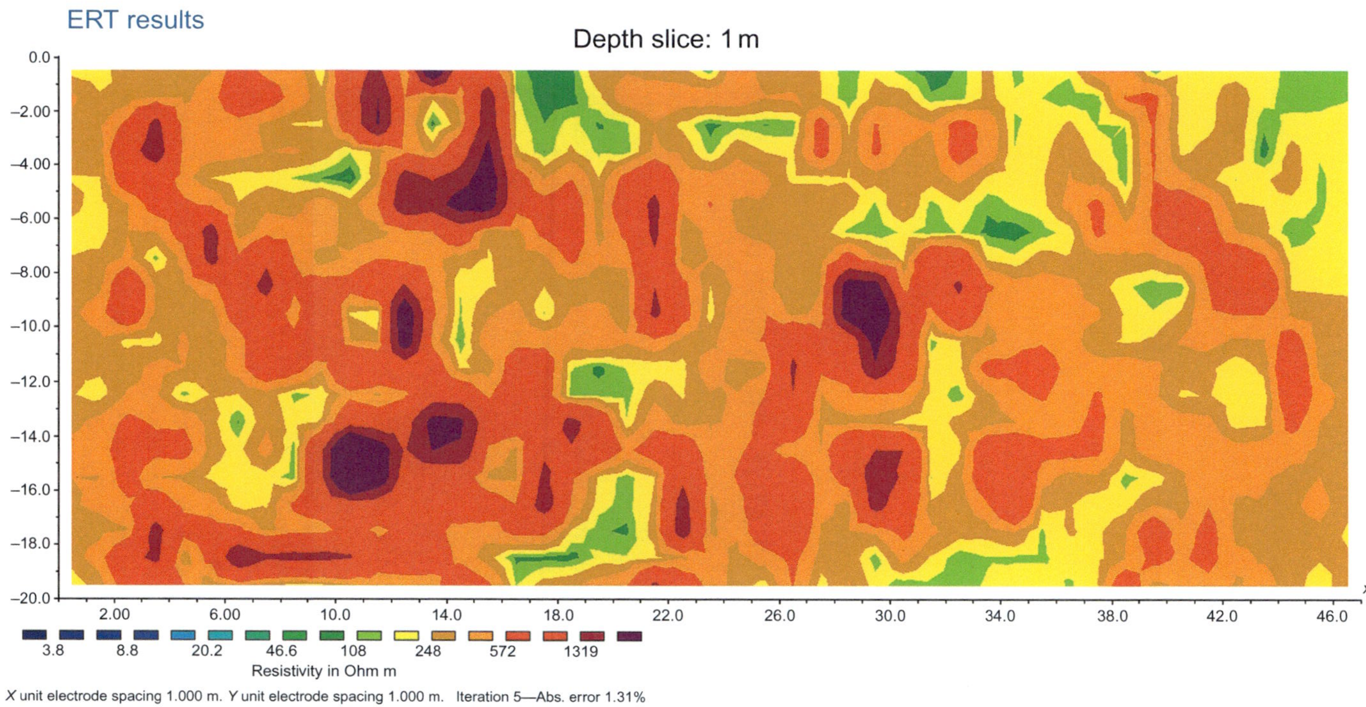

FIG. 19

ERT data results relative to a depth of about 1.0 m.

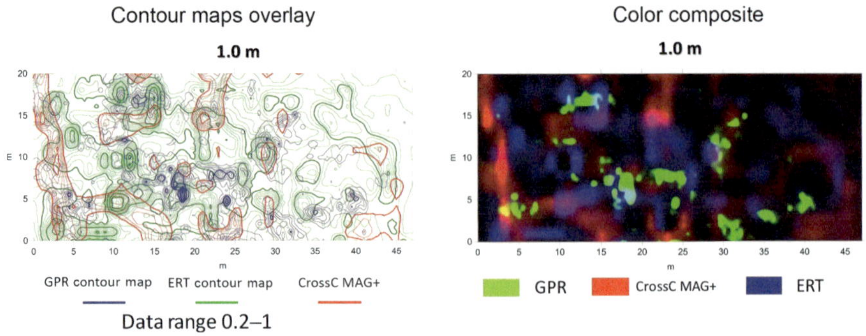

FIG. 20

Contour map overlay and RGB color composite integration for Cerveteri case study.

methods. The RGB gives instead good results since despite that it represents the combination of three layers, it is still possible to identify the anomalies relative to the different methods, as shown in Fig. 20.

8.2 DISCRETE QUANTITATIVE INTEGRATION (BINARY SUM AND CLUSTER ANALYSIS)

Regarding the discrete integration results, as for the other examples, these in general are useful only in order to produce maps representing in a unique image the main anomalies represented after the appropriate threshold has been selected by the user. The cluster analysis has been calculated in this case with $k=4$ where basically three of the four classes represent the main robust features of the single methods and the fourth class representing background subtler data, as shown in Fig. 21.

8.3 CONTINUOUS QUANTITATIVE INTEGRATION (SUM, PRODUCT AND PRINCIPAL COMPONENT ANALYSIS)

Also in this case, acceptable results are given by the sum of the three normalized datasets where all anomalies detected are represented in the final image. All these anomalies are probably related to the presence of the drainage tunnel network in the subsurface. The results of the product integration only show few features corresponding to the most robust anomalies detected by all the methods. Also in this case, the PCA was performed using the normalized results for the three geophysical methods where the overall correlation coefficient between the input variables remains very low with the highest value of r of about 0.2. As a result, the first principal component, the second, and the third simply represent the three input variables, magnetometry, resistivity, and GPR, respectively, so this type of analysis, also in this case, remains questioned, as shown in Fig. 22.

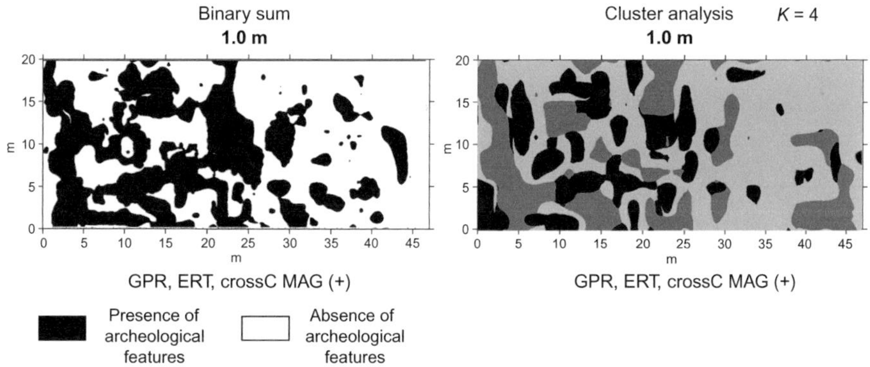

FIG. 21

Binary sum and cluster analysis for the Cerveteri case study.

9 HERAION ARCHEOLOGICAL SITE (TURKEY)

9.1 SITE CHARACTERISTIC

Heraion Teichos is a Thracian ancient city located in Tekirdag province in Marmara region of Turkey. Scientific excavations are continuing under the chairmanship of Prof. Dr. Nese Atik from Namik Kemal University, Department of Archeology, and with the support of Tekirdag Museum.

The results of ongoing excavations prove that the city was inhabited from third millennium BCE to AD XIII century. The city lived its most brilliant periods in classical and archaic eras. Even though fewer studies, Thracians are well known as the master for weaving, producing potteries in these periods. Considering the foundlings, it can be revealed that Heraion Teichos is also one of the important industrial ports in this area. The port walls can be seen surrounding around the acropolis that is built by using ashlar from Roman Period [54].

One of the important buildings of the Heraion Teichos is the Acropolis Gate. In the acropolis, small excavations were done in fortification wall, and in the gate again in 2000 and many different kinds of coins, figurines have been found related to the Archaic-Hellenistic Period [54,55]. In 2001, inside the acropolis fortifications and in the southwest part of North Gate, backyard and rooms surrounding this backyard have been determined during the excavation. Here, foundations built by rubble stone and roofs maiden by tiles have been leaped out clearly. Other foundlings prove that walls of these rooms are mostly made by mud bricks. One of the important foundlings in this part of the archaeologic site is the medical instruments. In 2004, a kiln that is used for medicine productions has been found around the backyard. Considering these interesting information and foundlings, it can be seen that this area has been used also as wellness-healing center and also the center for oblation [54].

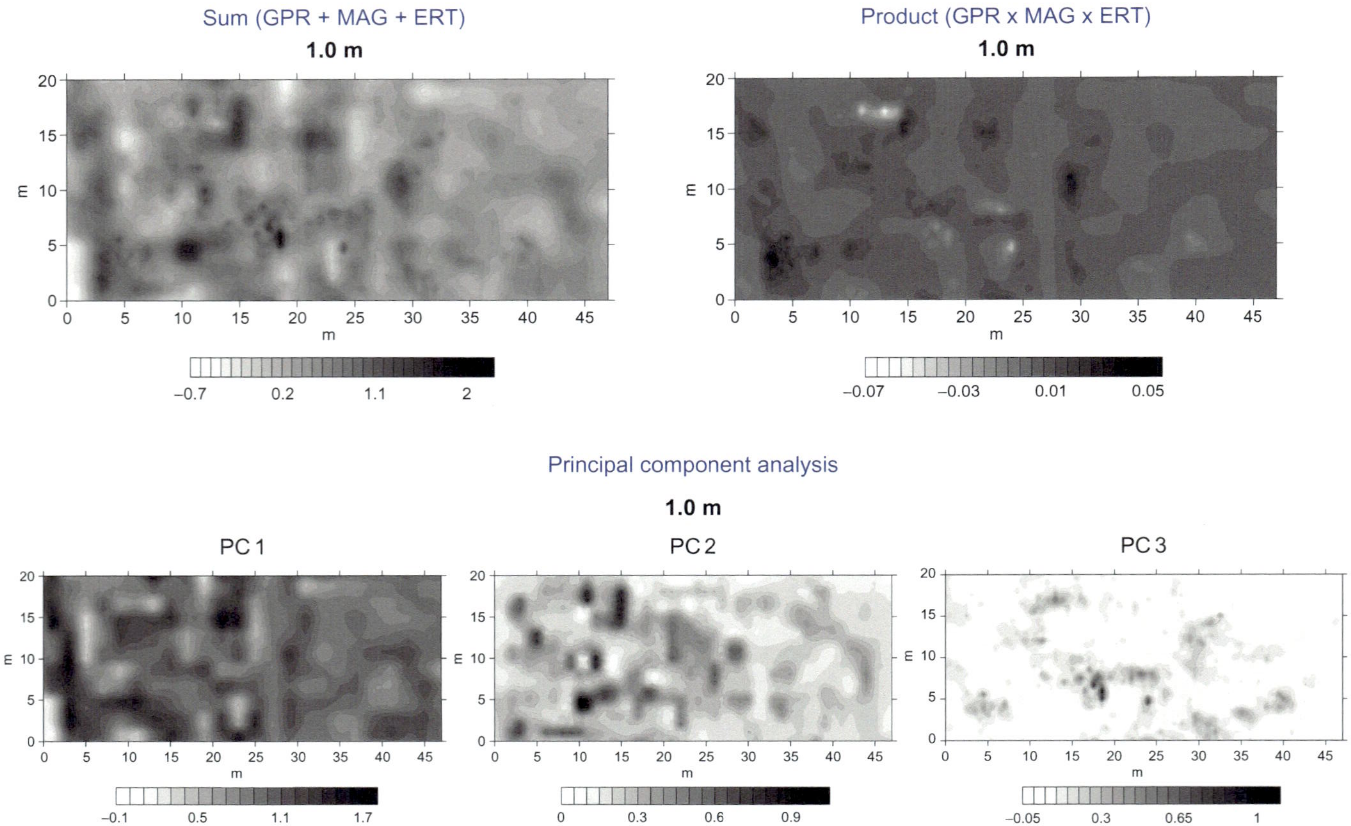

FIG. 22

(A) Maps of the sum and product integration using the three methods (GPR, ERT, and magnetometry). (B) The three components given by the principal components analysis are represented.

9.2 GEOPHYSICAL SURVEYS

Geophysical measurements were done in Heraion Teichos acropolis on several squared grid size 20 × 20 m. Measurements were collected along parallel profiles with an interval of 0.50 m.

GPR measurements were carried out with GSSI SIR 3000 GPR system equipped with a bistatic constant offset 400 MHz antenna. All profiles were collected alternatively in forward and reverse directions employing the GSSI cart system equipped with odometer.

Magnetometry measurements were collected employing Geometrics G858 cesium magnetometer with a gradiometric configuration. The distance between the ground and bottom sensor is 0.50 m; the distance between upper and bottom sensor is 1 m length.

9.3 DATA PROCESSING AND RESULTS

9.3.1 GPR data

All the GPR profiles were processed with GPR-SLICE v7.0 ground-penetrating radar imaging software [53]. The basic radargram signal processing steps included (i) postprocessing pulse regaining, (ii) DC drift removal, (iii) data resampling, (iv) band-pass filtering, (v) migration, and (vi) background filter. With the aim of obtaining a planimetric vision of all possible anomalous bodies, the time-slice representation technique was applied using all processed profiles [44]. Time slices for a depth of about 1.00 m are represented in Fig. 23.

9.3.2 Magnetic data

After despiking, filtering, and rearranging, the data were assembled in a contour of the gradient of the total magnetic field as shown in Fig. 24. The contour map representing the results of the magnetic method is characterized by the presence of many dipole anomalies that can be related to the presence of a portion of walls.

Magnetometry results seem more successful in this area possibly because of the nature of the target objects buried under the ground. The target structures to be detected are mostly the foundations and walls in this area, but in general, walls are masonry, so most of their parts are destroyed and scattered. It is not easy to follow the geometry of these walls only by using GPR measurements.

10 DATA INTEGRATION METHODS

All dataset was standardized and normalized for the integration approaches.

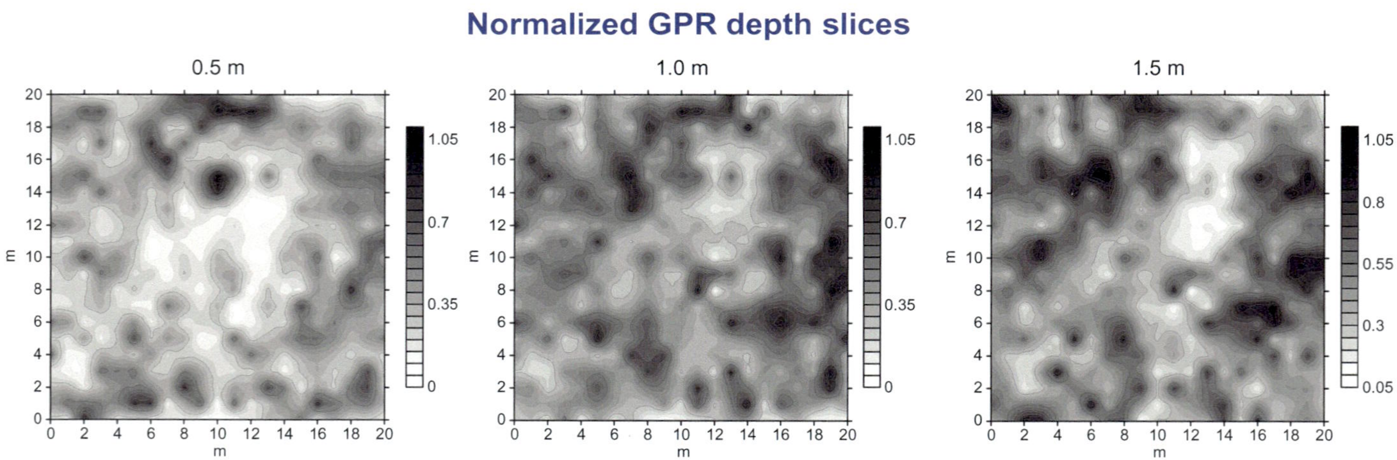

FIG. 23

Heraion Teichos. Normalized GPR depth slices.

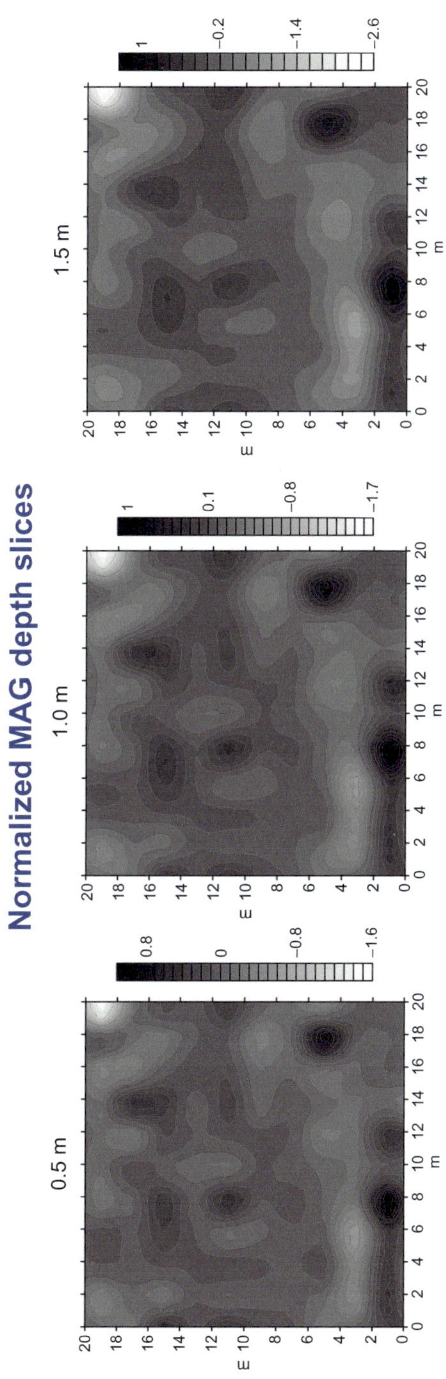

FIG. 24

Heraion Teichos. Normalized MAG depth slices.

10.1 QUALITATIVE INTEGRATION (CONTOUR MAPS OVERLAYS AND RGB)

By analyzing the graphical integration, also in this case, the GPR and magnetic method yield somewhat different results. The magnetic method detected an anomaly probably related to the portion of a wall that is not visible in the GPR results. This might be due to the fact that the contrast in the physical property measured by the GPR, between the wall and the surrounding material, is so low that the anomalous source is too subtle and therefore not visible as clearly as shown in the magnetic data results. The integration in this case is also useful to recover all the information that can be used for the final archeological interpretation. The RGB color composite method allows us to have a clearer idea about the correlation between the anomalies detected by the two methods, as shown in Fig. 25.

10.2 DISCRETE QUANTITATIVE INTEGRATION (BINARY SUM AND CLUSTER ANALYSIS)

In this case, the binary sum represents all the features within the normalized range between 0.5 and 1 and helps us to visualize those that are considered to be the main anomalous sources related to the presence of the structures in the subsurface. The cluster analysis has been calculated using a $k=4$ that best represents the main anomalies detected against the background data, as shown in Fig. 26.

10.3 CONTINUOUS QUANTITATIVE INTEGRATION (SUM, PRODUCT AND PRINCIPAL COMPONENT ANALYSIS)

Also in this case, the sum of the two normalized datasets allows us to have a new continuous dataset representing the indicators of source occurrence detected by the two methods. The results of the product integration only show a limited number of features that are part of the archeological structures in the subsurface as represented in Fig. 27.

As in the previous examples, correlation between the datasets was very low, and the two components just represent the two separated input variables, magnetometry and GPR, respectively (Fig. 28).

11 CONCLUSIONS

The integration of different geophysical methods and data analysis has provided an enhancement of different types of data, new analysis, and visualization capabilities for future interpretations of archeological and geophysical features at different investigated sites.

As expected, the integration of different geophysical datasets allows us a high potential to improve the knowledge of the subsurface. As known, a single

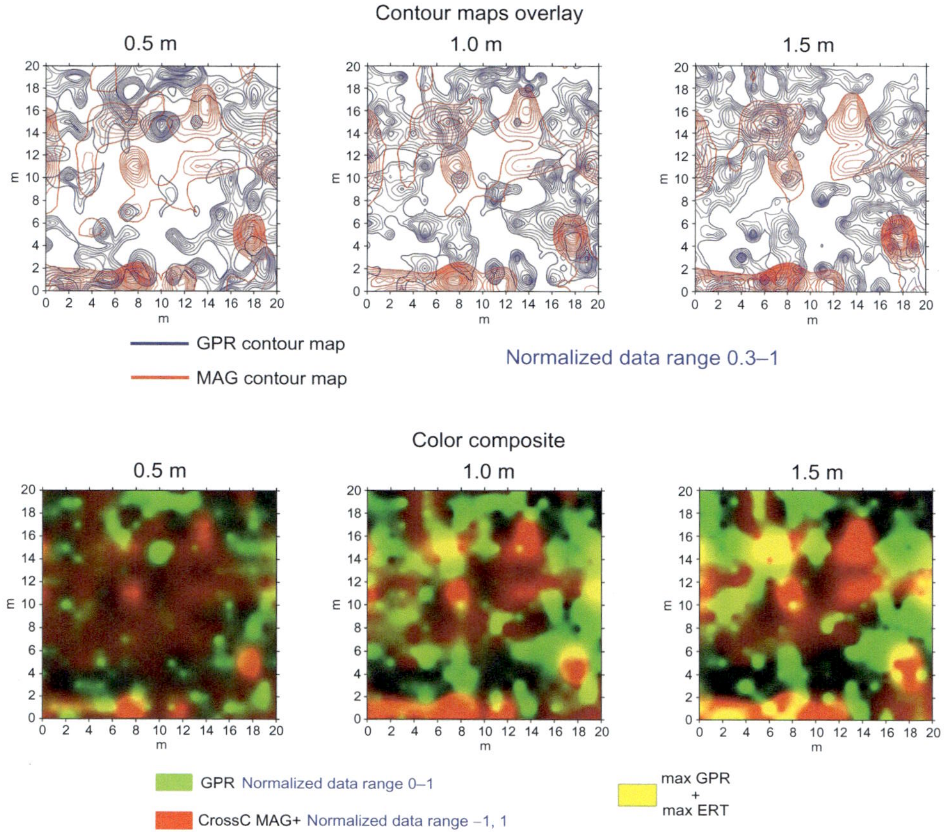

FIG. 25

Contour map overlay and RGB color composite integration for Heraion case study.

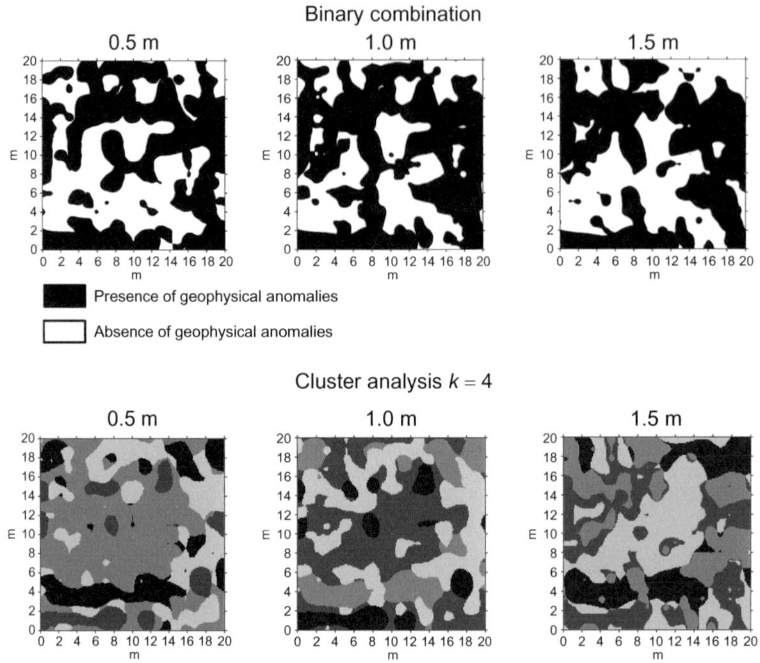

FIG. 26

Binary sum and cluster analysis for Heraion case study.

geophysical method might reveal only a portion of the searched buried building. Integrated geophysical datasets may show relationships between the different physical parameters and their contrast between the searched bodies and the surrounding subsoil.

Graphical combinations for data integration are easy to employ, and they show information from different sources into possible interpretable image, but their limitation is the relatively low dimensionality; only two datasets can be effectively represented. Therefore, these methods are descriptive, and no new datasets can be obtained for further analyses.

Discrete integrating methods allow application of statistical algorithms to a significant number of geophysical datasets.

Continuous data integrating methods generally produce a new dataset. This is characterized by the simultaneous presence of all anomalies individuated by the different geophysical methods. In these fused datasets, the anomalous conditions, frequently much less visible in a single dataset, can be pointed and analyzed.

The improvement of the processing techniques through an integration of different geophysical methods enables one to better define position, extension, depth,

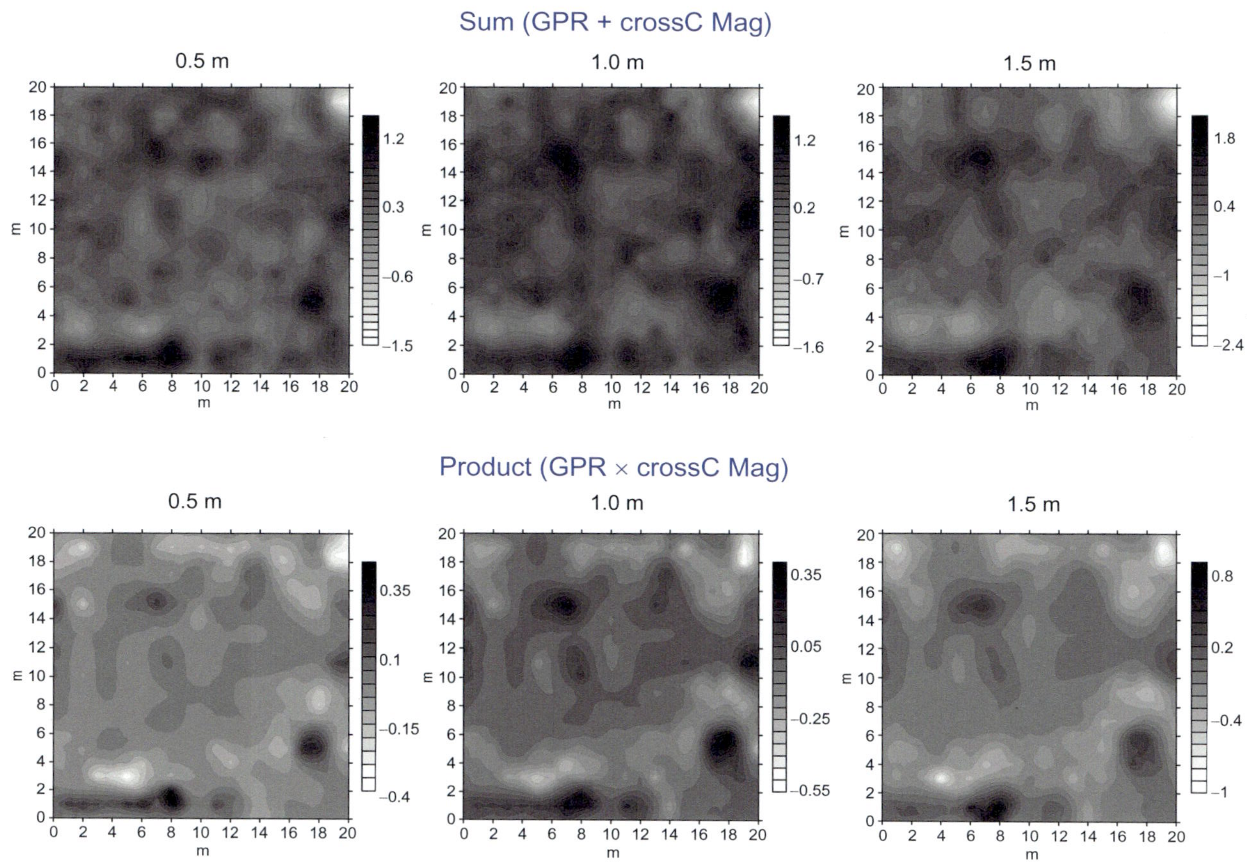

FIG. 27

Continuous quantitative integration applied to the Heraion Teichos (Turkey) case study. On the top, the sum and the product of the datasets are represented.

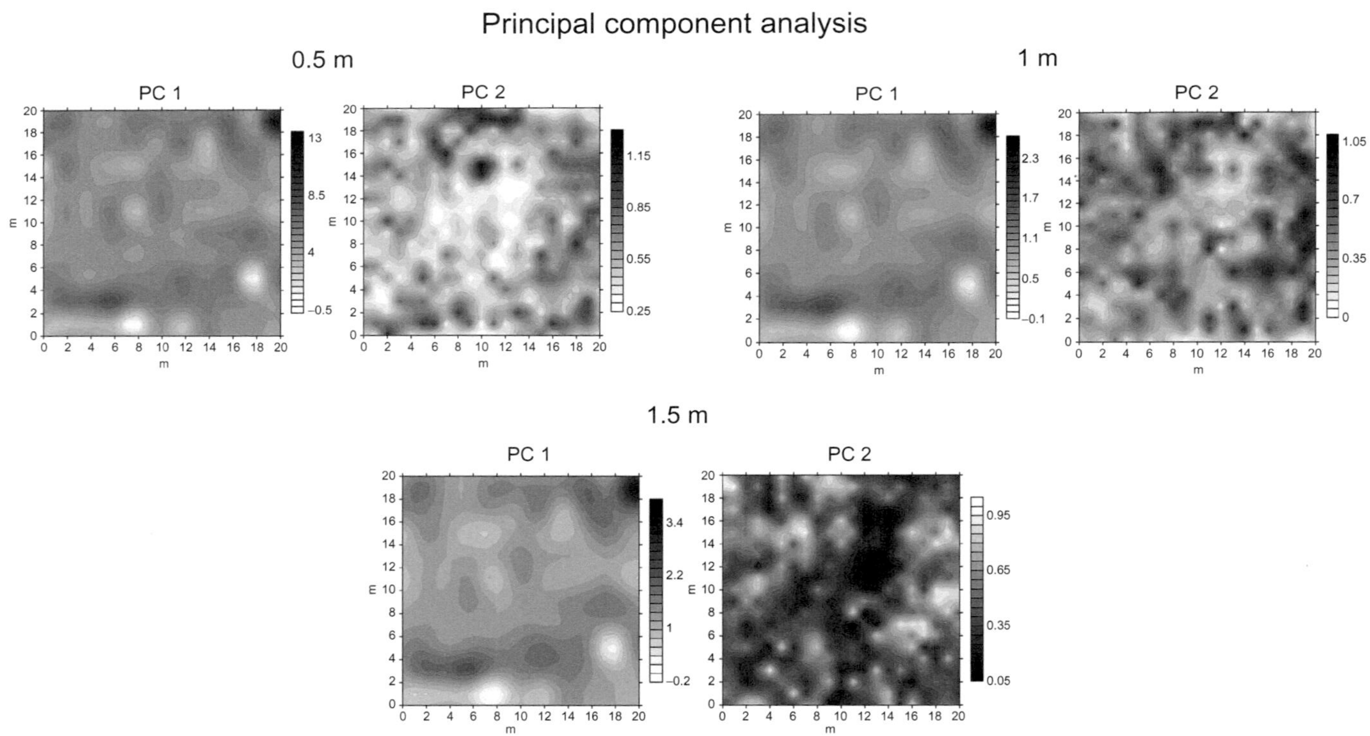

FIG. 28

Continuous quantitative integration applied to the Heraion Teichos (Turkey) case study. On the bottom, the two principal components for the three represented depths are represented.

thickness, and physical characteristics of any anomalous body within its geological context. Also the discrimination power between close but distinct archeological structures is notably enhanced. There is no doubt that if more geophysical methods are introduced arising from the same cultural source demand, their potentialities are interwoven. Their single side indeterminate aspects are greatly entangled to the extent of the target; therefore, its total property nearly always emerges by means of some kind of interrelationship.

REFERENCES

[1] E. Bozzo, S. Lombardo, F. Merlanti, M. Pavan, Geoelectric and electromagnetic measurements within an organised archaeological framework: the Marzabotto example, Ann. Geofis. 37 (Suppl. 5) (1994) 1199–1213.

[2] E. Brizzolari, F. Ermolli, L. Orlando, S. Piro, L. Versino, Integrated geophysical methods in archaeological surveys, J. Appl. Geophys. 29 (1992) 47–55.

[3] C. Camerlynk, M. Dabas, C. Panissod, Comparison between GPR and four electromagnetic methods for stone features characterization. An example, Archaeol. Prospect. 1 (1994) 5–17.

[4] C. Gaffney, V. Gaffney, Non-invasive investigations at Wroxeter at the end of twentieth century, Archaeol. Prospect. 7 (2) (2000) 65–80.

[5] S. Piro, Integrated geophysical prospecting at Ripa Tetta Neolithic site (Lucera, Foggia-Italy), Archaeol. Prospect. 3 (2) (1996) 81–88.

[6] I. Scollar, A. Tabbagh, A. Hesse, I. Herzog, Archaeological Prospecting and Remote Sensing, Cambridge University Press, Cambridge, 1990.

[7] J.W. Weymouth, Geophysical methods of archaeological sites surveying, in: M.B. Schiffer (Ed.), Advances in Archaeological Method and Theory, vol. 9, Academic Press, New York, 1986, pp. 311–395.

[8] M. Bernabini, E. Brizzolari, S. Piro, Improvement of signal-to-noise ratio in resistivity profiles, Geophys. Prospect. 36 (1988) 559–570.

[9] E. Brizzolari, S. Piro, L. Versino, Monograph on the geophysical exploration of the Selinunte Archaeological Park. Foreword, Boll. Geofis. Teor. Appl. 34 (1992).

[10] N.G. Papadopoulos, P. Tsourlos, G.N. Tsokas, A. Sarris, Two-dimensional and three-dimensional resistivity imaging in archaeological site investigation, Archaeol. Prospect. 13 (2006) 163–181.

[11] S. Piro, A. Samir, L. Versino, Position and spatial orientation of magnetic bodies from archaeological magnetic surveys, Ann. Geofis. 41 (3) (1998) 343–358.

[12] S. Piro, D. Peloso, R. Gabrielli, Integrated geophysical and topographical investigation in the territory of Ancient Tarquinia (Viterbo, Central Italy), Archaeol. Prospect. 14 (2007) 191–201.

[13] G.N. Tsokas, C.B. Papazachos, A. Vafdis, M.Z. Loukoyiannakis, G. Vargemezis, K. Tzimeas, The detection of monumental tombs buried in tumuli by seismic refraction, Geophysics 60 (6) (1995) 1735–1742.

[14] G.N. Tsokas, P. Tsourlos, Transformation of the resistivity anomalies from archaeological sites by inversion filtering, Geophysics 62 (1) (1997) 36–43.

[15] G.N. Tsokas, O. Hamsen, On the use of complex attribute and the inferred source parameter estimates in the exploration of archaeological sites, Archaeol. Prospect. 7 (1) (2000) 17–30.

[16] F. Cammarano, P. Mauriello, D. Patella, S. Piro, Geophysical methods for archaeological prospecting: a review, Science and Technology for Cultural Heritage 6 (2) (1997) 151–173.

[17] F. Cammarano, P. Mauriello, D. Patella, S. Piro, F. Rosso, L. Versino, Integration of high resolution geophysical methods. Detection of shallow depth bodies of archaeological interest, Ann. Geofis. 41 (3) (1998) 359–368.

[18] S. Campana, S. Piro (Eds.), Seeing the Unseen. Geophysics and Landscape Archaeology, CRC Press, Taylor & Francis Group, London, UK, 2008.

[19] M. Ciminale, H. Becker, D. Gallo, Integrated technologies for archaeological investigation: the Celone Valley project, Archaeol. Prospect. 14 (2007) 167–181.

[20] N.G. Diamanti, G.N. Tsokas, P.I. Tsourlos, A. Vafidis, Integrated interpretation of geophysical data in the archaeological site of Europos (Northern Greece), Archaeol. Prospect. 12 (2) (2005) 79–91.

[21] E. Finzi, N. Praticelli, L. Vettore, A. Zaja, Multi-temporal geophysical survey of a Roman bath complex in Montegrotto Terme (Padova, northern Italy), Archaeol. Prospect. 14 (3) (2007) 182–190.

[22] V. Gaffney, H. Patterson, S. Piro, D. Goodman, Y. Nishimura, Multimethodological approach to study and characterise Forum Novum (Vescovio, Central Italy), Archaeol. Prospect. 11 (2004) 201–212.

[23] K.L. Kvamme, Geophysical surveys as landscape archaeology, Am. Antiq. 68 (3) (2003) 435–456.

[24] E. Cardarelli, F. Fischanger, S. Piro, Integrated geophysical survey to detect buried structures for archaeological prospecting. A case-history at Sabine Necropolis (Rome, Italy), Near Surf. Geophys. 6 (1) (2008) 15–20.

[25] K.L. Kvamme, Integrating multidimensional geophysical data, Archaeol. Prospect. 13 (2006) 57–72.

[26] S. Piro, F. Cammarano, P. Mauriello, Quantitative integration of geophysical methods for archaeological prospection, Archaeol. Prospect. 7 (4) (2000) 203–213.

[27] W. Neubauer, A. Eder-Hinterleitner, Resistivity and magnetics of the Roman Town Carnuntum, Austria. An example of combined interpretation of prospection data, Archaeol. Prospect. 4 (1997) 179–189.

[28] E. Brizzolari, E. Cardarelli, M. Feroci, S. Piro, L. Versino, Vertical electric soundings and inductive electromagnetism used to investigate the calcarenitic layer in the Selinunte Archaeological Park, Boll. Geofis. Teor. Appl. 34 (1992) 109–119.

[29] E. Brizzolari, L. Orlando, S. Piro, L. Versino, Ground probing radar in the Selinunte Archaeological Park, Boll. Geofis. Teor. Appl. 34 (1992) 181–192.

[30] E. Brizzolari, E. Cardarelli, M. Feroci, S. Piro, L. Versino, Magnetic survey in the Selinunte Archaeological Park, Boll. Geofis. Teor. Appl. 34 (1992) 157–168.

[31] W. Neubauer, A. Eder-Hinterleitner, 3D-interpretation of post-processed archaeological magnetic prospection data, Archaeol. Prospect. 4 (1997) 191–205.

[32] S. Piro, R. Gabrielli, Multimethodological approach to investigate chamber tombs in the Sabine Necropolis at Colle del Forno (CNR, Rome, Italy), Archaeol. Prospect. 16 (2) (2009) 111–124.

[33] J. Ogden, S. Keay, G. Earl, K. Strutt, S. Kay, Geophysical prospection at Portus: an evaluation of an integrated approach to the interpretation of subsurface archaeological features, in: F. Bernard, J. Webb Crawford, D. Koller (Eds.), Making History Interactive. Computer Applications and Quantitative Methods in Archaeology (CAA). Proceedings of the 37th International Conference, Williamsburg, Virginia, United States of America, Archaeopress, Oxford, 2009, pp. 1–18.

[34] L.V. Eppelbaum, Geophysical observations at archaeological sites: estimating informational content, Archaeol. Prospect. 21 (2) (2014) 125–138.
[35] K.L. Kvamme, Archaeological prospection in fortified Great Plains villages: new insights through data fusion, visualization, and testing, in: M. Doneus, A. Eder-Hinterleitner, W. Neubauer (Eds.), Archaeological Prospection: 4th International Conference on Archaeological Prospection, Austrian Academy of Sciences Press, Vienna, 2001, pp. 141–143.
[36] J.C. Davis, Statistics and Data Analysis in Geology, third ed., New York, Wiley, 2002.
[37] Accordi, G., Carbone, F., Civitelli, G., Corda, L., De Rita, D., Esu, D., Funiciello, R., Kotsakis, T., Mariotti, G., Sposato, A. (1988). Note Illustrative alla Carta delle litofacies del Lazio—Abruzzo ed aree limitrofe. C.N.R—P. F. Geodinamica: sottoprogetto 4. *Quad. Ric. Scient.*, 114, vol. 5, 223 p.
[38] F. Carbone, Cretaceous depositional system of the evolving mesozoic carbonate platform of central appennine thrust belt, Italy, Geol. Romana 29 (1992) 31–53.
[39] A. Praturlon, Calcareous algae from Jurassic-Creataceous limestones of central Appennine (Southern Latium-Abruzzi), Geol. Romana 3 (1964) 171–202.
[40] L.B. Conyers, Ground-Penetrating Radar for Archaeology, AltaMira, Walnut Creek, CA, 2004.
[41] D. Goodman, S. Piro, Y. Nishimura, H. Patterson, V. Gaffney, Discovery of a 1st century Roman Amphitheater and Town by GPR, J. Environ. Eng. Geophys. 9 (1) (2004) 35–41.
[42] D. Goodman, S. Piro, Ground Penetrating Radar (GPR) surveys at Aiali (Grosseto), in: S. Campana, S. Piro (Eds.), Seeing the Unseen. Geophysics and Landscape Archaeology, CRC Press, Taylor & Francis Group, Oxon, pp. 297–302, ISBN: 978-0-415-44721-8.
[43] D. Goodman, S. Piro, Y. Nishimura, K. Schneider, H. Hongo, N. Higashi, J. Steinberg, B. Damiata, GPR archaeometry, in: H.M. Jol (Ed.), Ground Penetrating Radar Theory and Applications, Elsevier, Amsterdam, 2008, pp. 479–508.
[44] D. Goodman, S. Piro, GPR Remote Sensing in Archaeology, Springer, Berlin, 2013.
[45] J. Leckebusch, Ground-penetrating radar: a modern three-dimensional prospection method, Archaeol. Prospect. 10 (4) (2003) 213–240.
[46] J. Leckebusch, Semi-automatic feature extraction from GPR data for archaeology, Near Surf. Geophys. 6 (2008) 75–84.
[47] G. Leucci, N. Masini, R. Persico, F. Soldovieri, GPR and sonic tomography for structural restoration: the case of the Cathedral of Tricarico, J. Geophys. Eng. 8 (3) (2011) S76–S92.
[48] N. Lindford, From hypocaust to hyperbola: ground penetrating radar surveys over mainly Roman remains in the U.K, Archaeol. Prospect. 11 (4) (2004) 237–246.
[49] W. Neubauer, A. Eder-Hinterleitner, S. Seren, P. Melichar, Georadar in the Roman Civil town Carnuntum, Austria. An approach for archaeological interpretation of GPR data, Archaeol. Prospect. 9 (2002) 135–156.
[50] S. Piro, D. Goodman, Integrated GPR data processing for archaeological surveys in urban area. The case of Forum (Roma, Italy), 12th International Conference on Ground Penetrating Radar, June 16–19, 2008, Birmingham, UK, 2008. Proceedings Extanded Abstract Volume.
[51] S. Piro, S. Campana, GPR investigation in different archaeological sites in Tuscany (Italy). Analysis and comparison of the obtained results, Near Surf. Geophys. 2012 (10) (2012) 47–56.
[52] S. Piro, D. Goodman, Y. Nishimura, The study and characterisation of Emperor Traiano's Villa (Altopiani di Arcinazzo – Roma) using high-resolution integrated geophysical surveys, Archaeol. Prospect. 10 (2003) 1–25.

[53] Goodman, D. (2015). GPR-Slice 7.0, Manual, http://gpr-survey.com.
[54] Atik, N., Ve Işin M.A., 2004, Tekirdağ/Karevlialtı (2000), (2002), (2004). yılları Kazı Çalışmaları, *T.C. Kültür ve Turizm Bakanlığı Yayınları*, Yayın no: 115-1, syf: 47–58.
[55] Atik, N., (2003). Tekirdağ Karaevli (Antik Heraion Teichos) Hellenistik Devir Çanak Çömlekleri, *III*. Uluslararası Eskişehir Toprak Sempozyumu, Eskişehir, s. 298–305.

FURTHER READING

[56] E. Bozzo, F. Merlanti, G. Ranieri, L. Sambuelli, E. Finzi, EM-VLF soundings on the eastern hill of the archaeological site of Selinunte, Boll. Geofis. Teor. Appl. 34 (1991) 132–140.
[57] E. Bozzo, F. Merlanti, Magnetic and geoelectrical measurements on the eastern hill of the archaeological site of Selinunte, Boll. Geofis. Teor. Appl. 34 (1992) 145–156.
[58] E. Bozzo, S. Lombardo, F. Merlanti, Geophysical surveys at the Poliochni archaeological site (Lemnos Island, Greece): preliminary results, Archaeol. Prospect. 2 (1995) 1–13.
[59] E. Brizzolari, E. Cardarelli, S. Piro, L. Versino, Detection of subsurface magnetic anomalies of archaeological interest: computation of tridimensional magnetic anomalies and interpretation using bidimensional cross-correlation, in: A. Vogel (Ed.), Geophysical Exploration of Archaeological Sites, Series Theory and Practice of Applied Geophysics, vol. 7, Vieweg Publishing, Wiesbaden, 1993, pp. 3–16.
[60] D. Goodman, K. Schneider, S. Piro, Y. Nishimura, A.G. Pantel, Ground penetrating radar advances in subsurfaces imaging for archaeology, in: J. Wiseman, F. El-Baz (Eds.), Remote Sensing in ArchaeologySpringer, 2007, pp. 367–386 Chapter 15. ISBN 978-0-387-44453-6.
[61] M.H. Loke, R.D. Barker, Rapid least-squares inversion of apparent resistivity pseudosections using quasi-Newton method, Geophys. Prospect. 48 (1996) 131–152.
[62] M.H. Loke, R.D. Barker, Practical techniques for 3D resistivity surveys and data inversion, Geophys. Prospect. 44 (1996) 499–523.

Index

Note: Page numbers followed by *f* indicate figures, and *t* indicate tables.

D

N

O

P

T

Made in the USA
Monee, IL
15 September 2026

63317828R00319